D0023495

Kurt Gödel
COLLECTED WORKS
Volume III

Portrait of Kurt Gödel, ca. 1962. Copyright © Alfred Eisenstaedt.

Kurt Gödel

COLLECTED WORKS

Volume III
Unpublished essays and lectures

EDITED BY

Solomon Feferman
(Editor-in-Chief)

John W. Dawson, Jr.
Warren Goldfarb
Charles Parsons
Robert M. Solovay

Prepared under the auspices of the
Association for Symbolic Logic

New York   Oxford
OXFORD UNIVERSITY PRESS
1995

Oxford University Press

Oxford   New York
Athens   Auckland   Bangkok   Bombay
Calcutta   Cape Town   Dar es Salaam   Delhi
Florence   Hong Kong   Istanbul   Karachi
Kuala Lumpur   Madras   Madrid   Melbourne
Mexico City   Nairobi   Paris   Singapore
Taipei   Tokyo   Toronto

and associated companies in
Berlin   Ibadan

Published by Oxford University Press, Inc.,
200 Madison Avenue, New York, New York 10016

Oxford is a registered trademark of Oxford University Press, Inc.

Library of Congress Cataloging-in-Publication Data
(Revised for v. 3)
Gödel, Kurt.
Collected works.
German text, parallel English translation.
Includes bibliographies and indexes.
Contents: v. 1. Publications 1929–1936
v. 2. Publications, 1938–1974
v. 3. Unpublished essays and lectures.
1. Logic, Symbolic and mathematical.
I. Feferman, Solomon. II. Title.
QA9.G5313   1986   511.3   85-15501
ISBN: 0-19-503964-5 (v. 1)
ISBN: 0-19-503972-6 (v. 2)
ISBN: 0-19-507255-3 (v. 3)

9 8 7 6 5 4 3 2 1
Printed in the United States of America
on acid-free paper

Preface

This is the third volume of a comprehensive edition of the works of Kurt Gödel. Volumes I and II comprised all of his publications, ranging from 1929 to 1936 and from 1937 to 1974, respectively. The present volume consists of a selection of unpublished papers and texts for individual lectures found in Gödel's *Nachlass*. It opens with an overview of the contents of the *Nachlass* followed by a description of the Gabelsberger shorthand system that Gödel employed in many of his scientific notebooks and in a few of the items represented in this volume. A succeeding volume is to contain a selection of Gödel's scientific correspondence together with a full inventory of his *Nachlass*.

Our general aim here and in the other volumes has been to provide an edition which is comprehensive in scope yet selective with regard to previously unpublished material. In this volume, our primary criteria for judging an item to be worthy of inclusion were: (1) the manuscript had to be sufficiently coherent to permit editorial reconstruction; (2) the text was not to duplicate other works substantially in both content and tone (though treatments of similar topics aimed at different audiences or differing in degrees of detail might warrant inclusion); (3) the material had to possess intrinsic scientific interest.

Additional justification for our selections came from two lists prepared by Gödel himself, both preserved in his *Nachlass* and entitled "Was ich publizieren könnte" ("What I could publish"). Though undated, their contents imply that they were written fairly late in his life.[a] Some items on these lists are too vague to identify, while others refer to a body of work scattered throughout his notebooks. Of the identifiable items, *1972* and *1972a* were previously reproduced in Volume II of these *Works*; all others satisfying the criteria above (namely, *193?*, *1946/9*, *1951*, *1953/9* and *1970*) are published here for the first time. In addition, two of the less clearly identifiable items may refer to *1941* and to *1939b* together with *1940a*.

Despite their presence on these lists, we recognize that Gödel did not consider any of the texts presented here to be in final form; before submitting them for publication, he would no doubt have made a number of stylistic and, in some cases, substantive changes. Indeed, in one case (*1970a*) an item that had been intended for publication was withdrawn when a key argument was found to be in error. It has been included here

---

[a]In 1970, when he feared he might die, Gödel communicated a similar list to his friend and colleague, Oskar Morgenstern.

on the grounds that, even so, there is much to be learned from Gödel's ideas there (and in the related manuscripts *1970b, c*).

Multiple drafts of two articles, namely *1946/9* (in five versions) and *1953/9* (in six versions), posed particular problems of selection. In both cases Gödel evidently failed to arrive at a final text that he considered fully satisfactory. Two distinct versions of each, having been deemed sufficiently coherent and relatively finished in form, are reproduced here.

Taken as a whole, the texts presented in this volume substantially enlarge our appreciation of Gödel's scientific and philosophical thought, and in a number of cases they add appreciably to our understanding of his motivations.

The plan for this volume is essentially the same as for Volumes I and II of these *Works*. Each article, lecture text or closely related group of such is preceded by an introductory note which elucidates it and places it in historical context. These notes (varying greatly in length) have been written by members of the editorial board together with a number of outside experts. Articles originally written in German (transcribed where necessary from Gabelsberger script) are accompanied by English translations on facing pages. The unpolished nature of the items in this volume has necessitated more editorial comment and intervention than in the previous two volumes. Nonetheless, since this is a critical but not strictly documentary edition, editorial emendations, references and footnotes are handled so as to maximize readability and minimize editorial "barbed wire". Detailed information for the reader, subsuming that for Volumes I and II, is provided in a special section directly following this preface.

Once more our endeavor has been to make the full body of Gödel's work—including items previously available only to a few scholars—as accessible and useful to as wide an audience as possible, without compromising the requirements of historical and scientific accuracy. As with the preceding volumes, this one is expected to be of interest and value to professionals and students in the areas of logic, mathematics, computer science, philosophy and even physics, as well as to many non-specialist readers with a broad scientific background. Naturally, even with the assistance of the introductory notes, not all of Gödel's work can be made equally accessible to such a variety of readers; nonetheless, the general reader should be able to gain some appreciation of what Gödel accomplished in each case.

We continue to be indebted to the National Science Foundation and the Sloan Foundation, whose grants have made possible the production of these volumes, and to the Association for Symbolic Logic, which has sponsored our project and, with the able assistance of its Secretary-Treasurer, C. Ward Henson, has administered those grants. Our publisher, Oxford University Press, has been very accommodating to our overall plans and specific wishes.

We are grateful to the Institute for Advanced Study for its cooperation in giving us a free hand in our selection of manuscripts, and to its head librarian, Elliott Shore, and its archivist, Mark Darby, for making Gödel's books available to us for study of his annotations, particularly those in Husserl's works. We also received help from the staff of the Rare Books and Manuscripts Division of Firestone Library at Princeton University, where Gödel's *Nachlass* currently resides. Among those we must thank for their help with access to Gödel's papers are Jean Preston and Donald C. Skemer, and, most especially, Anne van Arsdale, Jane Moreton and Alice Clark for their patience with our seemingly endless requests for copies of papers and photographs.

The editorial board for this volume differs from that of Volumes I and II of these *Works*, due first to the death of Jean van Heijenoort and subsequently to the resignation of Gregory Moore (in order to devote full time to the Russell project) and Stephen Kleene (on account of ill health). All were substantial losses for us, but we are pleased to have been able to replace them with Warren Goldfarb and Charles Parsons. Furthermore, the work that Moore had done as managing editor and copy editor has been taken over by Cheryl Dawson.

We are indebted to the following experts who joined the members of the editorial board in preparing introductory notes for various of the texts in this volume: Robert M. Adams, George Boolos, Martin Davis, Dagfinn Føllesdal, Israel Halperin, David B. Malament, Wilfried Sieg, Howard Stein and A. S. Troelstra. Without their full engagement and willingness to take account of editorial suggestions, this volume would never have taken its present shape.

Transcription of Gödel's Gabelsberger shorthand, wherever needed here, was provided by Cheryl Dawson. In 1985, she and Tadashi Nagayama joined efforts in the long-term project of transcribing from Gödel's scientific and philosophical notebooks the material written in that script. That project is continuing, but regrettably, Nagayama had to leave the work before it was completed in order to return to his homeland. We are also grateful to Eckehart Köhler, to Stefan Bauer-Mengelberg and, most of all, to Wilfried Sieg for consultation and suggestions that shed light upon many dilemmas in the draft transcriptions of items presented in this volume.

John Dawson carried overall responsibility for translations from German to English. On various individual items he received assistance from Stefan Bauer-Mengelberg, William Craig, Eckart Förster, Patricia LaHay, Wilfried Sieg and Hans Sluga; their essential help was greatly appreciated. More specific credits for translation of individual items are given following the introductory notes pertaining thereto.

As in the first two volumes, responsibility for permissions and photographs also lay with John Dawson. Yasuko Kitajima of Aldine Press in Los Altos continued to provide the fine typesetting service in the TEX system. In the preparation of the comprehensive list of references and the index for this volume, as well as with the proof-reading of the volume as a whole, we have been ably assisted by Peter Johnson (who was in turn aided by Kathleen Healey). We also came to take for granted the efficiency and organization of the clerical staff at Penn State York, especially Carole V. Wagner. The Department of Mathematics at Stanford provided work space and some staff time; I wish particularly to thank my secretaries, Priscilla Feigen and Kathy Richards.

Finally, I wish to express my personal gratitude to my co-editors for their total engagement and dedication to this project, as well as to Cheryl Dawson for the excellence, care and thoroughness which she brought to the jobs of managing editor and copy editor; without them and without her, none of our plans would have come to fruition. One feature of our work together in recent years deserves special mention: namely, the use of the electronic mail system, which has radically transformed and facilitated our communications and decision-making, and has produced a new art of nagging. How did we ever live without it?

Solomon Feferman

Information for the reader

Dating of Gödel's works. We have continued to designate Gödel's works by date whenever we could determine the date of composition. To enable the reader to distinguish at a glance between those of Gödel's works published during his lifetime and those being published posthumously here, the date designators for the latter are preceded by an asterisk; for example, the lecture that Gödel apparently gave at the Conference on Epistemology of the Exact Sciences at Königsberg on 6 September 1930 has been designated *1930c.* In some cases, a date on the manuscript itself allowed us to date it; in other cases we relied on annotations on the copytext, internal evidence or contemporaneous correspondence. For details of our method of dating specific works, see either the corresponding introductory or textual notes. In cases where we know only a general time period in which Gödel worked on specific items, the date designator is given as a range of dates. When the date is uncertain and represents our best educated guess we have appended a question mark. When we are unsure when Gödel stopped work (e.g., *1961/?*), we have used a question mark for the end of the range.

Introductory notes. The purpose of the notes described in the preface above is (i) to provide a historical context for the items introduced, (ii) to explain their contents to a greater or lesser extent, (iii) to discuss further developments which resulted from them and (iv) in some cases to give a critical analysis. Each note was read in draft form by the editorial board, and then modified by the respective authors in response to criticisms and suggestions, the procedure being repeated as often as necessary in the case of very substantial notes. No attempt, however, has been made to impose uniformity of style, point of view, or even length. While the editorial board actively engaged in a critical and advisory capacity in the preparation of each note and made the final decision as to its acceptability, primary credit and responsibility for the notes rest with the individual authors.

Introductory notes are distinguished typographically by a running vertical line along the left- or right-hand margin and are boxed off at their end.[a] The authorship of each note is given in the Contents and at the end of the note itself.

References. This volume contains a comprehensive references section which includes a complete bibliography of Gödel's own published work as well as all items referred to in this volume.

---

[a]A special situation occurs when the note ends in mid-page before facing German and English text. Then the note extends across the top half of the facing pages and is boxed off accordingly.

In the list of references, each item is assigned a date with or without a letter suffix, e.g., "1930", "1930a", "1930b", etc.[b] The date is that of publication, where there is a published copy, or of presentation, for unpublished items such as a speech. A suffix is used when there is more than one publication in that year. (The ordering of suffixes does not necessarily correspond to order of publication within any given year.) For the sake of consistency throughout these volumes, we have continued the sequences of suffixes used in Volumes I and II rather than begin anew, even though the list of references here pertains only to this volume. A complete reference list for the whole set of volumes should thus be regarded as composed of that from the current volume plus all previous volumes in the series. It should be noted that in citations of Immanuel Kant's works, we have altered our format in order to conform to the usual practice in writings on Kant. Citations of the form A m/B n are, as usual, to the *Kritik der reinen Vernunft*, first edition (*Kant 1781*), page m/ second edition (*1787*), page n; citations merely of the form B n refer only to the second edition. Page references to other works of Kant are to the appropriate volume of the Academy edition of Kant's works (*Kant 1902–*).

Within the text of our volumes, all references are supplied by citing author(s) and date in italics, e.g., *Gödel 1930* or *Hilbert and Bernays 1934*. Where no name is specified or determined by the context, the reference is to Gödel's bibliography, as, e.g., in "Introductory note to *1929, 1930* and *1930a*". Examples of the use of a name to set the context for a reference are: "Lévy's proof was criticized by Steinitz (*1913*) ...", "as Hilbert does in *1926* ...", and "... proof was given by Sierpiński in *1947*". References to page numbers in Gödel's works are to those of the textual source, i.e., to the original page numbers in previously published works and to the manuscript page numbers of the works in this volume. The one exception is in citations of **1953/9-III*, where it proved to be more efficient and precise to cite the numbers of the very short sections. References to other items in these *Works* are cited by title, volume number and page number within the volume.

To make the references as useful as possible for historical purposes, authors' names are there supplied with first and/or middle names as well as initials, except when the information could not be determined. Russian names are given both in transliterated form and in their original Cyrillic spelling. In some cases, common variant transliterations of the same author's name, attached to different publications, are also noted.

Editorial annotations and textual notes. Editorial annotations within any of the original texts or their translations or within items quoted from

---

[b] "199?" is used for articles whose date of publication is not yet known.

other authors are signaled by double square brackets: [[ ]]; editorial emendations are discussed either in editorial footnotes (flagged by letters and
located below the horizontal line at the bottom of the page) or in the textual notes. Editorial footnotes are also used to emphasize an occasional
important textual problem of which the reader should be apprised. Single square brackets [ ] are used to incorporate corrections supplied by
Gödel. He also used square brackets occasionally, rather than parentheses; since we were not certain that these did not carry some significance
different from that of parentheses, we have not replaced them by parentheses. Each volume has, in addition, a separate list of textual notes
in which other corrections are supplied. Finally, the following kinds of
changes are made uniformly in the original texts: (i) footnote numbers are
raised above the line as simple numerals, e.g., ^{2} instead of $^{2)}$; (ii) spacing
used for emphasis in the original German is here replaced by italics, e.g.,
e r f ü l l b a r is replaced by *erfüllbar*, and underlining in manuscripts has
also been rendered in italics; (iii) references are replaced by author(s) and
date, as explained above; (iv) initial sub-quotes in German are raised, e.g.,
„engeren" becomes "engeren". In addition, spelling and capitalization have
been corrected without mention. Moreover, we have not attempted to indicate as such either deletions or insertions made by Gödel, but have simply
inserted or deleted text as indicated in the manuscript. Abbreviations
have been expanded throughout, except for "p." and "pp." in footnotes.
In most cases the intended word was entirely obvious. Where it was not,
the problem is discussed either in an editorial footnote or in the textual
notes to the individual work.

Logical symbols. The logical symbols used in Gödel's original articles
are here presented intact, even though these symbols may vary from one article to another. Authors of introductory notes have in some cases followed
the notation of the article(s) discussed and in other cases have preferred
to make use of other, more current, notation. Finally, logical symbols are
sometimes used to abbreviate informal expressions as well as formal operations. No attempt has been made to impose uniformity in this respect. As
an aid to the reader, we provide the following glossary of the symbols that
are used in one way or another in these volumes, where 'A', 'B' are letters
for propositions or formulas and '$A(x)$' is a propositional function of x or
a formula with free variable 'x'.

Conjunction ("A and B"): $A.B$, $A \wedge B$, $A \& B$
Disjunction ("A or B"): $A \vee B$
Negation ("not A"): $\overline{A}$, $\sim A$, $\neg A$
Conditional, or *Implication* ("if A then B"): $A \supset B$, $A \rightarrow B$
Biconditional ("A if and only if B"): $A \supset\subset B$, $A \equiv B$, $A \sim B$, $A \leftrightarrow B$
Universal quantification ("for all x, $A(x)$"): $(x)A(x)$, $\Pi x A(x)$, $x\Pi(A(x))$,
    $(\forall x)A(x)$

Existential quantification ("there exists an x such that $A(x)$"): $(Ex)A(x)$, $\Sigma x A(x)$, $(\exists x)A(x)$

Unicity quantification ("there exists a unique x such that $A(x)$"): $(E!x)A(x)$, $\Sigma!x A(x)$, $(\exists!x)A(x)$

Necessity operator ("A is necessary"): $\Box A$, NA

Possibility operator ("A is possible"): $\Diamond A$, MA

Minimum operator ("the least x such that $A(x)$"): $\epsilon x(A(x))$, $\mu x(A(x))$

Provability relation ("A is provable in the system S"): $S \vdash A$

Note: (i) The "horseshoe" symbol is also used for set-inclusion, i.e., for sets X, Y one writes $X \subset Y$ (or $Y \supset X$) to express that X is a subset of Y. (ii) Dots are sometimes used in lieu of parentheses, e.g., $A \supset . B \supset A$ is written for $A \supset (B \supset A)$.

Typesetting.   These volumes have been prepared by the TEX computerized mathematical typesetting system (devised by Donald E. Knuth of Stanford University), as described in the preface to Volume I. The resulting camera-ready copy was delivered to the publisher for printing. The computerized system was employed because: (i) much material, including the introductory notes and translations, needed to undergo several revisions; (ii) proof-reading was carried on as the project proceeded; (iii) in the case of previously published items, the papers could be prepared in a uniform, very readable form, instead of being photographed from the original articles. Choices of the various typesetting parameters were made by the editors in consultation with the publisher. Primary responsibility for preparing copy for the typesetting system in this volume lay with Cheryl A. Dawson, and the typesetting itself was carried out by Yasuko Kitajima.

For the aid of the reader who might wish to consult the original manuscripts, we have given Gödel's own pagination in the outside margins, with vertical bars in the text where the indicated page begins. The first page number is always omitted. However, it should be noted that Gödel occasionally added a page at the beginning, numbering it as page 0; in such cases, the beginning of his page 1 *is* indicated in the margin. Moreover, in *1953/9-III*, Gödel had specified many numbered sections; there we elected to give only the section numbers, as in most cases these give a more precise location than page numbers. Where the work is originally in German, we have not carried the page numbers across to English translations, since the word order in the translation does not always correspond closely enough to the German to permit giving a precise location.

Footnotes.   We use a combination of numbering and lettering, as follows. All of Gödel's own footnotes for his texts and their translations are numbered. Where Gödel numbered his footnotes in the order in which he wrote them, we have renumbered them in order of occurrence; in cases where he used his own non-numeric (and sometimes idiosyncratic) footnote symbols we have replaced them by numbering, except in Appendix B. (For treatment of footnotes there, see the introduction thereto.) To aid the reader who wishes to consult the original manuscripts, for those articles in which a large number of footnotes were renumbered, we have provided a concordance in the textual notes. For all the other material in this volume, footnotes are indicated by lower-case roman letters. See also *editorial annotations and textual notes* above.

Photographs. Primary responsibility for securing these lay with John Dawson. Their various individual sources are credited in the Permissions section, which follows directly.

Copyright permissions

The editors are grateful to the Institute for Advanced Study, Princeton, literary executors of the estate of Kurt Gödel, for permission to photocopy, transcribe, publish and translate manuscripts by Gödel found in his *Nachlass*. In addition, we wish to thank the following individuals for photographs used as illustrations in this volume:

Life Picture Sales, for the frontispiece portrait of Gödel by Alfred Eisenstaedt, copyright Time Warner Inc.

Mrs. Dorothy Morgenstern Thomas of Princeton, for the photo of Gödel and Einstein and that of Gödel standing before the entrance to the Firestone Library, both taken by her late husband, Oskar Morgenstern.

Dr. Veli Valpola of Espoo, Finland, for the picture of Gödel seated at his office desk.

Contents

Volume III

List of illustrations

Kurt Gödel
COLLECTED WORKS
Volume III

To the memory

of

Stephen C. Kleene

(5 January 1909–25 January 1994)

His fundamental contributions—especially in the areas of recursive function theory, intuitionism, and the foundations of computer science—have left an indelible imprint on mathematical logic. Our field has lost an enormous presence; we have lost a most valued friend and colleague.

The *Nachlass* of Kurt Gödel: an overview

Introduction

The *Nachlass* of Kurt Gödel was bequeathed by him to his wife Adele, who donated it to the Institute for Advanced Study in his memory prior to her death in 1981. Under terms of her will, literary rights to the papers are also vested in the Institute for Advanced Study.

The papers comprise documents relating to all periods of Gödel's life, including scientific correspondence, notebooks, drafts, unpublished manuscripts, academic, legal, and financial records and all manner of loose notes and memoranda. Family correspondence is notably absent, as are financial records after Gödel's emigration to the U.S. in 1940. Of the manuscript material, a substantial part is in Gabelsberger shorthand; a textbook for that obsolete German system is included among Gödel's books. Ancillary materials include an assortment of preprints and offprints sent to Gödel and some 700 books from Gödel's personal library. A bibliography of the latter is on file at the Institute for Advanced Study.

Cataloguing of the Gödel *Nachlass* was undertaken by the present writer in 1982 and completed in 1984. In 1985 the papers were transferred on indefinite loan to the manuscript archive in the Firestone Library at Princeton University. They are filed there as collection 282.

Scope and Content

The papers of Kurt Gödel include documents spanning the years 1905–1980, with the bulk of the material falling between 1930 and 1970. Of greatest extent and significance are Gödel's scientific correspondence (series 01), his notebooks (series 03), and numerous drafts, manuscripts and galleys of his articles and lectures, published and unpublished (series 04).

Prior to its arrangement, the Gödel *Nachlass* was stored in filing cabinets and moving cartons in a cage in the basement of the Historical Studies Library at the Institute for Advanced Study. Although gathered together in some haste and disarray following Gödel's death, most of the items were found in envelopes labelled by Gödel himself; on that basis, an attempt was made to retain or, where necessary, restore Gödel's original order. An exception was the division of the correspondence into two series (01, personal/scientific, and 02, institutional/commercial/incidental) for convenience of scholarly access. The former is arranged alphabetically by correspondent, the latter by subject. Folders and documents are numbered sequentially within each series, the first two digits serving as series designation. Thus folder 03/40 is the fortieth in series 03, and item 11013 the thirteenth in series 11. Envelopes bear the same number as their contents (or the first item among them, if more than one).

The correspondence in series 01 bulks between 1950 and 1975 but includes earlier items from such correspondents as Paul Bernays, Rudolf Carnap, Jacques Herbrand, Arend Heyting, Karl Menger, Emil Post, Oswald Veblen, John von Neumann and Ernst Zermelo. Other major correspondents include William Boone, Paul J. Cohen, Gotthard Günther, Georg Kreisel, Oskar Morgenstern, Abraham Robinson, Paul A. Schilpp, Dana Scott, Gaisi Takeuti, Jean van Heijenoort and Hao Wang. Approximately two-thirds of the correspondence is incoming. Family correspondence is virtually absent, but about 1000 pages of Gödel's letters to his mother are separately preserved in the *Neue Stadtbibliothek*, Vienna.

Early records in the Gödel *Nachlass* include patent correspondence of Gödel's father, birth and baptismal certificates, and Gödel's notebooks and report cards from elementary and secondary schools. Some university course notebooks are also preserved, but there are no enrollment or grade records from the University of Vienna.

Financial records (series 09) are quite detailed for the period 1930–1939 but are totally absent after Gödel's emigration in 1940. They include account books, cancelled checks and deposit slips, ledgers and various bills and receipts.

Series 05 and 06 comprise an assortment of loose manuscript notes and memoranda, including reading notes, library request slips, bibliographic memoranda, computation sheets (especially concerning Gödel's work in relativity theory) and personal notes on diverse subjects, including American history, languages, philosophy and theology. Especially prominent are voluminous notes on the works of Leibniz.

Gödel's personal notes and notebooks (series 03, 05 and 06) are largely in Gabelsberger shorthand, as are also some drafts of letters and lectures (series 01 and 04). Manuscript items are almost entirely in pencil.

Smaller categories include medical records (series 10), photographs (series 11) and ephemera (series 12). A few folders of correspondence from other sources have also been incorporated as addenda.

Of the assorted preprints and offprints sent to Gödel by others, three groups were segregated for retention: presentation copies, items accompanied by correspondence and items bearing annotations or with accompanying notes. Items within each of these groups are filed alphabetically by author but are not numbered or otherwise indexed. They were not transferred to Firestone Library with the other materials, but are presently housed together with Gödel's books in the basement of the Historical Studies Library at the Institute for Advanced Study.

A few non-documentary items, donated with the papers, were transferred for safekeeping to the Director's office at the Institute for Advanced

Study; they include Gödel's briefcase, door plate and National Medal of Science (medallion and lapel pin).

Series Descriptions

Series 01: Personal and scientific correspondence, 1927–1978

Boxes 1A–3C. Filed alphabetically by correspondent, and therein chronologically. Approximately 3500 items. Bulk dates: 1930–1975.

Incoming letters, and copies of outgoing letters, with friends and scientific colleagues; also letters of recommendation. Virtually no family correspondence. Major correspondents include Paul Bernays, William Boone, Paul J. Cohen, Gotthard Günther, Arend Heyting, Georg Kreisel, Karl Menger, Oskar Morgenstern, Abraham Robinson, Paul A. Schilpp, Dana Scott, Gaisi Takeuti, Jean van Heijenoort, John von Neumann and Hao Wang.

Series 02: Institutional, commercial, and incidental correspondence

Boxes 4A–5A. Filed by subject (type of correspondence). Approximately 1600 items.

Includes requests for biographical information, charitable solicitations, lecture invitations, editorial correspondence, honors bestowed, internal I.A.S. correspondence, literary solicitations, offprint and permissions requests and correspondence with professional societies. Also includes unsolicited correspondence from cranks and autograph seekers.

Series 03: Topical notebooks

Boxes 5B–7A. Filed by subject or title. Approximately 150 items. Mostly in Gabelsberger shorthand.

Includes school exercise books, university course notes, vocabulary notebooks, notes for Gödel's lectures at Vienna and Notre Dame and several series of notebooks on mathematical logic, philosophy and current events. Among the latter are sixteen mathematical workbooks (*Arbeitshefte*), fourteen labelled "Allg[emeine] Bild[ung]", nine history notebooks, six designated "Logic and foundations", four titled "Results on foundations" and fifteen philosophical notebooks containing material from before May 1941 until the end of Gödel's life. The philosophical notebooks were designated by Gödel as "Max" or "Phil", numbered

0–XV, of which volume XIII is missing; the second of three theological notebooks is also lost.

Series 04: Drafts and offprints of Gödel's lectures and papers

Boxes 7A–9B. Filed chronologically. Approximately 500 items.

Drafts, manuscript and typescript *Reinschriften*, galleys and offprints of Gödel's articles, lectures and reviews, published and unpublished, in English, German and Gabelsberger.

Series 05: Bibliographic notes and memoranda

Boxes 9B–11A. Filed by subject, and therein by date (where known). About 250 numbered groups (slips not all separately numbered).

Diverse notes and memoranda slips, including reading notes, library request slips, bibliographic excerpts and memoranda and memoranda books. Extensive notes on history, philosophy, and theology, especially the works of Leibniz (primary and secondary sources). Largely in Gabelsberger shorthand.

Series 06: Other loose manuscript notes

Boxes 11B–12. Filed by subject. Approximately 800 items.

Includes computation sheets, reference lists of formulas, miscellaneous mathematical notes and fragmentary drafts, and notes accompanying books and papers of others.

Series 07: Academic records and notices

Box 13A. Filed by type of document. Approximately 250 items.

Includes elementary and secondary school report cards, course announcements and enrollment slips for courses taught by Gödel, homework graded by Gödel, and administrative correspondence and announcements from the University of Vienna, Notre Dame and the I.A.S.

Series 08: Legal and political documents

Box 13A. Filed by type of document. Approximately 200 items.

Includes apartment rental agreements, birth, baptismal, marriage and citizenship certificates, copyright and publishing agreements, patent documents and correspondence of Gödel's father, passports and powers of attorney.

Series 09: Financial records, 1930–1939

Box 13B. Filed by type of document. Approximately 1100 items.

Account books, bank statements, currency exchange vouchers, cancelled checks, deposit slips, ledgers, securities transactions and various bills and receipts.

Series 10: Medical records

Box 14A. Filed by type of document. Approximately 150 items.

Dosage records, medical and dietary memoranda, lists of doctors, prescriptions and temperature records.

Series 11: Photographs

Boxes 14A–B, 15, and extra large container. Filed by subject. Approximately 200 photographic prints.

Snapshots of Gödel alone and with family, friends and colleagues. Some formal portraits. Photographs of the Gödel home in Princeton at 145 Linden Lane. Some photographs of Gödel's ancestors and his wife's parents. N.A.S.A. Mars and Apollo photos.

Series 12: Ephemera

Box 14C. Filed by type of document. Approximately 250 items.

Material not falling in any of the other series; includes advertisements, newspaper articles and clippings, annotated envelopes, concert programs, publisher's catalogs and Nazi proclamations issued to University of Vienna faculty.

John W. Dawson, Jr.

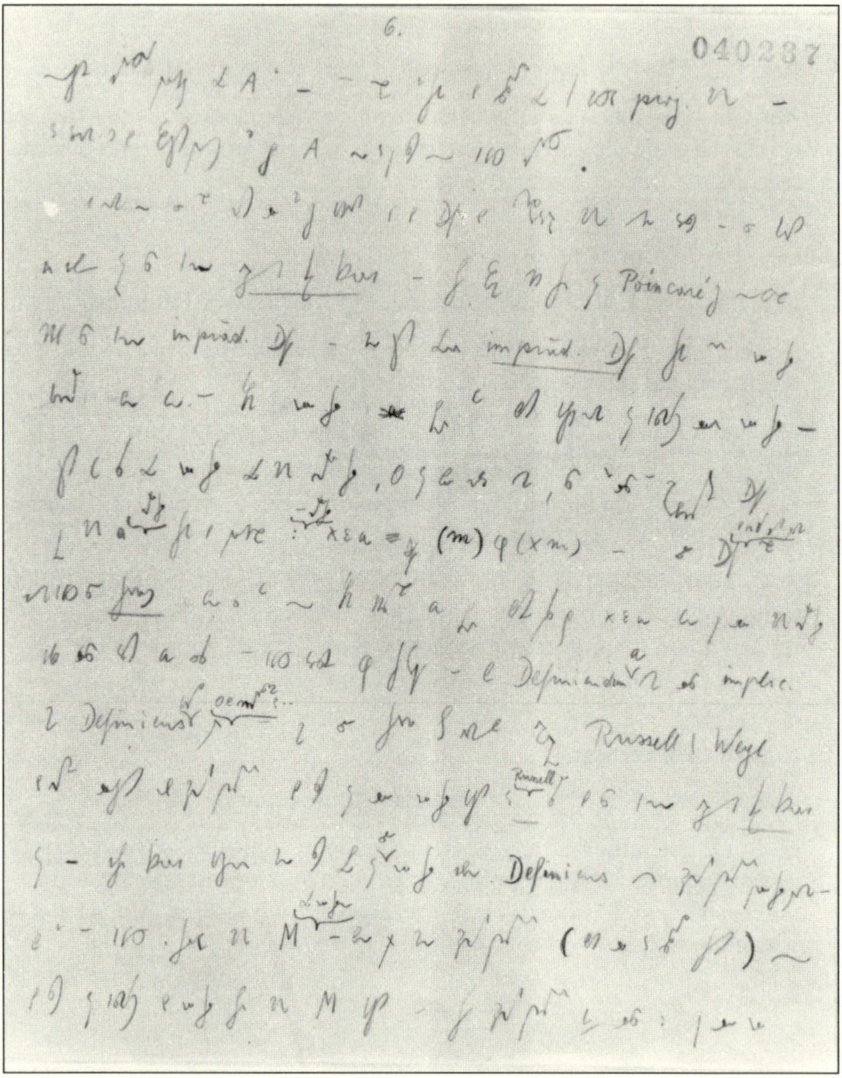

An example of Gödel's Gabelsberger shorthand (from **1939b*)

Gödel's Gabelsberger shorthand

Issues. Discussions of the *Nachlass* of Kurt Gödel often lead to questions about the shorthand he used. Quite a few stories have grown up around it: e.g., that it was a personal code that must literally be deciphered, or that Gödel used it to conceal what he wrote from other people. The myth of a personal code can be dispelled quite easily. The system he used, called "Gabelsberger", was once a standard system in Austria and Bavaria, indeed in most of southern Germany. Gödel learned it in his *Gymnasium* in the school years 1919–21, just before it was supplanted by the *Einheitskurzschrift*, a unification of several shorthand systems which had been competing for domination in Germany and Austria. The official date of the changeover in Austria was 1926, and, since Germany was even earlier than Austria in adopting the *Einheitskurzschrift*, at the present time there remain very few people who learned Gabelsberger as a "living" system. Thus we were extremely fortunate to enlist the aid of H. Landshoff, a retired photographer from New York City, who had learned Gabelsberger as a student in Munich. During the first two years of our project, he came weekly from New York to Princeton and provided invaluable assistance in transcribing all manner of documents. Without his interest and abilities, the time span of this project would have been considerably lengthened.

Because of the close relationship of the *Einheitskurzschrift* to the Gabelsberger, confusion has sometimes arisen, since the forms of the consonants in the later system were carried over from Gabelsberger but do not necessarily denote the same sounds they stood for in that system. So, at first glance, Gödel's script might very well look readable to someone who reads the modern system. But any attempt to impose the modern reading of consonants on the Gabelsberger system results in such nonsense that it *seems* like a code, or at least a system of very personal special signs.[a]

The second story, that Gödel used shorthand out of a desire for privacy, is perhaps more believable, but here again, the public nature of the Gabelsberger system reduces the credibility of this idea. While it is not unlikely that Gödel appreciated the privacy afforded by a system that few at the Institute for Advanced Study could read, secrecy would hardly seem

---

[a]In addition, Gabelsberger was adapted for many other languages in much the same way, using the sound frequencies of each particular language to determine what sounds would be represented by which symbols—for example, in the U.S.A. there was an English version called "Richter shorthand". Indeed, in the New York Public Library (before they phased out the old card catalog), one could find drawer after drawer of reference cards for books on shorthand systems, with at least one drawer devoted to German systems alone. A language that had an established shorthand would likely as not have "Gabelsberger" as part of the name of the shorthand.

to have been his primary reason for learning it. The only example we know where he obviously used shorthand for secrecy is in the midst of his English longhand notes for the logic course he once taught at Notre Dame, where he used it to conceal his examination questions.

Why then *would* a budding intellectual trouble to learn shorthand? At the present time in the U.S.A., there is a tendency to think of shorthand as useful primarily for business and legal purposes; however that has not always been the case—even today it is used elsewhere for many other purposes. For example, Woodrow Wilson used Graham shorthand extensively in taking lecture notes, in keeping personal diaries and in composing drafts of his writings. He once wrote (in shorthand) at the head of one of his diaries, "To save time is to lengthen life." He was very interested in shorthand systems and corresponded at length with various experts. Although Wilson found it necessary to learn shorthand on his own during a hiatus in his formal education, in Germanic countries shorthand seems to have been widely accepted as a standard part of a *Gymnasium* education. In addition to Gödel (whose "notebooks" are not really diaries, but mostly intellectual writings), others who used it for much the same purposes include Schrödinger, Waismann, Husserl, Heidegger (all of whom used Gabelsberger) and Carnap (who used the Stolze-Schrey system, the major competitor of Gabelsberger's system), as well as Zermelo. It was even used for correspondence by some; for example, Husserl addressed letters written in shorthand to Heidegger and others.[b] Moreover, shorthand saves not only time, but space as well, making it possible to copy large amounts of material much more efficiently than does longhand. This efficiency was especially advantageous before the advent of photocopying machines. There is a price, however, because skimming material written in shorthand becomes significantly harder after even a short delay, so that shorthand notes tend to have a certain temporal quality. When the material is fresh in the memory, reading shorthand is not much harder than reading longhand, but when the memory has faded, the task of reading even one's own notes becomes much more difficult. Indeed, when Gödel was asked about some of his early work on the independence of the Axiom of Choice, he recalled that his methods were closely related to those of Dana Scott, less so to Cohen's; but, he said, it would be too much effort to reconstruct the details from his shorthand notes.[c]

Usage.   Stenographic systems may be divided into families according to their form, and within that, according to their rules. In form they are either geometric or cursive, or a mixture of the two. Geometric shorthand

---

[b]Letter from S. IJsseling of the Husserl Archive to J. Dawson, 7 February 1983.

[c]Letter from Gödel to W. Rautenberg, published in *Mathematik in der Schule* 6, p. 20; quoted in *Dawson 1984a*, footnote 8.

systems use primarily straight lines, circles, dots or arcs, with little or no connection between elements, whereas cursive systems adapt parts of the elements of longhand cursive writing to represent specific sounds.[d] The earliest system of true stenography arose in Rome at the time of Cicero and was essentially geometric. Geometric systems reigned until 1818, when Franz Xaver Gabelsberger began to develop a cursive system, believing that cursive script would flow much more easily and therefore be much faster, as well as more aesthetic. (His *Anleitung zur deutschen Redezeichenkunst oder Stenographie*, first published in 1834, was followed and imitated by a long list of other cursive systems. The battles among proponents of the various systems became quite heated, at times even reaching the pitch of a moral issue.) In either case, a well-designed shorthand system minimizes redundancy and thus writing must be extremely accurate. This lack of redundancy also slows down the reading of the resulting text and may lead to problems in transcription.

A further division of shorthand systems into subfamilies is made on the basis of their treatment of the vowels. Earlier "tachygraphic" systems were systems of special signs for particular words, but were not particularly phonetic. In contrast, later systems were based much more on the sound patterns of the language. All of these systems have specific symbols for vowels, but some, including Gabelsberger, have other less time-consuming ways of indicating vowels as well. In particular, Gabelsberger normally uses a system of consonant-modification to indicate the occurrence of vowels before, between or after consonants. Thus an *a* may be denoted by a consonant symbol written darker on the downstroke (easily done with an old quill pen simply by bearing down harder), or it may be spelled out with a special symbol (when neither the preceding nor the succeeding consonant has a downstroke which can be shaded). Other vowels are denoted by raising or lowering the following consonant, or by altering the shape of the symbol slightly. If none of these modifications occur, the *e* is assumed. (If there is no word with *e* in such a context, one may use *ä*.) Since the writer of the shorthand thus has such latitude in deciding how to write a word, an unusual word may not be written consistently from one occurrence to another, because the writer may not yet have made up his mind how he wants to "spell" it. In addition, because Gödel wrote his notebooks in pencil, his manuscripts have faded over time, leading in some instances to uncertainty as to the exact reading of his text. Where this was a serious problem, we have flagged it with an editorial footnote.

In some methods of teaching the Gabelsberger system, a division is made into three levels. At the first level, everything is written out quite

---

[d]The Pitman system still in use in Great Britain and in parts of the northeastern U.S.A. is a geometric system, while the newer Gregg system is a mixture of cursive and geometric.

completely, with few elisions or abbreviations. At the second level, the
Verkehrsschrift ("commercial script"), certain standard elements may be
left out, to be filled in during transcription, while at the third level, the
Satzkürzung ("phrase abbreviation"), the stenographer is free to establish
further abbreviations specific to the field in which the dictation is being
taken. In the textbook used by Gödel,[e] the first two levels are melded into
one, with the assertion that if one intends to attain that level of skill, it
will be more efficient to learn it all at once, while the third level is put
off as being needed only by those taking large amounts of dictation profes-
sionally. Gödel learned and used the second level without going on to the
third and, fortunately for our transcription endeavor, he adhered strictly
to the system. In general, his shorthand is clearly written, with very few
personal abbreviations or special symbols. However, there are other inherent sources of difficulty in the transcription. Compound words need
not be concatenated or even hyphenated. Moreover, little if any punctuation is mandated, it being intended that the transciber add punctuation
as needed in transcription, without which there would be run-on sentences
and phrases. In the interests of efficiency in taking dictation, everything at
all superfluous is omitted. There are no capital letters, and thus that very
useful crutch for spotting German nouns is not available in the shorthand.
Most grammatical endings (on plural nouns, adjectives or regular verbs)
are omitted in writing, to be filled in as the material is transcribed. There
should be (and normally there is) only one way to fill everything in so as
to be consistent with all other information—but one does sometimes run
the risk of circularity. Furthermore, the person of the verb is often left
to the transcriber to determine. For example, the words *kann* and *kannst*
are both denoted by the same symbol, as are *wird*, *werde* and *werden*; but
kann and *können* are not. Similarly, while *will* and *wollen* have the same
symbol, *mag* and *mögen* do not.

   Though the system is generally quite phonetic, there are many special
signs, or *Sigeln*, which are used for common words or parts of words. These
sometimes take the form of a single shorthand letter, which, when used
alone, denotes an entire word. Thus the symbol for *f*, used alone, stands for
für, but when used at the end of a word, it could represent either a simple *f*
or one of the suffixes *-fach*, *-falls* or *-folge*. Particularly troublesome is the
prefix *un-*, which is denoted by the same symbol as the word *und*—since
a prefix is written close to but not necessarily attached to the stem of the
word, it is not always clear which is intended. Another very characteristic
way of making a symbol do multiple chores is to allow it to represent
different phonemes or morphemes according to its position above, on or

---

[e] *Weizman 1915*. Gödel's copy of the textbook was found among his books during
the cataloging process and was to be preserved with his *Nachlass*.

below The Line. (The Line is determined, in a rather indistinct fashion, by envisioning a line drawn through certain points on the symbols themselves.) So it is very easy for the writer of Gabelsberger shorthand to make an error, simply by forgetting to raise a symbol, by mistakenly raising it, or by raising it only slightly. The last is the most troublesome. For example, a simple dash, if raised sufficiently above the line, stands for *eine*; but if it is only slightly elevated, or slants a little, it is not clear whether it represents *eine*, might be the symbol for *sei* or really does represent an actual dash, which Gödel tended to use instead of periods. (This is one of his few idiosyncrasies.) Another particularly problematical example is the symbol for *ein*, because when elevated the same symbol may also stand for the verb *sein*, although the possessive adjective *sein* and each of its inflected forms has its own symbol. Both *ein* and *sein* are such frequently used words that a simple mistake in placement may render a reading totally nonsensical unless the transcriber is willing to charge the writer with an error; but when the writer is Kurt Gödel, such a presumption needs to be carefully justified.

There are also several special symbols whose forms are rather similar to symbols already in use elsewhere. For example, the symbol for the word *von* bears a fair resemblance to the Greek letter α, although, given the contexts in which they usually occur, this has been less of a problem than one might anticipate. The symbol for *aber* is identical to the mathematical symbol $<$, and this too must be determined from context. Even a simple dot may do double duty: as a period at the end of a sentence, or as the prefix *ab-*, *except* at the end of a sentence (where the prefix is written out).[f] In some cases, the original manuscript, rather than a photocopy or microfilm, must be inspected to know whether a given dot is in Gödel's hand, a flaw in the paper or an artifact of the copying process. Even on an original, deciding whether the dot is graphite or straw sometimes comes down to a question of the shininess of the dot.

Among Gödel's few personal *Sigeln*, the clearest is the shorthand *fu* for the word *Funktion*. Others are a bit more cryptic. The sign for *b* followed by an upward slanting line appears in contexts where *bedeuten* or *bezeichnen* would make sense. Since the upward slanting line can stand for either *ei* or *eu* (*if* there is no word with *ei* in the same context—if there is, the line must have a *u* attached at the end), and since Gödel also sometimes spelled both words out completely in shorthand, we have no basis for preferring one word over the other. Therefore we have simply filled in the choice that made sense in each particular occurrence of this symbol.

---

[f]Here Gödel seems to have modified the system for himself, using a dash instead of a period to avoid this difficulty.

While we cannot pretend there are no problems in transcribing Gödel's shorthand, neither do we intend to give the impression that it is a massive decoding job. In almost all cases, careful consultation with individuals expert in the subject matter of the manuscript has led to a good reading perfectly consistent with the shorthand, with reasonable allowance for human error in writing. In those cases where it has not, we have adopted the suggestions of our experts. All such places are noted, either in editorial footnotes or in the textual notes, if deemed of lesser importance.

We have decided not to indicate what parts of the text were in shorthand and what parts in longhand, even though Gödel often resorted to a few longhand/symbols in the midst of his shorthand. Names, in particular, occur in longhand in some places and in shorthand in others. In our original planning, we thought this distinction might be of some importance to the reader, but as transcription proceeded, it became apparent that there was no special significance to the hand in which a given word was written; it seems to have been simply a matter of Gödel's convenience at the time of writing. To make an issue of the distinction would have resulted in a much less readable text—and probably a less reliable one as well, because proofreading for font or special marks along with all the other variables would have been extremely time-consuming and difficult. We hope that the reader will find here a consistent, readable and reliable text.

Cheryl A. Dawson

Introductory note to *1930c

This item is, apparently, the lecture Gödel delivered on 6 September 1930 at the Conference on Epistemology of the Exact Sciences in Königsberg. In it Gödel presents his completeness theorem for first-order quantification theory, which he had proved in his doctoral dissertation *1929* (these *Works*, Volume I, pages 60–101) and published as *1930* (Volume I, pages 102–123). A one-paragraph abstract of the lecture was published as *1930a* (Volume I, pages 124–125). Further information on the Königsberg Conference and Gödel's participation in it can be found in the introductory note to *1931a*, Volume I, pages 196–199.

Gödel opens the lecture with an exposition of the completeness problem roughly similar to that in the first paragraph of *1930*, with no mention of the foundational issues discussed in his introduction to *1929*. First, he frames the general problem of showing that a logical system—in particular, that of *Principia mathematica*—can prove every logical truth. (Here, no doubt mindful of the members of the Vienna Circle in the audience, Gödel uses the term "tautology" for "logical truth"; this is not terminology he employs elsewhere.) He remarks that the problem has been solved affirmatively for a fragment, namely, the propositional calculus, where "logical truth" is taken to mean truth-functional validity, and situates his result as solving the problem for the next larger fragment, in which "logical truth" is taken to mean quantificational validity. Whereas the former proof depends on the decidability of truth-functional logic, Gödel notes, his proof for quantification theory goes through in the absence of knowledge about its decidability.

Throughout the lecture, Gödel considers only the "restricted functional calculus", that is, pure quantification theory without identity. He gives a concise description of its syntax and semantics, and goes on to sketch the body of the proof of *1929* and *1930*. After reformulating the theorem to state that every quantificational formula is either satisfiable or refutable, Gödel shows how this can be inferred for all formulas if it holds for formulas in Skolem normal form, that is, in prenex form with universal quantifiers preceding existential quantifiers. Using a simple example, he then illustrates how to generate from any formula F in Skolem normal form an infinite sequence of conjunctions of instantiations of F with the following properties: F provably implies the existential closure of each such conjunction; each such existential formula is either refutable or satisfiable; if each existential formula is satisfiable, then there is an interpretation making all of the conjunctions in the infinite sequence true at once; and any such interpretation immediately provides a model for F over the integers. All this, of course, implies that F is either refutable or satisfiable.

Naturally, in this sketch Gödel omits many details. For example, the reduction to Skolem normal form requires that various equivalences be provable in the axiom system. Gödel argues that the equivalences are correct and merely mentions that they also need to be shown provable. The passage from the satisfiability of each existential formula to that of an infinite class of formulas rests on the compactness of truth-functional logic. This step is elaborated in *1929* (these *Works*, Volume I, pages 84–87); in *1930* Gödel cites "familiar arguments" (his page 356), nowadays referred to as König's Lemma. In the present lecture Gödel says only, it "follows in a way I shall not go into here".

After presenting the proof, Gödel turns to what he calls an "application" of the result, namely, the derivation of a connection between the notions of categoricity and syntactic completeness. That is, by completeness, if a first-order theory T is categorical, then it is syntactically complete. The proof is straightforward, although not given by Gödel: if T is not syntactically complete, then there is a sentence F neither provable nor refutable in it; but then $T \cup \{F\}$ and $T \cup \{\sim F\}$ are both consistent, whence by completeness they both have models, which obviously cannot be isomorphic, so that T is not categorical. Two points are worth noting. First, if the argument is to apply to theories T with infinitely many non-logical axioms, then the completeness theorem in a stronger form than Gödel enunciates in the lecture must be assumed, namely, the form that applies to denumerably many formulas. This stronger form is proved in *1929* and *1930*. Second, in light of the upward and downward Löwenheim–Skolem Theorems, the result has no nontrivial applications, since no first-order theory that admits an infinite model can be categorical in Gödel's sense. However, those theorems yield a sharpening of the result: if a first-order theory T that has infinite models but no finite models is categorical in any infinite cardinal, then it is syntactically complete. (This seems to have been first observed in 1954, independently by J. Łoś and R. L. Vaught.[a] The sufficient condition for completeness is often called "Vaught's test". At the time of Gödel's writing, the downward but not the upward Löwenheim–Skolem Theorem was known.)

Finally, Gödel considers categoricity and syntactic completeness in the setting of higher-order logics. If such a logic L is proved complete—in the sense that any denumerable set of formulas in L is either satisfiable or else inconsistent (with respect to some axiom system)—then any theory framed with L as its underlying logic will be syntactically complete if it is categorical. Noting then that Peano Arithmetic is categorical—

---

[a]See *Łoś 1954*, p. 60 and *Vaught 1954*, p. 468.

where by Peano Arithmetic he means the second-order formulation—
Gödel infers that if higher-order logic is complete, then there will be
a syntactically complete axiom system for Peano Arithmetic. At this
point, he announces his incompleteness theorem: "The Peano axiom
system, with the logic of *Principia mathematica* added as superstructure, is not syntactically complete". He uses the result to conclude that
there is no (semantically) complete axiom system for higher-order logic.

There is an oddity here, in that the conclusion follows only given a
stronger claim, namely, that *no* sound axiom system for higher-order
logic, when added to the Peano axiom system, will yield a syntactically
complete system. Of course, his argument for incompleteness (as published in *1931*) does establish this, so the lacuna is one in exposition
only. Incidentally, Gödel's discussion of the interconnections among (semantic) completeness, syntactic completeness, and categoricity harkens
back to the end of the first paragraph of *1929*; indeed, this discussion
can serve to clarify the earlier, somewhat obscure, considerations (Volume I, pages 60–63; see also page 49).

Gödel's remarks are, clearly, the first public announcement of his incompleteness result. The next day, in the course of a roundtable discussion at the conference, Gödel expresses the result thus: "... one can even
give examples of propositions ... that, while contentually true, are unprovable in the formal system of classical mathematics" (*1931a*, his page
148). As this announcement was embedded in a discussion of whether
consistency is a sufficient "criterion of adequacy" for formal theories,
the nature of Gödel's remarkable result may not have been apparent. In
contrast, the statement of the result in the present lecture is direct and
emphatic. As is pointed out in Volume I, page 199, however, there seems
to have been little reaction at the conference to Gödel's announcements.

Warren Goldfarb

The translation is based on a draft translation by Jean van Heijenoort;
it was subsequently revised by John Dawson, William Craig and Warren
Goldfarb.

Translator's note: In this lecture, Gödel is careful to distinguish two
different notions of completeness, to wit: (1) that of the provability of
every valid sentence, and (2) that of the provability, for any sentence A
of the system, of either A or its negation. The corresponding German
adjectives he uses are, respectively, "vollständig" and "entscheidungsdefinit". To preserve the distinction, in this translation the former is
rendered by "complete", the latter by "syntactically complete".

Vortrag über Vollständigkeit
des Funktionenkalküls
(*1930c)

Das Thema meines Vortrages knüpft an die axiomatische Begründung der Logik an, wie sie z. B. in den *Principia mathematica* von Whitehead und Russell vorliegt. Das Wesen dieser Methode besteht bekanntlich darin, daß gewisse Axiome und Schlußregeln an die Spitze gestellt und in allen weiteren Entwicklungen keine anderen Hilfsmittel verwendet werden als eben diese Axiome und Schlußregeln. Das Problem, von dem ich hier einen Ausschnitt behandeln möchte, ist die Frage nach der Tragweite dieser Hilfsmittel, d. h. genauer: Ist wirklich jeder wahre Satz der Logik (jede Tautologie) mittels der Axiome und Schlußregeln der *Principia mathematica* in endlich vielen Schritten ableitbar? Dabei muß natürlich der Begriff der Tautologie, von dem hier die Rede ist, noch näher präzisiert werden, was im Folgenden auch geschehen wird. Das Problem, von dem ich eben sprach (ich will es kurz "das Vollständigkeitsproblem" nennen), ist bisher nur für den einfachsten Fall, nämlich den Aussagenkalkül gelöst. Dieser wird in den *Principia mathematica* auf fünf Axiomen begründet, von denen sich eines als überflüssig erwiesen hat. Die restlichen vier sind hier angeschrieben [[(wo $X \rightarrow Y$ die Formel $\overline{X} \vee Y$ abkürzt):

(I)
$\quad$ (i) $X \vee X \rightarrow X$ $\quad$ (iii) $X \vee Y \rightarrow Y \vee X$

$\quad$ (ii) $X \rightarrow X \vee Y$ $\quad$ (iv) $(X \rightarrow Y) \rightarrow (Z \vee X \rightarrow Z \vee Y)$]].

Als Schlußregeln gelten: Die Einsetzungsregel (für Aussagevariable darf eine beliebige Formel eingesetzt werden) und die Implikationsregel (aus A und $A \rightarrow B$ darf B geschlossen werden). Der Begriff der Tautologie ist hier sehr leicht in der bekannten Weise zu präzisieren: Eine Aussageformel heißt "tautologisch", wenn sie bei jeder Verteilung von Wahrheitswerten auf die in ihr vorkommenden Aussagevariablen immer den Wahrheitswert "wahr" ergibt. Das Vollständigkeitsproblem lautet also hier: Ist jede tautologi-

2 sche Aussageformel aus den obigen vier Axiomen nach den | angeführten Schlußregeln ableitbar? Daß man diese Frage beantworten kann (und zwar fällt die Antwort bejahend aus), beruht wesentlich darauf, daß für den Aussagenkalkül das Entscheidungsproblem gelöst ist, d. h., daß man Verfahren kennt, um in endlich vielen Schritten zu entscheiden, ob eine vorgelegte Aussageformel tautologisch ist oder nicht. Der Weg, der von hier aus zur Lösung des Vollständigkeitsproblems führt, ist kurz der folgende: Eines der Entscheidungsverfahren besteht darin, daß man durch Anwendung der De Morganschen Regeln und Ausmultiplizieren nach dem distributiven Gesetz

Lecture on completeness
of the functional calculus
(*1930c)

The theme of my lecture concerns the axiomatic grounding of logic, as it is presented, for example, in Whitehead and Russell's *Principia mathematica*. As is well known, the essence of this method consists in giving at the outset certain axioms and rules of inference, and in using in all further developments no devices other than just those axioms and rules of inference. The problem, a segment of which I would like to treat here, is the question of the power of these devices, that is, more exactly: Is, in fact, every true proposition of logic (every tautology) derivable in finitely many steps by means of the axioms and rules of inference of *Principia mathematica*? Here of course the concept of tautology that is at issue must be made more precise, as will also be done in what follows. The problem just mentioned (I shall call it for short "the completeness problem") has so far been solved only for the simplest case, namely the propositional calculus. In *Principia mathematica* the latter is based on five axioms, one of which has turned out to be redundant. The other four are listed here [[(where $X \to Y$ abbreviates $\overline{X} \vee Y$):

(I)   (i) $X \vee X \to X$ (iii) $X \vee Y \to Y \vee X$
     (ii) $X \to X \vee Y$ (iv) $(X \to Y) \to (Z \vee X \to Z \vee Y)$]].

As inference rules we have the substitution rule (for a propositional variable any formula may be substituted) and the implication rule (from A and $A \to B$, B may be inferred). Here, in the well-known manner, the concept of tautology is quite easily made precise: a propositional formula is said to be "tautological" if, under every assignment of truth values to the propositional variables that occur in it, it always receives the truth value "true". The completeness problem therefore takes the form: Is every tautological propositional formula derivable from the four axioms given above by means of the aforementioned rules of inference? That we can answer this question (with the answer turning out positive) rests essentially on the fact that the decision problem for the propositional calculus has been solved; that is, we know procedures for deciding in finitely many steps whether any given propositional formula is tautological or not. The path that leads from that to the solution of the completeness problem is, briefly, the following: One of the decision procedures consists in the construction, by use of the De Morgan rules and by multiplying out according to the distributive law,

eine gewisse Normalform herstellt, der man dann ohneweiters ansieht, ob sie eine tautologische ist. Da man bei diesem Entscheidungsverfahren also im Wesentlichen nur die De Morganschen Regeln und das distributive Gesetz verwendet und da diese beiden Sätze, wie man sich leicht überzeugt, aus den obigen Axiomen folgen, so läßt sich dieses Entscheidungsverfahren ins Formale übertragen, d. h. es liefert für jede tautologische Formel zugleich einen Beweis aus den obigen Axiomen und Schlußregeln.

Geht man im Aufbau der Logik einen Schritt über den Aussagenkalkül hinaus, so gelangt man zum sogenannten engeren Funktionenkalkül. Für diesen ist das Entscheidungsproblem keineswegs gelöst. Trotzdem gelingt es, wie ich im Folgenden zeigen will, auch für diesen Bereich den Vollständigkeitssatz zu beweisen. Ich erinnere zunächst daran, was man nach Hilbert und Ackermann [*1928*] unter "engerem Funktionenkalkül" versteht. Es gehören dazu diejenigen Formeln, welche sich aus den [an der Tafel] hierstehenden Grundzeichen aufbauen, d. h. Aussagevariable, Funktionsvariable, mit Individuen als Argumente, und den logischen Operationen (x), $\lor$, $\overline{}$, wobei (x) sich nur auf Individuen, nicht auf Funktionen, beziehen darf. Die übrigen logischen Konstanten (Ex), $\to$, & usw. lassen sich natürlich aus den obigen definieren. | Hier sind zur Erläuterung einige Formeln des engeren Funktionenkalküls angeschrieben.

$\llbracket$ (1) $(x)(y)(z)[F(x,y) \,\&\, F(y,z) \to F(x,z)]$

(2) $(w)(Ez)(x_1)(y_1)(Ex_2)(Ey_2)[F(x_1,y_1,x_2,y_2,z) \,\&\, G(x_2,y_2,w)]$

(3) $(x)(Ey)(z)(Eu)A(x,y,z,u)$

(4) $(v)(Ew)F(v,w) \,\&\, (x)(y)[F(x,y) \to (z)(Eu)A(x,y,z,u)]$

(4a) $\overline{(v)(Ew)F(v,w) \,\&\, (x)(y)[F(x,y) \to (z)(Eu)A(x,y,z,u)]}$

(4b) $\overline{(v)(Ew)(z)(Eu)A(v,w,z,u)}$
$\&\,(x)(y)[(z)(Eu)A(x,y,z,u) \to (z)(Eu)A(x,y,z,u)]$

(5) $(v)(x)(y)(z)(Ew)(Eu)\{F(v,w) \,\&\, [F(x,y) \to A(x,y,z,u)]\}\rrbracket$

Ich gehe nun dazu über, den Begriff der Tautologie für den engeren Funktionenkalkül zu definieren. Jede Formel des engeren Funktionenkalküls enthält freie Variable, d. h. solche, die nicht durch ( ) oder (E) Zeichen gebunden sind (Aussage- und Funktionsvariable, eventuell auch Individuenvariable). Setzt man für diese freien Variablen entsprechende Gegenstände ein, d. h. für die Aussagevariable Aussagen, für die Funktionsvariable Relationen mit der entsprechenden Stellenzahl [und] für die Individuenvariable (falls solche vorkommen) Individuen, wobei man natürlich einen bestimmten Individuenbereich zugrundelegen muß, so entsteht ein Satz, der entweder wahr oder falsch ist. Nimmt man z. B. als Individuenbereich die reellen Zahlen und setzt in der Formel (1) [oben] für F die Relation $>$ ein, so besagt der entstehende Satz, daß $>$ transitiv ist, ist also wahr. Nimmt man dagegen als Individuenbereich die natürlichen Zahlen und für F die Relation "Nachfolger von", so ist der entstehende Satz falsch. Nun

of a certain normal form, for which it is then immediately evident whether it is a tautology. Since this decision procedure thus uses essentially only the De Morgan rules and the distributive law, and since—as one readily sees—these two principles follow from the axioms above, the decision procedure can be cast into the formal framework: in other words, for every tautological formula it furnishes at the same time a proof from the axioms and rules of inference above.

If in the development of logic we go one step beyond the propositional calculus, we come to the so-called restricted functional calculus. For it the decision problem has by no means been solved. Nevertheless, as I shall show in what follows, for this part [of logic] too the completeness theorem can be proved. First let us recall what, according to Hilbert and Ackermann [*1928*], we are to understand by the "restricted functional calculus". To it belong those formulas that are constructed from the basic signs standing here [on the blackboard], that is, propositional variables, functional variables (with individuals as arguments), and the logical operations (x), $\lor$, and $\overline{}$, where (x) may be applied only to individuals, not to functions. The remaining logical constants, (Ex), $\rightarrow$, &, and so on, can of course be defined from those just given. As an illustration, a few formulas of the restricted functional calculus are written down here.

$$\llbracket \;\; (1) \;\; (x)(y)(z)[F(x,y) \, \& \, F(y,z) \rightarrow F(x,z)]$$
$$(2) \;\; (w)(Ez)(x_1)(y_1)(Ex_2)(Ey_2)[F(x_1,y_1,x_2,y_2,z) \, \& \, G(x_2,y_2,w)]$$
$$(3) \;\; (x)(Ey)(z)(Eu)A(x,y,z,u)$$
$$(4) \;\; (v)(Ew)F(v,w) \, \& \, (x)(y)[F(x,y) \rightarrow (z)(Eu)A(x,y,z,u)]$$
$$(4\mathrm{a}) \;\; \overline{(v)(Ew)F(v,w) \, \& \, (x)(y)[F(x,y) \rightarrow (z)(Eu)A(x,y,z,u)]}$$
$$(4\mathrm{b}) \;\; \overline{(v)(Ew)(z)(Eu)A(v,w,z,u)}$$
$$\& \, (x)(y)[(z)(Eu)A(x,y,z,u) \rightarrow (z)(Eu)A(x,y,z,u)]$$
$$(5) \;\; (v)(x)(y)(z)(Ew)(Eu)\{F(v,w) \, \& \, [F(x,y) \rightarrow A(x,y,z,u)]\} \; \rrbracket$$

I now go on to define the concept of tautology for the restricted functional calculus. Every formula of the restricted functional calculus contains free variables, that is, variables not bound by the signs ( ) or (E) ([namèly,] propositional and functional variables, possibly also individual variables). If for these free variables we substitute corresponding objects—that is, propositions for the propositional variables, relations with the proper number of places for the functional variables, and individuals for the individual variables (if such variables occur), where, of course, a definite domain of individuals must be chosen as a basis—then we obtain a proposition that is either true or false. If, for example, we take the real numbers as the domain of individuals, and if in formula (1) [above] we substitute the relation $>$ for F, then the resulting proposition states that $>$ is transitive, and thus is true. If, on the other hand, one takes the natural numbers as the domain of individuals and the relation "successor of" for F, then the

heißt eine Formel "tautologisch", wenn sie bei jeder beliebigen Einsetzung bei Zugrundelegung jedes beliebigen Individuenbereichs einen wahren Satz liefert.

Die Axiome, auf welche der engere Funktionenkalkül in den *Principia mathematica* aufgebaut wird, sind hier angeschrieben und mit (i) bis (vi) bezeichnet.

⟦ (i) – (iv) wie in (I)
  (v) $(x)F(x) \to F(y)$
  (vi) $(x)[X \lor F(x)] \to X \lor (x)F(x)$⟧

Als Schlußregeln gelten die Implikationsregel und die Einsetzungsregel für Aussage- und Funktionsvariable und ferner noch die folgende: Enthält eine Formel $A(x)$ die freie Individuenvariable x, so darf daraus die Formel $(x)A(x)$ geschlossen werden. Das Vollständigkeitsproblem lautet nun wieder: Ist jede Tautologie des engeren Funktionenkalküls mittels der eben angeführten Axiome und Schlußregeln ableitbar? Um diesen Satz zu beweisen, spreche ich ihn zunächst in einer etwas anderen Form aus und führe dazu zwei neue Begriffe ein, nämlich "erfüllbar" und "widerlegbar". "Tautologisch" hieß eine Formel, wenn sie bei jeder Einsetzung einen wahren
4   Satz liefert. "Erfüllbar" heißt sie dann, wenn | das für mindestens eine Einsetzung der Fall ist. "Widerlegbar" soll eine Formel heißen, wenn ihre Negation (d. h. die Formel, welche durch Vorsetzen eines Negationszeichens entsteht) beweisbar ist. Dabei bedeutet natürlich "beweisbar" hier und im Folgenden immer, mittels obiger sechs Axiome und drei Schlußregeln beweisbar. Jetzt können wir den Vollständigkeitssatz auch in der hier mit II bezeichneten Form aussprechen, nämlich so:

(II) Jede Formel des engeren Funktionenkalküls ist entweder erfüllbar oder widerlegbar.

Nehmen wir an, dieser Satz sei richtig. Ich will zeigen, daß dann jede tautologische Formel beweisbar ist. Sei A eine beliebige tautologische Formel, d. h., A ergibt bei jeder Einsetzung einen wahren Satz. Dann ergibt die Negation von A bei keiner Einsetzung einen wahren Satz, d. h. sie ist nicht erfüllbar; dann ist sie aber nach II widerlegbar. Wenn aber die Negation von A widerlegbar ist, ist A selbst beweisbar. A war aber eine beliebige tautologische Formel. Aus II folgt also die frühere Fassung und es gilt auch das Umgekehrte, worauf ich nicht näher eingehe.

Wir haben also zu zeigen, daß jede Formel des engeren Funktionenkalküls die Eigenschaft hat, entweder widerlegbar oder erfüllbar zu sein. Diese Eigenschaft will ich kurz mit E bezeichnen. Um den Beweis leichter zu führen definiere ich zunächst zwei neue Begriffe, nämlich den der "Normalformel" und den des "Grades" einer Normalformel. Normalformel soll eine Formel heißen, bei der sämtliche Präfixe, d. h. ( ) und (E) Zeichen zu

resulting proposition is false. Now a formula is said to be "tautological" if, under any substitution whatever based on any ⟦non-empty⟧ domain of individuals whatever, it yields a true proposition.

The axioms on which the restricted functional calculus is built in *Principia mathematica* are listed here and numbered (i) to (vi).

⟦ (i) – (iv) as above
  (v) $(x)F(x) \rightarrow F(y)$
  (vi) $(x)[X \vee F(x)] \rightarrow X \vee (x)F(x)$ ⟧

As rules of inference we have the implication rule, the substitution rule for propositional and functional variables, and moreover the following: If a formula $A(x)$ contains the free individual variable x, then from it the formula $(x)A(x)$ may be inferred. The completeness problem now takes the form: Is every tautology of the restricted functional calculus derivable by means of the axioms and rules of inference just cited? To prove this theorem, I will first express it in a somewhat different form, and toward that end I introduce two new concepts, namely, "satisfiable" and "refutable". A formula was said to be tautological if it yields a true proposition under every substitution. It shall be said to be satisfiable if that is the case for at least one substitution. A formula shall be said to be refutable if its negation (that is, the formula that results by prefixing a negation sign) is provable. Of course, here and in what follows, "provable" always means provable by means of the six axioms and three rules of inference given above. Now we can also express the completeness theorem in the following form, here denoted by II:

(II) Every formula of the restricted functional calculus is either satisfiable or refutable.

Let us suppose that this theorem is correct. I shall show that then every tautological formula is provable. Let A be any tautological formula whatever, so that A yields a true proposition under any substitution. Then under no substitution does the negation of A yield a true proposition, that is, it is not satisfiable; but then, by II, it is refutable. If, however, the negation of A is refutable, then A itself is provable. But A was any tautological formula whatever. Hence the earlier version follows from II. The converse also holds, but I shall not go into details.

We thus have to show that every formula of the restricted functional calculus has the property of being either refutable or satisfiable. For brevity I shall denote this property by E. In order to carry out the proof more easily, I first define two new concepts, namely, that of "normal formula" and that of the "degree" of a normal formula. A formula shall be called a normal formula if all the quantifiers, that is, the signs ( ) and (E), occur at the

Anfang stehen [und sich auf die ganze Formel beziehen], und zwar an
erster Stelle ein All- und an letzter Stelle ein E-Zeichen. Hier [Formel
(2) oben] ist als Beispiel eine Normalformel angeschrieben. Unter Grad
einer Normalformel verstehe ich die Anzahl der Komplexe von Allzeichen,
also zwei in obigem Beispiel. Diese Zahl hängt natürlich mit der Anzahl
der Wechsel zwischen All- und E-Zeichen nahe zusammen. Der Beweis
des Vollständigkeitssatzes wird sich in drei Schritten vollziehen, die hier
angegeben sind. [Es] wird zu zeigen sein:

5 | 1. Wenn jede Normalformel die Eigenschaft E hat (d. h. entweder wider-
legbar oder erfüllbar ist), dann gilt dasselbe für jede Formel.

2. Wenn jede Normalformel k-ten Grades die Eigenschaft E hat, dann
auch jede $(k + 1)$-ten Grades.

3. Jede Normalformel ersten Grades hat die Eigenschaft E.

Wenn diese drei Sätze bewiesen sind, es wird also das Problem zunächst auf
das entsprechende für Normalformeln reduziert und dann vollständige Induktion nach dem Grad angewendet. Punkt 1 ist rasch erledigt. Man kann
ja bekanntlich jede Formel auf die Normalform bringen, d. h. eine andere
angeben, deren Präfixe sämtlich am Anfang stehen und die mit der ersten
äquivalent ist, und zwar ist diese Äquivalenz nicht nur richtig, sondern auch
beweisbar. Durch einen kleinen Kunstgriff, den ich hier übergehe, kann
man auch immer eine Normalform finden, deren Präfix mit einem Allzeichen beginnt und mit einem E-Zeichen endet, die also eine Normalformel
in unserem Sinn ist. Die Erfüllbarkeit bzw. Widerlegbarkeit dieser Normalformel zieht nun offenbar die der ursprünglichen Formel nach sich, da ja die
Äquivalenz zwischen beiden sogar beweisbar ist, und daraus folgt natürlich,
daß wenn diese Normalformel entweder erfüllbar oder widerlegbar ist, dasselbe auch für die vorgelegte Formel gilt, womit Punkt 1 erledigt ist.

Sätze 2 und 3 zeige ich nach der Methode, die Skolem zum Beweise
des Löwenheimschen Satzes verwendet hat. Zunächst Punkt 2. Der Beweis beruht hier darauf, daß man zu jeder Formel $(k + 1)$-ten Grades eine
k-ten Grades angeben kann, deren Erfüllbarkeit die Erfüllbarkeit der ursprünglichen und deren Widerlegbarkeit die Widerlegbarkeit der ursprüng-

6 lichen zur Folge hat. Ich erläutere dies an einem einfachen Beispiel. | Die
vorgelegte Formel $(k + 1)$-ten Grades habe die hier [oben] angeschriebene
Gestalt (3), also den zweiten Grad. $A(x, y, z, u)$ ist dabei aus Funktionsvariablen allein mittels der Operationen des Aussagenkalküls aufgebaut zu
denken (also ohne (x) und (Ex)). Nun betrachten Sie die in der nächsten
Zeile angeschriebene Formel (4). Der darin vorkommende Buchstabe F
bedeutet eine zweistellige Funktionsvariable. Für diese Formel (4) gilt folgendes:

i. Ihre Normalformel (sie ist hier [oben] als Formel 5 angeschrieben) hat
einen um 1 geringeren Grad als die vorgelegte Formel, nämlich den ersten.

ii. Aus ihrer Erfüllbarkeit bzw. Widerlegbarkeit folgt die Erfüllbarkeit
bzw. Widerlegbarkeit der vorgelegten Formel.

beginning [and their scope extends to the right end of the formula] and if, moreover, a universal quantifier comes first and an existential quantifier last. Here [formula (2) above] an example of a normal formula is written down. By the degree of a normal formula I understand the number of blocks of universal quantifiers—thus two in the example above. This number is of course closely related to the number of alternations between universal and existential quantifiers. The proof of the completeness theorem will be carried out in three steps, which are stated here. It is to be shown:

1. If every normal formula has property E (that is, is either refutable or satisfiable), then so does every formula.

2. If every normal formula of degree k has property E, so does every one of degree $k + 1$.

3. Every normal formula of degree 1 has property E.

The problem is thus first reduced to the corresponding problem for normal formulas, and then mathematical induction is applied to the degree. Point 1 is quickly disposed of. After all, as is well known, every formula can be brought into normal form, that is, we can specify another formula, all of whose quantifiers occur at the beginning, which is equivalent to the first; and indeed, this equivalence is not only correct, but also provable. By a little artifice, which I pass over here, we can also always find a normal form whose prefix begins with a universal quantifier and ends with an existential one, and which is therefore a normal formula in our sense. The satisfiability (or the refutability) of this normal formula now evidently entails that of the original formula, since the equivalence between the two is indeed provable. From this it follows, of course, that if this normal formula is either satisfiable or refutable, then the same also holds for the given formula. Point 1 is thus disposed of.

I will establish propositions 2 and 3 by the method Skolem used to prove Löwenheim's theorem. First, point 2. Here the proof rests on the fact that for every formula of degree $k+1$ we can specify a formula of degree k whose satisfiability (or whose refutability) entails that of the original formula. I illustrate this by a simple example. Let the given formula of degree $k + 1$ have the form (3) written here [above], hence degree 2. In it, $A(x, y, z, u)$ is to be thought of as built up from functional variables solely by means of the operations of the propositional calculus (hence without (x) or (Ex)). Now consider formula (4) [above], written on the next line. The letter F occurring in it denotes a two-place functional variable. For this formula (4) the following holds:

i. Its normal formula (which is written here [above] as formula (5)) has a degree one less than that of the given formula, namely, degree 1.

ii. From its satisfiability (or refutability) follows the satisfiability (or refutability) of the given formula.

Ich zeige das zunächst für die Widerlegbarkeit. Nehmen wir an, die Formel (4) sei widerlegbar. Dann ist ihre Negation (d. h. Formel (4a)) beweisbar. Führen wir nun in dieser Formel die darunter ⟦an der Tafel⟧ angeschriebene Substitution aus, d. h., setzen wir an Stelle von F ⟦(mit, in jedem Falle, den bezeichneten Individuenvariablen)⟧ diejenige Formel, welche aus der ursprünglichen Formel dadurch entsteht, daß man die ersten beiden Zeichen des Präfixes wegläßt ⟦und die entsprechenden Individuenvariablen durch die bezeichneten ersetzt⟧. Dadurch ergibt sich Formel (4b), welche also auch beweisbar ist. Sie ist die Negation einer Konjunktion, besagt also, daß mindestens eines der beiden Konjunktionsglieder falsch ist. Nun ist aber das zweite richtig (und sogar beweisbar), daher ist das erste widerlegbar. Das erste Glied ist aber nichts anderes, als ⟦eine alphabetische Variante der⟧ vorgelegten Formel (3). Aus der Widerlegbarkeit von (4) folgt also die von (3). Man überzeugt sich ebensoleicht, daß aus der Erfüllbarkeit von (4) die Erfüllbarkeit von (3) folgt, was ich hier nicht näher ausführe. Das hier skizzierte Verfahren läßt sich ganz allgemein durchführen und führt
7   immer zu | einer Formel niedrigeren Grades, deren Erfüllbarkeit die Erfüllbarkeit und deren Widerlegbarkeit die Widerlegbarkeit der ursprünglichen zur Folge hat, und damit ist Punkt 2 des obigen Gedankenganges erledigt.

Nun zu Punkt 3, d. h. zu dem Satz, daß jede Formel ersten Grades entweder erfüllbar oder widerlegbar ist. Ich erläutere das Beweisverfahren wieder an einem einfachen Beispiel. Ich nehme an, die vorgelegte Formel ersten Grades, von der zu zeigen ist, daß sie die Eigenschaft E hat, habe die hier angegebene einfache Gestalt.

⟦ (6) $(x)(Ey)(Ez)A(x, y, z)$⟧

$A(x, y, z)$ ist dabei ebenso wie früher nur mittels der Operationen des Aussagenkalküls ($\lor$, Negation, usw.) aufgebaut zu denken. Nun betrachten Sie, bitte, die hier angeschriebene Reihe von Formeln.

⟦ (7) $A(x_0, x_1, x_2), A(x_1, x_3, x_4), A(x_2, x_5, x_6), \ldots, A(x_n, x_{2n+1}, x_{2n+2})$⟧

Sie entstehen aus $A(x, y, z)$ dadurch, daß man an Stelle von x, y, z anders bezeichnete Variablen einsetzt, und zwar an Stelle von x, welches hier ⟦in Formel (6)⟧ durch ein Allzeichen gebunden ist, der Reihe nach die Variablen x_0, x_1, x_2 usw., an Stelle von y, z jeweils zwei neue Variablen, d. h. solche, die in den vorhergehenden Formeln noch nicht vorkommen. Die Konjunktion (&-Verknüpfung) der ersten n so gebildeten Formeln bezeichne ich mit A_n und mit $(P_n)A_n$ die Formel, welche daraus entsteht, wenn sämtliche in A_n vorkommende Variable durch E-Zeichen gebunden werden. Nun ist $(P_n)A_n$ offenbar eine Folgerung der obigen Formel (6), die ich mit $(P)A$ abkürze, d. h. die hier stehende Formel (8)

⟦ (8) $(P)A \to (P_n)A_n$⟧

I show this first for refutability. Let us assume that formula (4) is refutable. Then its negation (that is, formula (4a)) is provable. Let us now carry out on this formula the substitution written down underneath it, that is, in place of F [(with, in each instance, the indicated individual variables)] let us put the formula that we obtain from the original formula [(3)] by deleting the first two quantifiers [and replacing the corresponding individual variables by those indicated]. Thus we obtain formula (4b), which is therefore provable too. It is the negation of a conjunction, and so states that at least one of the two conjuncts is false. But now the second one is true (and indeed provable), hence the first one is refutable. The first conjunct, however, is nothing but [an alphabetic variant of] the given formula (3). Hence from the refutability of (4) that of (3) follows. Just as easily we may convince ourselves that from the satisfiability of (4) that of (3) follows; I won't go into further details here. The procedure sketched here can be carried out quite generally and always leads to a formula of lower degree whose satisfiability (refutability) entails that of the original formula; and with that point 2 in the above argument is disposed of.

Now to point 3, that is, to the proposition that every formula of degree 1 is either satisfiable or refutable. Once again I illustrate the proof procedure by a simple example. I assume that the given formula of degree 1, which is to be shown to have property E, has the simple form given here.

[(6) $(x)(Ey)(Ez)A(x, y, z)$]

Here, just as before, $A(x, y, z)$ is to be thought of as built up solely by means of the operations of the propositional calculus ($\lor$, negation, and so on). Now consider, if you will, the sequence of formulas written here.

[(7) $A(x_0, x_1, x_2), A(x_1, x_3, x_4), A(x_2, x_5, x_6), \ldots, A(x_n, x_{2n+1}, x_{2n+2})$]

They result from $A(x, y, z)$ when $x, y,$ and z are replaced by different variables. Specifically, in place of x, which is here [in formula (6)] bound by a universal quantifier, we put successively the variables $x_0, x_1, x_2,$ and so on, and in place of y and z put in each case two new variables, that is, variables that do not occur in the preceding formulas. I denote by A_n the conjunction of the first n formulas thus formed, and by $(P_n)A_n$ the formula that results when all the variables occurring in A_n are bound by [initial] existential quantifiers. Now $(P_n)A_n$ is obviously a consequence of formula (6) above, which I abbreviate by $(P)A$; that is, formula (8) here

[(8) $(P)A \rightarrow (P_n)A_n$]

ist tautologisch (allgemeingültig). Denn aus dem Bestehen von $(P)A$ folgt
zunächst die Existenz von Individuen x_0, x_1, x_2, für welche $A(x_0, x_1, x_2)$
gilt (denn gemäß $(P)A$ gibt es ja sogar für jedes x_0 zwei Individuen x_1, x_2,
für die das gilt). Für das Individuum x_1 gibt es aber nach $(P)A$ wieder zwei
8   Individuen x_3, x_4, so daß $A(x_1, x_3, x_4)$ gilt und in dieser Weise | schließt
man weiter. Es zeigt sich nun, daß die hier angeschriebene Implikation (8)
nicht nur tautologisch sondern auch beweisbar ist, was ich hier nicht näher
ausführen kann.

Die Formeln $(P_n)A_n$ haben eine besonders einfache Bauart. Ihr Präfix
besteht nur aus E-Zeichen. Für solche Formeln sind alle Fragen lösbar; insbesondere kann man leicht zeigen, daß jede solche Formel entweder widerlegbar oder erfüllbar ist. Da also jedes einzelne $(P_n)A_n$ entweder widerlegbar oder erfüllbar ist, sind nur zwei Möglichkeiten denkbar: Entweder
sind alle $(P_n)A_n$ erfüllbar oder mindestens eine ist widerlegbar. Nun werde
ich zeigen, daß aus der Widerlegbarkeit einer einzigen der Formeln $(P_n)A_n$
die Widerlegbarkeit der vorgelegten Formel $(P)A$ und aus der Erfüllbarkeit
aller die Erfüllbarkeit von $(P)A$ folgt. Damit ist natürlich gezeigt, daß $(P)A$
entweder erfüllbar oder widerlegbar ist, was zu beweisen war. Daß aus der
Widerlegbarkeit eines $(P_n)A_n$ die von $(P)A$ folgt, ergibt sich natürlich sofort aus dieser Implikation (8), genauer gesagt aus der Beweisbarkeit dieser
Implikation. Daß aus der Erfüllbarkeit aller $(P_n)A_n$ die Erfüllbarkeit von
$(P)A$ folgt, zeigt man etwa so: A_n ist ja definiert als die &-Verknüpfung
der ersten n hier angeschriebenen Formeln ([und $(P_n)A_n$ ist A_n] mit einem
Präfix aus E-Zeichen). Aus der Erfüllbarkeit aller $(P_n)A_n$ folgt daher,
daß die Formeln jedes endlichen Abschnittes der obigen Reihe gleichzeitig erfüllbar sind. Daraus aber folgt in einer hier nicht näher anzugebenen
Weise die gleichzeitige Erfüllbarkeit aller abzählbar vielen Formeln der obigen Reihe. Das System von Funktionen und Individuen, welches das leistet,
d. h. alle Formeln der obigen Reihe zu wahren Sätzen macht, ergibt aber
auch in die Formel $(P)A$ eingesetzt einen wahren Satz. Denn jedes Individuum x kommt in dieser Reihe irgendwo an erster Stelle vor und es
gibt folglich zwei weitere y, z, für die $A(x, y, z)$ gilt. Gerade das aber behauptet $(P)A$. Das eben skizzierte Verfahren läßt sich auf jede beliebige
Formel ersten Grades anwenden und liefert so den Beweis, daß jede solche
Formel entweder erfüllbar oder widerlegbar ist und dies war der letzte
9 | Punkt des obigen Gedankenganges. Damit ist die Vollständigkeit der
Axiome und Schlußregeln des engeren Funktionenkalküls nachgewiesen,
d. h. es ist gezeigt, daß in diesem Bereich jede tautologische Formel
beweisbar ist oder daß die Begriffe Tautologie und beweisbare [Formel]
umfangsgleich sind.

Ich möchte noch auf eine Anwendung aufmerksam machen, die man von
dem Bewiesenen auf die allgemeine Theorie der Axiomensysteme machen
kann. Sie betrifft die Begriffe "entscheidungsdefinit" und "monomorph".

is tautological (logically valid). From the fact that $(P)A$ is the case, there follows, first of all, the existence of individuals x_0, x_1, x_2 for which $A(x_o, x_1, x_2)$ holds (for, according to $(P)A$, there are indeed, for every x_0, two individuals x_1 and x_2 for which it holds). But for the individual x_1 there are, according to $(P)A$, again two individuals, x_3 and x_4, such that $A(x_1, x_3, x_4)$ holds, and one can go on arguing in this way. Now it can be shown that implication (8), written here [above], is not only tautological but also provable; but I cannot go into further detail about that here.

The formulas $(P_n)A_n$ have a particularly simple structure. Their prefix consists only of existential quantifiers. For such formulas all questions are solvable; in particular, it can easily be shown that every such formula is either satisfiable or refutable. Since therefore each one of the $(P_n)A_n$ is either refutable or satisfiable, only two possibilities are conceivable: Either all $(P_n)A_n$ are satisfiable or at least one of them is refutable. I shall now show that from the refutability of a single one of the formulas $(P_n)A_n$ there follows the refutability of the given formula $(P)A$, and from the satisfiability of all of them there follows the satisfiability of $(P)A$. With that, of course, it is shown that $(P)A$ is either satisfiable or refutable, as was to be proved. That from the refutability of a $(P_n)A_n$ there follows that of $(P)A$ is, of course, an immediate consequence of the implication (8), or, more exactly, of the provability of this implication. That from the satisfiability of all $(P_n)A_n$ there follows the satisfiability of $(P)A$ can be shown along the following lines: A_n is defined as the conjunction of the first n formulas written here [in (7) above] ([and $(P_n)A_n$ is A_n] with a prefix consisting of existential quantifiers). From the satisfiability of all $(P_n)A_n$ it therefore follows that the formulas of every finite segment of the sequence above are simultaneously satisfiable. But from that there follows, in a way I shall not go into here, the simultaneous satisfiability of all denumerably many formulas of the sequence above. Now, the system of functions and individuals which achieves that, that is, which turns all formulas of the sequence above into true propositions, also yields a true proposition when substituted in the formula $(P)A$; for every individual x has a first occurrence somewhere in this sequence, and consequently there are two further individuals y and z for which the formula $A(x, y, z)$ holds. But that is just what $(P)A$ asserts. The procedure just sketched can be applied to any formula whatever of degree 1, and so furnishes the proof that every such formula is either satisfiable or refutable; that was the last point of the line of argument outlined above. The completeness of the axioms and rules of inference of the restricted functional calculus is thus demonstrated, that is, it is shown that in this domain every tautological formula is provable, or that the concepts of tautology and of provable formula have the same extension.

I would furthermore like to call attention to an application that can be made of what has been proved to the general theory of axiom systems. It concerns the concepts "syntactically complete" and "monomorphic".

Ein Axiomensystem heißt bekanntlich entscheidungsdefinit, wenn jeder einschlägige Satz aus den Axiomen entscheidbar, d. h. entweder er selbst oder seine Negation in endlich vielen Schritten ableitbar ist. Dagegen heißt es monomorph, wenn je zwei seiner Realisierungen isomorph sind. Es läßt sich vermuten, daß zwischen diesen beiden Begriffen ein enger Zusammenhang besteht, doch läßt sich ein solcher bisher nicht allgemein angeben. Man kennt ja sehr viele monomorphe Axiomensysteme, von denen man keineswegs weiß, ob sie entscheidungsdefinit sind, z. B. das der euklidischen Geometrie. Auf Grund der hier vorgebrachten Entwicklungen läßt sich nun zeigen, daß für eine spezielle Klasse von Axiomensystemen, nämlich die, deren Axiome sich im engeren Funktionenkalkül ausdrücken lassen, aus der Monomorphie immer die Entscheidungsdefinitheit folgt, genauer die Entscheidungsdefinitheit für solche einschlägige Sätze, die sich im engeren Funktionenkalkül ausdrücken lassen. Dabei läßt sich im engeren Funktionenkalkül ein Satz (Axiom) dann ausdrücken, wenn darin nicht von Mengen oder Folgen von Grunddingen die Rede ist, z. B. alle Axiome der Geometrie mit Ausnahme der Stetigkeitsaxiome. Könnte man den Vollständigkeitssatz auch für die höheren Teile der Logik (den erweiterten Funktionenkalkül) beweisen, so würde sich ganz allgemein zeigen lassen, daß aus der Monomorphie die Entscheidungsdefinitheit folgt und da man z. B. weiß, daß das Peanosche Axiomensystem monomorph ist, würde daraus die Lösbarkeit jedes in den *Principia mathematica* ausdrückbaren Problems der Arithmetik und Analysis folgen.

10 | Indessen ist eine solche Ausdehnung des Vollständigkeitssatzes, wie ich ihn in letzter Zeit bewiesen habe, unmöglich, d. h. es gibt mathematische Probleme, die sich in den *Principia mathematica* zwar ausdrücken aber mit den logischen Hilfsmitteln der *Principia mathematica* nicht lösen lassen. Dabei werden als Axiome das Reduzibilitätsaxiom, das Unendlichkeitsaxiom (in der Fassung: Es gibt genau abzählbar [viele] Individuen), sogar das Auswahlaxiom zugelassen. Man kann das auch so ausdrücken: Das Peanosche Axiomensystem mit der Logik der *Principia mathematica* als Überbau ist nicht entscheidungsdefinit. Doch würde es zu weit führen, auf diese Dinge näher einzugehen.

As is well known, an axiom system is said to be syntactically complete if every proposition in the underlying language is decidable on the basis of the axioms, that is, either it itself or its negation is derivable in finitely many steps. On the other hand, it is said to be monomorphic [categorical] if any two of its realizations are isomorphic. One would suspect that there is a close connection between these two concepts, yet up to now such a connection has eluded general formulation. Indeed, quite a few monomorphic axiom systems are known—Euclidean geometry, for example—for which we have no idea whether they are decidable. In view of the developments presented here it can now be shown that, for a special class of axiom systems, namely, those whose axioms can be expressed in the restricted functional calculus, syntactical completeness always follows from monomorphicity [categoricity] (more exactly, syntactical completeness for those propositions of the underlying language that can be expressed in the restricted functional calculus). Here a proposition (axiom) can be expressed in the restricted functional calculus if there is no mention in it of sets or sequences of basic objects (for example, each axiom of geometry with the exception of the axiom of continuity). If the completeness theorem could also be proved for the higher parts of logic (the extended functional calculus), then it could be shown in complete generality that syntactical completeness follows from monomorphicity; and since we know, for example, that the Peano axiom system is monomorphic, from that the solvability of every problem of arithmetic and analysis expressible in *Principia mathematica* would follow.

Such an extension of the completeness theorem is, however, impossible, as I have recently proved; that is, there are mathematical problems which, though they can be expressed in *Principia mathematica*, cannot be solved by the logical devices of *Principia mathematica*. Here the axiom of reducibility, the axiom of infinity (in the form: There are exactly denumerably many individuals), and even the axiom of choice are admitted as axioms. This fact can also be expressed thus: The Peano axiom system, with the logic of *Principia mathematica* added as superstructure, is not syntactically complete. It would take us too far afield, however, to go into these things in more detail.

Introductory note to *1931?

This paper gives an excellent overview of the results in *1931* without entering into the overwhelming amount of detail on which they rest. On the back of the manuscript Gödel wrote "(Vortrag?)" [Lecture?]; we surmise that the paper was intended as a popular treatment of the results of *1931*. An early communication by him of only a part of those results took place at Königsberg on 5–7 September 1930 and is reported in *1931a*. At that time he had not yet discovered his second incompleteness theorem (which *is* described in the last paragraph of *1931?*), nor had he yet formulated his undecidable propositions number-theoretically (see these *Works*, Volume I, page 137). After *1931* was written, Gödel undoubtedly received invitations to present his results in a single lecture. We can presume that the content of

[Über unentscheidbare Sätze]
(*1931?*)

Ein formales System S heißt bekanntlich vollständig (entscheidungsdefinit), wenn jeder in den Symbolen von S ausdrückbare Satz p aus den Axiomen von S entscheidbar, d. h. entweder p oder non-p in endlich vielen Schritten nach den Schlußregeln des Logikkalküls aus den Axiomen ableitbar ist. Im folgenden wird ein Verfahren skizziert, welches nicht nur zeigt, daß sämtliche bisher aufgestellten formalen Systeme der Mathematik (*Principia mathematica*, Axiomensysteme der Mengenlehre, Systeme der Hilbertschen Schule) unvollständig sind, sondern darüber hinaus ganz allgemein zu beweisen gestattet: *Jedes formale System mit endlich vielen Axiomen, das die Arithmetik der natürlichen Zahlen enthält, ist unvollständig.* Dasselbe gilt auch für Systeme mit unendlich vielen Axiomen, vorausgesetzt, daß die Axiomenregel (d. h. das Gesetz, nach dem die unendliche Menge von Axiomen erzeugt wird) konstruktiv ist (in einem genau präzisierbaren Sinn[1]). Man kann für jedes formale System, das den

---

[1]Die Axiomenregel kann man sich etwa in Form eines Gesetzes, das jeder natürlichen Zahl n ein Axiom zuordnet, gegeben denken. Man wird sie im allgemeinsten Sinn "konstruktiv" nennen, wenn das Gesetz ein Verfahren liefert, für jede Zahl n das zugehörige Axiom effektiv hinzuschreiben. Der im obigen Satz zugrundegelegte Begriff von "konstruktiv" ist zwar dem Wortlaut nach enger, aber es ist kein Gesetz bekannt, das im einen Sinn konstruktiv wäre und im anderen nicht.

1931? is more or less what he said in several such lectures. In particular, we know that he lectured on the topic to the Schlick Circle on 15 January 1931, at Bad Elster on 15 September 1931, and at Washington, D.C., on 20 April 1934, but for none of those occasions do we have an identifiable lecture text. (A separate, fragmentary manuscript for his lecture of 18 April 1934 to the New York Philosophical Society does exist in the *Nachlass*, along with his own notes for the long series of lectures on the topic that he gave from February to May 1934 at the Institute for Advanced Study, Princeton, N.J. For the notes on the latter taken by Stephen C. Kleene and J. Barkley Rosser, see *Gödel 1934.*)

Stephen C. Kleene

The translation of *1931?* is by Stephen C. Kleene, with subsequent revisions by John Dawson and William Craig.

⟦On undecidable sentences⟧
(*1931?*)

As is well known, a formal system S is called complete (in the syntactic sense) if every sentence p expressible in the symbols of S is decidable from the axioms of S, that is, either p or not-p is derivable in finitely many steps from the axioms ⟦of the system⟧ by means of the rules of inference of the calculus of logic. In what follows, a procedure is sketched which not only shows that all the previously established formal systems of mathematics (*Principia mathematica*, axiom systems of set theory, systems of the Hilbert school) are incomplete, but which, beyond that, allows one to prove quite generally: *Every formal system with finitely many axioms that contains the arithmetic of the natural numbers is incomplete.* The same holds also for systems with infinitely many axioms, provided that the rule for the axioms (that is, the law according to which the infinite set of axioms is generated) is constructive (in a sense that can be made quite precise[1]). For

---

[1]The rule for the axioms can be thought of as given, say, in the form of a law that associates an axiom with each natural number n. It will be called "constructive" in the most general sense if the law furnishes a procedure whereby, for each number n, the associated axiom can effectively be written down. The concept of "constructive" used in the ⟦incompleteness⟧ statement above is, to be sure, narrower in a literal sense, but no law is known that is constructive in the one sense and not in the other.

2   genannten Voraussetzungen genügt, einen unentscheidbaren Satz effektiv
     angeben, und die so kon|struierten Sätze gehören der Arithmetik der natür-
     lichen Zahlen an. Dabei werden zur *Arithmetik* diejenigen Begriffe und
     Sätze gerechnet, welche sich allein mittels der Begriffe der Addition und der
     Multiplikation und der logischen Verknüpfungen ("nicht", "oder", "und",
     "alle", "es gibt") ausdrücken lassen, wobei "alle" und "es gibt" sich nur
     auf natürliche Zahlen beziehen dürfen.

     Das Beweisverfahren, welches dieses Resultat liefert, verläuft folgen-
     dermaßen: Man numeriere zunächst die Formeln des vorgelegten Systems
     (es gibt natürlich nur abzählbar viele) in einer beliebigen aber ein für
     allemal festgehaltenen Weise. Dadurch ist jedem Begriff, der sich auf
     Formeln bezieht (jedem metamathematischen Begriff) ein bestimmter Be-
     griff, der sich auf natürliche Zahlen bezieht, zugeordnet, z. B. entspricht der
     metamathematischen Relation: "Die Formel a ist aus der Formel b nach
     den Schlußregeln ableitbar" die folgende Relation R zwischen natürlichen
     Zahlen: R besteht zwischen den Zahlen m und n dann und nur dann, wenn
     die Formel mit der Nummer m aus der Formel mit der Nummer n ableit-
     bar ist. Dem Begriff "beweisbare Formel" entspricht die Klasse derjenigen
     Zahlen, welche Nummern von beweisbaren Formeln sind usw. Es zeigt sich
     nun (wie eine eingehende Untersuchung lehrt), daß die so auf dem Umweg
     über die Metamathematik definierten Relationen zwischen Zahlen (bzw.
     Klassen von Zahlen) sich keineswegs prinzipiell von den sonst in der Arith-
     metik vorkommenden Relationen und Klassen (z. B. "Primzahl", "teilbar"

3   usw.) unterschei|den, sondern sich ohneweiters auf diese durch Definition
     zurückführen lassen, ohne erst den Umweg über die metamathematischen
     Begriffe zu gehen; d. h. genauer: Diese Relationen erweisen sich als arith-
     metisch im oben präzisierten Sinn. Daß dies der Fall ist, beruht letzten
     Endes darauf, daß die metamathematischen Begriffe lediglich auf gewisse
     kombinatorische Verhältnisse der Formeln Bezug nehmen, die sich in den
     zugeordneten Zahlen (bei geeigneter Wahl der Zuordnung) direkt wider-
     spiegeln.

     Wir betrachten nun die Gesamtheit der in dem vorgelegten formalen Sys-
     tem enthaltenen Aussagefunktionen mit einer freien Variablen und denken
     uns diese in eine Folge geordnet:

$$\varphi_1(x), \varphi_2(x), \ldots, \varphi_n(x), \ldots . \tag{1}$$

Dann definieren wir eine Klasse K von natürlichen Zahlen folgendermaßen:

$$n \in K \equiv {\sim} Bew\, \varphi_n(n), \tag{2}$$

wobei *Bew x* bedeuten soll: x ist eine beweisbare Formel. Die Klasse K
wurde auf dem Umweg über metamathematische Begriffe ("Aussagefunktion", "beweisbar" etc.) definiert. Dieser Umweg ist aber, wie oben be-

every formal system that satisfies the stated assumptions one can effectively specify an undecidable sentence, and the sentences so constructed belong to the arithmetic of the natural numbers. Here those concepts and sentences are to be reckoned as belonging to arithmetic which can be expressed solely by means of the concepts of addition and multiplication and the logical connectives ("not", "or", "and", "all", "there is"), where "all" and "there is" are allowed to refer only to natural numbers.

The method of proof that furnishes this result runs as follows: At the outset one numbers the formulas of the given system (there are of course only countably many [of them]) in an arbitrary manner, kept fixed, however, once and for all. With every concept that involves reference to formulas ([that is,] with every metamathematical concept), a definite concept that involves reference to natural numbers is thereby associated; for example, to the metamathematical relation "The formula a is derivable from the formula b by the rules of inference" corresponds the following relation R between natural numbers: R holds between the numbers m and n if and only if the formula with the number m is derivable from the formula with the number n. To the concept "provable formula" corresponds the class of those numbers that are numbers of provable formulas, etc. Now it turns out (as one learns from a detailed investigation) that the relations between numbers or classes of numbers which are thus defined by [taking] a detour through metamathematics are in principle not at all different from the relations and classes that occur elsewhere in arithmetic (for example, "prime number", "divisible", etc.); rather, they can be reduced to them in a straightforward manner by means of definition[s], without first taking the detour via metamathematical concepts. In other words, more accurately, these relations turn out to be arithmetical in the precise sense adopted above. That this is the case rests, in the end, on the circumstance that these metamathematical concepts involve only certain combinatorial relationships among the formulas, [relationships] which (under a suitable choice of the association [between formulas and numbers]) are directly mirrored among the associated numbers.

We consider now the totality of the propositional functions with one free variable that are contained in the formal system under consideration and think of these as ordered in a sequence

$$\varphi_1(x), \varphi_2(x), \ldots, \varphi_n(x), \ldots. \tag{1}$$

Then we define a class K of natural numbers as follows:

$$n \in K \equiv {\sim}Bew\ \varphi_n(n), \tag{2}$$

where *Bew x* shall mean: x is a provable formula. The class K has been defined via a detour through metamathematical concepts ("propositional function", "provable", etc.). But, as remarked above, this detour is avoid-

merkt, vermeidbar, d. h. man kann eine (im obigen Sinn) arithmetische Klasse angeben, die mit K umfangsgleich ist. Daß dies tatsächlich möglich ist, ist der springende Punkt für das folgende und muß natürlich ausführlich bewiesen werden, worauf aber hier nicht eingegangen werden kann. Da nach Voraussetzung die Arithmetik in dem zugrun|degelegten System S enthalten ist, so gibt es in S und daher in der Reihe (1) eine Aussagefunktion $\varphi_k(x)$, welche mit K umfangsgleich ist, für die also gilt:

$$\varphi_k(n) \equiv {\sim} Bew\ \varphi_n(n). \tag{3}$$

Setzt man an Stelle von n k ein, so folgt:

$$\varphi_k(k) \equiv {\sim} Bew\ \varphi_k(k). \tag{4}$$

Aus (4) ergibt sich aber, daß weder $\varphi_k(k)$ noch ${\sim}\varphi_k(k)$ beweisbar sein kann, denn aus der Annahme, $\varphi_k(k)$ sei beweisbar, würde folgen, daß $\varphi_k(k)$ richtig und daher wegen (4) nicht beweisbar wäre. Aus der Annahme, daß ${\sim}\varphi_k(k)$ beweisbar ist, folgt, daß ${\sim}\varphi_k(k)$ richtig und daher wegen (4) $\varphi_k(k)$ beweisbar ist, was ebenfalls mit der Annahme im Widerspruch steht. Im Beweise wurde stillschweigend vorausgesetzt, daß jeder in S beweisbare Satz richtig ist. Diese Voraussetzung kann, wie eine nähere Untersuchung zeigt, durch eine viel schwächere ersetzt werden, die nur wenig mehr verlangt als Widerspruchsfreiheit des betrachteten formalen Systems.

Das eben skizzierte Verfahren liefert für jedes System, das den oben genannten Voraussetzungen genügt, einen in diesem System unentscheidbaren arithmetischen Satz. Dieser Satz ist aber keineswegs absolut unentscheidbar; man kann vielmehr immer zu "höheren" Systemen übergehen, in denen der betreffende Satz entscheidbar wird (natürlich bleiben trotzdem andere Sätze unentscheidbar). Im besonderen ergibt sich, daß z. B. die Analysis ein in diesem Sinn höheres System ist als die Zahlen|theorie und das Axiomensystem der Mengenlehre wiederum höher als die Analysis. Es folgt also z. B., daß es zahlentheoretische Probleme gibt, die sich nicht mit zahlentheoretischen sondern nur mit analytischen bzw. mengentheoretischen Hilfsmitteln lösen lassen.

Aus den obigen Untersuchungen ergibt sich auch ein Resultat betreffend Widerspruchsfreiheitsbeweise. Die Aussage, daß ein System widerspruchsfrei ist, ist eine metamathematische Aussage und ist daher nach dem obigen Verfahren durch eine arithmetische (daher in demselben System ausdrückbare) Aussage ersetzbar. Es zeigt sich nun, daß diese Aussage immer in dem System, dessen Widerspruchsfreiheit sie behauptet, unbeweisbar ist (unter den selben Voraussetzungen wie oben). D. h., die Widerspruchsfreiheit eines formalen Systems (der oben charakterisierten Art) läßt sich niemals mit geringeren (oder denselben) Schlußweisen zeigen, die in dem betreffenden System formalisiert sind; vielmehr braucht man dazu immer irgendwelche Schlußweisen, die über das System hinausgehen.

able, that is, one can specify a class that is arithmetical (in the sense above) and that is coextensive with K. That this is in fact possible is the crucial point for what follows, and it must of course be proved in detail; however, I cannot go into that here. Since, by assumption, arithmetic is included in the system S under consideration, there is in S, and therefore in the sequence (1), a propositional function $\varphi_k(x)$ that is coextensive with K, for which, therefore, ⟦the following⟧ holds:

$$\varphi_k(n) \equiv \sim Bew \; \varphi_n(n). \tag{3}$$

If one substitutes k for n, it follows that

$$\varphi_k(k) \equiv \sim Bew \; \varphi_k(k). \tag{4}$$

From (4) it follows, however, that neither $\varphi_k(k)$ nor $\sim\varphi_k(k)$ can be provable, because from the supposition that $\varphi_k(k)$ is provable it would follow that $\varphi_k(k)$ is true, and therefore, by (4), is not provable. From the supposition that $\sim\varphi_k(k)$ is provable, it follows that $\sim\varphi_k(k)$ is true, and therefore, by (4), that $\varphi_k(k)$ is provable, which likewise stands in contradiction to the supposition. In the proof it has been tacitly assumed that every sentence provable in S is true. This assumption can, as a closer investigation shows, be replaced by a much weaker one, which requires only a little more than consistency of the formal system being considered.

The procedure just sketched furnishes, for every system that satisfies the aforementioned assumptions, an arithmetical sentence that is undecidable in that system. That sentence is, however, not at all absolutely undecidable; rather, one can always pass to "higher" systems in which the sentence in question is decidable. (Some other sentences, of course, nevertheless remain undecidable.) In particular, for example, it turns out that analysis is a system higher in this sense than number theory, and the axiom system of set theory is higher still than analysis. It therefore follows, e.g., that there are number-theoretic problems that cannot be solved with number-theoretic, but only with analytic or, respectively, set-theoretic methods.

From the investigations above there emerges also a result concerning consistency proofs. The proposition that a system is consistent is a metamathematical proposition, and therefore, by the procedure above, is replaceable by an arithmetical proposition (hence one expressible in the same system). Now it turns out that this ⟦arithmetical⟧ proposition is always unprovable in the system whose consistency it asserts (given the same assumptions as above ⟦about the system⟧). That is, the consistency of a formal system (of the kind characterized above) can never be established by methods of proof more stringent than (or the same as) those formalized in the system in question; rather, for that one always needs some methods of proof that transcend the system.

Introductory note to *1933o

This rich and, in certain respects, remarkable article is drawn from the handwritten text for an invited lecture which Gödel delivered (under the same title) to a meeting of the Mathematical Association of America, held jointly with the American Mathematical Society in Cambridge, Massachusetts, 29–30 December 1933. A report of the meeting is to be found in the *American Mathematical Monthly*, Volume 41, 1934, pages 123–131. According to that report, Gödel's paper was supposed to appear in a later issue of the *Monthly*. However, we have no evidence that the paper was ever actually prepared for publication, let alone submitted.[a]

Gödel had arrived on his first trip to the United States in October 1933, for his visit to the Institute for Advanced Study in Princeton, which was to last until May 1934. In the months February to May 1934 he gave a course of lectures (*Gödel 1934*) at the Institute on the incompleteness results.[b] According to these dates the present article may well represent the first public lecture which Gödel delivered in English. Only one version of the text was found in the Gödel *Nachlass*; the manuscript is clearly written and there are no ambiguities. The English itself is very readable and the style is close to that of later publications by Gödel in English. There is no evidence of any editorial assistance by native speakers, though it is of course possible that colleagues at the Institute provided some help.

The aim of Gödel's lecture is announced in his first paragraph with admirable clarity: "The problem of giving a foundation for mathematics ... can be considered as falling into two different parts. At first [the] methods of proof [actually used by mathematicians] have to be reduced to a minimum number of axioms and primitive rules of inference, ... and then secondly a justification in some sense or other has to be sought for these axioms ...".

Gödel says that the first apparent reduction of all of mathematics to a few axioms and rules of inference, as carried out by Frege,[c] was faulty in that it led to contradictions, and some restrictions in dealing with infinite sets (or "aggregates") are necessary. The way such restrictions must

---

[a]Incidentally, the other speakers at the same session at which Gödel delivered this lecture were A. N. Whitehead on "Logical definitions of extension, class and number", and Alonzo Church on "The Richard paradox"; only the latter was subsequently published in the *Monthly*. (See *Church 1934*.)

[b]See the Gödel chronology by John W. Dawson, Jr. in these *Works*, Vol. I, p. 39.

[c]Peano is also mentioned in this connection.

be made "seems to be essentially uniquely determined by the two requirements of avoiding the paradoxes and retaining all of mathematics",
including the theory of sets. Gödel then claims that the first part of
the problem of foundations for mathematics has been solved in a completely satisfactory way by means of formalization in the simple theory
of types (here abbreviated STT)[d] when all "superfluous restrictions" are
removed.

According to Gödel, there are three such restrictions in STT: (1) Just
pure types 0, 1, 2, ... are admitted, where the objects of type $(n + 1)$
admit only objects of type n as members; (2) thence "$a \in b$" is taken
to be meaningless when a, b are not of successive type levels; (3) only
finite type levels are admitted. For STT the objects of type level n are
interpreted as ranging over sets T_n, where $T_{n+1} = \mathcal{P}(T_n)$ is the set of all
subsets of T_n. In place of (1) Gödel considers the inessential modification of replacing the pure type levels T_n by the cumulative type levels,
informally interpreted by the collections S_n, where $S_{n+1} = S_n \cup \mathcal{P}(S_n)$.
Then in place of (2), he assumes that "$a \in b$" is meaningful for all a, b in
the universe of sets under consideration, with "$a \in b$" taken to be false
whenever b is of the same or lower cumulative type level than a. Finally,
this permits in place of 3 a natural extension to transfinite levels S_α,
with $S_\alpha = \bigcup_{\beta < \alpha} S_\beta$ for limit α and $S_{\alpha+1} = S_\alpha \cup \mathcal{P}(S_\alpha)$ for all α. As
Gödel states, this kind of generalization of the theory of types is formu-
lated axiomatically in theories of sets (or "aggregates") due to Zermelo,
Fraenkel and von Neumann. He refers specifically to *von Neumann 1929*
for the motivation to remove the restrictions 1–3.[e]

However, Gödel does not fix a specific formal system of set theory as
the result of removing the restrictions 1–3 from STT. And indeed there
is an ambiguity in his description (pages 8–9) of the transfinite levels α
to be admitted, for which he seems to suggest that α must already be
"defined" in an earlier system S_β with $\beta < \alpha$. It is not clear what notion
of definability Gödel has in mind, nor whether he is thinking about S_α as
a collection of sets or as a formal system true at level α in the cumulative
hierarchy.[f]

[d]See the introductory note to *1930b*, *1931* and *1932b*, these *Works*, Vol. I, p. 129
and *Gödel 1931*, pp. 176–178 for Gödel's *1931* version of STT.

[e]The basic papers are, of course, *Zermelo 1908*, *Fraenkel 1922* and *Skolem 1923a*.
Gödel could have referred to *Zermelo 1930* for the first clear statement of the informal
interpretation of axiomatic set theory in the cumulative hierarchy.

[f]Under the former reading Gödel may have intended a formulation related to
Fraenkel's Replacement Axiom, namely, that one has a formula $F(x, y)$ which defines
a function from members x of a set $a \in S_\beta$ to ordinals y whose range $\alpha = \{y | \exists x (x \in a \land F(x, y))\}$ is closed under predecessor and where β is previously obtained. The
latter reading in terms of formal systems suggests a notion like that of autonomous
progressions of theories, in the sense of *Feferman 1962*. Charles Parsons has suggested to me that on neither reading is the procedure strong enough to justify the ZF

Though one is freed of the "superfluous restrictions" 1–3 in the manner described (Gödel continues), it should not be considered that one has reached an end with this [unspecified] system of axioms for set theory, since the process of "creating" ever higher types can be continued beyond those dealt with in any given system S. While we set out to find a single formal system for mathematics, instead we have found an infinity of such systems. In practice one can confine oneself to a system for the first few type levels, where all the mathematical methods and results developed up to now can be formally represented. But, says Gödel, the unending process of construction of formal systems for ever higher types is necessary in principle, as is evidenced by the incompleteness results of his paper *1931*: For each consistent system S based on the theory of types, there are true (and simple) arithmetic propositions which cannot be proved in S but which become provable if one adds axioms to S for the next higher type. A special case of this general theorem is that "... there are arithmetic propositions which can be proved only by analytical methods and, further, that there are arithmetic propositions which cannot be proved even by analysis but only by methods involving extremely large infinite cardinals and similar things."[g] Here Gödel harks back to the well-known footnote 48a of his 1931 paper and anticipates his later advocacy, in the *1964* supplement to *1947* and elsewhere, of the need for large cardinal axioms to settle open arithmetical and set-theoretical problems, including the continuum hypothesis.

Gödel turns from this (on page 15) to the second part of the foundational problem for mathematics, namely, that of providing a justification for the axioms and rules arrived at in the first part. Here "... it must be said that the situation is extremely unsatisfactory." There is no problem about treating the formalism in a purely formal way as a kind of game with symbols, "... but as soon as we come to attach a meaning to our symbols serious difficulties arise." These difficulties are essentially of three kinds: (1°) the non-constructive notion of existence, (2°) the "still more serious" general notion of class (of arbitrary type), and (3°) the axiom of choice.

As to (1°), Gödel points out that it is possible to infer the existence of an object (even an integer) satisfying a property P, without actually

---

axioms. According to *Wang 1974*, p. 186, apparently Gödel thought more general considerations are necessary to arrive at the Replacement Axiom.

[g]Note that on p. 13, Gödel calls a proposition "arithmetic" if it states that a decidable property holds for all integers. This kind of proposition is nowadays said to belong to the class Π_1^0 (cf. the introductory note to *1930b, 1931* and *1932b*, these *Works*, Vol. I, p. 135), while "arithmetical" is reserved for propositions which fall somewhere in the hierarchy of Π_m^0 and Σ_m^0 classes. Gödel did use "arithmetical" in a sense equivalent to the latter in *1931*, §3 (translation in these *Works*, Vol. I, p. 181).

being able to produce (or describe a procedure to produce) a specific instance of P: he rightly traces this back to the assumption of the law of excluded middle in the basic logic of the formalism.

Concerning (2°), Gödel finds problematic the use of impredicative definitions for specifying classes (of any given type), for example to define a set of integers consisting of all n satisfying a property P, where P is defined by reference to "all" properties of integers (or "all" classes of integers as their extensions), among them P itself. This kind of definition can be considered as unobjectionable only if one assumes that there is a pre-existent totality of all properties (or all classes) of integers (and more generally of all properties of a given type).[h] But we have a vicious circle if properties are regarded as "generated by our definitions". The trouble with such a predicative (definitionist) conception of properties is that adhering to it "... makes impossible an adequate theory of real numbers, many of whose fundamental theorems appear to depend essentially on the non-predicative definitions."

Gödel says practically nothing about (3°), the axiom of choice, as the third "weak spot" in the axioms, on the grounds that "... it is of less importance for the development of mathematics." This casual remark is a bit surprising in view of the intense controversy over the legitimacy of the axiom of choice (and its equivalents such as the well-ordering principle) which figured prominently in foundational discussions in the first third of this century (cf. *Moore 1982*). However, it may be that Gödel had already started thinking about the relative consistency of the axiom of choice (with the axioms of Zermelo-Fraenkel set theory) and had evolved a plan as to how this might be established.[i]

What is most surprising is Gödel's next statement (page 19): "The result of the preceding discussion is that our axioms, if interpreted as meaningful statements, necessarily presuppose a kind of Platonism, which cannot satisfy any critical mind and which does not even produce the conviction that they are consistent." This does not seem to square with Gödel's unequivocal assertions, quoted in letters of 1967 and 1968 to Hao Wang and reproduced in *Wang 1974*, pages 8–11, that he had held a Platonistic philosophy of mathematics since his student days in Vienna.[j] These are reinforced by Gödel's statement, in an unpublished

---

[h]On p. 17, Gödel says that "the notion of 'class of integers' is essentially the same as that of 'property of integers' ...". Much later (in *1944*) he distinguished between these two notions, but found it unobjectionable to assume a totality of all properties (or all concepts) as well as a totality of all classes of a given type; cf. *Gödel 1944*, pp. 138 ff. and the discussion by Charles Parsons thereto (these *Works*, Vol. II, pp. 106 ff.).

[i]Gödel established the consistency of AC with ZF in 1935 (though not announced until 1938); cf. these *Works*, Vol. I, pp. 9 and 21–22.

[j]See also *Wang 1978*, p. 183.

(and unsent!) response of 1975 to a questionnaire from Burke D. Grand-jean, that he had held such views since 1925.[k] There is certainly no ques-tioning Gödel's unremitting espousal in print of full-fledged Platonism, beginning with his 1944 article on Russell's mathematical logic and con-tinuing (especially in his 1947 article on Cantor's continuum problem) until his death. Of course, one may take Gödel's retrospective assertions about his earlier views *cum grano salis*, colored as they are by his later firm stance. Alternatively, one might seek to make his statement in this article coherent with the rest of the evidence concerning his views by distinguishing varieties of Platonism. For example, it is reasonable to say that one holds such a position as regards integers, but not as regards sets; conceivably Gödel could have started as a moderate Platonist in this sense and only later shifted to the stronger position.[1] Another pos-sibility is that Gödel had, at the time of this lecture, a temporary period of doubt about set-theoretical Platonism. Unless further evidence from the *Nachlass* comes to light that we are not presently aware of, all of this must, unfortunately, remain speculative.

In any case, within the context of this article, Gödel's assertion of the unacceptability of the Platonist position as justification for the axioms of set theory must be taken at face value. He does go on to say (pages 19–20) that, at least, it is very likely that these axioms are consistent, since their consequences have been developed extensively without arriv-ing at an inconsistency. Moreover, the question of consistency is a purely combinatorial one about manipulation of symbols according to specified rules, so one might hope to establish the consistency of these axioms by "unobjectionable methods". This leads into a description of Hilbert's foundational program, which was to be carried out by finitary[m] consis-tency proofs for formal systems for mathematics. However, according to Gödel (page 25) there is no hope of carrying out Hilbert's program even

---

[k]See *Feferman 1984a*, pp. 549–552 for this and further information about Gödel's retrospective claims.

[1]Cf. the suggestion by Martin Davis (quoted in these *Works*, Vol. I, p. 31, fn. 21), that Gödel's Platonism regarding sets may have evolved more gradually than his later statements would suggest; Davis pointed particularly to the remark at the end of *Gödel 1938* as evidence for this.

[m]Gödel equates finitism with intuitionism in its most restricted sense; he denotes by "A" a system in which just such methods are allowed. While he does not specify A in any detail, he does describe its essential features.

According to *Bernays 1967*, p. 502, the prevailing view in the Hilbert school at the beginning of the 1930s equated finitism with intuitionism (cf. also *Sieg 1990*, p. 272). This was shaken in part by Heyting's formalization of intuitionistic arithmetic and Gödel's simple reduction (in his *1933e*) of classical arithmetic to that system. Wilfried Sieg has suggested to me that the kinds of considerations presented by Gödel in the lecture under discussion helped logicians to recognize finitist mathematics as a proper part of intuitionistic mathematics.

for classical arithmetic (now referred to as Peano arithmetic, PA), let alone for set theory, "in view of some recently discovered facts". Namely, by Gödel's second incompleteness theorem (*Satz* XI of *1931* as strengthened in *1932b*), no formal system S which contains PA can prove its own consistency, unless it is inconsistent. Gödel goes on to say (page 26) that

> ... all the intuitionistic [finitary] proofs complying with the requirements of the system A [of finitistically allowable methods] which have ever been constructed can easily be expressed in the system of classical analysis and even in the system of classical arithmetic, and there are reasons for believing that this will hold for any proof which one will ever be able to construct.
>
> So it seems that not even classical arithmetic can be proved to be non-contradictory by the methods of the system A ....

This last is especially interesting in view of the cautious statement Gödel had made near the close of *1931* concerning the significance of the second incompleteness theorem for Hilbert's program, in which he said: "I wish to note expressly that Theorem XI ... [does] not contradict Hilbert's formalistic viewpoint ... [which] presupposes only the existence of a consistency proof in which nothing but finitary means of proof is used, and it is conceivable that there exist finitary proofs that *cannot* be expressed in the formalism of P ... ."[n] Here "P" denotes Gödel's system of simple type theory (STT), with the axioms of classical arithmetic taken to hold for the objects of lowest type. Clearly, Gödel had convinced himself in the interim that all consistency proofs using methods that were clearly finitary, as then recognized, could readily be formalized in PA, and that there was nothing in sight that suggested finitary methods would be capable of going beyond arithmetic, and certainly not beyond analysis. Hence it would be hopeless to carry out Hilbert's program for arithmetic unless some radically new finitary methods not expressible in arithmetic could be found. But concerning this last possibility, as we saw, Gödel added even more strongly that he had "reasons to believe" that no such methods could ever be produced. He gives no indication what such reasons are, and in fact he never provided such in print until *1958*, in the introductory discussion—which, however, makes essential reference to consistency proofs by ordinals as first provided in *Gentzen 1936*.

It should be noted that Gödel does refer to partial successes of Hilbert's program, "the most far-reaching of which is the following the-

[n]Translation from the original; cf. these *Works*, Vol. I, p. 195.

orem proved by Herbrand" which provides a general method for consistency proofs of certain systems with axioms whose existential quantifiers can be realized constructively.[o] However, this can be used to account only for a fragment of classical arithmetic, namely, that in which induction is restricted to quantifier-free formulas.[p]

The article concludes with a discussion of the possible use of intuitionistic methods in the wider sense, as developed by the Brouwerian school. This goes beyond the system A of finitary methods by, among other things, permitting negation to be applied to general propositions, as, e.g., $\neg(x)F(x)$, which can (in certain cases) be established by intuitionistic methods without producing an x such that $\neg F(x)$. For, the meaning given to $\neg p$ is that one has a (constructive) demonstration that any proof of p reduces to absurdity $\bot$, which is the same as the conditions for $p \supset \bot$. Gödel mentions that for the "... axioms of the intuitionistic mathematics as stated by Heyting ... we find that for the notion of absurdity exactly the same propositions hold as do for negation in ordinary mathematics—at least, this is true within the domain of arithmetic" (page 29). Presumably Gödel is here referring to his results in *1933e*, which provided a translation of classical arithmetic (PA) into its intuitionistic counterpart (HA) which preserves formulas that do not contain the disjunction or existence operation symbols. But he objects to the intuitionistic meaning of negation (and hence, more generally, of implication) because it refers to the totality of all possible (intuitionistic) proofs of a proposition, and this violates what he considers to be a basic principle of constructivity, namely, that "... the word 'any' can be applied only to those totalities for which we have a finite procedure for generating all their elements." Since the supposed totality of intuitionistic proofs is too vague to meet this criterion, Gödel concludes that "... this foundation of classical arithmetic by means of the notion of absurdity is of doubtful value." Still he hopes that in the future "one may find other and more satisfactory methods of construction beyond the limits of the system A ..." to "found" classical arithmetic and analysis. As we see from Troelstra's introductory note to *1958* and *1972* in these *Works*, Volume II, pages 217–241, this was something Gödel struggled to achieve until late in his life, but he never succeeded to his own satisfaction, even for arithmetic.[q]

---

[o] As pointed out to me by Warren Goldfarb, the theorem stated by Gödel is never stated this way by Herbrand, but it is an easy consequence of a theorem Herbrand formulated in his *1930* (cf. *Herbrand 1971*, p. 179) and again in *1931* (cf. *Herbrand 1971*, p. 289).

[p] Cf. *Herbrand 1931*; that case had previously been treated in a rather complicated special way by von Neumann (*1927*).

[q] See also Troelstra's introductory note to *1941* in the present volume.

No doubt Gödel would have worked on improving this article were he ever to have brought it forward to publication, but it could hardly have been bettered for its time. What is impressive about the piece is how well it holds up after more than half a century's subsequent intensive development of those areas of mathematical logic that most bear on Gödel's foundational concerns, in set theory, proof theory and constructivity. In only two (related) respects would he perhaps want to modify certain points to take account of technical developments in recent years. The first has to do with the claimed impossibility of an adequate theory of real numbers (and hence of analysis) without the use of impredicative definitions (page 19). Here Gödel could have referred to Weyl's landmark work *Das Kontinuum (1918)*[r] for its pioneering effort to demonstrate the viability of a system of mathematical analysis, based on strictly predicative principles, that could account for the bulk of nineteenth-century analysis. Following World War II, a body of work produced over a period of years by Lorenzen, Kreisel, Takeuti, and others (including the author) has led to the conclusion that much of twentieth-century analysis can also be given a predicative foundation.[s] Indeed, it follows from work of *Feferman 1977, Takeuti 1978a,* and *Feferman 1985a* that this portion of analysis can already be formalized in a system S which is proof-theoretically (finitarily) reducible to classical arithmetic PA (and which is a conservative extension of PA). Moreover, it appears that such an S serves to account for all scientifically applicable portions of analysis. In a way, this outcome was anticipated by Gödel's remark on page 11 (already noted above) that "all mathematics hitherto developed" can be carried out in much weaker systems than (Zermelo–Fraenkel or von Neumann) axiomatic set theory, since the first few type levels suffice. Indeed, that is correct, but it does not go far enough: what is crucial is Weyl's recognition (in his *1918*) that *sequential (Cauchy) completeness* is adequate for much of analysis where *set-theoretical (Dedekind cut) completeness* is ordinarily applied; the latter is essentially impredicative, whereas the former permits a strictly predicative development.

This ties up with the second respect in which Gödel might have modified his discussion in the light of recent technical work, in this case that carried out by Friedman, Simpson and their followers under the rubric of "reverse mathematics". According to their results, of which an informative survey is to be found in *Simpson 1988,* much of nineteenth- and twentieth-century analysis and algebra can already be formalized in a system S_0 that is conservative over the fragment PRA (primitive

---

[r]Now translated into English, as *Weyl 1987.*

[s]Cf. *Feferman 1964* for an earlier survey, and *Feferman 1988* for a more up-to-date (though less readily available) formulation and assessment of Weyl's program.

recursive arithmetic) of PA. It is generally accepted that PRA is a finitistically acceptable system. Moreover, by *Sieg 1985* one has a clear finitistic reduction of S_0 to PRA (incidentally, obtained mainly by making use of basic proof-theoretical techniques due to Herbrand and Gentzen). It is in this sense that partial realizations of Hilbert's program are much more successful than could have been recognized at the time of Gödel's lecture.

As for the current picture of the more general issues raised by Gödel, one may say that on the one hand, set-theoretical foundations have by and large won the day among working mathematicians, and that on the other hand, the need to provide some sort of constructive foundations for mathematics has not receded among the relatively small group of "critical minds" who reject the Platonist philosophy underlying set theory. Efforts directed at such constructive foundations have taken a great variety of forms, in particular through the massive body of work in proof theory on extensions of Hilbert's program, but there is no settled opinion as to what has been achieved thereby. The interested reader should consult such sources as *Buchholz et alii 1981*, *Feferman 1988a*, *Kreisel 1968*, *Schütte 1977* and *Takeuti 1987* (including the appendices by Feferman, Kreisel, Pohlers, and Simpson).

Solomon Feferman[t]

[t]I wish to thank M. Davis, J. W. Dawson, P. de Rouilhan, W. Goldfarb, C. Parsons, and W. Sieg for their helpful comments on a draft of this note.

The present situation in the foundations
of mathematics
(*1933o)

The problem of giving a foundation for mathematics (and by mathematics I mean here the totality of the methods of proof actually used by mathematicians) can be considered as falling into two different parts. At first these methods of proof have to be reduced to a minimum number of axioms and primitive rules of inference, which have to be stated as precisely as possible, and then secondly a justification in some sense or other has to be sought for these axioms, i.e., a theoretical foundation of the fact that they lead to results agreeing with each other and with empirical facts.

The first part of the problem has been solved in a perfectly satisfactory way, the solution consisting in the so-called "formalization" of mathematics, which means that a perfectly precise language has been invented, by which it is possible to express any mathematical proposition by a formula. Some of these formulas are taken as axioms, and then certain rules of inference | are laid down which allow one to pass from the axioms to new formulas and thus to deduce more and more propositions, the outstanding feature of the rules of inference being that they are purely formal, i.e., refer only to the outward structure of the formulas, not to their meaning, so that they could be applied by someone who knew nothing about mathematics, or by a machine. [This has the consequence that there can never be any doubt [as] to what cases the rules of inference apply [to], and thus the highest possible degree of exactness is obtained.]

The important fact that all [of] mathematics can be reduced to a few formal axioms and rules of inference was discovered by Frege and Peano. But when it was first attempted to set up such a formal system for mathematics, i.e., a system of axioms and rules of inference, one serious difficulty arose, namely that, if the axioms and rules of inference were formulated in the way that | seemed to be suggested at first sight, they led to obvious contradictions, and it became clear that certain restrictions had to be made in dealing with infinite aggregates. How these restrictions must be made seems to be essentially uniquely determined by the two requirements of avoiding the paradoxes and retaining all of mathematics (including the theory of aggregates). At least hitherto only one solution which meets these two requirements has been found, although more than 30 years have elapsed since the discovery of the paradoxes. This solution consists in the theory of types. (I mean the simple theory of types, not the complicated form, which leads to the axiom of reducibility.)

It may seem as if another solution were afforded by the system of axioms for the theory of aggregates, as presented by Zermelo, Fraenkel | and von

2

3

4

Neumann; but it turns out that this system of axioms is nothing else but a natural generalization of the theory of types, or rather, it is what becomes of the theory of types if certain superfluous restrictions are removed. This appears very clearly, for instance, from von Neumann's paper "Über eine Widerspruchsfreiheitsfrage in der axiomatischen Mengenlehre" [1929] and comes about as follows: The restriction on the logical rules, introduced by the theory of types, consists essentially in this, that the general notion of a class or aggregate is discarded and replaced by an infinite series of different notions of class. That is to say: In order to speak of classes at all, it is required that first a system of things (called individuals) be given (you may, for instance, regard the integers as individuals); then you can form
5   the notion of a class of those individuals and | speak about all such classes. Then you can go one step further and form the notion of a class whose elements are these classes of individuals, i.e., of a class of classes of individuals (called classes of the second type) and speak about all such classes. So you can go on indefinitely in this hierarchy, without ever being able to form the most general notion of class or to speak about all classes whatsoever. But on this hierarchy of types the following unnecessary restrictions have been put in *Principia mathematica*, unnecessary from the point of view of setting up a formal system which avoids the logical paradoxes and retains all [of] mathematics, the only question with which we are dealing now.

1. Only the so-called pure types have been admitted in *Principia mathematica*, i.e., no class can be formed which contains classes of different type among its elements.

2. Propositions of the form $a \in b$ are regarded as meaningless (i.e., neither true nor false) if a and b are not of the appropriate types—if,
6   for instance, | a is of a higher type than b. This complication can be removed simply by stating that $a \in b$ is to be false if a and b are not of the appropriate types.

The removal of these first two restrictions is not very essential; it can easily be seen that no contradiction can arise from it [and that for each proposition provable in the new system there is an equivalent in *Principia mathematica*]. The situation is quite different with the third restriction, which I am going to explain now. In Russell's theory the process of going over to the next higher type—for instance, from classes of individuals to classes of classes of individuals—can be repeated only a finite number of times; i.e., to each class occurring in the system of *Principia mathematica*, there is assigned a finite number n, indicating in how many steps the class
7   under consideration can be reached, starting from | the level of individuals. This number n may be arbitrarily large, but it must be finite.

Now there is no reason whatever to stop the process of formation of types at this stage [as has been pointed out, e.g., by Hilbert]. You can, e.g., form the class of *all* classes of finite type, which, of course, is not of finite type, but may be called of type ω. [The general definition of type ω would be that a class belongs to it if it contains only classes of finite type among its elements, but for each arbitrarily large integer n contains elements of a type higher than n.] It is clear how this process can be continued indefinitely. You can make the class of all classes of finite type play the role of the class of individuals, i.e., take it as a basis for a new hierarchy of types and thus form classes of type $\omega + 1$, $\omega + 2$, and so on for each transfinite ordinal.

There is one objection to this process | of forming classes of infinite   8
type, which may have been one of the reasons why Russell refrained from it; namely, that in order to state the axioms for a formal system, including all the types up to a given ordinal α, the notion of this ordinal α has to be presupposed as known, because it will appear explicitly in the axioms. On the other hand, a satisfactory definition of transfinite ordinals can be obtained only in terms of the very system whose axioms are to be set up. I don't think that this objection is serious for the following reason: The first two or three types already suffice to define very large ordinals. So you can begin by setting up axioms for these first types, for which purpose no ordinal whatsoever is needed, then define a transfinite ordinal α | in terms   9
of these first few types and by means of it state the axioms for the system, including all classes of type less than α. (Call it S_α.) To the system S_α you can apply the same process again, i.e., take an ordinal β greater than α which can be defined in terms of the system S_α and by means of it state the axioms for the system S_β including all types less than β, and so on.

| The place which the system of axioms for the theory of aggregates oc-   9a
cupies in this hierarchy can be characterized by a certain closure property as follows: There are two different ways of generating types, the first consisting in going over from a given type to the next one and the second in summing up a transfinite sequence of given types, which we did, e.g., in forming the type ω. Now the statement made by the axioms of the theory of aggregates is essentially this, that these two processes do not lead you out of the system if the second process is applied only to such sequences of types as can be defined within the system itself. [That is to say: If M is a set of ordinals definable in the system and if to each ordinal of M you assign a type contained in the system, then the type obtained by summing up those types is also in the system.] But it would be a mistake to suppose that with this system of axioms for the theory of sets we should have | 10
reached an end to the hierarchy of types. For all the classes occurring in this system can be considered as a new domain of individuals and used as a starting point for creating still higher types. There is no end to this process [and the totality of all systems thus obtained seems to form a totality of similar character to the set of ordinals of the second ordinal class].

So we are confronted with a strange situation. We set out to find a formal system for mathematics and instead of that found an infinity of systems, and whichever system you choose out of this infinity, there is one more comprehensive, i.e., one whose axioms are stronger. In practice

we can safely confine ourselves to one of those systems (e.g., the system for the theory of aggregates) because all the mathematical methods and
11   proofs hitherto developed are in this system, and, apart from | certain theorems of the theory of aggregates, all mathematics hitherto developed is contained even in much weaker systems, which include just a few of the first types. Nevertheless the situation created by the existence of an infinity of systems, each of which can be extended by further notions and axioms, may be considered as unsatisfactory and as discrediting the theory of types, which led to this situation.

   But, as a matter of fact, this character of our systems turns out to be a strong argument in favor of the theory of types. For it is in perfect accord with certain facts which can be established quite independently. It can be shown that any formal system whatsoever—whether it is based on the theory of types or not, if only it is free from contradiction—must
12   necessarily be deficient in its methods | of proof. Or to be more exact: For any formal system you can construct a proposition—in fact a proposition of the arithmetic of integers—which is certainly true if the system is free from contradiction but cannot be proved in the given system. Now if the system under consideration (call it S) is based on the theory of types, it turns out that exactly the next higher type not contained in S is necessary to prove this arithmetic proposition, i.e., this proposition becomes a provable theorem if you add to the system S the next higher type and the axioms concerning it.

   This fact is interesting also from another point of view; it shows that the construction of higher and higher types is by no means idle, but is necessary for proving theorems even of a relatively simple structure, namely
13   arith|metic propositions, whereby I mean the following: I call a proposition "arithmetic" if it states that a certain property P belongs to all integers, where P is such a property as is decidable for each particular integer by a general procedure. The theorem of Goldbach, which states that each even number is the sum of two prime numbers, would be an example of an arithmetic proposition in this sense. A special case of the general theorem about the existence of undecidable propositions in any formal system is that there are arithmetic propositions which can be proved only by analytical methods and, further, that there are arithmetic propositions which cannot be proved even by analysis but only by methods involving extremely large infinite cardinals and similar things.

14   Now, | coming back to the theory of types, it seems to me that there is a widespread feeling among logicians that something is wrong with this theory and that there must exist another more satisfactory way of avoiding the paradoxes. I think this feeling is justified as to the form of the theory presented in *Principia mathematica*. But if the superfluous restrictions which I mentioned above are dropped, most of the objections which have been brought against it do not hold any longer. [E.g., the necessity of

stating the logical axioms for each type separately disappears, because you can introduce a variable running through any given set of types if you drop the restriction concerning the purity of types.] As I mentioned above, the theory of types, if we understand it in the more general form which I explained, is until now the only solution to the problem of restricting the rules of the so-called naive logic so as to avoid the logical paradoxes and retain all of mathematics, and it is very likely | to remain so. All   15 other solutions that have been presented up to now either remained vague promises, i.e., have not been followed up to the point of setting up a formal system, or led to contradictions.

I come now to the second part of our problem, namely, the problem of giving a justification for our axioms and rules of inference, and as to this question it must be said that the situation is extremely unsatisfactory. Our formalism works perfectly well and is perfectly unobjectionable as long as we consider it as a mere game with symbols, but as soon as we come to attach a meaning to our symbols serious difficulties arise. There are essentially three kinds of such difficulties.

| The first is connected with the non-constructive notion of existence.   16 That is to say: Under the axioms of our systems we are permitted, e.g., to form a proposition stating "There is an integer which has a certain property P", and although we may have no means of ascertaining whether such an integer exists or not, we apply the law of excluded middle to this proposition, just as if in some objective realm of ideas this question were settled quite independently of any human knowledge. This treatment has strange results, as may be expected in advance; e.g., that we can often prove the existence of an integer with a given property without anyone's being able actually to name such an integer | or even to describe a procedure   17 by the application of which we could obtain such an integer (the so-called non-constructive existence proofs).

The second weak spot, which is still more serious, is connected with the notion of class. As I explained above, the general notion of class has been eliminated from our systems and split up into an infinite series of notions of classes of different type. But take an arbitrary notion out of this series, for instance the notion of "a class of the first type", i.e., a class of integers. As a class of integers, at least if it is infinite, can be given only by a characteristic property belonging to its members, the notion of "class of integers" is essentially the same as that of "property of integers", and this notion appears as a primitive idea in our systems. Not only that, but the words "all" and "there exists" are applied to properties of integers in just the same way as to integers using, e.g., the law of excluded middle. This ⟦way of⟧ proceeding is particularly objectionable in connection with properties | because it gives rise not only to non-constructive existence propositions   18 but also to the so-called method of impredicative definition, which consists essentially in this, that a property P is defined by a statement of the following form: An integer x shall possess the property P if for all properties (including P itself) some statement about x is true. Again, as in the case of the law of excluded middle, this process of definition presupposes that the totality of all properties exists somehow independently of our knowledge and our definitions, and that our definitions merely serve to pick out certain of these previously existing properties. If we assume this, the method of non-predicative definition is perfectly all right (as has been pointed out by Ramsey), for there is certainly nothing objectionable in characterizing a particular element of a previously given totality by reference to the whole totality; we do this if, e.g., we speak of the tallest building in a city.

But the situation becomes entirely different if we regard the properties as *generated* by our definitions. For it is certainly a vicious circle to generate an object by reference to a totality in which this very object is supposed
19   to be | present already. Russell, in order to avoid this vicious circle, found himself compelled to split up the notion of property of a given type into an infinite number of subtypes [in such a way that a property involving a reference to a totality of properties never belongs to this totality]. This device avoids the vicious circles but it also makes impossible an adequate theory of real numbers, many of whose fundamental theorems appear to depend essentially on the non-predicative definitions.

The third weak spot in our axioms is connected with the axiom of choice, which however I do not want to go into the details of because it is of less importance for the development of mathematics.

The result of the preceding discussion is that our axioms, if interpreted as meaningful statements, necessarily presuppose a kind of Platonism, which cannot satisfy any critical mind and which does not even produce the conviction that they are consistent. Nevertheless it is for other reasons extremely unlikely that they should actually involve contradiction. For the consequences of the objectionable methods, such as impredicative defini-
20   tion, have been followed up in all directions, especially | in the theory of aggregates and of real functions, without ever reaching any inconsistency. Thus the conjecture arises that, although we have failed to give an unobjectionable meaning to the symbols of our formal systems, we might at least be able to prove their freedom from contradiction by unobjectionable methods. And it seems reasonable to expect that this is possible, because the statement to be proved—I mean the statement that a given formal system is non-contradictory—is of a very simple character and does not
21   involve | any of those objectionable notions such as "property of integers". Indeed, freedom from contradiction simply means that, if we start with certain formulas (called axioms) and carry out on them as many times as we wish certain operations (given by the rules of inference), we shall never obtain two contradictory formulas, i.e., two formulas of which one is the negation of the other. In our proof for freedom from contradiction we don't have to worry about the meaning of the symbols of our systems because

the rules of inference never refer to the meaning, and so the whole matter becomes merely a combinatorial question about the handling of symbols according to given rules.

Of course, the chief point in the desired proof of freedom from contradiction is that it must be conducted by perfectly unobjectionable | methods; i.e., it must strictly avoid the non-constructive existence proofs, non-predicative definitions and similar things, for it is exactly a justification for these doubtful methods which we are seeking. Now, what remains of mathematics if we discard these methods [and retain only things that can be constructed and operations that can actually be carried out] is the so-called intuitionistic mathematics, and the domain of this intuitionistic mathematics is by no means so uniquely determined as it may seem at first sight. For it is certainly true that there are different notions of constructivity and, accordingly, different layers of intuitionistic or constructive mathematics. As we ascend in the series of these layers, we are drawing nearer to ordinary non-constructive mathematics, and at the same time the methods of proof and construction which we admit are becoming less satisfactory and less convincing. The lowest of these layers, i.e., the strictest form of constructive mathematics, can be described roughly by the following two characteristics:

| 1. The application of the notion of "all" or "any" is to be restricted to those infinite totalities for which we can give a finite procedure for generating all their elements (as we can, e.g., for the totality of integers by the process of forming the next greater integer and as we cannot, e.g., for the totality of all properties of integers).

2. Negation must not be applied to propositions stating that something holds for all elements, because this would give existence propositions. Or to be more exact: Negatives of general propositions (i.e., existence propositions) are to have a meaning in our system only in the sense that we have found an example but, for the sake of brevity, do not state it explicitly. I.e., they serve merely as an abbreviation and could be entirely dispensed with if we wished.

From the fact that we have discarded the notion of existence and the logical rules concerning it, it follows that we are left with essentially only one method for proving general propositions, namely, complete induction applied to the generating process of our elements.

| And finally, we require that we should introduce only such notions as are decidable for any particular element and only such functions as can be calculated for any particular element. Such notions and functions can always be defined by complete induction, and so we may say that our system [I will call it A] is based exclusively on the method of complete induction in its definitions as well as in its proofs. This method possesses a particularly high degree of evidence, | and therefore it would be the most desirable thing if the freedom from contradiction of ordinary non-constructive math-

ematics could be proved by methods allowable in this system A. And, as a matter of fact, all the attempts for a proof for freedom from contradiction undertaken by Hilbert and his disciples tried to accomplish exactly that. But unfortunately the hope of succeeding along these lines has vanished entirely in view of some recently discovered facts. It can be shown quite generally that there can exist no proof of the freedom from contradiction of a formal system S which could be expressed in terms of the formal system
26 S itself, i.e., which would proceed by such methods of proof as are | expressible in the system S itself. Now all the intuitionistic proofs complying with the requirements of the system A which have ever been constructed can easily be expressed in the system of classical analysis and even in the system of classical arithmetic, and there are reasons for believing that this will hold for any proof which one will ever be able to construct.

So it seems that not even classical arithmetic can be proved to be non-contradictory by the methods of the system A, because this proof, if complying with the rules of the system A, would be expressible in classical arithmetic itself, which is impossible. Nevertheless, interesting partial results have been obtained, the most far-reaching of which is the following theorem proved by Herbrand: If we take a theory which is constructive in the sense that each existence assertion made in the axioms is covered by a construction, and if we add to this theory the non-constructive notion of existence and all the logical rules concerning it, e.g., the law of excluded
27 middle, we shall never get into any contradiction. | One would think that this was all we wanted. But unfortunately in classical arithmetic we do more than merely apply the rules of logic (e.g., the law of excluded middle) to expressions involving the non-constructive notion of existence. We also apply complete induction to such expressions, i.e., we form properties of integers by means of the non-constructive notion of existence and, in order to prove that these properties belong to all integers, apply complete induction; and that is the point where Herbrand's proof fails. Herbrand's methods might be generalized also for systems which adopt Russell's subdivision of types into subtypes, but, as I mentioned before, for larger systems containing the whole of arithmetic or analysis the situation is hopeless if you insist upon giving your proof for freedom from contradiction by means of the system A.
28 Now if we look at the intuitionistic mathematics as developed | by Brouwer and his followers, we become aware that they by no means confine themselves to the system A. The first place where its limits are transcended is the notion of "absurdity". In our system A we have forbidden any negation of general propositions or, I should rather say, we have admitted it only in the sense that we have actually found a counterexample. Brouwer, however, introduces a different form of negation of general propositions, called "absurdity", and by the assertion that a proposition p is absurd he means that one has succeeded in deriving a contradiction from p (of course

by intuitionistic methods of proof). Now it may happen, and actually does happen, that you can derive a contradiction from the proposition "for every x, $F(x)$ is true" by intuitionistic methods of proof without anyone's being able to give a counterexample, i.e., an x for which $F(x)$ is false, so that we have a perfect substitute for non-constructive existence theorems. And much more than this is true. If we investigate the | axioms of the   29
intuitionistic mathematics as stated by Heyting, a disciple of Brouwer, we find that for the notion of absurdity exactly the same propositions hold as do for negation in ordinary mathematics—at least, this is true within the domain of arithmetic. So we have succeeded, by means of the notion of absurdity, in giving an interpretation and hence also a proof for freedom from contradiction for classical arithmetic, which was impossible by means of the system A alone. The character of the axioms assumed by Heyting about the notion of absurdity can be seen from the following example: $p \supset \neg\neg p$, which means: If p has been proved, then the assumption $\neg p$ leads to a contradiction. This is obvious, because p and $\neg p$ already constitute the contradiction. Axioms of this kind do not violate the main principle of constructive mathematics[[, to wit,]] that you can speak with sense only about things | which you can actually construct and operations which you   30
can actually carry out. For the statements made in Heyting's axioms have in each case this form, that out of a proof or reductio ad absurdum of a certain proposition you can construct a proof or reductio ad absurdum of a certain other proposition, and the axioms are such that these constructions can always be carried out. So Heyting's axioms concerning absurdity and similar notions differ from the system A only by the fact that the substrate on which the constructions are carried out are proofs instead of numbers or other enumerable sets of mathematical objects. But by this very fact they do violate the principle, which I stated before, that the word "any" can be applied only to those totalities for which we have a finite procedure for generating all their elements. For the totality of all possible proofs certainly does not | possess this character, and nevertheless the word "any"   31
is applied to this totality in Heyting's axioms, as you can see from the example which I mentioned before, which reads: "Given *any* proof for a proposition p, you can construct a reductio ad absurdum for the proposition $\neg p$". Totalities whose elements cannot be generated by a well-defined procedure are in some sense vague and indefinite as to their borders. And this objection applies particularly to the totality of intuitionistic proofs because of the vagueness of the notion of constructivity. Therefore this foundation of classical arithmetic by means of the notion of absurdity is of doubtful value. But there remains the hope that in future one may find other and more satisfactory methods of construction beyond the limits of the system A, which may enable us to found classical arithmetic and analysis upon them. This question promises to be a fruitful field for further investigations.

Introductory note to *1933?

This manuscript gives a compact, correct proof of the theorem of Paul Lévy and Ernst Steinitz which generalizes to n-dimensional real vector space a classical theorem of Bernhard Riemann on conditionally convergent series.

The manuscript was found with many blanks for formulae not filled in and without bibliography; it is not clear whether Gödel meant this manuscript for ultimate publication or whether he had actually read the relevant references. The manuscript describes the proof given as "based on ideas of Steinitz"; in fact it is essentially a rearrangement of material in *Steinitz 1913* and *1914*. Gödel's proof is compact in that it uses the theorem, due to Steinitz, that the convex hull of a compact set in a finite-dimensional real vector space is closed.

The theorem of Lévy[a] and Steinitz asserts that for any countable collection of vectors in a finite-dimensional real vector space, the sum-set—i.e., the set of sums of those series which (i) have as terms the vectors in the given collection and (ii) are convergent—is a linear set (i.e., a subspace displaced by the addition of some fixed vector). A strengthened form of this theorem identifies the sum-set, and, in particular, its dimension.

This theorem was first formulated by Lévy in his *1905* (page 509), where he introduced the important concept of "principal vector", called by him *point d'indétermination*, later by Steinitz *Hauptrichtung* (*1914*, page 31), and later by Gödel *Hauptvektor*. Lévy's proof was criticized by Steinitz (*1913*) as difficult to follow for the two-dimensional case and definitely incomplete for higher dimensions. Steinitz then gave an entirely different proof (*1913*) based on a very interesting polygonal rearrangement inequality, which asserts for a real normed linear vector space of finite dimension N: There exists a real finite constant K such that for any finite family of vectors, $v_1, v_2, \ldots, v_m$, each of length less than or equal to 1, and with sum 0, there is a rearrangement, say $u_1, u_2, \ldots, u_m$, such that the partial sums $u_1 + \ldots + u_i$ have length less than or equal to K, for all i.

In *1914* Steinitz pointed out the important gap in the proof by Lévy and completed that proof. The papers by Steinitz give a very detailed exposition of the properties of convex sets, and it takes considerable patience to extract from them the parts necessary to prove the theorem of Lévy and Steinitz.

---

[a]Note that in the manuscript Gödel misspells "Lévy".

Wilhelm Groß (*1917*) gave a short elementary proof, by induction, of the polygonal rearrangement theorem of Steinitz (with a much larger value for K), and used it to give a short proof of the Lévy-Steinitz theorem in the weaker form. Other authors, among them Abraham Wald (*1933*), published a proof of the theorem which was close to the proof of Gödel, and it may well be that publication of Wald's proof lessened Gödel's interest in publication of his own manuscript.

During World War II, an example, by Marcinkiewicz, to show that the theorem failed in Hilbert space was recorded in the original Polish "Scottish Book" (Warsaw). (See *Mauldin 1981*, page 188.)[b] Using a deep theorem of Aryeh Dvoretzky (*1961*) on approximate imbedding of finite-dimensional Euclidean space in an arbitrary normed linear space of sufficiently high dimension, Vladimir M. Kadets (*1986*) imbedded the Marcinkiewicz example in every Banach space. Then M. I. Kadets and Krzystof Wozniakowski (*1989*), and, independently, P. A. Kornilov (*1988*), constructed an example in Hilbert space for which the sum-set consisted of exactly two values; the technique of V. M. Kadets (*1986*) is sufficient to enable the authors also to imbed this example in an arbitrary Banach space.

That the theorem of Lévy and Steinitz does hold in the infinite-dimensional linear topological (but not Banach) space of all real sequences with topology defined by componentwise convergence was shown independently by Wald (*1933a*), Stanimir Troyanski (*1967*) and Yitzhak Katznelson and O. Carruth McGehee (*1974*), and that it holds in every metrizable nuclear space by Wojciech Banaszczyk (*1990*).

The best possible value for the Steinitz constant K is known only for the two-dimensional Euclidean case, where $K = \sqrt{5/4}$, as shown by V. Bergström (*1931*) and by W. Banaszczyk (*1987*); and for the space of vectors $x = (x_1, x_2)$ with real x_i and $\|x\| = |x_1| + |x_2|$, where $K = 3/2$, which may be shown by combining the work of Grinberg and Sevast'yanov (*1980*) and W. Banaszczyk (*1987*). Estimates for K in the N-dimensional case were obtained in various papers: $K \leq 2N$ by Steinitz (*1913*); $K \leq 2^N - 1$ by Groß (*1917*); $K < N$ by F. A. Behrend (*1954*); $K \leq N$ (with a remarkably simple proof) by Grinberg and Sevast'yanov (*1980*); and $K \leq N - 1 + 1/N$ by W. Banaszczyk (*1987*).

Israel Halperin

The translation of *1933?* is by John Dawson, with suggestions from Israel Halperin and William Craig.

---

[b] An account of work done by others on the theorem of Lévy and Steinitz has been given in *Halperin 1986*. For a bibliography of related work, see *Halperin and Ando 1989*.

Vereinfachter Beweis eines Steinitzschen Satzes
(*1933?)

Ist $\mathfrak{A} = v_1, v_2, \ldots$ eine bedingt konvergente Reihe von Vektoren des $\mathbb{R}_n$, so bildet die Menge derjenigen Vektoren, die man durch Umordnen der Glieder von $\mathfrak{A}$ als Reihensumme erhalten kann, nach P. Lévy [[*1905*]] und E. Steinitz [[*1913, 1914,* und *1916*]] eine lineare Mannigfaltigkeit des $\mathbb{R}_n$. Für diesen Satz wird im folgenden ein auf die Steinitzschen Methoden sich stützender aber einfacherer Beweis gegeben.

Ein Einheitsvektor e des $\mathbb{R}_n$ heiße "Hauptvektor", wenn für jedes positive ϵ die Summe der absoluten Beträge der Vektoren, welche mit e einen Winkel $< \epsilon$ einschließen, unendlich ist.

Hilfssatz 1: *Es gibt endlich viele Hauptvektoren $e_1, \ldots, e_s$ ($s > 0$) und positive Zahlen $p_1, \ldots, p_s$ (alle $p_k > 0$) derart, daß $p_1 e_1 + \cdots + p_s e_s = 0$.*

Beweis:[a] Die Menge der Endpunkte der vom Nullpunkt aus aufgetragenen Hauptvektoren heiße $\mathfrak{N}$, ihre konvexe Hülle (kleinste $\mathfrak{N}$ enthaltende konvexe Menge) sei $\mathfrak{N}^c$.

$\mathfrak{N}$ und daher auch $\mathfrak{N}^c$ sind kompakt.[b]

Der Nullpunkt gehört zu $\mathfrak{N}^c$, denn wäre dies nicht der Fall, so würde es einen Stütz durch den Nullpunkt geben (er heiße H), so daß $\mathfrak{N}^c$ und daher alle Hauptvektoren auf einer Seite von H liegen. (Sie heiße die "$\mathfrak{N}$ Seite von H".) Die Summe der Absolutbeträge der|jenigen Vektoren von $\mathfrak{A}$, welche [nicht] auf der $\mathfrak{N}$ Seite[c] von H liegen, wäre dann endlich, daher auch die Summe ihrer Projektionen auf eine zu H senkrechte Gerade l, während die Summe der Projektionen der auf der $\mathfrak{N}$ Seite liegenden Glieder unendlich wäre, was der Konvergenz von $\mathfrak{A}$ widerspricht.[d]

Die Menge der Endpunkte derjenigen Vektoren, die sich als Linearkombinationen von Hauptvektoren mit positiven Koeffizienten darstellen lassen, bildet offenbar eine konvexe $\mathfrak{N}$ enthaltende Menge $\mathfrak{M}$. Also ist $\mathfrak{N}^c \subset \mathfrak{M}$ und daher $0 \in \mathfrak{M}$, womit die Behauptung bewiesen ist.

Wir nennen eine Folge S aus den Folgen $S_1, \ldots S_s$ "zusammengesetzt",[e] wenn $S_1, \ldots, S_s$ Teilfolgen von S sind, von der Art, daß jedes Glied von S in genau einer der Folgen S_k vorkommt.

---

[a]Lemma 1 was essentially proved by Steinitz (*1914*, pp. 18,19,36,37) with $2 \leq s \leq n + 1$.

[b]$\mathfrak{N}$ is not empty, since *conditional* convergence of A was assumed. That $\mathfrak{N}^c$ is closed follows from the theorem of Steinitz (*1913*, p. 154) that if $w = \Sigma_{i=1}^{p} c_i v_i$ with $0 \leq c_i \leq 1$ for all i, $\Sigma_{i=1}^{p} c_i = 1$, and if all v_i are in $\mathfrak{N}$, then there are d_i, with $0 \leq d_i \leq 1$ and $\Sigma_{i=1}^{n} d_i = 1$, and w_i in $\mathfrak{N}$, for all i, such that $w = \Sigma_{i=1}^{n} d_i w_i$.

56

Simplified proof of a theorem of Steinitz
(*1933?)

If $\mathfrak{A} = v_1, v_2, \ldots$ is a conditionally convergent sequence of vectors of $\mathbb{R}_n$, then according to [work of] P. Lévy [1905] and E. Steinitz [1913, 1914 and 1916] the set of those vectors that can be obtained as the sum of a series through rearrangement of the terms of $\mathfrak{A}$ forms a linear manifold of $\mathbb{R}_n$. In what follows, a proof of this theorem is given that is based on Steinitz's methods but is simpler.

Call a unit vector e of $\mathbb{R}_n$ a "principal vector" if, for every positive ϵ, the sum of the absolute values of the vectors that form an angle $< \epsilon$ with e is infinite.

Lemma 1: *There are finitely many principal vectors $e_1, \ldots, e_s (s > 0)$ and positive numbers $p_1, \ldots, p_s$ (all $p_k > 0$) such that $p_1 e_1 + \ldots + p_s e_s = 0$.*

Proof:[a] Let the set of endpoints of the principal vectors starting from the origin be called $\mathfrak{N}$, and let its convex hull (the smallest convex set containing $\mathfrak{N}$) be $\mathfrak{N}^c$.

$\mathfrak{N}$, and therefore also $\mathfrak{N}^c$, is compact.[b]

The origin belongs to $\mathfrak{N}^c$, because if that were not the case, there would be a support[ing hyperplane] through the origin (call it H) such that $\mathfrak{N}^c$, and therefore all principal vectors, would lie on one side of H. (Call it the "$\mathfrak{N}$ side of H".) The sum of the absolute values of those vectors of $\mathfrak{A}$ which [do not] lie on the $\mathfrak{N}$ side[c] of H would then be finite, hence so would the sum of their projections onto a line l perpendicular to H, whereas the sum of the projections of the terms that lie on the $\mathfrak{N}$ side would be infinite, which contradicts the convergence of $\mathfrak{A}$.[d]

The set of endpoints of those vectors that may be represented as linear combinations, with positive coefficients, of principal vectors obviously forms a convex set $\mathfrak{M}$ that contains $\mathfrak{N}$. Thus $\mathfrak{N}^c \subset \mathfrak{M}$, and therefore $0 \in \mathfrak{M}$, so that the assertion is proved.

We say a sequence S is "composed" from[e] the sequences $S_1, \ldots, S_s$ if $S_1, \ldots, S_s$ are subsequences of S such that every term of S occurs in exactly one of the sequences S_k.

---

[c]By definition a side of H excludes all points of H.

[d]This argument was used by Steinitz (1914, pp. 36,37).

[e]The construction *zusammengesetzte Reihen* was used by Steinitz (1914, pp. 37,38) under the name *durch Ineinanderschieben*.

Hilfssatz 2: *Seien* $\mathfrak{a}_1, \ldots, \mathfrak{a}_s$ [*von null verschiedene*] *Vektoren in* $\mathbb{R}_n$, *die nicht schon in einem* [*echten*] *linearen Teilraum von* $\mathbb{R}_n$ *liegen und für die* $\mathfrak{a}_1 + \cdots + \mathfrak{a}_s = 0$. *Sei ferner jedem* $\mathfrak{a}_k$ *eine Reihe,* S_k *von mit* $\mathfrak{a}_k$ *richtungsgleichen Vektoren mit* $S_k = \mathfrak{w}_1^k, \mathfrak{w}_2^k, \ldots$; $\lim_{i\to\infty} \mathfrak{w}_i^k = 0$, $\sum_{i=1}^{\infty} |\mathfrak{w}_i^k| = \infty$, $k = 1, \ldots, s$, *zugeordnet und sei* S_0 *eine Reihe von Vektoren mit* $S_0 = w_1, w_2, \ldots$; $\lim_{i\to\infty} w_i = 0$. *Dann gibt es eine aus* $S_0, S_1, \ldots, S_s$ *zusammengesetzte Reihe* S, *die einen beliebigen Vektor* w *des* $\mathbb{R}_n$ *als Summe hat.*

Beweis: Es genügt, den Satz für $w = 0$ zu beweisen, da man im allgemeinen Fall $-w$ zur Reihe S_0 hinzufügen und dann den Satz für 0 anwenden kann.

Zu jedem w_i bestimme man s Zahlen w_i^k, $k = 1, \ldots, s$ (die "Koordinaten" von w_i) in der Weise, daß $w_i = \sum_{k=1}^{s} w_i^k \mathfrak{a}_k$ und $| \lim_{i\to\infty} w_i^k = 0$.

Unter den "Koordinaten" von $\{\mathfrak{w}_i^k\}$ verstehe man das s-tupel von Zahlen, dessen k-tes Glied $\dfrac{|\mathfrak{w}_i^k|}{|\mathfrak{a}_k|}$ [ist] und dessen übrige Glieder 0 sind.

Für die i-te Partialsumme $\mathfrak{S}^i$ der zu konstruierenden Reihe verstehe man unter ihren "Koordinaten" $\mathfrak{S}_1^i, \mathfrak{S}_2^i, \ldots, \mathfrak{S}_s^i$ die Summen der entsprechenden Koordinaten ihrer Glieder, so daß immer gilt: $\mathfrak{S}^i = \sum_{k=1}^{s} \mathfrak{S}_k^i \mathfrak{a}_k$.

S^i bezeichne das Maximum der Zahlen $\mathfrak{S}_1^i, \mathfrak{S}_2^i, \ldots, \mathfrak{S}_s^i$. Die zu konstruierende Reihe bilde man wie folgt: ihr erstes Glied sei w_1 (also $\mathfrak{S}^1 = w_1$), dann nehme man aus jeder der Reihen $\{\mathfrak{w}_i^k\}$ so viele Glieder, daß für die neue Partialsumme $\mathfrak{S}^{n_1}$ gilt: $\mathfrak{S}_k^{n_1} > S^1 (k = 1, 2, \ldots, s)$; dann folgt w_2;[f] usw.

Mit Rücksicht darauf, daß wegen $\mathfrak{w}_i^k \to 0$, $w_i \to 0$ auch $\mathfrak{S}_k^i - \mathfrak{S}_1^i \to 0$, $k = 1, \ldots, s$, und daß wegen $\mathfrak{a}_1 + \cdots + \mathfrak{a}_s = 0$, bestätigt man leicht, daß $\sum_S v = 0$.

Hilfssatz 2 gilt auch noch, wenn die Reihen S_k nicht parallel zu $\mathfrak{a}_k$ sind, sondern die Reihen $\mathfrak{w}_i^k - (\mathfrak{w}_i^k)'$ konvergieren (wobei $(\mathfrak{w}_i^k)'$ die Projektion von $\mathfrak{w}_i^k$ auf die Richtung $\mathfrak{a}_k$ bedeutet).

Denn sind μ^k, $k = 1, \ldots, s$, die Summen dieser Reihen, so bilde man nach Hilfssatz 2 die aus $\{(\mathfrak{w}_i^k)'\}$, $k = 1, \ldots, s$, zusammengesetzte Reihe, welche $w - (\mu^1 + \cdots + \mu^s)$ als Summe hat. Ersetzt man in dieser Reihe $(\mathfrak{w}_i^k)'$ durch $\mathfrak{w}_i^k$, so konvergiert die so entstehende Reihe gegen w.

Es ist ferner ersichtlich, daß man zu jedem Hauptvektor e eine Teilfolge $\{\mathfrak{w}_i\}$ von $\mathfrak{A}$ bilden kann, so daß $\sum_i |\mathfrak{w}_i| = \infty$ und $\{\mathfrak{w}_i - (\mathfrak{w}_i)'\}$ absolut konvergiert (wobei $(\mathfrak{w}_i)'$ die Projektion von $\mathfrak{w}_i$ auf e bedeutet) (Hilfssatz 3).

Zum Beweise braucht man nur eine Folge positiver Zahlen $| 0 < \epsilon_i < \frac{\pi}{2}$ mit beschränkter Summe anzunehmen und für jedes ϵ_i innerhalb des Kegels

---

[f]Then take from each of the remaining terms of the sequences $\{\mathfrak{w}_i^k\}$ so many terms that $S_k^{n_2} > S^{n_1} (k = 1, \ldots, s)$; next take w_3.

Lemma 2: Let $a_1, \ldots, a_s$ be [non-zero] vectors in $\mathbb{R}_n$ that do not lie in a [proper] linear subspace of $\mathbb{R}_n$ and for which $a_1 + \ldots + a_s = 0$. Furthermore, to each a_k let there be assigned a sequence S_k of vectors with the same direction as a_k, with $S_k = \mathfrak{w}_1^k, \mathfrak{w}_2^k, \ldots$; $\lim_{i \to \infty} \mathfrak{w}_i^k = 0$, $\sum_{i=1}^{\infty} |\mathfrak{w}_i^k| = \infty$, $k = 1, \ldots, s$; and let S_0 be a sequence of vectors, with $S_0 = w_1, w_2, \ldots$, $\lim_{i \to \infty} w_i = 0$. Then for an arbitrarily given vector w of $\mathbb{R}_n$ there is a sequence S composed from $S_0, S_1, \ldots, S_s$ that has w as its sum.

Proof: It suffices to prove the theorem for $w = 0$, since in the general case one can adjoin $-w$ to the sequence S_0 and then apply the theorem for 0.

For every w_i one can specify s numbers w_i^k, $k = 1, \ldots, s$ (the "coordinates" of w_i) in such a way that $w_i = \sum_{k=1}^{s} w_i^k a_k$ and $\lim_{i \to \infty} w_i^k = 0$.

By the "coordinates" of $\{\mathfrak{w}_i^k\}$ is to be understood the s-tuple of numbers whose k^{th} term [is] $\dfrac{|\mathfrak{w}_i^k|}{|a_k|}$ and whose remaining terms are 0.

By the "coordinates" $\mathfrak{S}_1^i, \mathfrak{S}_2^i, \ldots, \mathfrak{S}_s^i$ of the ith partial sum $\mathfrak{S}^i$ of the sequence to be constructed are to be understood the sums of the corresponding coordinates of its terms, so that one always has $\mathfrak{S}^i = \sum_{k=1}^{s} \mathfrak{S}_k^i a_k$.

Let S^i denote the maximum of the numbers $\mathfrak{S}_1^i, \mathfrak{S}_2^i, \ldots, \mathfrak{S}_s^i$. The sequence to be constructed is to be formed as follows: Let its first term be w_1 (so that $\mathfrak{S}^1 = w_1$); then from each of the sequences $\{\mathfrak{w}_i^k\}$ take enough terms so that for the new partial sum $\mathfrak{S}_k^{n_1}$ we have $\mathfrak{S}_k^{n_1} > S^1 (k = 1, 2, \ldots, s)$. Next follows w_2, and so on.[f]

Recalling that $\mathfrak{S}_k^i - \mathfrak{S}_1^i \to 0$, $k = 1, \ldots, s$ (since $\mathfrak{w}_i^k \to 0$ and $w_i \to 0$), and considering that $a_1 + \ldots + a_s = 0$, one easily confirms that $\sum_S v = 0$.

Lemma 2 still holds also when the sequences S_k are not parallel to a_k, but the sequences $\mathfrak{w}_i^k - (\mathfrak{w}_i^k)'$ converge (where $(\mathfrak{w}_i^k)'$ denotes the projection of $\mathfrak{w}_i^k$ in the direction a_k).

For if μ^k, $k = 1, \ldots, s$, are the sums of these sequences, then, according to Lemma 2, one may form the sequence composed from $\{(\mathfrak{w}_i^k)'\}$, $k = 1, \ldots, s$, which has $w - (\mu^1 + \ldots + \mu^s)$ as its sum. If, in this sequence, $(\mathfrak{w}_i^k)'$ is replaced by $\mathfrak{w}_i^k$, then the resulting sequence converges to w.

Furthermore the following is evident (Lemma 3): For each principal vector e one can form a subsequence $\{\mathfrak{w}_i\}$ of $\mathfrak{A}$ such that $\sum_i |\mathfrak{w}_i| = \infty$ and $\{\mathfrak{w}_i - (\mathfrak{w}_i)'\}$ converges absolutely (where $(\mathfrak{w}_i)'$ denotes the projection of $\mathfrak{w}_i$ onto e).

For the proof, one need only take a sequence of positive numbers $0 < \epsilon_i < \frac{\pi}{2}$ with bounded sum, and, for each ϵ_i, choose from within the

mit der Achse e und dem Öffnungswinkel ϵ_i so viele Glieder $v_{i_1}, \ldots, v_{i_m}$ aus $\mathfrak{A}$ herauszugreifen, daß ihre Summe $1 < \sum_r |v_{i_r}| < 2$ [erfüllt].

Nehmen wir nun an, der Steinitzsche Satz sei für Räume von niedrigerer als n-ter Dimension schon bewiesen und sei M der von den nach Hilfssatz 1 existierenden Vektoren $e_1, \ldots, e_s$ aufgespannte lineare Teilraum und $M^\perp$ der dazu orthogonale; ferner v_M, bzw. $v_{M^\perp}$, die Projektionen von v in M, bzw. $M^\perp$, und sei L der Summenbereich von $\{v_{M^\perp} | v \in \mathfrak{A}\}$, welcher nach induktiver Annahme eine lineare Mannigfaltigkeit ist.

Dann gehört ein Vektor μ dann und nur dann zum Summenbereich von $\mathfrak{A}$, wenn $\mu_{M^\perp} \in L$.[g]

Die Notwendigkeit der Bedingung ist trivial. Sei also [μ ein Vektor, sodaß] $\mu_{M^\perp} \in L$; dann gibt es eine Anordnung der gegebenen Reihe (wir nehmen an, $\mathfrak{A}$ sei schon eine solche), so daß $\sum_{\mathfrak{A}} v_{M^\perp} = \mu_{M^\perp}$. Zu jedem e_k bestimme man nach Hilfssatz 3 eine Teilfolge S_k von $\mathfrak{A}$ in der dort angegebenen Weise und zwar so, daß keine zwei der Reihen S_k gemeinsame Glieder haben; weil $\sum_{S_k} |v_{M^\perp}| < \infty$, ist $\sum_{S_k} v_{M^\perp}$ absolut konvergent. Mit S_0 bezeichne man die Teilfolge von $\mathfrak{A}$, welche übrig bleibt, wenn man sämtliche Folgen $S_1, \ldots, S_s$ aus $\mathfrak{A}$ wegläßt. Nach Hilfssatz 2 kann man aus $S_0, S_1, \ldots, S_s$ eine Folge S zusammensetzen, so daß $\sum_S v_M = \mu_M$.

Dann ist auch $\mathfrak{A}$ aus $S_0, S_1, \ldots, S_s$ zusammengesetzt. Andererseits sind auch aus denselben Reihen zusammengesetzt die Reihen $\sum_{\mathfrak{A}} v_{M^\perp}, \sum_S v_{M^\perp}$. $\sum_{S_k} v_{M^\perp}, k = 1, \ldots, s$, sind konvergent, daher auch $\sum_{S_0} v_{M^\perp}$ konvergent und zwei Reihen, die aus denselben konvergenten Reihen zusammengesetzt sind, ergeben offenbar dieselbe Summe. Es ist also $\sum_S v_{M^\perp} = \sum_{\mathfrak{A}} v_{M^\perp} = \mu_{M^\perp}$.[h]

---

[g]And hence this sum domain is a linear manifold.

cone with axis e and opening-angle ϵ_i enough terms $v_{i_1}, \ldots, v_{i_m}$ from $\mathfrak{A}$ so that their sum satisfies $1 < \sum_r |v_{i_r}| < 2$.

Let us now assume that Steinitz's Theorem has already been proved for spaces of dimension lower than n. Let M be the linear subspace spanned by the vectors, $e_1, \ldots, e_s$, whose existence is guaranteed by Lemma 1, and let $M^\perp$ be the [subspace] orthogonal to it. Furthermore, let v_M (respectively, $v_{M^\perp}$) be the projection of v onto M (respectively, $M^\perp$) and let L be the sum domain of $\{v_{M^\perp} | v \in \mathfrak{A}\}$, which, by the inductive assumption, is a linear manifold.

[We shall show that] a vector μ then belongs to the sum domain of $\mathfrak{A}$ if and only if $\mu_{M^\perp} \in L$.[g]

The necessity of the condition is trivial. So [for a proof of the sufficiency] let [μ be a vector with] $\mu_{M^\perp} \in L$; then there is an arrangement of the given sequence (we assume $\mathfrak{A}$ to be one such) such that $\sum_{\mathfrak{A}} v_{M^\perp} = \mu_{M^\perp}$. According to Lemma 3, to each e_k one may, in the way indicated there, specify a subsequence S_k of $\mathfrak{A}$ so that no two of the sequences S_k have common terms; because $\sum_{S_k} |v_{M^\perp}| < \infty$, $\sum_{S_k} v_{M^\perp}$ is absolutely convergent. Denote by S_0 the subsequence of $\mathfrak{A}$ that remains when all the sequences $S_1, \ldots, S_s$ are omitted from $\mathfrak{A}$. According to Lemma 2, a sequence S can be composed from $S_0, S_1, \ldots, S_s$ so that $\sum_S v_M = \mu_M$.

Then $\mathfrak{A}$ is also composed from $S_0, S_1, \ldots, S_s$. On the other hand, from those same sequences the sequences $\sum_{\mathfrak{A}} v_{M^\perp}$, $\sum_S v_{M^\perp}$ are composed as well. [The sequences] $\sum_{S_k} v_{M^\perp}$, $k = 1, \ldots, s$, are convergent, hence $\sum_{S_0} v_{M^\perp}$ is also convergent; and two sequences that are composed from the same convergent sequences obviously yield the same sum. Thus $\sum_S v_{M^\perp} = \sum_{\mathfrak{A}} v_{M^\perp} = \mu_{M^\perp}$.[h]

---

[h] And so $\Sigma_S v = \mu$.

Introductory note to *1938a*

The text that follows consists of notes for a lecture Gödel delivered in Vienna on 29 January 1938 to a seminar organized by Edgar Zilsel. The lecture presents an overview of possibilities for continuing Hilbert's program in a revised form. It is an altogether remarkable document: biographically, it provides, together with *1933o* and *1941*, significant information on the development of Gödel's foundational views; substantively, it presents a hierarchy of constructive theories that are suitable for giving (relative) consistency proofs of parts of classical mathematics (see §§2–4 of the present note); and, mathematically, it analyzes Gentzen's 1936 proof of the consistency of classical arithmetic in a most striking way (see §7). A surprising general conclusion from the three documents just mentioned is that Gödel in those years was intellectually much closer to the ideas and goals pursued in the Hilbert school than has been generally assumed (or than can be inferred from his own published accounts).

1. The setting.

Edgar Zilsel (1891–1944) had been connected with the Vienna Circle. By 1938 his main interest was history and sociology of science.[a] The concrete stimulus for Gödel's preparing the lecture was Zilsel's question whether anything new had happened in the foundations of mathematics and his request that Gödel should describe the "status of the consistency question" to the seminar. This is reported in Gödel's notes[b] concerning

---

[a] Cf. *Zilsel 1976*, which collects in German translation articles published in English after Zilsel's emigration. The English versions are collected in *Zilsel 199?*. Zilsel had taught physics and philosophy at the *Volkshochschule* in Vienna, but as a result of the Dollfuss coup in 1934, he was dismissed from or eased out of this position (see *Behrmann 1976* and *Dvořák 1981*, pp. 23–25) and thereafter worked as a *Gymnasium* teacher. He emigrated in 1938 to England and moved on to America the next year. In 1944, while teaching at Mills College in California, he committed suicide.

Testimony differs as to whether Zilsel was a member of the Vienna Circle or merely someone sympathetic to their views who attended some of the sessions; see *Dvořák 1981*, pp. 30–31.

[b] The notes (*Zusammenkunft bei Zilsel*) are in Gödel's *Nachlass* (document no. 030114) and were transcribed from Gabelsberger by Cheryl Dawson. In editing the text and preparing this note, we have also used a document (no. 040147) entitled *Konzept* (i.e., draft), evidently an earlier draft of Gödel's notes for this lecture. On our use of this document, see the textual notes.

the organizational meeting of the seminar on 2 October 1937 at Zilsel's home; it was at this organizational meeting that Zilsel made his request and that Gödel, after some reflection, agreed to speak on the consistency question.

As Zilsel did not have a university position, the seminar was probably an informal, private affair and continued to meet at Zilsel's home.[c] Gödel's notes do not give a clear view of the seminar's intended theme, except that it was rather general. The list of names of those present or mentioned as possible participants confirms this; they spread over several fields and include no one, except for Gödel, who was much involved in mathematical logic.[d] In his immediate response to Zilsel's request Gödel suggested presenting a German version of the lecture *1933o, given in Cambridge, Mass. (published in this volume); but on reflection he added that that talk was "zu prinzipiell", which can be translated roughly as "too general".

The Zilsel lecture gives, as we remarked, an overview of possibilities for a revised Hilbert program. The central element of that program was to prove the consistency of formalized mathematical theories by finitist means. Gödel's 1931 incompleteness theorems have been taken to imply that for theories as strong as first-order arithmetic this is impossible, and indeed, so far as Gödel ventures to interpret Hilbert's finitism, that is Gödel's view in the present text as well as earlier in *1933o (though not in 1931) and later in *1941, 1958 and 1972. The crucial questions then are what extensions of finitist methods will yield consistency proofs, and what epistemological value such proofs will have.

Two developments after *Gödel 1931* are especially relevant to these questions. The first was the consistency proof for classical first-order arithmetic relative to intuitionistic arithmetic obtained by Gödel (*1933e*). The proof made clear that intuitionistic methods went beyond finitist ones (cf. footnote j below). Some of the issues involved had been discussed in Gödel's lecture *1933o, but also in print, for example in *Bernays 1935* and *Gentzen 1936*. Most important is Bernays' emphasis on the "abstract element" in intuitionistic considerations.[e] The second development was Gentzen's consistency proof for first-order arithmetic using as the additional principle—justified from an intuitionistic

---

[c]Karl Popper writes (*1976*, p. 84) of having given a paper to a gathering there several years earlier.

It is likely that the seminar did not continue long after Gödel's talk, since the Anschluß took place only six weeks later, and not long after that Zilsel emigrated (cf. fn. a).

[d]We are indebted to Katalin Makkai for researching this matter.

[e]Significantly, Gödel refers to that earlier discussion in *1958*.

standpoint—transfinite induction up to ϵ_0. Already in *1933o* (page 31) Gödel had speculated about a revised version of Hilbert's program using constructive means that extend the limited finitist ones without being as wide and problematic as the intuitionistic ones:

> But there remains the hope that in future one may find other and more satisfactory methods of construction beyond the limits of the system A [capturing finitist methods], which may enable us to found classical arithmetic and analysis upon them. This question promises to be a fruitful field for further investigations.

The Cambridge lecture does not suggest any intermediate methods of construction; by contrast, Gödel presents in the Zilsel lecture two "more satisfactory methods" that provide bases to which not only classical arithmetic but also parts of analysis might be reducible: quantifier-free theories for higher-type functionals and transfinite induction along constructive ordinals. Before looking at these possibilities, we sketch the pertinent features of the Cambridge talk, because they give a very clear view not only of the philosophical and mathematical issues Gödel addresses, but also of the continuity of his development.[f]

2. Relative consistency.

Understanding by mathematics "the totality of the methods of proof actually used by mathematicians", Gödel sees the problem of providing a foundation for these methods as falling into two distinct parts (page 1):

> At first these methods of proof have to be reduced to a minimum number of axioms and primitive rules of inference, which have to be stated as precisely as possible, and then secondly a justification in some sense or other has to be sought for these axioms, i.e., a theoretical foundation of the fact that they lead to results agreeing with each other and with empirical facts.

The first part of the problem is solved satisfactorily through type theory and axiomatic set theory, but with respect to the second part Gödel considers the situation to be extremely unsatisfactory. "Our formalism", he contends, "works perfectly well and is perfectly unobjectionable as long as we consider it as a mere game with symbols, but as soon as

---

[f]Cf. Solomon Feferman's detailed introductory note to *1933o* in this volume, p. 36.

we come to attach a meaning to our symbols serious difficulties arise" (page 15). Two aspects of classical mathematical theories (the non-constructive notion of existence and impredicative definitions) are seen as problematic because of a necessary Platonist presupposition "which cannot satisfy any critical mind and which does not even produce the conviction that they are consistent" (page 19). This analysis conforms with that given in the Hilbert school, for example in *Hilbert and Bernays 1934, Bernays 1935* and *Gentzen 1936*.[g] Gödel expresses the belief, again as the members of the Hilbert school did, that the inconsistency of the axioms is most unlikely and that it might be possible "to prove their freedom from contradiction by unobjectionable methods".

Clearly, the methods whose justification is being sought cannot be used in consistency proofs, and one is led to the consideration of parts of mathematics that are free of such methods. Intuitionistic mathematics is a candidate, but Gödel emphasizes (page 22) that

> the domain of this intuitionistic mathematics is by no means so uniquely determined as it may seem at first sight. For it is certainly true that there are different notions of constructivity and, accordingly, different layers of intuitionistic or constructive mathematics. As we ascend in the series of these layers, we are drawing nearer to ordinary non-constructive mathematics, and at the same time the methods of proof and construction which we admit are becoming less satisfactory and less convincing.

The strictest constructivity requirements are expressed by Gödel (pages 23–25) in a system A that is based "exclusively on the method of complete induction in its definitions as well as in its proofs". That implies that the system A satisfies three general characteristics:[h] (**A1**) Universal quantification is restricted to "infinite totalities for which we can give a finite procedure for generating all their elements"; (**A2**) Existential statements (and negations of universal ones) are used only as abbreviations, indicating that a particular (counter-)example has been found without—for brevity's sake—explicitly indicating it; (**A3**) Only decidable notions and calculable functions can be introduced. As the method of complete induction possesses for Gödel a particularly high degree of evidence, "it would be the most desirable thing if the freedom from contradiction of ordinary non-constructive mathematics could be proved by methods allowable in this system A" (page 25).

---

[g]Of these writings *Bernays 1935* is more ready to defend Platonism, with certain qualifications.

[h]The designations (**A1**)–(**A3**) are introduced by us for ease of reference.

Gödel infers that Hilbert's original program is unattainable from two claims: first, *all* attempts for finitist consistency proofs actually undertaken in the Hilbert school operate within system A; second, *all* possible finitist arguments can be carried out in analysis and even classical arithmetic. The latter claim implies jointly with the second incompleteness theorem that finitist consistency proofs cannot be given for arithmetic, let alone analysis. Gödel puts this conclusion here quite strongly: "...unfortunately the hope of succeeding along these lines ⟦using only the methods of system A⟧ has vanished entirely in view of some recently discovered facts" (page 25). But he points to interesting partial results and states the most far-reaching one, due to *Herbrand 1931* in a beautiful and informative way (page 26):

> If we take a theory which is constructive in the sense that each existence assertion made in the axioms is covered by a construction, and if we add to this theory the non-constructive notion of existence and all the logical rules concerning it, e.g., the law of excluded middle, we shall never get into any contradiction.

Gödel conjectures that Herbrand's method might be generalized to treat Russell's "ramified type theory", i.e., we assume, the theory obtained from system A by adding ramified type theory instead of classical first-order logic.[i]

There *are*, however, more extended constructive methods than those formalized in system A; this follows from the observation that system A is too weak to prove the consistency of classical arithmetic together with the fact that the consistency of classical arithmetic can be established relative to intuitionistic arithmetic.[j] The relative consistency proof is made possible by the intuitionistic notion of absurdity, for which "exactly the same propositions hold as do for negation in ordinary mathematics—at least, this is true within the domain of arithmetic"

---

[i] In *Konzept*, p. 0.1, Gödel mentions Herbrand's results again and also the conjecture concerning ramified type theory. The obstacle for an extension of Herbrand's proof is the principle of induction for "transfinite" statements, i.e., formulae containing quantifiers. Interestingly, as discovered in *Parsons 1970*, and independently by Mints (*1971*) and Takeuti (*1975*, p. 175), the induction axiom schema for purely existential statements leads to a conservative extension of A, or rather its arithmetic version, primitive recursive arithmetic. How Herbrand's central considerations can be extended (by techniques developed in the tradition of Gentzen) to obtain this result is shown in *Sieg 1991*.

[j] In his introductory note to *1933e* (these *Works*, Vol. I, p. 284), Troelstra mentions relevant work also of Kolmogorov, Gentzen and Bernays. Indeed, as reported in *Gentzen 1936*, p. 532, Gentzen and Bernays discovered essentially the same relative consistency proof independently of Gödel. According to Bernays (*1967*, p. 502), the above considerations made the Hilbert school distinguish intuitionistic from finitist methods. Hilbert and Bernays (*1934*, p. 43) make the distinction without referring to the result discussed here.

(page 29). This foundation for classical arithmetic is, however, "of doubtful value": the principles for absurdity and similar notions (as formulated by Heyting) employ operations over *all* possible proofs, and the totality of all intuitionistic proofs cannot be generated by a finite procedure; thus, these principles violate the constructivity requirement (**A1**).

Despite his critical attitude towards Hilbert and Brouwer, Gödel dismisses neither in **1933o* when trying to make sense out of Hilbert's program in a more general setting,[k] namely, as a challenge to find consistency proofs for systems of "transfinite mathematics" relative to "constructive" theories. And he expresses his belief that epistemologically significant reductions may be obtained.

3. Layers of constructivity.

In his lecture at Zilsel's, Gödel explores options for such relative consistency proofs and goes beyond **1933o* by considering—in detail—three different and, as he points out, *known*[l] ways of extending the arithmetic version of system A: the first one uses higher-type functionals, the second introduces absurdity and a concept of consequence, and the third adds the principle of transfinite induction for concretely defined ordinals of the second number class. The first way is related to the hierarchy of functionals in *Hilbert 1926* and is a precursor of the *Dialectica* interpretation; the very brief section IV of the text is devoted to it. Of course, the second way is based on intuitionistic proposals, whereas the last way is due to Gentzen. The second and third way are discussed extensively in sections V and VI.

The broad themes of the preceding discussion are taken up in sections I through III, putting the notion of reducibility first; a theory T is called *reducible to a theory* S if and only if either

$$S \text{ is a subsystem of } T \text{ and } S \text{ proves } Wid\ S \rightarrow Wid\ T$$

or

$$S \text{ proves } Wid\ T.$$

As an example for the first type of reducibility Gödel alludes to his own relative consistency proof for the axiom of choice; as an example

---

[k]This is in contrast to the lecture at Zilsel's, where he makes some (uncharacteristically polemical) remarks against the members of the Hilbert school and against Gentzen in particular; cf. fn. ss below.

[l]When introducing the three ways of extending the basic system, Gödel starts out by saying: Drei Wege sind bisher bekannt (i.e., Three ways are known up to now).

for the second type of reducibility he mentions the reduction of analysis to logic (meaning, we assume, simple type theory). Concerning the epistemological side of the problem, Gödel emphasizes that a proof is satisfactory in the first case only if S is a proper part of T, and in the second case only if S is more evident, more reliable than T. Though he admits that the latter criterion is subjective, he points to the fact that there is general agreement that constructive theories are better than non-constructive ones, i.e., those that incorporate "transfinite" existential quantification.

Acknowledging the vagueness of the notion of constructivity, Gödel formulates in section II what he calls a *Rahmendefinition* that incorporates the requirements (**A1**) through (**A3**) which, in *1933o*, had motivated the system A. Now there are four conditions:[m] (**R1**) restates (**A3**), namely, that the primitive operations must be computable and that the basic relations must be decidable; (**R2**) combines aspects of (**A1**) and (**A2**) to restrict appropriately the application of universal and existential quantifiers; (**R3**) is an open list of inference rules and axioms that includes defining equations for primitive recursive functions, axioms and rules of the classical propositional calculus, the rule of substitution, and ordinary complete induction (for quantifier-free formulae); (**R4**) indicates what was in *1933o* the positive motivation for restricting universal quantification, namely, the finite generation of objects: "Objects should be surveyable (that is, denumerable)."

This point of view is modified in *1941*, where we find (on pages 2 and 3) versions of the first two conditions. The list of basic axioms and rules is introduced later in a very similar way, though not as part of the *Rahmendefinition*, but rather as part of the specification of the system Σ for finite-type functionals. There is no analogue to (**R4**). These changes (in exposition) are most interesting, as both (**R3**) and (**R4**) are viewed as "problematic" in the lecture at Zilsel's. As to (**R3**), the restriction to induction for just natural numbers is viewed as problematic, because induction for certain transfinite ordinals is also evident; (**R4**) is problematic "because of the concept of function". Here, it seems, is the germ of Gödel's analysis in *1958* of the distinction between (strictly) finitist and intuitionistic considerations.

Gödel focuses in the present text immediately on number theory. The theory that corresponds to system A is obviously *primitive recursive arithmetic*, PRA, and it is viewed as the fundamental system in the hierarchy of constructive systems described briefly in section III.

---

[m]For convenience in our further discussions, we use (**Ri**) to refer to these conditions, which were simply numbered by Gödel.

4. Higher-type functionals.

The first kind of extension of PRA consists in the introduction of (defining equations for) functionals of finite type. Gödel suggests continuing, as Hilbert does in *1926*, the introduction of types into the transfinite, "if it is demanded that types only [be admitted] for those ordinal numbers which have been defined in an earlier system". The mathematical details for the basic finite type theory, as well as its transfinite extension, are sketchy. Nevertheless, this extension is most interesting for at least three reasons: It is the only extension that, according to Gödel, satisfies *all* requirements of the "Rahmendefinition"; it is connected to Hilbert and Ackermann's hierarchies of recursive functionals; and, finally, it is the first known articulated, even if very rudimentary, step in the evolution of Gödel's *Dialectica* interpretation.

Gödel indicates only by an example how to extend PRA by a recursion schema for functionals; the example he gives is as follows:

$$\Phi(f, 1, k) = f(k)$$
$$\Phi(f, n + 1, k) = f(\Phi(f, n, k)).$$

This definition of the iteration functional is almost identical to the ones given in *Hilbert 1926*, page 186, and *Ackermann 1928*, page 118 (*van Heijenoort 1967*, pages 388 and 495 respectively), where the iteration functional is used for defining the Ackermann function.[n] The latter function was, as will be recalled, the first example of a calculable function that cannot be defined by ordinary primitive recursion. However, it can be defined by primitive recursion if higher-type objects are allowed. Hilbert (*1926*, page 186; *van Heijenoort 1967*, page 389) formulated the full definition schema as follows:

$$\rho(\mathfrak{g}, \mathfrak{a}, 0) = \mathfrak{a}$$
$$\rho(\mathfrak{g}, \mathfrak{a}, n + 1) = \mathfrak{g}(\rho(\mathfrak{g}, \mathfrak{a}, n), n);$$

$\mathfrak{a}$ is of arbitrary type (and ρ and $\mathfrak{g}$ must satisfy the obvious type restrictions). It is not entirely clear, but certainly most plausible, that Gödel envisioned using some form of the full schema (for what he calls here "closed systems").

Gödel claims that all his requirements are satisfied. Equality of

---

[n]This "iterator" suffices to define the recursor functional, which directly yields the schema of definition by primitive recursion. See *Diller and Schütte 1971* or *Troelstra 1973*, theorem 1.7.11 (p. 56).

higher-type objects could pose a problem for (**R1**); on this point see below. The claim that (**R4**) is satisfied would seem to mean that Gödel thought of the higher-type variables as ranging over the primitive recursive functionals themselves, which are given by terms. But then it is puzzling that Gödel earlier says that (**R4**) is "problematic ... because of the concept of function" (page 3).

Various claims and conjectures are formulated in subsections IV, 4 and 5. We shall make some frankly speculative remarks, which, together with (the introductory note to) *1941*, may help the reader to speculate further on Gödel's statements. However, before discussing the assertions in subsection 4, we try to clarify the character of the extension procedure that is claimed to "contain" certain additions. The procedure is described more directly in a remark of *Konzept* (quoted in full below—see editorial note n to the text); the recursion schema mentioned there is the schema depicted above for defining the iteration functional:

> In general this recursion schema [is used] for the introduction of Φ_i [with the help of] earlier f_i and the functions obtained by substituting in one another. ... This hierarchy can be continued by introducing functions whose arguments are such Φ and admitting once again recursive definitions according to a numerical parameter. And that can even be extended into the transfinite.

Already in *Ackermann 1928*, pages 118–119 (*van Heijenoort 1967*, page 495), there is a discussion of a hierarchy A_i of classes of functionals. The elements of A_i are, in van Heijenoort's terminology (*ibid.*, page 494), functionals of level i, i.e., their arguments are at most of level $(i - 1)$ and their values are natural numbers. Gödel does not explicitly envisage equations of terms of higher type; if he indeed does not, then his hierarchy corresponds to Ackermann's. Otherwise, one is led quite directly to a hierarchy of classes H_i, restricted to functionals of type level $\leq i$, where numbers are of type level 0, and functionals with arguments of type σ and values of type τ are of type level $\max[\text{level}(\sigma) + 1, \text{level}(\tau)]$.[o]

Gödel began section III with the statement, "The finitary systems form a hierarchy." In section VI he uses S_1 and S_ω, evidently for systems in this hierarchy. A reasonable conjecture is that they are the systems for functionals of higher type in either the hierarchy A_i or H_i.

---

[o]This reading would, on the conjecture stated immediately below, make the systems S_i correspond roughly to the subsystems T_i of the **T** of *Gödel 1958* described in *Parsons 1972*, section 4. Parsons uses the term "rank".

It is, however, more likely that Gödel envisaged systems with equations only of numbers, thus closer to that of *Spector 1962* than to the **T** of *Gödel 1958*.[P] Otherwise one would expect some mention of the problems involved in interpreting equations of higher type so as to be decidable. (Hilbert also did not discuss the interpretation of such equations.) Such equations do occur in *1941*, pages 15 and 17, but from the notes of Gödel's lectures on intuitionistic logic at Princeton at the time it is clear that this use was informal, and the Σ of *1941* had only equality of numbers as a primitive.[q]

Although it would be natural today to take S_ω as the union of the S_i and thus roughly the Σ of *1941* (or perhaps the **T** of *1958*), this is hard to reconcile with Gödel's later statement (page 13) that recursion on ϵ_0 may be obtainable in S_ω. We will understand S_ω as a system containing functionals of lowest transfinite type (what on the other understanding would be $S_{\omega+1}$).

The "addition of recursion on several variables", as 4.1 asserts, is contained in the procedure since the schema allows nested recursion. The claim in 4.2, "addition of the statement *Wid*", seems to be clear, as S_{i+1} proves the consistency of S_i; that also fits well with the discussion of the provability predicates B_i on page 10. Finally, in 4.3 the "addition of Hilbert's rule of inference" is claimed to be contained in the procedure. In *1931c* Gödel reviews Hilbert's paper in which the infinitary rule was introduced, and in *1933e* he considers Herbrand's formulation of classical arithmetic, which includes that rule in the following form (*Herbrand 1931*, page 5; *Herbrand 1971*, page 291):

> Let $A(x)$ be a proposition without apparent variables [i.e., a quantifier-free formula]; if it can be proved by intuitionistic procedures that this proposition, intuitionistically considered, is true for every x, then we add $(x)A(x)$ to the hypotheses.

As, for Herbrand, intuitionistic and finitistic considerations are coextensive, that is exactly Hilbert's formulation. A mathematically precise version of Gödel's claim follows from *Rosser 1937*; cf. also Feferman's introductory note to *Gödel 1931c* (these *Works*, Volume I, page 208).

[P]In his introductory note to *1958* and *1972*, Troelstra points out that a subsystem $\mathbf{T}_0$ of **T** is sufficient for Gödel's interpretation of first-order arithmetic; this subsystem is common to **T** and the extensional variant. Thus, to obtain the Gödel interpretation in the latter, extensionality is needed only to derive Troelstra's "replacement schemas" (3) (these *Works*, Vol. II, p. 224).

On the problems of higher-type equality, see further the informative remarks in the same note, pp. 227–229.

[q]See the introductory note to *1941* in this volume, p. 186.

There is no indication in this section as to how the system for functionals (of finite type) is to be used in consistency proofs of classical or, respectively, intuitionistic number theory; there is not even an explicit claim that the consistency of number theory can be established relative to $\cup S_n$ (or in S_ω; see above). In subsection 5 of the present section, Gödel formulates only negative claims and conjectures. In 5.1, he says that "with finite types one cannot prove the consistency of number theory". We assume that he intended to say what he formulated at the end of the Yale lecture, namely, the system that goes up only "to a given finite type" is not sufficient, instead the system for all finite types is needed. This can be rephrased in our terminology: No system S_i, $i < \omega$, will suffice; their union is needed for relative consistency. The negative conjecture in 5.2 parallels that for transfinite induction in VI, 13: Even when the extension procedure is iterated transfinitely, along ordinals satisfying the restrictive condition mentioned above, one will not be able to prove the consistency of analysis. This conjecture is in stark contrast to that for the "modal-logical" route, which, according to Gödel in section V, 11, "leads furthest" and by means of which the consistency of analysis is "probably obtainable".

5. The modal-logical route.

In section V, Gödel turns to the detailed examination of the "modal-logical" route, that is, giving consistency proofs relative to intuitionistic systems of the sort that had been first formalized by Heyting. Intuitionistic mathematics, as a framework for such proofs, does not satisfy the conditions Gödel laid down at the outset, because of the free application of negation and the conditional. Thus the now well-known translation of classical into intuitionistic arithmetic (from *Gödel 1933e*) gives a simple relative consistency proof. And Gödel was quite correct in conjecturing the possibility of extending this route to stronger classical theories: Both analysis and a (carefully formulated) set theory can be shown to be consistent relative to their versions with intuitionistic logic.[r]

Gödel reformulates his proof from *1933e*, to show that full intuitionistic logic is not used. He proves (page 7) that for every formula of number theory containing only the conditional and universal quantifica-

---

[r]For references and brief discussion, see A. S. Troelstra's introductory note to *1933e*, these *Works*, Vol. I, pp. 284–285. To the work relevant to his introductory note to *Gödel 1933f* (these *Works*, Vol. I, pp. 296–298), Troelstra has suggested adding *Flagg 1986*, *Flagg and Friedman 1986* and *Shapiro 1985*.

tion, $\neg\neg A \supset A$ is provable in an intuitionistic system. The details are not entirely clear, but Gödel wants to emphasize that one does not use the principle $\neg A \supset (A \supset B)$. The logical axioms he states are all principles of positive implicational logic, except for the rule of generalization (C, page 7) and axiom B7, which states for elementary formulae, also with free variables, $p \supset q . \supset C . \neg p \vee q$.[s] The role of this axiom in Gödel's intended argument is not clear.

The proof of $\neg\neg A \supset A$ proceeds by induction on the construction of A, and the atomic case is simply said to be "clear". Nothing beyond minimal logic is used in the induction step. Thus it would be for the atomic case that B7 is used.[t] The likely interpretation of Gödel's intention is that he assumes (as part of finitist arithmetic, in line with (**R3**)) some of classical truth-functional logic applied to quantifier-free formulae. One might reason as follows for atomic A: $\neg A \vee A$ follows from B5 and B7. Then we can reason by dilemma: Assuming $\neg A$, we infer $\neg\neg\neg A$ and therefore $\neg\neg\neg A \vee A$; assuming A, we infer $\neg\neg\neg A \vee A$; by disjunction-elimination we have $\neg\neg\neg A \vee A$ and by B7 $\neg\neg A \supset A$. Thus B7 is the only assumption used beyond minimal logic, but this argument does use the introduction and elimination rules for disjunction not mentioned by Gödel.[u]

In fact minimal logic is sufficient, and this use of B7 redundant, given that $\neg A$ has been defined as $A \supset 0 = 1$. For $0 = 1 \supset A$ is derivable (in fact for all A) by induction on the construction of A. In the atomic case, where A is an equation $s = t$, we use primitive recursion to define a function ϕ such that $\phi(0) = s$ and $\phi(Sx) = t$, so that $\phi(1) = t$, and then $0 = 1 \supset s = t$ follows.[v] It is doubtful that Gödel had this argument in mind, since if so there would be no reason for the presence of B7.

[s] We use Gödel's symbols for connectives, including the unusual '$\supset C$' for the biconditional.

[t] If one sets out to prove $\neg\neg(x = y) \supset x = y$ in intuitionistic arithmetic, one will normally use $\neg A \supset (A \supset B)$ or some equivalent logical principle, and this seems to be essential. Consider $\neg\neg(Sx = 0) \supset Sx = 0$. This could not be proved by minimal logic from the axioms of intuitionistic arithmetic, because minimal logic is sound if one interprets the connectives other than negation classically, and $\neg A$ as true for any truth-value of A. But on that interpretation $\neg\neg(Sx = 0) \supset Sx = 0$ is false. But note that the equivalence of $\neg A$ and $A \supset 0 = 1$ fails on this interpretation.

[u] Why does Gödel assume B7 instead of directly assuming $\neg\neg A \supset A$ for atomic A? Possibly he thought it more evident when $\neg A$ is defined as $A \supset 0 = 1$.

Gödel does remark that conjunction and disjunction are definable from negation and the conditional (p. 6), but the context indicates that this would be after classical logic has been derived for the restricted language. However, if in the atomic case we define $p \vee q$ as $\neg p \supset q$, then B7 reduces to $p \supset q . \supset C . \neg\neg p \supset q$, which yields $\neg\neg p \supset p$ by putting p for q and applying axiom 5. Possibly that is what Gödel had in mind by calling the atomic case "clear".

[v] We owe this observation to A. S. Troelstra.

If A contains only negation, the conditional and universal quantification, and if every atomic formula is negated, then the provability of $\neg\neg A \supset A$ in minimal logic follows from theorem 3.5 of *Troelstra and van Dalen 1988*.[w] Thus for a formula satisfying Gödel's conditions, this will be true for the formula A^g obtained by replacing each atomic subformula P by $\neg\neg P$. Since $P \supset \neg\neg P$ is provable in minimal logic, the above reasoning using axiom B7 enables us to prove $P \leftrightarrow \neg\neg P$ for atomic P, whence minimal logic suffices to prove $A^g \leftrightarrow A$. Gödel remarks further that this proof does not require any nesting of applications of the conditional to universally quantified statements. What seems to be the case is that it is not needed essentially more than is directly involved in the construction of the formula A itself.

Gödel comments that the proof goes so easily because "*Heyting's system violates all essential requirements on* constructivity" (page 8). This is an example of a somewhat disparaging attitude toward intuitionistic methods, at least as explained by Heyting and presumably Brouwer, when applied to the task at hand. But it is clear that Gödel's requirement (**R2**) is violated, as he asserts at the beginning of the section. He claims that requirement (**R3**) is violated because "certain propositions are introduced as evident" (page 4). Probably he has in mind the logical axioms applied to formulae with quantifiers, and possibly also induction applied to such formulae, since these are the non-finitary axioms of intuitionistic arithmetic. But he does not elaborate. We can understand why he thought requirement (**R4**) violated by turning to *1933o*. There he argues that intuitionistic mathematics uses methods going beyond the system A, and says (page 30), when commenting on the logical principle "$p \supset \neg\neg p$":

> So Heyting's axioms concerning absurdity and similar notions differ from the system A only by the fact that the substrate on which the constructions are carried out are proofs instead of numbers or other enumerable sets of mathematical objects. But by this very fact they do violate the principle, which I stated before, that the word "any" can be applied only to those totalities for which we have a finite procedure for generating all their elements.

The idea that intuitionistic mathematics has proofs as basic objects is central to his later analysis in *1958* of the distinction between finitist and intuitionistic mathematics.

---

[w] *Troelstra and van Dalen 1988*, pp. 62–68, gives a general treatment of provability by minimal logic of "negative translations", based on *Leivant 1985*.

In the present text Gödel considers an interpretation of intuitionistic logic starting from the idea that the conditional is to be understood in terms of absolute derivability, i.e., provability by arbitrary correct means, not limited to the resources of a single formal system. It is not clear how much Gödel was influenced by Heyting's early formulations of the "BHK-interpretation" of logical constants, which in *1930b* and *1931* are very sketchy and incomplete.[x] In his writings of the 1930's, Gödel does not comment on Heyting's conception of a mathematical proposition as expressing an "expectation" or "intention" whose fulfillment is given by the proof of the proposition.[y] Gödel remarks only that he used the notion of derivability to interpret intuitionistic logic already in *1933f*; but there he does not analyze it further, and here he states that the earlier work did not put any weight on constructivity. Intuitionistic propositional logic was interpreted in the result of adding the operator **B** to classical propositional logic (in effect, in a version of the modal logic S4).

In order to obtain a constructive system, Gödel proposes replacing the provability predicate **B** with a three-place relation $z\mathbf{B}p, q$, meaning "z is a derivation of q from p", which he says can "with enough good will" be regarded as decidable. He then formulates some axioms, but in a confusing way, since he sometimes uses **B** as two-place. The axioms as Gödel writes them are as follows:

(1) $z\mathbf{B}p, q \ \& \ u\mathbf{B}q, r \ \rightarrow \ f(z, u)\mathbf{B}p, r$

(2) $z\mathbf{B}\varphi(x, y) \ \rightarrow \ \varphi(x, y)$

(3) $u\mathbf{B}v \ \rightarrow \ u'\mathbf{B}(u\mathbf{B}v)$

He suggests further a rule of inference:

(4) If q has been derived with proof a, infer $a\mathbf{B}q$.

---

[x] *Heyting 1930* and *1930a* present his intuitionistic formal systems without discussing questions of interpretation at all. But *Heyting 1931* at least was surely known to Gödel before his work on intuitionistic logic and arithmetic. In *1933f*, fn. 1, Gödel does refer to *Kolmogorov 1932*, which presents the interpretation of intuitionistic logic as a calculus of "problems"; cf. Troelstra's comment, these *Works*, Vol. I, p. 299.

[y] *Heyting 1931*, p. 113; cf. *1930b*, pp. 958–959. In neither of these texts does Heyting directly give an explanation of the conditional, but see *1934*, p. 14, which appeared after the remarks in **1933o* but before the present text. In fact Gödel saw earlier drafts of much of *Heyting 1934*, which was the result of Heyting's work on a survey of the foundations of mathematics that was to be written jointly with Gödel. Gödel, however, never finished his part, which was to include a discussion of logicism. Heyting sent him a version of his section on intuitionism with a letter of 27 August 1932.

It seems reasonable to conjecture that this is to be a system based on classical propositional logic, with the presupposition that formulae are decidable. In particular, the $\varphi(x, y)$ of (2) is evidently not an arbitrary formula of, say, intuitionistic arithmetic, but presumably one constructed by finitistically admissible means plus **B**. Gödel does not make clear how the possible second (or second and third) arguments of **B** are to be constructed, although it is clear that they can contain logical operators not admissible elsewhere, such as universal quantifiers. Let us suppose that the language contains the symbol $\top$ for an unanalyzed tautology or other trivial truth. Then we read the two-place "$z\mathbf{B}q$" as an abbreviation for "$z\mathbf{B}\top, q$".[z]

Gödel asserts that these axioms are sufficient to prove, for some a,

(5) $a\mathbf{B}[(u){\sim}u\mathbf{B}(0 = 1)]$.

Clearly from (2) we have

(6) $u\mathbf{B}(0 = 1) \to 0 = 1$,

and, assuming enough arithmetic to prove ${\sim}(0 = 1)$,

(7) ${\sim}u\mathbf{B}(0 = 1)$.

Evidently the step to (5) is to use rule (4), but Gödel gives no indication how the universal quantifier is to be introduced. Gödel may well have understood the rule, in application to a case like this where a formula with free variable has been derived, as introducing a symbol a for the general proof; in that case there would only be a notational difference between "$a\mathbf{B}[{\sim}u\mathbf{B}(0 = 1)]$" and "$a\mathbf{B}[(u){\sim}u\mathbf{B}(0 = 1)]$" as conclusion.[aa]

Gödel inquires whether the system he has sketched is constructive in the sense he has explained. His answer (page 9) is that the violation of requirement (**R2**) is avoided, since to the right of **B**, where "forbidden" logical operations occur, the formula occurs "in quotes". But requirements (**R3**) and (**R4**) are still not satisfied. This defect might be removed, if one interpreted **B** as referring to proofs of the system itself. In the usual proof of Gödel's second incompleteness theorem, it is shown that formulae corresponding to axioms (1) and (3), and a version of rule (4), are derivable in the system. But then of course the conclusion has to be drawn that (2) is *not* derivable, and just for the case $0 = 1$ that occurs in the present argument. In view of the remark, "Essentially not

---

[z]A notation in *Konzept*, p. 7, indicates that Gödel thought of the two-place "$z\mathbf{B}p$" as an abbreviation for "$z\mathbf{B}Ax, p$", but he gives no explanation of what axioms the expression "Ax" refers to. In that same place (3) is stated as "$u\mathbf{B}v \to g(u)\mathbf{B}(u\mathbf{B}v)$", which makes clear that a primitive function giving the proof of $u\mathbf{B}v$ in terms of the given one of v is being assumed.

[aa]P. 7 of *Konzept* contains other formulae in which the universal quantifier occurs, some of them crossed out, but no suggestion as to how an introduction such as that we have discussed is to go.

the underlined—that is essentially the consistency of the system",[bb] it seems that Gödel saw this; so it is not clear what he thought was the value of his suggestion. By the "introduction of types" he achieves (now using $\mathbf{B}$ for "provable") that only $\mathbf{B}_{n+1} \sim \mathbf{B}_n (0 = 1)$ holds; this seems to be the natural direction in which the interpretation of $\mathbf{B}$ as referring to formal proofs would go.[cc]

Gödel's remarks present in a very sketchy way an idea that was pursued in subsequent work by G. Kreisel and others (apparently without knowledge of Gödel's earlier discussion), of developing formal theories of constructions and proofs, with a basic predicate like Gödel's $z\mathbf{B}p$. Gödel seems not to attack systematically the question that arises already at the beginning of this work in *Kreisel 1962a* of interpreting intuitionistic logic by giving a definition of "$z\mathbf{B}A$", where A is an arbitrary formula of first-order logic or of some intuitionistic theory. The idea that one should use the clauses of the BHK-interpretation of the intuitionistic connectives to give an inductive definition of $z\mathbf{B}A$ is a very natural one and may well have occurred to Gödel at this time. Without it, it is hard to see how a theory on the lines Gödel sketches could serve the purpose that seems to be intended for it of being a vehicle for consistency proofs. Gödel, however, seems not concerned to develop the idea very far for this purpose, but rather to exhibit where it falls short of meeting his constraints.

Kreisel in *1962a* and *1965* treated the proof relation as decidable, in agreement with Gödel's suggestion. This led to complications in the inductive definition of $z\mathbf{B}A$ for formulae of intuitionistic logic. The obvious clause for $z\mathbf{B}(A \to C)$ would be $\forall u[u\mathbf{B}A \to z(u)\mathbf{B}C]$ (thinking of z as a function), and this appears not to be decidable. Kreisel thus altered the definition to:

(∗) $z\mathbf{B}(A \to C)$ iff z is a pair $\langle z_1, z_2 \rangle$ and $z_2\mathbf{B}\forall u[u\mathbf{B}A \to z_1(u)\mathbf{B}C]$.[dd]

The same needed to be done for universal quantification. With an axiom like Gödel's (2), (∗) and $z\mathbf{B}(A \to C)$ imply $u\mathbf{B}A \to z_1(u)\mathbf{B}C$. Thus Kreisel's definiens implies the above-mentioned obvious one.

There were considerable difficulties in developing a theory along these lines. Kreisel's ideas were naturally developed in the framework of the type-free lambda-calculus, but then the resulting theory is inconsistent;

[bb] "The underlined" seems to be the formula expressing consistency; see editorial note y to the text below.

[cc] Possibly $\mathbf{B}_n$ is intended to mean provability in S_n; see §4 above.

[dd] *Kreisel 1962a*, p. 205; *1965*, p. 128.

see *Goodman 1970*, §9.[ee] Development of a theory along these lines modified so as to avoid paradox was never carried much beyond the interpretation of arithmetic; see *Goodman 1970* and *1973*. A lucid treatment of the basic issues is *Weinstein 1983*. An alternative approach to a theory of constructions involved abandoning the requirement of the decidability of **B**. This led to various typed theories of which that of Per Martin-Löf is best known; see for example *Martin-Löf 1975, 1984* and the accounts in *Beeson 1985* and *Troelstra and van Dalen 1988*. *Sundholm 1983* criticizes from this point of view the motivation of Kreisel's approach, and thus indirectly Gödel's suggestion of decidability.

6. Transfinite induction and recursion.

Gödel mentions repeatedly that the modal-logical route we just discussed was the heuristic viewpoint guiding Gentzen's 1936 consistency proof for classical arithmetic. However, the characteristic principle by means of which Gentzen went beyond finitist mathematics (if that is thought of as being formalized in the system S_1) is the principle of transfinite induction for both proofs and definitions of functions. And it is to the number-theoretic formulation of these principles that Gödel turns immediately—a task, incidentally, that was not taken up explicitly by Gentzen. First of all it has to be clarified how to grasp (in a mathematically expressible way) specific countable ordinals α from a finitist standpoint. That can be achieved in the system S_1 with function parameters, by considering definable linear orderings $\prec$ of the natural numbers such that, for a definable functional Φ and for all functions f, S_1 proves:

$$\sim \{ f(\Phi(f) + 1) \prec f(\Phi(f)) \};$$

i.e., $\prec$ is provably well-founded. Such an ordering is said to represent α if it is order-isomorphic to α. Two points should be noted. First, there is obviously no function quantification in S_1; universal statements concerning functions (as well as numbers) are expressed just using free-variable statements. Second, the connection between α and $\prec$ is not formulated within finitist mathematics (and is also not needed for the further systematic considerations).

[ee] Although Goodman's statement (p. 109) is more cautious, his argument seems to prove the inconsistency of the "starred theory" of *Kreisel 1962a* as it stands. *Kreisel 1965* envisages the typed λ-calculus, but then it is not clear what type can be assigned to z_2 in (*). Goodman's own solution involves a stratification of constructions into levels; see *1970*, §§10, 13, and the criticism in *Weinstein 1983*, pp. 265–266.

Gödel observes[ff] that S_1 proves the transfinite induction principles for such $\prec$. The *principle of proof* by transfinite induction is formulated as follows: If one can prove $E(a)$ from the assumption $(x)(x \prec a \rightarrow E(x))$, then one is allowed to infer $(x)E(x)$. This inference is represented by the rule

$$\frac{(x)(x \prec a \rightarrow E(x)) \rightarrow E(a)}{(x)E(x).}$$

The corresponding *principle of definition* by transfinite induction (what would nowadays be called a schema of transfinite *recursion*) for $\prec$ is formulated in this way:[gg] if g_i, $1 \leq i \leq n$, are functions with $g_i(x) \prec x$ when x is not minimal, and A is any term (in the language of a definitional extension of S_1), then there is a unique solution to the functional equation

$$\varphi(x) = A(\varphi(g_1(x)), \dots , \varphi(g_n(x))).$$

That can be done for (orderings representing) ordinals like $\omega + \omega$, ω^2, ω^3, etc. Gödel points out that ω^ω is a precise limit for S_1, since S_1 proves the proof principle of transfinite induction for quantifier-free formulae up to any ordinal $\alpha < \omega^\omega$, but not up to ω^ω. It seems, amazingly, that this result was rediscovered and established in detail only more than twenty years later in *Church 1960* and *Guard 1961*.

In current terminology ω^ω is the *proof-theoretic ordinal* of S_1. The analysis of formal theories in terms of their proof-theoretic ordinals has been a major topic in proof theory ever since Gentzen's consistency proof and his subsequent analysis of the (un-)provability of transfinite induction in number theory. Returning to the present text, Gödel asserts that what can be done in the system S_1, namely, prove induction inferences from other axioms, can also be done in number theory for even larger ordinals. And, as in the case of S_1, there are ordinals for which this cannot be done in number theory. One such ordinal is the first epsilon-number ϵ_0. That ordinal is definable as the limit of the sequence

$$\alpha_1 = 2^{\omega+1} \text{ and } \alpha_{n+1} = 2^{\alpha_n},$$

where exponentiation[hh] is given by

[ff]Treatments of transfinite induction in PRA (i.e., S_1 without function parameters) are given in *Kreisel 1959c* and *Rose 1984*, pp. 165 ff.

[gg]A systematic treatment of so-called ordinal recursive functions, with references to the literature, in particular to the work of Péter, Kreisel and Tait, is found in *Rose 1984*.

[hh]Cf. editorial note bb to the text.

$$2^1 = 2, 2^\beta = \sum_{x<\beta} 2^x.$$

I.e., we associate with each element the sum of the preceding ones; Gödel views this as a "very intuitive construction procedure".[ii] ϵ_0 is obtained by countably iterating the transition from α to 2^α, and if this transition were given, Gödel states in section VI, 9, ϵ_0 would be given. How could one obtain this transition formally in arithmetic? One would assume that α is already represented in the sense above; then one would have to *define* an ordering that represents 2^α and *prove* that that ordering is a well-founded relation (using a suitable functional). Clearly, one may use both the proof and definition principle of transfinite induction for α.

In section VI, 6, Gödel gives a straightforward constructive *definition* of an ordering $\prec$ representing ϵ_0 and explains in the next subsection the obstacle against carrying out the *usual proof* of its well-foundedness.[jj] The proof proceeds by transfinite induction and uses the impredicative induction property "being an ordinal". This property can't be formulated in the language of arithmetic since it requires genuine universal quantification over functions (to be used in the induction schema). Gentzen's consistency proof together with Gödel's second incompleteness theorem is needed to show more than the failure of the usual argument, namely, that there is *no* proof in number theory at all. Gödel does not remark, as he did for ω^ω and S_1, that ϵ_0 is the proof-theoretic ordinal of arithmetic. That for any arithmetic statement transfinite induction up to any ordinal less than ϵ_0 is indeed provable in arithmetic is shown in *Hilbert and Bernays 1939* (page 366); it is also shown in *Gentzen 1943*, where the unprovability of transfinite induction up to ϵ_0 is established without appeal to the second incompleteness theorem. For a modern and very beautiful presentation of these mathematical considerations (incorporating advances due mostly to Schütte and Tait) see *Schwichtenberg 1977*.

In spite of the fact that the transition from α to 2^α (and thus ϵ_0) is not given in arithmetic, Gödel emphasizes (page 12) with respect to the epistemological side:

> ... one will not deny a high degree of intuitiveness to the inference by induction on ϵ_0 thus defined, as in general to the procedure of *defining an ordinal by induction on ordinals* (even though this is an impredicative procedure).

[ii] He emphasized that also in *Konzept*, p. 10, where he added "... und eine Reduktion darauf erscheint mir als wertvoll" ("... and a reduction [to that construction] seems to me to be valuable").

[jj] This (usual) argument is carefully presented in Supplement V of *Hilbert and Bernays 1970*, pp. 534–5.

Consequently, it is natural for Gödel to consider the system obtained from S_1 by adding the principle of transfinite induction up to ϵ_0 (for quantifier-free E). He asks as the first important, philosophical question, whether this theory is still constructive.[kk] Indeed, all requirements are satisfied except for (**R3**); that condition demands the *exclusive* use of ordinary induction. But in a sense transfinite induction is just a generalization of ordinary induction, and Gödel thinks that therefore "... the deviation from the requirement 3 [our (**R3**)] is perhaps not such a drastic one" (page 13). Gödel's reason for accepting induction up to ϵ_0 is not the special combinatorial character of ϵ_0, but rather the fact that ϵ_0 falls into a broader class of ordinals *definable by recursion on already defined ordinals*. And to this procedure, though impredicative, Gödel "will not deny a high degree of intuitiveness". Such broadened views of "constructivity" underlie developments in proof theory described, e.g., in *Feferman 1981* and *1988a*, but also in *Feferman and Sieg 1981*, where the use of generalized inductive definitions is emphasized. As to the use of large constructive ordinals in current proof theory, we refer to *Buchholz 1986*, *Pohlers 1989* and *Rathjen 1991*.

The second important, mathematical question is: How far does one get with extensions of S_1 by adding the inference rule for induction on ordinals that are obtained by ordinal recursive procedures along already obtained ordinals? As to ϵ_0, Gentzen did prove the consistency of number theory and, Gödel adds, "probably also of Weyl's *Kontinuum*".[ll] As the precise theory underlying the development of analysis in *Weyl 1918* is (in one interpretation) a conservative extension of number theory, Gödel is indeed right; cf. *Feferman 1988*. He even thinks that with sufficiently large ordinals one can establish the consistency of analysis and of parts of set theory, as Gentzen had hoped. But he doubts that ordinals satisfying his principle of definability will be sufficiently large. Gödel conjectures at the end of subsection 13 and in 14 that the addition of transfinite induction for such ordinals may not lead to stronger theories than the S_i, and that transfinite induction up to ϵ_0 is already provable in S_ω (see §4 above).[mm]

[kk] A detailed analysis of such quantifier-free theories is given in *Rose 1984*, chapters 6 and 7.

[ll] Gödel presumably meant that the method of Gentzen's proof would yield a proof of the consistency of Weyl's *Kontinuum*, not that Gentzen had literally proved this.

[mm] In *Tait 1968* it is shown that the **T** of *Gödel 1958* is closed under recursion on standard orderings of type less than ϵ_0; from *Kreisel 1959c* it then follows that induction on these orderings is also derivable. This would confirm Gödel's conjecture for a suitable formulation of S_ω (as interpreted in §4 above).

7. Interpreting Gentzen's consistency proof.

Subsections 16 through 19 are a remarkable *tour de force*: on less than two pages Gödel analyzes, with a surprising twist, the essence of Gentzen's consistency proof for classical arithmetic and indicates precisely where in the proof the ordinal exponentiation step occurs that forces the use of all ordinals below ϵ_0. As we mentioned above, Gödel repeatedly points out that the modal-logical route, as a way of assigning a finitist meaning to transfinite statements, was the heuristic viewpoint guiding Gentzen's proof. The latter point was made already in *Konzept*[nn] and is in complete accord with Gentzen's intentions; in section 13 (page 536) of his *1936* (see also *Gentzen 1969*, page 173) Gentzen writes:

> The concept of the *"stability of a reduction rule (Reduziervorschrift)"* for a sequent, to be defined below, serves as a formal substitute for the contentual *concept of correctness*; it gives a special *finitist interpretation* of statements, which replaces the *in-itself conception* of them.[oo]

At the very end of his paper (page 564; *1969*, page 201) Gentzen points back to the definition of *Reduziervorschrift* in §13 and claims that the *most crucial part* of his consistency proof consists in providing a finitist meaning to the theorems of classical arithmetic:

> For every arbitrary statement, so long as it has been proved, a *reduction rule* according to 13.6 *can be stated*, and this fact represents the finitist sense of the statement in question, which is gained precisely through the consistency proof.

Gödel claims that the finished proof is only remotely connected to the modal-logical one and maintains that Gentzen proves of each theorem a double-negation translation "... in a different sense from the modal-logical". Gödel formulates the different sense in a mathematically and conceptually perspicuous way: it turns out to be the sense provided by the "no-counterexample interpretation" introduced by Kreisel in *1951*!

[nn]In *Konzept*, page iii, we read: "Die transfiniten Aussagen erhalten einen finiten Sinn." (I.e., the transfinite statements obtain a finitist meaning.)

[oo]In this and in the succeeding quotation, the translation in *Gentzen 1969* is revised. The "in-itself conception" (*an-sich Auffassung*, §9.2) is what we would call realist or Platonist.

These matters are formulated paradigmatically in subsection 16 by considering a formula

(1) $(x)(Ey)(z)(Eu)A(x,y,z,u)$,

in prenex normal form with a decidable matrix A. Proving the negation of this formula constructively means presenting a number c, a unary function f, and a proof of

(2) $\sim A(c,y,f(y),u)$.

A proof of the double negation of (1) consists then in a proof that such a c and f cannot exist: for each f and c one can find functionals $y_{f,c}$ and $u_{f,c}$ such that

(3) $A(c,y_{f,c},f(y_{f,c}),u_{f,c})$;

thus, there cannot be counterexamples c and f.[PP] The functionals y and u are called by Gödel a *reduction*. In subsections 17 through 19 Gödel sketches how to find reductions for theorems in number theory from their formal proofs. (And it is for the treatment of modus ponens that ordinal exponentiation comes in.) Here is not the place to show that Gödel captures the mathematical essence of Gentzen's proof, as that would require a somewhat detailed description of that proof.

Gödel's analysis and presentation are surprising indeed. What accounts partially for the dramatic difference between Gentzen's and Gödel's presentations is the latter's free use of functionals and, to be sure, neglect of all formal details. Functionals do occur in Gentzen's presentation and also in Bernays' description of Gentzen's unpublished consistency proof in *Hilbert and Bernays 1970*, but only in a cautious way to express that the *Reduziervorschriften* are independent of arbitrary choices (see, e.g., pages 536 and 537 in *Gentzen 1936*). The difficult and cumbersome presentation (and what is perceived as an unmotivated manner of associating ordinals to derivations) resulted in a quite general dismissal of Gentzen's first consistency proof; it is the second consistency proof in *Gentzen 1938a* that has been at the center of proof-theoretic research. A widely shared attitude of logicians towards (the "most crucial part" of) Gentzen's first proof can be gleaned from a remark in *Kreisel 1971*. With respect to that most crucial part, i.e., the finitist sense given to logically complex theorems by the *Reduziervorschriften*, Kreisel writes (page 252):

> He [Gentzen] has reservations about his own proposal of expressing this [finitist] sense in terms of the reductions used in

---

[PP]The reasoning behind the necessary shift of quantifiers is made explicit in *1941*, p. 9.

his proof because the proposed sense is only "loosely connected"
with the form of the theorem considered (and, it might be added,
the connection is so tortuous that one couldn't possibly remember it).[qq]

That is particularly striking when contrasted with Gödel's uncovering of
the no-counterexample interpretation in his (clearly more sympathetic)
reading in late 1937 and early 1938.

8. Concluding remarks.

Gödel commented to Zilsel, as reported in his notes on the organizational meeting, that Gentzen's result is of only mathematical interest,
"... ist nur mathematisch interessant"; and this judgment was not un-
informed: Gödel had read (at least the unpublished version of) the consistency proof carefully and had discussed it extensively with Bernays.[rr]
On this point he obviously modified his views when preparing the lecture for the Zilsel seminar. When discussing the finitist character of
the system obtained from PRA by the principle of transfinite induction,
Gödel points out, as the reader may recall, that it violates only (**R3**).
He continues: "This [new] inference can be considered as a generalization of ordinary induction, and in this respect the deviation from the
requirement 3 is perhaps not such a drastic one." In his concluding section VII Gödel evaluates the epistemological significance of consistency
proofs relative to the systems he considered; with respect to Gentzen's
proof he states, "one will not be able to deny of Gentzen's proof that it
reduces operating with the transfinite E to something more evident (the
first ϵ-number)".

Gentzen's consistency proof meets, consequently, the general condition Gödel formulated for a "satisfying" relative consistency proof,
namely, that such a proof should reduce to something that is more evident. In comparison with a reduction to the basic finitist system Gödel
considers the epistemological significance to be "very much diminished".

[qq] It should be mentioned briefly that Kreisel misrepresents Gentzen's remark
on p. 564 (to which he alludes in this quotation): according to Gentzen, the loose
connection to the form of the theorem is *not* due to the reductions, but rather to the
initial (standard) double negation translation, so that, for example, an existential
statement does not have its strong finitist meaning, but only the weaker one of its
translation.

[rr] As to this episode, see *Kreisel 1987*, pp. 173–174.

But then we have to realize that Gödel expressed in this lecture a rather high regard for Hilbert's original program; if that could have been carried out, "that would have been without any doubt of enormous epistemological value". When comparing Gödel's philosophical remarks with those of Bernays (e.g., in *1935*) or with the reflective considerations of Gentzen (e.g., in *1936* and *1938*) one still finds a marked affinity of their general views.[ss] It is the absolutely unencumbered mathematical analysis that most distinguishes Gödel's presentation from theirs.

Wilfried Sieg and Charles Parsons[tt]

The text of *1938a* was transcribed from Gabelsberger shorthand by Cheryl Dawson and edited by Cheryl Dawson, Charles Parsons and Wilfried Sieg. The translation is by Charles Parsons, revised using suggestions of John and Cheryl Dawson and Wilfried Sieg.

[ss]Despite Gödel's rather critical remarks concerning Gentzen, e.g., on p. 13, "But here again the drive of Hilbert's pupils to derive something from nothing stands out." It is difficult to justify this remark either narrowly, as applying to Gentzen's paper, or more broadly, as applying, e.g., also to Bernays.

[tt]We are grateful to Robert S. Cohen, Cheryl Dawson, John Dawson, Warren Goldfarb, Nicolas Goodman and especially Solomon Feferman and A. S. Troelstra for information, suggestions, and/or comments.

Vortrag bei Zilsel
(*1938a)

I

[Von Widerspruchsfreiheit kann man nur reden mit bezug auf Teilsysteme der Mathematik. Deswegen ist die Frage aber nicht weniger wichtig, weil es Systeme gibt, die so umfassend sind.—Es handelt sich um eine mathematische Frage: jeder Widerspruchsfreiheitsbeweis ist selbst mathematisch und ist daher in bestimmten mathematischen Systemen mit bestimmten Schlußweisen durchführbar.][a] Ich glaube, man muß bei dieser Frage zunächst feststellen:

1. Der Widerspruchsfreiheitsbeweis hat Sinn nur im Sinne einer Reduktion T [ist] reduzierbar [auf] S:[b]

       1. *Wid S* → *Wid T* beweisbar [in] S *Teilsystem* [von T]

       2.   *Wid T*   beweisbar [in] S

2 ist vernünftig. Beispiele: Geometrie, Analysis—Logik, Auswahlaxiom *Teilsystem*!

2. Triviale Bemerkung, aber nicht überflüssig. Hilbert [*1928*, Seite 85]:

---

[a]This passage is taken—almost literally—from *Konzept*, p. 1. See the textual notes.

[b] "Reduzierbar" is used in the text in both senses indicated here; cf. I 4 A and B below, but also the argument in III 3. As to the examples, one finds a more detailed statement in *Konzept*, p. 1: "In diesem Sinne kann man sagen, die Widerspruchsfreiheit

Lecture at Zilsel's
(*1938a)

I

[One can only talk of consistency in relation to partial systems of mathematics. The question is no less important for that reason, since there are systems that are so encompassing. We have to do with a mathematical question: every consistency proof is itself mathematical and can therefore be carried out in definite mathematical systems with definite modes of inference.][a] I believe that in treating this question one must first note:

1. The consistency proof is meaningful only in the sense of a reduction:
   T is reducible to S:[b]

    1. *Wid S* → *Wid T* provable [in] S (a *subsystem* of T)
    2.   *Wid T*   provable [in] S

2 is reasonable. Examples: geometry, analysis—logic, axiom of choice *subsystem*!

2. A trivial observation [that is] not superfluous. Hilbert says [*1928*, page 85; translation from *van Heijenoort 1967*, page 479]:

Widerspruchsfreiheit der Geometrie sei auf die der Analysis, die der Analysis auf die der Logik reduziert (nicht im anderen Sinne)." ["In this sense one can say that the consistency of geometry is reduced to that of analysis, that of analysis to that of logic (not in the other sense).") We assume that Gödel means by "Logik" here the simple theory of types, so that the sense he has in mind is sense 2; clearly, "Auswahlaxiom, *Teilsystem*" is meant to give an example of reducibility in sense 1. On these matters see also section 3 of the introductory note.

Schon jetzt möchte ich als Schlußergebnis die Behauptung aus-
sprechen: die Mathematik ist eine voraussetzungslose Wissenschaft. Zu ihrer Begründung brauche ich weder den lieben Gott
... noch die Annahme einer besonderen auf das Prinzip der
vollständigen Induktion abgestimmten Fähigkeit unseres Ver-
standes, wie Poincaré, noch die Brouwersche Urintuition und
endlich auch nicht, wie Russell und Whitehead, Axiome der Un-
endlichkeit [und] Reduzierbarkeit ....

Es wird also so getan, als könne man die Widerspruchsfreiheit aus gar
keinen Voraussetzungen ableiten, was nicht einmal dann richtig wäre, wenn
2 | sich das ursprüngliche Hilbertsche Programm durchführen ließe.
3. Die Frage der Widerspruchsfreiheit zerfällt [für die gesamte Mathematik]
also in Teilfragen[: Kann man das System T mittels des Systems S als
widerspruchsfrei erweisen?—Und hier gibt es teils positive, teils negative
Antworten.]^c In der Aufzählung der Teilantworten erschöpft sich die mathematische Seite der Frage.
4. Die Frage hat aber auch eine erkenntnistheoretische Seite. *Man will ja
einen Widerspruchsfreiheitsbeweis zum Zwecke der besseren Fundierung der
Mathematik (Tieferlegung der Fundamente)* [führen], und es kann mathematisch sehr interessante Beweise geben, die das nicht leisten ([wie etwa]
Tarski[s für die] Analysis^d). Befriedigend [ist] ein Beweis [nur], wenn er

> A. *auf einen echten Teil reduziert* [oder]
> B. *auf etwas zurückführt, was zwar nicht Teil, aber was evi-
> denter, zuverlässiger, etc. ist, so daß dadurch die Überzeugung
> gestärkt wird.*

A bedeutet zweifellos einen objektiven Fortschritt (Überflüssigmachung von
Voraussetzungen [ist] fast dasselbe wie [ein] Beweis). B [ist] zunächst
problematisch, weil subjektiv verschieden, aber *de facto* nicht so schlimm;
denn es besteht im allgemeinen Übereinstimmung, daß die konstruktiven
Systeme besser sind als die, welche mit dem Existential "es gibt" arbeiten.
Und auch historisch handelt es sich darum, die nichtkonstruktive Mathematik auf die konstruktive zurückzuführen.

II

Die Schwierigkeit, diese Frage (ob das geht oder nicht) in positivem oder
negativem Sinn zu entscheiden, liegt an der Verschwommenheit des *Begriffs*

---

^{c}These sentences, as well as the addition "für die gesamte Mathematik", are taken
from *Konzept*, p. 1, #3.

Already at this time I should like to assert what the final outcome will be: mathematics is a presuppositionless science. To
found it I do not need God ... or the assumption of a special
faculty of our understanding attuned to the principle of mathematical induction, as does Poincaré, or the primal intuition of
Brouwer, or, finally, as do Russell and Whitehead, axioms of
infinity [or] reducibility ...

Thus one acts as if consistency could be derived from no presuppositions
at all, which would not even be correct if the original Hilbert program could
be carried out.
3. The question of consistency [for the whole of mathematics] thus di-
vides into partial questions. [Can one show the system T to be consistent
by means of the system S?—And here there are partly positive, partly
negative answers.][c] The enumeration of the partial answers exhausts the
mathematical side of the question.
4. The question has, however, also an epistemological side. *After all we
want a consistency proof for the purpose of a better foundation of mathematics (laying the foundations more securely),* and there can be mathematically very interesting proofs that do not accomplish that (as, for example,
Tarski's for analysis[d]). A proof is only satisfying if it either

A. *reduces to a proper part* or
B. *reduces to something which, while not a part, is more evident,
reliable, etc., so that one's conviction is thereby strengthened.*

A signifies without doubt an objective step forward (making assumptions
superfluous is almost the same as a proof). B is at the outset problematic
because subjectively different—but de facto not so bad, since there exists
general agreement that constructive systems are better than those that
work with the existential "there is". And also historically the task has
been to reduce non-constructive to constructive mathematics.

II

The difficulty in deciding this question (whether or not that works) positively or negatively is due to the haziness of the *concept "constructive"*.

---

[d]Gödel refers presumably to truth definitions for languages of finite type, as presented, e.g., in section 4 of *Tarski 1935*, and their use in consistency proofs.

"konstruktiv". Ich beginne mit einer Rahmendefinition,[e] welche wenigstens notwendige, wenn nicht schon hinreichende Bedingungen angibt ([die] auch nicht ganz scharf, aber immerhin recht brauchbar zu sein scheinen).

1. Die Grundoperationen und Relationen müssen berechenbar und entscheidbar sein. (D. h., [man benötigt ein] Verfahren; hier [ist] ein Punkt, [der] wohl nicht scharf [ist].[f]) *Daraus folgt die Anwendbarkeit des Aussagenkalküls* [und] insbesondere *rekursive Definitionen*.

3 |2. Einschränkung in der Verwendung von ( ) und (E). Ich formuliere zunächst möglichst scharf. (Die Lesarten ergeben sich dann von selbst.) E soll überhaupt nicht unter den Grundzeichen vorkommen, und die Operationen des Aussagenkalküls sollen auf ( ) Aussagen nicht angewendet werden. Bei "¬" [ist das] klar, [und auch für die Implikation "⊃",] sonst [hätte man] aus der Definition

$$\neg(x)F(x) \equiv (x)F(x) \supset 0 = 1.$$

Daraus folgt: [man behandelt] *nur Aussagen mit freien Variablen*, also von der Form: es seien irgendwelche Objekte vorgelegt etc. Ein konstruktives E [ist] dann als Abkürzung einführbar, indem man erlaubt zu schließen:

$$F(a) \ldots (Ex)F(x).$$

$(x)(Ey)A(x,y)$ darf aus $A(x,f(x))$ geschlossen werden, und mit solchen Regeln [ist] das nur dann beweisbar. Da [E] nur Abkürzung, [al]so unwesentlicher Bestandteil des Systems [ist], welcher weggelassen werden kann, [wird] im Folgenden davon abgesehen. [Es] *bleibt also dabei, daß alle [Variablen] nur als freie Variablen auftreten.*

3. *Schlußregeln und Axiome*: jedenfalls die [rekursiven] Definitionen, [der] Aussagenkalkül, [die] Einsetzungsregel und vollständige Induktion—aber vielleicht noch weitere.[g] Das lasse ich offen.

4. Objekte sollen überblickbar sein (d. h. abzählbar).[h]

1 [und] 2 unbedingt. 3 [und] 4 problematisch; 4 z. B. wegen [des] Funktionsbegriffs; 3 wegen [der] Rekursion nach transfiniten Ordinalzahlen.

---

[e]The "Rahmendefinition" is provided by 1; that is supported by the text below, e.g., V 1.

[f]The interpretive emendations are based on *Konzept*, section C, p. 3 and section 2, p. 2. It seems that Gödel indicates here (as well as in the corresponding statements of *Konzept*) that he views the notion of computability as "nicht ganz scharf" when considered in the context of constructive mathematics. For a later explicit formulation

I begin with a framework definition,[e] which at least gives necessary if not sufficient conditions (which also seem to be not completely sharp, but nevertheless quite usable).

1. The primitive operations and relations must be computable and decidable (that is, one needs a procedure for computing functions and deciding relations; this is a point that is surely not sharp[f]). *From this follows the applicability of the propositional calculus* and in particular *recursive definitions.*

2. Restriction in the application of ( ) and (E). I formulate as sharply as possible at first. (Variant readings then follow of themselves.) E should not occur at all among the primitive signs, and the operations of the propositional calculus should not be applied to ( ) statements. In the case of "¬" that is clear, [and also for implication "⊃";] otherwise one would have from the definition

$$\neg(x)F(x) \equiv (x)F(x) \supset 0 = 1.$$

From this follows: we allow *only statements with free variables*, thus of the form: Let certain arbitrary objects be given, etc. A constructive E can then be introduced as an abbreviation by allowing the inference:

$$F(a)\ldots(Ex)F(x).$$

$(x)(Ey)A(x,y)$ may be inferred from $A(x, f(x))$, and is only provable with such rules [i.e., only in that way]. Since E is only an abbreviation, hence an inessential part of the system, which can be left out, it will be disregarded in the sequel. *Thus all variables continue to occur only as free variables.*

3. *Rules of inference and axioms*: in any case [recursive] definitions, the propositional calculus, the rule of substitution, and ordinary complete induction—but perhaps still further [axioms and rules].[g] I leave that open.

4. Objects should be surveyable (that is, denumerable).[h]

1 and 2 definitely. 3 and 4 [are] problematic, 4 for example because of the concept of function, 3 because of recursion on transfinite ordinal numbers.

---

of his concerns with the correctness of Church's Thesis for intuitionistic computability, cf. *van Heijenoort 1985*, pp. 114–116, and also *Wang 1974*, pp. 81–99, in particular, pp. 87–89.

[g]The addition "rekursiven", in the sense of "primitive recursive", seems to be implied by the first comment below (section III 1).

[h]In *Konzept*, p. 2, #1, Gödel formulated: "Die Gesamtheit der Objekte (Individuen, Relationen, Funktionen), von denen im System die Rede ist, muß überblickbar sein (abzählbar sein)." ["The totality of objects (individuals, relations, functions) that are spoken of in the system must be surveyable (denumerable)."] There are complementary remarks in *Gödel *1933o*, pp. 8 and 10; but cf. also sections 2 and 3 of the introductory note.

III

1. Die finiten Systeme bilden eine Hierarchie. [[Zu]]unterst [[ist die]] *finite Zahlentheorie*; die Objekte sind Zahlen und durch die gewöhnliche Rekursion definierte Funktionen und solche [[Funktionen]], die daraus durch Einsetzung entstehen. Schon die *Brouwersche Mathematik* geht darüber erheblich hinaus. Aber ich glaube, *Hilbert wollte mit dieser den Beweis führen.*[i]

2. *Vielleicht sogar ein noch* schwächeres, denn [[die]] *vollständige Induktion* [[wurde]] erst [[durch]] von Neumann [[eingeführt]] (Hilbert abgeschwächt), 4 | außerdem *Bernays* elementar kombinatorisch und Hilbert selbst.[j]

> 1. Einer der Gesichtspunkte für den Widerspruchsfreiheitsbeweis: Es ist nötig, durchweg dieselbe Sicherheit des Schließens herzustellen, wie sie in der gewöhnlichen niederen Zahlentheorie vorhanden ist [[*Hilbert 1926*, Seite 170]].
>
> 2. Es ist möglich, auf rein anschauliche und finite Weise—gerade wie die Wahrheiten der Zahlentheorie—auch diejenigen Einsichten zu gewinnen, die die Zuverlässigkeit des mathematischen Apparates gewährleisten [[ib., Seite 171]].

Schließlich *zugegeben*, daß eine Erweiterung des Standpunkts nötig ist.

3. *Wie weit kommt man mit der finiten Zahlentheorie, bzw. wie weit kommt man nicht?*

Negativ: *A* [[ist]] nicht reduzierbar auf ein Teilsystem *T*, dessen Widerspruchsfreiheit in *A* beweisbar ist.[k] [[Aus diesen Voraussetzungen hat man sowohl, daß]]

$Wid(T)$ bew[[eisbar in]] A[[, als auch daß]]
$Wid(T) \to Wid(A)$ beweisbar in *T*, also in *A*. [[Es folgt:]]
$Wid(A)$ [[ist]] beweisbar in *A*.

Daraus folgt schon: die transfinite Arithmetik [[ist]] nicht mehr (Herbrand) Teil davon [[d. h. Teil der finiten Zahlentheorie]]; außerdem [[gilt das für das System von]] Russell–Whitehead ohne Reduzierbarkeitsaxiom.[l]

---

[i]It seems that for Gödel "finite Zahlentheorie" coincides with Primitive Recursive Arithmetic in the sense of Hilbert and Bernays. Cf. the system *A* in *Gödel *1933o*, pp. 23–25. A systematic argument for this identification is given in *Tait 1981*.

[j]The additions to the text are conjectural as to the meaning. They are based on the following historical facts: in the first steps towards his program in the early twenties, Hilbert thought that the consistency problem could be solved using a system without free variables; cf. *Bernays 1967*, p. 500. But it is clearly stated already in *Bernays 1922* that free variables and metamathematical induction are needed.—"von Neumann", we think, refers to *von Neumann 1927*.

III

1. The finitary systems form a hierarchy with *finitary number theory at the lowest level*; the objects are numbers and functions defined by ordinary recursion and functions that arise from them by substitution. *Brouwer's mathematics* already goes considerably beyond it. But I believe that Hilbert *wanted to carry out the proof* [of consistency] *with this.*[i]

2. *Perhaps even a still* weaker [system], for *complete induction* was first [introduced by] von Neumann (Hilbert weakened), moreover *Bernays* elementary combinatorial and Hilbert himself.[j]

> 1. One of the points of view for the consistency proof: It is necessary to make inference everywhere as reliable as it is in ordinary elementary number theory [*Hilbert 1926*, page 170].
>
> 2. It is possible to obtain in a purely intuitive and finitary way, just as with the truths of number theory, those insights that guarantee the reliability of the mathematical apparatus [*ibid.*, page 171, translations from *van Heijenoort 1967*, page 377, slightly revised].

Finally *granted*, that an extension of the standpoint is necessary:

3. *How far do we get, or fail to get, with finitary number theory?*

Negative: A is not reducible to a subsystem T, whose consistency is provable in A.[k] [Supposing that it is, we have both]

> $Wid(T)$ provable in A, and
> $Wid(T) \rightarrow Wid(A)$ provable in T, therefore in A. It follows that $Wid(A)$ is provable in A.

From this already follows: transfinite arithmetic is no longer (Herbrand) a part [of finitary number theory]. Moreover, [that holds for the system of] Russell-Whitehead without the axiom of reducibility.[l]

---

[k] A is taken to be some formal system that contains finitist arithmetic T as a proper part.

[l] Gödel presumably refers to *Herbrand 1931*. There Herbrand entertains the possiblity that all finitist arguments can be carried out already in elementary (Peano) arithmetic and claims: "If this were so, the consistency of ordinary arithmetic would already be unprovable" (p. 8; *Herbrand 1971*, p. 297). Clearly, Herbrand means in this sentence "unprovable by finitist means".—Gödel's second claim seems to be that the system of *Principia mathematica* without the axiom of reducibility goes beyond finitist mathematics. That is supported by *Konzept* (p. 0.1, #2), where that system is considered to be part of transfinite mathematics.

4. *Wie also erweitern?* (Erweiterung nötig.) Drei Wege [sind] bisher bekannt:

1. Höhere Typen von Funktionen (Funktionen [von] Funktionen von Zahlen, etc.).
2. Modalitätslogischer Weg (Einführung einer Absurdität auf Allsätze angewendet und eines "Folgerns").
3. Transfinite Induktion, d. h., es wird der Schluß durch Induktion für gewisse konkret definierte Ordinalzahlen der zweiten Klasse hinzugefügt.

Bemerkung. Nur die Systeme 1 genügen sämtlichen Forderungen. 2, 3 genügen nicht der Forderung 3 (gewisse Sätze [werden] als evident eingeführt), [Systeme] 2 [genügen] auch nicht der Forderung 4.[m]

IV

5 |1. [Die Erweiterung 1] besteht wesentlich in der Einführung einer Funktionsvariablen $f(x)$, welche nicht nur über die früher definierten Funktionen, sondern auch über die mit ihrer Hilfe zu definierenden läuft, und [im] Definitionsschema für Funktionen [von] Funktionen $\Phi(f, n, k)$ durch Induktion nach n, z. B.

$$\Phi(f, 1, k) = f(k)$$
$$\Phi(f, n+1, k) = f(\Phi(f, n, k)) \text{ Iteration}$$

Beweise [benutzen] nur [den] Aussagenkalkül, [die] Einsetzung[sregel] und [die gewöhnliche] Induktion. [Alle] *Forderungen* [sind] *erfüllt.*

2. Für f darf wieder eine mittels Φ definierte Funktion eingesetzt werden, aber [das ist] nicht zirkelhaft, weil Ausdrücke in eine Reihe [ge]ordnet werden [können].[n]

3. Fortsetzung für die Funktionen von Funktionen von Funktionen ... und schließlich auch ins Transfinite. Trotzdem [erhält man ein] abgeschlossenes System, wenn verlangt wird, daß nur Typen nach solchen Ordinalzahlen [zugelassen werden], welche in einem früheren System definiert [worden sind].

---

[m] Indeed, Gödel states at the beginning of the next section that systems under 2 do not satisfy the second requirement either.

[n] In *Konzept*, p. 4, Gödel writes more explicitly: "Allgemein [wird] dieses Rekursionsschema zur Einführung von Φ_i [mit Hilfe von] früheren f_i und den durch Einsetzung ineinander gewonnenen Funktionen [benutzt]. ... $\Phi(f, n, k) = f(f(f \dots f(k) \dots)$ [ist] berechenbar, wenn f [es] ist. Das kann allgemein gezeigt werden, indem man die [definierenden] Ausdrücke in eine bestimmte Reihenfolge bringt und zeigt, daß die

4. *How then shall we extend?* (Extension is necessary.) Three ways are known up to now:
   1. Higher types of functions (functions of functions of numbers, etc.).
   2. The modal-logical route (introduction of an absurdity applied to universal sentences and a [notion of] "consequence").
   3. Transfinite induction, that is, inference by induction is added for certain concretely defined ordinal numbers of the second number class.

Remark. Only the systems 1 satisfy all requirements. 2, 3 do not satisfy requirement 3 (certain propositions are introduced as evident), [the systems] 2 also do not satisfy requirement 4.[m]

IV

1. [Extension 1] consists essentially in the introduction of a function variable $f(x)$, which ranges not only over the functions defined earlier, but also over those to be defined with its help, and in the schema of definition for functionals $\Phi(f, n, k)$ by induction on n, for example

$$\Phi(f, 1, k) = f(k)$$
$$\Phi(f, n+1, k) = f(\Phi(f, n, k)) \text{ Iteration}$$

Proofs [use] only the propositional calculus, [the rule of] substitution, and [ordinary] induction—All *requirements are satisfied.*
2. For f, a function defined by means of Φ can again be substituted, but that is not circular, because expressions [can be] ordered in a series.[n]
3. Continuation for functions of functions of functions ... and ultimately into the transfinite. Nevertheless [one gets a] closed system, if it is demanded that types only [be admitted] for those ordinal numbers which have been defined in an earlier system.

Berechnung der späteren [Funktionen] auf die Berechnung der vorhergehenden zurückgeführt werden kann. Diese Hierarchie kann fortgesetzt werden, indem man Funktionen einführt, deren Argumente [solche] Φ sind, und [indem man] wiederum rekursive Definitionen nach einem Zahlparameter zuläßt. Und das kann sogar ins Transfinite erweitert werden." ("In general this recursion schema [is used] for the introduction of Φ_i [with the help of] earlier f_i and the functions obtained by substituting in one another. ... $\Phi(f, n, k) = f(f(f \ldots f(k) \ldots)$ is computable, when f is [computable]. That can be shown generally, by arranging the defining expressions in a certain sequence and showing that the computation of the later [functions] can be reduced to the computation of the earlier ones. This hierarchy can be continued by introducing functions whose arguments are such Φ and admitting once again recursive definitions according to a numerical parameter. And that can even be extended into the transfinite.")

4. In diesem Verfahren ⟦sind⟧ enthalten:
   1. Hinzufügung von Rekursionen nach mehreren Variablen,
   2. Hinzufügung der Aussage *Wid*,
   3. Hinzufügung der Hilbertschen Schlußregel.[o]
5. Wie weit kommt man damit?
   ⟦1.⟧ Negativ: Mit endlichen Typen kann man die Zahlentheorie nicht als widerspruchsfrei beweisen.
   2. Vermut⟦ung⟧: daß die Analysis bereits in keinem solchen System ⟦als widerspruchsfrei⟧ beweisbar sein wird.

6 |Das Ganze, ein interessantes offenes Problem (weil diese die einzigen Systeme sind, welche allen vier Forderungen genügen).

Nachweisen an den einzelnen Forderungen.

V

1. Ich komme jetzt zum zweiten Weg, dem modalitätslogischen. Heyting hat den Intuitionismus formalisiert. *Heytings System genügt außer der Rahmendefinition überhaupt nichts*, weil er $\neg$ ⟦und⟧ $\supset$ anwendet auf ( ) Aussagen.
2. *Alle die scheinbar schwächeren Annahmen (Satz vom ausgeschlossenen Dritten ⟦ist⟧ nicht ⟦vorausgesetzt⟧ und auch nicht allgemein beweisbar) konnte ⟦man⟧ vollkommen ersetzen, indem man beweisen kann*, daß für die aus $\neg$, $\supset$, ( ) und den arithmetischen Grundbegriffen aufgebauten Sätze alle klassisch beweisbaren Sätze gelten (und aus diesen Begriffen sind ja die übrigen &, $\vee$ definierbar).[p] Es gilt also auch der *Satz vom ausgeschlossenen Dritten*, allerdings nicht für das Heytingsche ⟦System⟧, aber für das aus diesem definierte. Aber das ⟦ist⟧ egal. (Man hat ein Modell.)
3. Dieses Resultat ⟦ist⟧ sogar schon aus geringeren Voraussetzungen als den Heytingschen ableitbar [bei Heyting problematisch: $\neg p . \supset p \supset q$]. ⟦Es ist⟧ *interessant mit wie geringen Voraussetzungen*, denn man ⟦hat nur⟧ einmal $\supset$ auf ( ) Aussagen angewendet.[q] Also:
   A. Als einziger neuer Grundbegriff zu der finiten Zahlentheorie wird $\supset$ hinzugefügt: $\neg p$ definiert durch $p \supset 0 = 1$ ($p \supset a$).
   B. Über das $\supset$ nur lauter anscheinend einwandfreie Annahmen, nämlich

---

[o] This is an altogether novel rule of inference formulated in *Hilbert 1931* (for an informative discussion see Feferman's introductory note to *Gödel 1931c*, these *Works*, Vol. I, pp. 208–213): Given a finitistic proof that each instance of the quantifier-free formula $A(x)$ is a correct numerical statement, the universal generalization $(\forall x)A(x)$ can be taken as an axiom of arithmetic. In *Herbrand 1931*, with which Gödel was thoroughly familiar and whose formalism he actually used in his *1933e*, Hilbert's rule is prominently

4. [The following] are contained in this procedure:
   1. Addition of recursion on several variables,
   2. Addition of the statement *Wid,*
   3. Addition of Hilbert's rule of inference.[o]
5. How far does one get with this?
   1. Negative: With finite types one cannot prove the consistency of number theory.
   2. Conjecture: that already the [consistency of] analysis will not be provable in any such system.

All of this [constitutes] an interesting open problem (because these are the only systems which satisfy all four requirements). To show for the individual requirements.

V

1. I now come to the second way—the modal-logical. Heyting has formalized intuitionism. *Heyting's system satisfies nothing at all beyond the framework definition,* because he applies $\neg$ and $\supset$ to ( )-statements.

2. *The apparently weaker assumptions (the law of the excluded middle is not assumed and is also not generally provable) could be totally replaced, in that one can prove* that, for the statements built up from $\neg$, $\supset$, ( ), and the basic arithmetical concepts, all classically provable statements hold (and the remaining [connectives] & and $\vee$ are of course definable from these).[p] Therefore the *law of the excluded middle* also holds, to be sure not for Heyting's [system], but for the system defined from it—but that doesn't matter. (One has a model.)

3. This result is already provable even from fewer presuppositions than Heyting's [in the case of Heyting, $\neg p . \supset p \supset q$ is problematic]. It is *interesting with how few presuppositions,* for $\supset$ has been applied to ( )-statements [only] once.[q] Therefore:
   A. $\supset$ is added to finitary number theory as the only new primitive concept: $\neg p$ [is] defined as $p \supset 0 = 1(p \supset a)$.
   B. About $\supset$, nothing but apparently unexceptionable assumptions, namely,

---

employed. The main theorem of *Herbrand 1931* asserts the consistency of the classical system with Hilbert's rule, but with induction restricted to quantifier-free formulas. For further discussion see the introductory note.

[p]Cf. *Gödel 1933e*, in particular, p. 37.

[q]On this remark see section 5 of the introductory note.

1. Transitivität: $p \supset q . \supset : q \supset r . \supset . p \supset r$
2. Vertauschung zweier Prämissen: $p \supset . q \supset r : \supset : q \supset . p \supset r$
|3. Weglassung einer doppelten Prämisse: $p \supset . p \supset q : \supset . p \supset q$
4. Modus ponendo ponens: $p \supset : p \supset q . \supset q$
5. Identität: $p \supset p$
6. Wahrer Satz folgt aus jedem: $q \supset . p \supset q$
7. Für elementare Formeln auch mit freien Variablen:

$$p \supset q . \supset \subset . \neg p \lor q$$

[C.] Für das ( ) die Schlußregel $A \supset F(x) : A \supset (x)F(x)$.

Lauter scheinbar unverdächtige Schlüsse. Das System *wahrscheinlich abundant* und vielleicht [erhält man] mit & ein noch eleganteres System.[r] Jedenfalls [ist] keine Spur eines Satzes vom ausgeschlossenen Dritten zu bemerken. *Problematisch* [ist] *vielleicht 6*, aber nicht wenn $\supset$ als beweisbar gedeutet [wird] [Łukasiewicz $p \supset q . \supset p : \supset p$].[s]

4. Für jeden Satz der Zahlentheorie[, der mit $\supset$ und ( ) aufgebaut ist, gilt:] $\neg\neg A \supset A$. [Der Beweis beruht auf dem aussagenlogischen Theorem $(p \supset q) \supset (((p \supset a) \supset a) \supset ((q \supset a) \supset a))$, das mit Hilfe von Axiom 1 leicht zu beweisen ist.][t]

1. Für elementare Formeln klar, auch wenn sie freie Variablen enthalten.
2. Wenn für $F(x)$ bewiesen [ist] $\neg\neg F(x) \supset F(x)$, so auch $\neg\neg(x)F(x) \supset (x)F(x)$:

$$(x)F(x) \supset F(y)$$
$$\neg\neg(x)F(x) \supset \neg\neg F(y)$$
$$\supset F(y)$$
$$\supset (y)F(y),$$

[3.] und ebenso [beweist man die] Übertragung von p, q auf $p \supset q$:

$$p \supset : p \supset q . \supset q$$
$$p \supset . \neg\neg(p \supset q) \supset \neg\neg q$$
$$\supset q$$
$$\neg\neg(p \supset q) . \supset . p \supset q$$

|5. *Daß das so leicht geht, liegt eben daran, daß das Heytingsche System alle wesentlichen Forderungen an* Konstruktivität *verletzt, aber vielleicht*

---

[r]The system of pure implicational logic Gödel describes here is indeed "abundant"; compare the contemporary presentation in *Hilbert and Bernays 1934*, section 3.c)3. In Supplement III of *Hilbert and Bernays 1939* a system that contains "&" in addition to implication is investigated; see their references to the work of Hertz and Gentzen.

1. Transitivity: $p \supset q . \supset : q \supset r . \supset . p \supset r$
2. Interchange of two premisses: $p \supset . q \supset r : \supset : q \supset . p \supset r$
3. Leaving out a doubled premiss: $p \supset . p \supset q : \supset . p \supset q$
4. Modus ponendo ponens: $p \supset : p \supset q . \supset q$
5. Identity: $p \supset p$
6. [A] true proposition follows from every [proposition]:
   $q \supset . p \supset q$
7. For elementary formulae, also with free variables:
   $p \supset q . \supset \subset . \neg p \lor q$

[C.] For ( ) the rule of inference $A \supset F(x) : A \supset (x)Fx$.

Nothing but apparently innocent inferences. The system is *probably abundant* and perhaps with & [one obtains] a still more elegant system.[r] In any case no trace of a law of excluded middle is to be noticed. *6 is perhaps problematic*, but not when $\supset$ is interpreted as provable [Lukasiewicz $p \supset q . \supset p : \supset p$].[s]

4. For every proposition of number theory [which is built up by $\supset$ and ( )] $\neg\neg A \supset A$ [holds. The proof rests on the theorem of propositional logic $(p \supset q) \supset (((p \supset a) \supset a) \supset ((q \supset a) \supset a))$, which is easy to prove with the help of Axiom 1.][t]

1. For atomic formulae clear, also if they contain free variables.
2. If for $F(x)$, $\neg\neg F(x) \supset F(x)$ has been proved, then also $\neg\neg(x)F(x) \supset (x)F(x)$:

$$(x)F(x) \supset F(y)$$
$$\neg\neg(x)F(x) \supset \neg\neg F(y)$$
$$\supset F(y)$$
$$\supset (y)F(y),$$

3. And likewise [one proves] the transition from p, q to $p \supset q$:

$$p \supset : p \supset q . \supset q$$
$$p \supset . \neg\neg(p \supset q) \supset \neg\neg q$$
$$\supset q$$
$$\neg\neg(p \supset q) . \supset . p \supset q$$

5. *That this goes so easily rests on the fact that Heyting's system violates all essential requirements on* constructivity, but perhaps it is still to be made

---

[s]The reference to Łukasiewicz might be to *Łukasiewicz and Tarski 1930*. In section 4 of that paper the authors discuss a semantically characterized, implicational fragment of sentential logic. They prove a completeness theorem (their Theorem 29) for a calculus whose axioms are Gödel's axioms 1,6 and this formula, i.e., Peirce's law.

[t]Here we follow *Konzept*, p. 6. See the textual notes.

[ist es] doch irgendwie konstruktiv [zu] machen.—Weswegen erkennen es
die Intuitionisten überhaupt an? Sie denken an eine konstruktive Deutung:

$$p \supset q \text{ bedeutet } \text{``}q \text{ ist aus } p \text{ ableitbar''},$$

"ableitbar" verstanden nicht in einem bestimmten System, sondern im absoluten Sinn (d. h., man kann es evident machen); und "ableitbar" im
konstruktiven Sinn verstanden, wie ich oben E eingeführt habe, d. h. man
hat eine Ableitung. Und bei dieser Interpretation [sind] tatsächlich [die]
obigen Axiome plausibel, aber nicht für "ableitbar" in einem bestimmten
System.

6. Auf dieser Idee basierend habe ich eine Interpretation [in *1933f*] mittels
B gegeben, ohne auf konstruktive Forderungen Wert zu legen[, und] den
gewöhnlichen Aussagenkalkül ergänzt[e ich] durch B ("ist beweisbar im
absoluten Sinn") und [durch] Axiome 1$^\mathrm{u}$ und 2. $Bp \rightarrow BBp$; 3. $Bp \rightarrow p$;
4. B kann hinzugefügt werden. *Intuitionismus* [ist] *daraus ableitbar.*$^\mathrm{v}$

7. Merkwürdiges Ergebnis, obwohl diese Axiome sämtlich außerordentlich
plausibel [sind], [sind] *trotzdem daraus Sätze über B ableitbar, welche
sicher für jedes definierte B falsch sind.* Nämlich $B \sim\!B(0 = 1)$.$^\mathrm{w}$ Auf
etwas ähnliches müssen wir auch jetzt gefaßt sein.

8. Damals [hatte ich] keinen Wert darauf gelegt, mittels *des B ein konstruktives System zu erhalten.* [Das] B ist ja nicht konstruktiv, und außerdem
[wurde] darauf der gewöhnliche Aussagenkalkül angewendet. *Aber das ist
vermeidbar:* [den] Grundbegriff zBp, q, [d. h.,] z ist eine Ableitung von
q aus p, [kann man] mit dem nötigen guten Willen als entscheidbar annehmen.

Axiome: z. B. Transitivität der Implikation: $zBp, q \,\&\, uBq, r \rightarrow f(z, u)Bp, r$.

Andere Axiome: $zB\varphi(x, y) \rightarrow \varphi(x, y)$, $uBv \rightarrow u'B(uBv)$; ferner, wenn
9   q bewiesen und a der Beweis, so daß ist anzuschreiben "aBq", | würde wie
oben beweisbar $aB[(u)\sim uB(0 = 1)]$.

9. *Es fragt sich nun, ist dieses System konstruktiv im obigen Sinn?*

    A. Es ist *kein Einwand, daß* auf alle Aussagen ja jetzt doch logische
        Operationen (nämlich das B) angewendet werden, was gerade ver-
        boten war, denn die Aussage tritt hier in der *suppositio materialis*
        als Gegenstand auf, unter Anführungszeichen.

$^\mathrm{u}$By "Axiom 1" Gödel must mean $Bp \rightarrow . B(p \rightarrow q) \rightarrow Bq$. Indeed, in his *1933f*,
Gödel used the axioms $Bp \rightarrow p$, $Bp \rightarrow . B(p \rightarrow q) \rightarrow Bq$, and $Bp \rightarrow BBp$; in addition,
he had the rule that allows the inference from p to Bp. The "interpretation" alluded to
above is given there by the following translation:

$$\neg p \quad \text{is translated as} \quad \sim\!Bp,$$
$$p \supset q \qquad\qquad\quad \text{as} \quad Bp \rightarrow Bq,$$
$$p \vee q \qquad\qquad\quad \text{as} \quad Bp \vee Bq$$
$$\text{and } p \wedge q \qquad\qquad \text{as} \quad p \,.\, q.$$

somehow constructive. For what reason do the intuitionists recognize it at all? They are thinking of a constructive interpretation:

$$p \supset q \text{ means "} q \text{ is derivable from } p \text{"},$$

[with] "derivable" understood not in a particular system, but in the absolute sense (that is, one can make it evident), and "derivable" understood in the constructive sense, as I introduced E above, that is, one has a derivation. And on this interpretation the *above axioms* are actually *plausible*, but not for "derivable" in a particular system.

6. Relying on this idea, I gave [in *1933f*] an interpretation by means of B, without laying stress on constructive requirements, and I supplemented the usual propositional calculus with B ("is provable in the absolute sense"), and Axioms 1^{u} and 2. $Bp \to BBp$; 3. $Bp \to p$; 4. B can be added. *Intuitionism is derivable from this.*[v]

7. A curious result, although these axioms are all extraordinarily plausible: *nevertheless propositions about B are derivable from them which are surely false for every defined B*: namely, $B \sim B(0 = 1)$.[w] Now as well, we have to be prepared for something similar.

8. At that time [I hadn't] put any value on *obtaining a constructive system by means of B*. B is not constructive, and moreover the usual propositional calculus was applied to it. *But that can be avoided*: the basic notion zBp, q, that is, z is a derivation of q from p, can be viewed as decidable with enough good will.

Axioms: for example transitivity of implication: $zBp, q \& uBq$, $r \to f(z, u)Bp, r$.

Other axioms: $zB\varphi(x, y) \to \varphi(x, y), uBv \to u'B(uBv)$; furthermore, if q has been proved and a is the proof, so that "aBq" is to be written down, then as above $aB[(u) \sim uB(0 = 1)]$ would be provable.

9. *The question now arises, is this system constructive in the above sense?*

    A. It is *no objection that* logical operations (namely B) are now applied to all statements, just what was forbidden, because the statement occurs here in *suppositio materialis* as an object, in quotation marks.

---

[v]The meaning of the remark "Intuitionismus ist daraus ableitbar" is not entirely clear. One possibility is that he means simply that on the interpretation given, the theorems of intuitionistic propositional logic are derivable. Another suggestion is that he is alluding to the converse of this result obtained by McKinsey and Tarski (*1948*, cf. these *Works*, Volume I, p. 296). Gödel conjectured this result in *1933f*.

[w]In *Konzept* Gödel argues also that these axioms—in spite of their plausibility from an intuitionistic standpoint—are false for each formalized provability predicate. He does this as follows: the first axiom gives $B(p\&\neg p) \supset p\&\neg p$; so we have by logic $\sim B(p\&\neg p)$, and thus with the rule $B \sim B(p\&\neg p)$. The same argument is indicated at the end of *1933f*, with $0 \neq 0$ replacing $p\&\neg p$.

B. [[Sind die]] Voraussetzungen erfüllt? So wie es steht, sind 3 und 4 sicher nicht erfüllt [aber sehr hohe Evidenz], denn Beweise [[sind]] unüberblickbar; und die Grundsätze sind nicht Definitionen und folgen auch nicht aus den oben angeführten Schlußregeln.
Die Voraussetzungen 1 und 2 sind bei nötigem guten Willen erfüllt.

C. Vielleicht aber ist es möglich, das System so zu präzisieren, daß tatsächlich alle Voraussetzungen bis auf 3 erfüllt [[sind]], indem man sich auf Beweise des Systems selbst beschränkt. Das [[ist]] vielleicht nicht widerspruchsvoll, denn die äquivalente arithmetische Aussage [[ist]] nicht beweisbar. Dann [[können]] die meisten von den Axiomen tatsächlich auf Definitionen zurückgeführt [[werden]].
Im Wesentlichen nicht das Unterstrichene[x]—Das ist im Wesentlichen die Widerspruchsfreiheit des Systems.

10. Ein anderer Weg, um zu versuchen, ob man aus diesem System etwas Vernünftiges bekommt, ist die Typeneinführung:[y]

$$B_{n+1} \sim B_n (0 = 1)$$

beweisbar. Aber [[solch ein Versuch macht]] große Schwierigkeiten, obwohl Gentzen angibt, daß dies der heuristische Gesichtspunkt [[für seinen Widerspruchsfreiheitsbeweis sei,]] jedenfalls *indem man es ins Transfinite fortsetzt.*

11. Von den drei Wegen [[ist]] dieser am schlechtesten, sogar vielleicht schlechter als der zu beweisende, daher führt er auch am weitesten (Analysis wahrscheinlich zu bekommen). Aber [[er ist]] durchaus nicht müßig, sondern vielleicht heuristisch sehr wertvoll. Daher so ausführlich obwohl außerhalb durch Gentzen...[z]

10 | VI

1. Schon innerhalb *der finiten Zahlentheorie* (oder wenigstens des Systems mit f) [[sind]] gewisse Ordinalzahlen der zweiten Klasse definierbar, und [[es ist]] nachweisbar, daß Beweise und Definitionen nach diesen Ordinalzahlen möglich sind, z. B. $\omega + \omega$, und beweisbar, daß Ordinalzahl ...

2. *Was heißt das, finit?* $\Phi(f)$ definierbar, so daß

$$\sim \{ f(\Phi(f) + 1) \underset{R}{\leq} f(\Phi(f)) \}.$$

---

[x]All text underlined in the manuscript has here been rendered in italics. We conjecture that "das Unterstrichene" may refer to the last line of 8 above, where in the original text each occurrence of "B" had a dot underneath.

B. Are the presuppositions satisfied? As it stands, 3 and 4 are surely not satisified [but [[there is]] very high evidence], for proofs are unsurveyable, and the principles are not definitions and also don't follow from the rules of inference cited above.

The presuppositions 1 and 2 are, with enough good will, satisfied.

C. Perhaps, however, it is possible to specify the system in such a way that actually all presuppositions except 3 are satisfied, by limiting oneself to proofs of the system itself. That is perhaps not contradictory, for the equivalent arithmetic statement is not provable. Then most of the axioms [[can]] actually be reduced to definitions.

Essentially not the underlined[x]—that is essentially the consistency of the system.

10. Another way to understand whether one gets something reasonable from this system is the introduction of types:[y]

$$B_{n+1} \sim B_n (0 = 1)$$

[[is]] provable. But [[such an attempt has]] great difficulties, although Gentzen claims that this is the heuristic point of view [[for his consistency proof]], at any rate *if one continues it into the transfinite.*

11. This is the worst of the three ways, even perhaps worse than the one to be proved; thus it also leads furthest (analysis is probably obtainable), but it is definitely not idle but perhaps heuristically very valuable, thus so detailed although except by Gentzen ...[z]

VI

1. Already within *finitary number theory* (or at least the system with f), certain ordinals of the second number class are definable, and it can be shown that proofs and definitions according to these ordinals are possible, for example $\omega + \omega$, and provable, that ordinal number ...

2. *What does this mean, [[in]] finitary [[terms]]?* $\Phi(f)$ [[is]] definable, so that

$$\sim\{f(\Phi(f) + 1) \underset{R}{\leq} f(\Phi(f))\}.$$

---

[y] In *Konzept*, p. 8, Gödel wrote more explicitly that B_n means "beweisbar im finiten System n-ten Typs" ("provable in the finitary system of nth type"). As to the latter notion, see section 5 of the introductory note.

[z] What follows "daher" in this sentence is evidently incomplete. It seems Gödel means that what he has just said is a reason for treating the approach more fully, but that only Gentzen has made use of its heuristic value.

Daraus folgt dann:

1. Schluß durch Induktion, d. h. wenn E eine Aussagenfunktion [ist],
   [und] es ist aus der Annahme $x \underset{R}{\le} a \to E(x)$ ableitbar $E(a)$ im System
   S_1, dann ist $(x)E(x)$ beweisbar im System S_1;

2. Definition durch Induktion: Wenn $g_1(x), \ldots, g_v(x)$ irgendwelche Funktionen sind, so daß $g_i(x) \underset{R}{\le} x$, und man schreibt [für] irgendeinen Ausdruck [A] ein

   $$\varphi(x) = A(\varphi(g_1(x)), \ldots, \varphi(g_v(x))),$$

   so gibt es ein φ[, das diese Gleichung erfüllt].

3. Dasselbe [ist] auch für höhere Ordinalzahlen, z. B. ω^2, ω^ω, ω^{ω^2}, noch
   möglich. Also [sind] Induktionsschlüsse beweisbar aus den anderen Axi-
   omen. *ω^ω schon nicht mehr in der finiten Zahlentheorie (aber alle [$< \omega^\omega$]
   definierbar in der finiten Zahlentheorie).*

4. Aber es gibt gewisse Zahlen, für welche das nicht möglich [ist] zu be-
   weisen und zwar nicht einmal in der *transfiniten Arithmetik*. Eine solche
   Zahl ist die erste ϵ-Zahl—definiert[aa] durch [$\alpha_1 = 2^{\omega+1}$ und] $\alpha_{n+1} = 2^{\alpha_n}$.
   Anschauliches Bild dieser Zahl. Zunächst [betrachte ich den] Prozeß 2^α:

$$2^1 = 2, \quad 2^\beta = \sum_{x < \beta} 2^x,$$

11 | d. h., jedem Element *wird die Summe der vorhergehenden zugeordnet*;
[das ist ein] sehr anschauliches Bildungsverfahren.[bb]

5. Schwierigkeit: *Warum das nicht in der Arithmetik formalisiert* werden kann und auch nicht in der durch endliche finite Systeme ergänzten?
Angenommen es sei schon bewiesen, α ist eine Ordinalzahl in dem schärfsten
Sinn, d. h., ein $\Phi(f)$ [ist] definiert, so daß [jede Teilfolge] abbricht.—Wir
wollen versuchen, zunächst eine *Anordnung von* natürlichen Zahlen in der
Arithmetik zu definieren, welche 2^α darstellt, und dann zu beweisen, daß
diese Anordnung eine Wohlordnung ist (d. h. ein Φ dazu definieren). Selbstverständlich [werden] beide durch Induktion nach der schon als Ordinalzahl
nachgewiesenen Anordnung α [definiert].

6. *Definition* geht: das sieht man am leichtesten, wenn man daran denkt,
daß 2^α erhalten werden kann durch absteigende n-Tupel von Zahlen $< \alpha$,
lexikographisch geordnet:

$$(\text{Beweis}: \beta < 2^\alpha \to \beta = 2^{\gamma_1} + \ldots + 2^{\gamma_k}, \quad \gamma_i < \alpha)$$

[aa]The base case for the definition of the fundamental sequence for ϵ_0 was omitted in
the text, but given by Gödel in *Konzept*, p. 9.

From that then follows:

1. Inference by induction, that is, if E is a propositional function, and $E(a)$ is derivable in the system S_1 from the assumption $x \underset{R}{<} a \rightarrow E(x)$,

   then $(x)E(x)$ is provable in the system S_1;
2. Definition by induction: If $g_1(x), \ldots, g_v(x)$ are any functions such that $g_i(x) \underset{R}{<} x$, and for any expression $[\![A]\!]$, one writes down

$$\varphi(x) = A(\varphi(g_1(x)), \ldots, \varphi(g_v(x))),$$

   then there is a φ [that satisfies this equation].

3. The same is also still possible for larger ordinals, for example ω^2, ω^ω, ω^{ω^2}. Therefore inferences by induction are provable from the other axioms. ω^ω *already no longer in finitary number theory (but all $[\![< \omega^\omega]\!]$ definable in finitary number theory)*.

4. But there are certain numbers for which it is not possible to prove that, not even in *transfinite arithmetic*. One such number is the first ϵ-number— defined[aa] by $[\![\alpha_1 \doteq 2^{\omega+1}$ and$]\!]$ $\alpha^{n+1} = 2^{\alpha_n}$. $[\![$An$]\!]$ *intuitive picture of this number*. First $[\![$I consider the$]\!]$ process 2^α:

$$2^1 = 2, \quad 2^\beta = \sum_{x<\beta} 2^x,$$

that is, to every element *is assigned the sum of the previous ones*. Very intuitive construction procedure.[bb]

5. Difficulty: *Why can that not be formalized in arithmetic*, also not if it is extended by finite finitary systems? Suppose it has already been proved that α is an ordinal number in the sharpest sense, that is, a $\Phi(f)$ is defined, so that [every subsequence] terminates. Next we will try to define in arithmetic an *ordering* of natural numbers which represents 2^α, and then to prove that this ordering is a well-ordering (that is, to define a Φ for it). Obviously both [will be defined] by induction according to the ordering α, already shown to be an ordinal.

6. The *definition* goes through: That is most easily seen if one considers the fact that 2^α can be obtained by decreasing n-tuples of numbers $< \alpha$, lexicographically ordered:

$$(\text{Proof: } \beta < 2^\alpha \rightarrow \beta = 2^{\gamma_1} + \ldots + 2^{\gamma_k}, \gamma_i < \alpha).$$

[bb]One might think from the end of *Konzept*, p. 9, that at this point Gödel left out the successor case. However, his following comment and the argument of 7 support the hypothesis that he has exactly what he intended. If we call the function defined by this recursion 2[exp]x, then for $1 < x < \omega$, 2[exp]$x = 2^{x-1}$; otherwise 2[exp]$x = 2^x$.

Nun kann man innerhalb der Arithmetik die n-Tupel abbilden auf die Zahlen (z. B. durch Primzahlen), und [diese] Anordnung [ist] auch definierbar.
7. *Beweis* geht nicht: denn wie ist *das gewöhnlich [zu] beweisen?* Man würde beweisen, 2^1 ist eine Ordinalzahl. Angenommen 2^γ ($\gamma < \beta$) ist eine Ordinalzahl, dann ist ja $2^\beta = \sum_{\gamma < \beta} 2^\gamma$. [Die] Summe einer wohlgeordneten Folge von Ordinalzahlen ist eine Ordinalzahl. Die Eigenschaft, nach der man Induktion anwendet, ist also das "*Ordinalzahlsein*", d. h. aber, *jede Teilmenge* hat ein erstes Element, *oder jede* Teilfolge bricht ab. [Diese Eigenschaft ist] also imprädikativ, also nicht mehr formulierbar in der Zahlentheorie (*Existentiale Operationen für Klassen* nötig und sogar [das] Reduzibilitätsaxiom), und das ist *wesentlich, wie aus dem Gentzenschen Beweis hervorgeht*.
8. Vor Gentzen zeigte es sich nur in dem Fehlschlag der Versuche, z. B. [des] Versuchs, ein Φ zu definieren. $\Phi(f)$ führt auf eine Rekursion nach α, aber in den g_i kommt das Φ selbst wieder vor.
12 | 9. Durch abzählbare Iteration dieses Übergangs von α zu 2^α entsteht dann ϵ_o. Diese [Ordinalzahl wäre] also ohne weiteres gegeben, falls dieser Übergang gegeben [wäre]. Nichtsdestoweniger wird man dem Induktionsschluß nach dieser so definierten Zahl ϵ_o einen hohen Grad von Anschaulichkeit nicht absprechen, wie überhaupt dem Verfahren, eine *Ordinalzahl durch Induktion nach Ordinalzahlen zu definieren* (obwohl [es] ein imprädikatives Verfahren [ist]).
10. Es macht, wie gesagt, keine Schwierigkeit, eine Anordnung vom Typus ϵ_o *innerhalb eines finiten Systems (sogar S_1) zu definieren*. Nur der Beweis der vollständigen Induktion ist unmöglich. Man kann also diese als neue Schlußregel hinzufügen, d. h. also folgendes: Sei R_o die Ordnung, und es sei gelungen aus $(x)[x \underset{R_o}{<} a \rightarrow E(x)]$ abzuleiten $E(a)$, dann darf man daraus $(x)E(x)$ schließen. Oder andere Systeme würde man erhalten, indem man das für andere Ordinalzahlen tut, die durch Induktion nach Ordinalzahlen definiert sind.
11. Ist dieses System finit? Es *genügt allen unseren Bedingungen* außer 3. Die hatte ja gelautet, es soll kein anderer Schluß als [die] gewöhnliche Induktion vorkommen. Dieser [neue] Schluß kann aber als eine Verallgemeinerung der gewöhnlichen Induktion aufgefaßt werden, und insofern ist die Abweichung von der Forderung 3 vielleicht keine so tiefgehende.
12. Ich möchte übrigens bemerken, daß Gentzen *einen "Beweis"* für diesen Schluß zu geben suchte, und [er] sag[te] sogar, daß dies der wesentliche Teil seines Widerspruchsfreiheitsbeweises sei. In Wirklichkeit handelt es sich dabei aber gar nicht um einen Beweis, sondern um eine Berufung auf Evidenz; was ja auch klar ist. Ich möchte vorlesen, was Gentzen selbst über diesen Beweis sagte.[cc] Ich glaube es hat mehr Sinn, ein Axiom präzis zu

Now one can, within arithmetic, map the n-tuples onto the numbers (for example by primes), and this ordering is also definable.

7. The *proof* does not go through: for how is *this usually to be proved?* One would prove that 2^1 is an ordinal. Supposing that 2^γ is an ordinal ($\gamma < \beta$), then so is $2^\beta = \sum_{\gamma < \beta} 2^\gamma$. The sum of a well-ordered sequence of ordinals is an ordinal. The property to which induction is applied is therefore that of "*being an ordinal*", that is that *every subset* has a first element, *or every subsequence breaks off*. [This property is] therefore impredicative, thus no longer formulable in number theory (*existential operators for classes* are necessary, and even the axiom of reducibility), and that is *essential, as is evident from Gentzen's proof.*

8. Before Gentzen this came to light only in the failure of attempts, for example the attempt to define a Φ. $\Phi(f)$ leads to a recursion on α, but Φ itself occurs again in the g_i.

9. By countable iteration of this transition from α to 2^α, ϵ_0 is generated. This [ordinal would be] therefore given immediately, once this transition is given. Nonetheless, one will not deny a high degree of intuitiveness to the inference by induction on ϵ_0 thus defined, as in general to the procedure of *defining an ordinal by induction on ordinals* (even though this is an impredicative procedure).

10. As we have said, there is no difficulty in defining an ordering of type ϵ_0 *within a finitary system (even S_1).* Only the proof of complete induction is impossible. Therefore one can add this as a new rule of inference, that is the following: Let R_0 be the ordering, and suppose one has succeeded in deriving $E(a)$ from $(x)[x \underset{R_o}{<} a \to E(x)]$, then from this one may conclude $(x)E(x)$. Or other systems would be obtained by doing this for other ordinal numbers that are defined by induction on ordinals.

11. Is this system finitary? It *satisfies all our conditions* except 3. Recall that this had stated that no other inference should occur than ordinary induction. This [new] inference can be considered as a generalization of ordinary induction, and in this respect the deviation from the requirement 3 is perhaps not such a drastic one.

12. I would like to remark by the way that Gentzen sought to give a *"proof"* of this rule of inference and even said that this was the essential part of his consistency proof. In reality, it's not a matter of a proof at all, but of an appeal to evidence—what is after all also clear. Let me read what Gentzen himself says about this proof.[cc] I think it makes more sense to formulate

---

[cc] At this point Gödel evidently read a passage from Gentzen which he does not identify, almost certainly from *Gentzen 1936*. Gentzen's proof of transfinite induction is in §15.4 of the paper, but possibly Gödel read from the comments on the proof in §16.11.

formulieren und zu sagen, daß [[es]] eben nicht weiter reduzierbar ist. Aber
13 es tritt wohl hier wieder | das Bestreben der Hilbert Schüler, aus nichts
etwas zu deduzieren, hervor.

13. *Wie weit kommt man damit?* Gentzen hat die Widerspruchsfreiheit der
Zahlentheorie [[bewiesen]], wahrscheinlich auch des Weylschen *Kontinuums*
[[*1918*]]. Gentzen hoffte ferner, die Analysis und Teile der Mengenlehre
und höher zu erhalten; bei genügend grossen Ordinalzahlen ist das wohl
nicht zu bezweifeln. Ob man aber mit diesem Prinzip, [[d. h.]] Definition
von *Ordinalzahlen durch* transfinite *Induktion nach bereits definierten Ordinalzahlen*, das ja immerhin einen hohen Grad von Anschaulichkeit be-
sitzt, auskommt, ist wohl fraglich; hier vielleicht negativ. [Insbesondere
$\epsilon_{[0]}$ vielleicht in S_ω.]

14. Verhältnis zu den ersten Erweiterungen. Vielleicht [[geht]] dieses ganze
Prinzip (es erscheint sehr anschaulich) nicht über die Systeme S_i hinaus.

15. Jetzt möchte ich noch einiges über den Gentzenschen Beweis sagen.
Zunächst muß man sagen, daß seine Arbeit sehr kompliziert [[und]] durchaus nicht ein Muster an Klarheit ist. Es gibt ferner bis jetzt anscheinend
nur sehr wenige Mathematiker, die sie kontrolliert haben. Es ist also die
Möglichkeit eines Fehlers nicht auszuschließen. Ich glaube das eigentlich
nicht, sondern ich habe den Eindruck, daß die Reduktion stimmt.

14 |16. Gentzen gibt an, er habe seinen Beweis in dem Bestreben, den modali-
tätslogischen zu präzisieren, gefunden. Die Verwandtschaft des fertigen
Beweises ist aber eine ziemlich entfernte. Es ist von jeder beweisbaren
Formel eine ¬¬ bewiesen, aber in einem anderen Sinn als dem modalitätslogischen. Nehmen wir eine zahlentheoretische Formel in der Normalform

$$\Phi = (x)(Ey)(z)(Eu)A(x, y, z, u),$$

wobei A ein elementarer (entscheidbarer) Ausdruck ist.—Was wäre ein konstruktiver Beweis einer solchen Formel? [[Für gewisse gegebene]] $f(x)$ [[und]]
$g(x, z)$ [[ein konstruktiver Beweis von]] $(x, z)A(x, f(x), z, g(x, z))$. Wie sieht
die Negation der obigen Formel aus, wenn man sie auf die Normalform
bringt? $(Ex)(y)(Ez)(u){\sim}A(x, y, z, u)$. Ein konstruktiver Beweis der Negation wäre als[[o]] die Angabe einer Zahl c und einer Funktion $f(y)$, so daß
für alle $(y, u){\sim}A(c, y, f(y), u)$. [[Ein Beweis von]] ¬¬$\Phi$ wäre der Beweis, daß
es kein solches f, c geben kann: man kann zu jedem f, c; y, u angeben, [[so
daß]] $(f, c)A(c, \underline{y_{f,c}}, f(y_{f,c}), \underline{u_{f,c}})$. Solche Funktionsfunktionen y, u nennt
er eine Reduktion.[1]

---

[1]Der uferlose Begriff des "Beweis" wird also hier ersetzt durch den ebenso ufer-
losen Begriff der Funktionsfunktion.—Das geht, wo das Ersetzen durch bestimmten
Beweis[[begriff]] eben nicht geht.

an axiom precisely and to say that it is just not further reducible. But here again the drive of Hilbert's pupils to derive something from nothing stands out.

13. *How far does one get with this?* Gentzen proved the consistency of number theory [and] probably also of Weyl's *Kontinuum* [*1918*]. Gentzen hoped further to obtain analysis and parts of set theory and higher; with sufficiently large ordinals, that is no doubt true. Whether one gets by with this principle, that is *definition of ordinals by* transfinite *induction on previously defined ordinals*, which admittedly still has a high degree of intuitiveness, seems questionable; here perhaps negative. [In particular $\epsilon_{[0]}$ perhaps in S_ω.]

14. Relation to the first extensions. Perhaps this whole principle (it seems very intuitive) doesn't go beyond the systems S_i.

15. Now I would like to say something about Gentzen's proof. First one must say that his paper is very complicated and not at all a model of clarity. There are, furthermore, so far apparently only very few mathematicians who have checked it. Thus the possibility of a mistake [in the proof] can't be ruled out. I don't really believe that; rather I have the impression that the reduction is correct.

16. Gentzen claims he found his proof in the endeavor to make the modal-logical one more precise. The affinity of the finished proof is, however, rather remote. A $\neg\neg$ of every provable formula is proved, but in a different sense from the modal-logical. Consider a number-theoretic formula in the normal form

$$\Phi = (x)(Ey)(z)(Eu)A(x, y, z, u)$$

where A is an elementary (decidable) expression.—What would a constructive proof of such a formula be? [For certain given] $f(x)$ and $g(x, z)$, [a constructive proof of] $(x, z)A(x, f(x), z, g(x, z))$. How does the negation of the above formula look, if one brings it into the normal form $(Ex)(y)(Ez)(u){\sim}A(x, y, z, u)$? Thus a constructive proof of the negation would be the giving of a number c and a function $f(y)$, so that for all $(y, u){\sim}A(c, y, f(y), u)$. [A proof of] $\neg\neg\Phi$ would be the proof that there can be no such f, c: one can for every f, c give y, u so that $(f, c)A(c, \underline{y_{f,c}}, f(y_{f,c}), \underline{u_{f,c}})$. He calls such functionals y, u a reduction.[1]

---

[1]The vast notion of "proof" is thus here replaced by the equally vast notion of functional.—That works, where the replacement by a definite [notion of] proof does not.

17. Es wird gezeigt, wie man aus dem Beweis für eine Formel eine Reduktion finden kann. Zunächst ist es ganz leicht, für die Axiome Reduktionen anzugeben. Auch für den Satz vom ausgeschlossenen Dritten [ist es] ganz leicht:[dd]

$$(Ex)(y)[\varphi(x) \vee \sim\varphi(y)] \quad (f)[\varphi(\overbrace{\psi(f)}^{x}) \vee \sim\varphi(f(\overbrace{\psi(f)}^{x}))] \qquad [\![*]\!]$$

$$\sim\varphi(x).\varphi(f(x)) \qquad \overbrace{\psi(f)}^{x} = f(0) \mid \varphi(f(0))$$

$$\sim\varphi(0).\underline{\varphi(f(0))} \qquad\quad [\![=]\!] \quad 0 \mid \sim\varphi(f(0))$$

$$\sim\underline{\varphi(f(0))}.\varphi(ff(0))$$

15 |18. Die Methode, wie dieses ψ definiert wird, ist die Probiermethode. Man setzt zuerst $\psi(f) = 0$. Geht es damit nicht, so muß $\varphi(f(0))$ gelten. Dann geht es mit $\psi(f) = f(0)$.

Schlußregeln. Die ganze Schwierigkeit des Beweises liegt in der Schlußregel $P \quad P \supset Q \; [\![/Q]\!]$. Hat man die, so ist insbesondere der Induktionsschluß klar. [Seien die Formeln]

$$F(0) \qquad F(n) \supset F(n+1)$$

bewiesen. Daher hat man für jedes $F(n)$ ein Reduktionsverfahren, [d. h. man hat] jedem n ein Reduktionsverfahren [für] $F(n)$ zugeordnet; [das] heißt, man hat ein Reduktionsverfahren für $(n)F(n)$.

19. Der Beweis für $P \supset Q$ geht folgendermaßen: den Funktionsfunktionen, welche finit definiert sind (d. h. für jedes konkret vorgelegte f berechenbar), kann man ja Ordinalzahlen der zweiten Klasse zuordnen (Souslinsches Schema). Die reduzierende Funktion für Q wird definiert durch transfinite Induktion nach der Ordinalzahl der reduzierenden Funktion für P, und wenn man sich die Ordinalzahl, welche der reduzierenden Funktion für Q zugeordnet ist, ausrechnet, so tritt die von P im Exponenten auf. Es ist also genau der Schluß, daß man eine gewisse neue Ordinalzahl einführt durch Rekursion nach der bereits als Ordinalzahl erkannten und dann nach dieser neuen wieder rekursive Definition anwendet.

[dd]That is, we set x (i.e, $\psi(f)$) $= f(0)$ if $\varphi(f(0))$ holds, $= 0$ otherwise. Thus if $\varphi(f(0))$ holds, the first disjunct of $[\![*]\!]$ holds, otherwise the second disjunct holds. That is, one

17. It is shown how one can find a reduction for a formula from the proof. First, it is quite easy to give reductions for the axioms. Even for the law of the excluded middle it is very easy:[dd]

$$(Ex)(y)[\varphi(x) \vee \sim\varphi(y)] \quad (f)[\varphi(\overbrace{\psi(f)}^{x}) \vee \sim\varphi(f(\overbrace{\psi(f)}^{x}))) \qquad [\![*]\!]$$

$$\sim\varphi(x).\varphi(f(x)) \qquad \overbrace{\psi(f)}^{x} = f(0) \mid \varphi(f(0))$$

$$\sim\varphi(0).\varphi(\underline{f(0)}) \qquad\qquad [\![=]\!] \quad 0 \mid \sim\varphi(f(0))$$

$$\sim\varphi(\underline{f(0)}).\varphi(ff(0))$$

18. The method by which this ψ is defined is the method of trial and error. First one sets $\psi(f) = 0$. If that doesn't work, then $\psi(f(0))$ must be true. Then $\psi(f) = f(0)$ works.

Rules of inference. The whole difficulty of the proof lies in the rule of inference $P \; P \supset Q [\![/Q]\!]$. If we have that then in particular the induction rule is clear. [Suppose that the formulae]

$$F(0) \qquad F(n) \supset F(n+1)$$

have been proved. Then we have a reduction procedure for every n, that is, we have correlated to each n a reduction procedure for $F(n)$; that means that we have a reduction procedure for $(n)F(n)$.

19. The proof for $P \supset Q$ goes as follows: We can assign ordinals of the second number class to the functionals that are defined in a finitary way (that is, computable for every concretely presented f) (Souslin's schema). The reducing function for Q is defined by transfinite induction on the ordinal of the reducing function for P, and if we compute the ordinal that is assigned to the reducing function for Q, then that for P occurs in the exponent. It is therefore exactly the inference of introducing a certain new ordinal by recursion on an ordinal already recognized as such and then again applying recursive definition on this new [ordinal].

---

of the two last formulas on the left must be false, thus ruling out f as a Skolem function for the negated prenex of the orginal formula. Essentially the same argument occurs in *Kreisel 1951*, p. 257. Cf. the discussion in section 7 of the introductory note.

16 | VII

Zum Schluß möchte ich auf die historische und erkenntnistheoretische
Seite der Frage zurückkommen und also fragen, (1) ob einem Widerspruchs-
freiheitsbeweis mittels der drei erweiterten Systeme ein Wert im Sinn einer
Tieferlegung der Fundamente zukommt; (2) was damit zusammenhängt,
ob das Hilbertsche Programm dadurch, daß über die finite Zahlentheorie
notwendig hinausgegangen wird, in einem wesentlichen Punkt vereitelt ist.
Dazu kann man zweierlei sagen: (1) Falls das ursprüngliche Hilbertsche
Programm durchführbar gewesen wäre, so wäre das zweifellos von unge-
heurem erkenntnistheoretischem Wert gewesen. Es wären nämlich beide
Forderungen erfüllt worden: (A) Die Mathematik wäre auf einen sehr
kleinen Teil von sich reduziert worden (also eine große Anzahl von un-
abhängigen Annahmen [wären] überflüssig geworden). (B) Es wäre wirklich alles auf eine konkrete Basis reduziert worden, auf die alle sich müssen
einigen können.[ee] [(2)] Bei den Beweisen mittels des erweiterten Finitismus
ist das erste gar nicht mehr der Fall, denn man muß ja immer, um etwas als
widerspruchsfrei [zu] beweisen, gewisse andere Annahmen an die Stelle der
als widerspruchsfrei [zu beweisenden] setzen, so daß man keine Reduktion
[im obigen Sinne][ff], sondern eine Ersetzung oder Verschiebung hat. Das
zweite (Reduktion auf die konkrete Basis, d. h. also Erhöhung des Evidenz-
grades) [ist] bei den verschiedenen Systemen in verschiedenen Graden der
Fall, also z. B. bei den modalitätslogischen gar nicht, bei den höheren Funk-
tionstypen am meisten, bei den transfiniten Ordinalzahlen—soweit man nur
17 das oben ausgesprochene Prinzip anwendet—auch | noch in einem ziemlich
hohen Grad. Man wird in diesem Sinn dem Gentzenschen Beweis nicht
absprechen können, daß er das Operieren mit dem transfiniten E auf etwas
evidenteres (die erste ϵ-Zahl) zurückführt. Auf jeden Fall scheint mir, daß
die erkenntnistheoretische Bedeutung, im Sinne einer besseren Fundierung,
dadurch daß sie [die verschiedenen Systeme] nicht in der finiten Zahlentheorie enthalten sind, sehr vermindert wird. [Davon] ganz unbeschadet
[ist] die mathematische Bedeutung dieser Untersuchung. Diese scheint
hier tatsächlich außerordentlich groß zu sein, und ich bin überzeugt, daß
die dabei verwendeten Methoden in der Grundlagenforschung und auch
außerhalb ihrer zu sehr interessanten Resultaten führen werden.

[ee]B is a paraphrase of *Hilbert 1926*, p. 180, ll. 9–10; see *van Heijenoort 1967*, p. 384
ll. 6–7.

VII

In conclusion, I would like to return to the historical and epistemological side of the question and then ask (1) whether a consistency proof by means of the three extended systems has a value in the sense of laying the foundations more securely; (2) what is closely related, whether the Hilbert program is undermined in an essential respect by the fact that it is necessary to go beyond finitary number theory.

To this we can say two things: (1) If the original Hilbert program could have been carried out, that would have been without any doubt of enormous epistemological value. The following requirements would both have been satisfied: (A) Mathematics would have been reduced to a very small part of itself (therefore a large number of independent assumptions would have become superfluous). (B) Everything would really have been reduced to a concrete basis, on which everyone must be able to agree.[ee] (2) As to the proofs by means of the extended finitism, the first is no longer the case at all, since in order to prove something consistent, one must always put other assumptions in the place of those [to be proved] consistent, so that one doesn't have a reduction [in the above sense],[ff] but rather a replacement or shifting. The second (reduction to the concrete basis, which means increase of the degree of evidence) obtains for the different systems to different degrees, thus for example for the modal-logical not at all, for the higher function types the most, for the transfinite ordinal numbers—insofar as one applies only the principle stated above—also to a rather high degree. In this sense, one will not be able to deny of Gentzen's proof that it reduces operating with the transfinite E to something more evident (the first ϵ-number). In any case, it seems to me that the epistemological significance, in the sense of a better foundation, is very much diminished by the fact that [the different systems] are not contained in finitary number theory. The mathematical significance of this investigation is totally unaffected. The latter seems to me in fact to be extraordinarily great, and I am convinced that the methods applied here will lead to very interesting results in foundational research and also outside it.

[ff]Without the addition there is (the possibility of seeing) a conflict with the subsequent remarks and also the discussion in section 1.

Introductory note to *Gödel *1939b* and *1940a*

1. Introduction

Gödel's *Nachlass* contains notes for four different lectures on the consistency of the continuum hypothesis. The editors have selected two of these for publication here, because they shed light on Gödel's motivations or present novel approaches to this topic, while the lectures not chosen cover substantially the same ground as the published work that appears in these *Works*, Volume II, pages 26–101.

1939b is the text of a lecture given at Göttingen. In my opinion it contains Gödel's clearest exposition of the ideas behind his proof of the relative consistency of the continuum hypothesis.

1940a is the text of a lecture given at Brown University. It contains two novel elements. First, Gödel attempts to motivate the proof in terms of the notion of "ordinal definability". Indeed his heuristic remarks make it seem as if he is confusing that notion with the notion of constructibility, though a remark at the end of the paper makes the distinction clear. Second, there is an alternate presentation of the notion of constructibility which is supposed to resemble the treatment in Hilbert's attempted proof of the continuum hypothesis in *Hilbert 1926*. This alternate treatment gives a very quick and direct description of the constructible subsets of an ordinal α.

In what follows I comment separately on each of the two lectures and then turn to Gödel's remarks in both that bear on the relationship of his work on L to earlier work of Russell. Finally, section 5 gives some details concerning the "Hilbert style" proof of the consistency of the continuum hypothesis given in *1940a*.

2. Remarks on *1939b*

2.1. Gödel begins this lecture with a comment that he completed the proof of the result on L in "June of last year" (presumably June, 1938). However, in a letter he wrote to Karl Menger on 15 December 1937 (quoted in part in *Menger 1981*, his unpublished memoir "Recollections of Kurt Gödel") Gödel confided that he had "finally succeeded [last summer] in proving the consistency of the continuum hypothesis (even the generalized form) with respect to general set theory", but he asked Menger not to tell anyone else about it. Evidence in Gödel's *Nachlass* also supports a June 1937 dating. (Cf. note s, Volume I, page 36 of these *Works*.)

2.2. On page 2, Gödel also comments concerning his consistency result "... consistency therefore holds in an absolute sense, insofar as it makes any sense at all today to speak of absolute consistency." We have already indicated in our introductory note to Gödel's main papers on L that this is not the case. (Cf. these *Works*, Volume II, page 22.) In fact, many large cardinal axioms (in particular, the assertion that measurable cardinals exist) refute the assertion that $V = L$.

2.3. On pages 13–14, Gödel writes "The question whether there are $\aleph_1$ or $\aleph_0$ constructible sets of natural numbers is to all appearances a proposition that is undecidable in set theory ...".

This shows remarkable prescience. Nowadays, of course, we can obtain models in which there are only countably many constructible reals by a forcing extension that collapses $\aleph_1^L$. (This was first shown in *Lévy 1970*.) On the other hand, large cardinals such as measurables *prove* that there are only countably many constructible reals.

2.4. Later on page 14, Gödel writes "For after all ... the axioms of reducibility, infinity and choice are the only axioms of classical mathematics that do not have [a] tautological character."

I struggled for a long time to give a meaning to this sentence under which it is not obviously false. Clearly the word "tautological" is being used somewhat loosely. Still, e.g., the power set axiom seems quite *non-tautological*, since the consistency strength of ZFC *minus* the power set axiom is the same as that of second-order number theory.

The best I could come up with is the following well-known result. Let M be a transitive class which is a model of the comprehension axiom and satisfies the principle:

$$(\forall x \subset M)(\exists y \in M)(x \subseteq y).$$

Then M is a model of ZF.

The force of this remark is that (under the modest side condition just stated) *all* the axioms of ZF holding in V also hold in M, provided M satisfies comprehension. Yet if this were what Gödel was driving at, why would he have mentioned the axiom of infinity?

Feferman has suggested an alternate explanation which is worth considering. He believes that Gödel has the logicist explanation of set-theoretical notions in mind. If one regards the system PM of *Principia mathematica* including the axiom of reducibility as *logical* (and in that sense *tautological*), then all of Zermelo's axioms are accounted for except those for infinity and choice. In particular, the power set axiom is built into the type structure.

2.5. Still further on, on page 14, Gödel writes: "the axiom of choice likewise holds, since it is an easy consequence of the generalized continuum hypothesis."

There are two results of this general character in the literature.

1. One is a result announced by Lindenbaum and Tarski in *1926* whose first published proof was given by Sierpiński in *1947*. This uses the following formulation of GCH: Let $\mathbf{m}$ be an infinite cardinal; then there is no cardinal $\mathbf{n}$ such that $\mathbf{m} < \mathbf{n} < 2^{\mathbf{m}}$.

The result in question asserts that this form of the GCH entails the axiom of choice. However, this does not suffice for Gödel's purpose, since he only proves the GCH in L for cardinals which are "alephs" (i.e., the cardinals of well-orderable sets).

2. The second result does not deduce the axiom of choice from the GCH *per se*, but from the weaker proposition that the power set of every well-ordered set is well-orderable.[a] The proof of this implication uses the axiom of foundation in an essential way.

This result *is* available to Gödel for his proof, since it is evident that L satisfies the axiom of foundation. However, actually spelling out this proof would be significantly more involved than the usual approach (which directly verifies the axiom of choice in L). Thus Gödel's remark seems the kind of thing one says when one doesn't want to give the details of an argument.

2.6. The closing sentences of this lecture are quite prophetic: "I am fully convinced that the assumption that non-constructible sets exist is also consistent. A proof of that would perhaps furnish the key to the proof of the independence of the continuum hypothesis from the other axioms of set theory. That would then yield the definitive result that one must really be content with a proof of the consistency of the continuum hypothesis, because then what would have been shown is exactly that a proof of the proposition itself [CH] does not exist."

3. Comments on *1940a*

3.1. In the early pages of the notes for this lecture, Gödel seems to be

---

[a]This theorem was announced by Herman Rubin in 1960. A proof may be found in *Rubin and Rubin 1985*, Part I, §5, Theorem 5.7 (p. 76).

talking about ordinal definability. Indeed he talks of "all possible definitions" and then says that one must consider not only the definable reals but "those definable in terms of ordinals". This is clearly only a way of motivating the proof. That Gödel is quite clear on the distinction between ordinal definability and constructibility (at the time of writing out his lecture) is shown by his comments on page 31.

In fact, it is known that the *only* thing one can prove about **HOD** (the class of hereditarily ordinal definable sets) is that it is a model of ZFC. More precisely, given any countable transitive model M of ZFC, there is a countable transitive model N of ZFC with the same ordinals such that $M = \textbf{HOD}^N$. Similarly, one can find models of ZFC in which there are $\aleph_2$ ordinal-definable reals.[b] Thus Gödel's implicit move from **HOD** to L is essential.

3.2. On page 31 of the lecture, Gödel writes: "... the consistency of A ⟦Gödel's name in this paper for the proposition that '$V = L$' or, alternatively, for the (different!) proposition that 'Every real is constructible'⟧ does not mean that for every real number definable, say, in the system of *Principia mathematica* one could prove that it is constructible .... On the contrary, it is possible actually to define certain real numbers which very likely cannot be proved to be constructible, although one can of course assume consistently that they are."

One can only speculate as to what sorts of definitions Gödel had in mind. The simplest plausible example that Gödel *might* have been thinking of is $\{n \in \omega \mid 2^{\aleph_n} = \aleph_{n+1}\}$. Of course, nowadays, one knows of Δ^1_3 definitions that can consistently be assumed (relative to the consistency of ZF) to define non-constructible sets,[c] but it seems highly unlikely that Gödel had these arcane definitions in mind. However, Gödel might simply have had in mind something like the complete Σ^1_3 subset of ω, feeling that going beyond the analytic hierarchy level of the definition of "constructible real" would yield some definition that consistently gives a non-constructible real.[d]

---

[b] Both these results are easily proved by the methods of *McAloon 1971*; the second is explicitly stated there. The first, which is somewhat harder, is not. I don't know who first noticed it, and don't remember from whom I first heard of it; at any rate, I have carried out a proof for myself.

[c] Cf. *Jensen and Solovay 1970*, p. 92, Theorem 4.1.

[d] The statement A (that every real is constructible) on p. 34 of *1940a* is known to be equivalent to a Π^1_3 sentence (one of the form $(\forall \alpha)(\exists \beta)(\forall \gamma)S(\alpha, \beta, \gamma)$ where the variables α, β and γ range over the set of functions from ω to ω and the predicate S is arithmetic). Moreover, this bound is sharp: A is not provably equivalent (in ZFC) to the negation of a Π^1_3 sentence unless ZFC is inconsistent.

3.3. In discussing reasons to believe that the existence of a non-constructible real should be consistent with the axioms of set theory, Gödel
writes (on page 33) "... an inconsistency of $\sim A$ [where here A is the
proposition that every real is constructible] would imply an inconsistency of the notion of a random sequence, where by a random sequence
I mean one which follows no mathematical law whatsoever, and it seems
very unlikely that this notion should imply a contradiction."

At first glance, this seems a foreshadowing of my notion of a real
being random over a transitive model of set theory. (Cf. *Solovay 1970.*)
In this latter notion, a real x is random over a transitive model of set
theory M iff x lies in no Borel set of Lebesgue measure zero coded by a
real of M. The analogous notion (of an absolutely random real) would
be a real that lies in no ordinal-definable set of measure zero. It is of
course consistent that such reals exist, as is shown, for example, by the
model of *Solovay 1970.*

Upon reflection, however, I doubt that this notion is what Gödel had
in mind. More likely, it seems to me that by "random" he meant a real
which is not ordinal definable. This seems to be what the phrase "no
mathematical law whatsoever" was intended to express.

4. Gödel's references to Russell's influence on his work.

4.1. In reading through these two lectures, I was struck by the paucity
of references to Zermelo and the subsequent developments leading to
modern set theory, especially in contrast to Gödel's frequent references
to Russell (and to a lesser extent to Hilbert).

Here are some examples:

1. In presenting the ramified hierarchy (in *1939b*, page 8), Gödel
   refers to it by saying that the "objects of which set theory speaks
   fall into a transfinite sequence of Russellian types." This seems
   to me rather foreign to Russell's notion of types. For Russell, one
   gets the impression that the types are necessarily disjoint from one
   another. (They are the domains of propositional functions rather
   than sets.) Cf. the discussion on pages 48–55 of *Whitehead and
   Russell 1925.* There Russell specifically states (page 53) that "since
   the orders of functions are only defined step by step, there can be
   no process of 'proceeding to the limit,' and functions of an infinite
   order cannot occur."

   On the contrary, if one is working in set theory where the bottom domain is a transitive set (for example, the empty set), then

the various stages of the finite hierarchy are nested (one within
the other) and nothing is more natural than to continue into the
transfinite. Thus it seems strange to refer here to Russell rather
than Zermelo.

2. The second place where the reference to Russell, rather than to
   Zermelo, seems strange is in the pride of place given to the axiom of reducibility over the axiom of separation of Zermelo. For
   example, on pages 14–15 of *1939b, Gödel writes: "To be sure,
   one must observe that the axiom of reducibility appears in different mathematical systems under different names and in different
   forms, for example, in Zermelo's system of set theory as the axiom
   of separation ..."

Now in Zermelo's sytem the axiom of separation merely records
one of the natural properties of the concept of set. In Russell's
system, where he is guided by his "vicious circle" principle to a
bewildering apparatus of types of all levels and ranks, the axiom
of reducibility is generally regarded as the grossest philosophical
expediency.[e] True, as Gödel suggests, it serves to make comprehension available (more precisely, it implies strong closure principles of the extensions of "predicative" functions), but it makes a
mockery of Russell's claim to do away with sets and to analyze
propositional functions into a series of levels and types. Therefore
it seems strange for Gödel to refer to the separation axiom of set
theory as merely a form of the reducibility axiom of Russell.

4.2. That Gödel was not ignorant of Zermelo's basic work on the axiomatization of set theory is evidenced by his reference to *Zermelo 1908*
in footnotes 1 and 11 of *1939a* and on page 2 of *1940*. Curiously, there
seem to be no other explicit references to that work in all of Gödel's
published writings on set theory. Perhaps Gödel took this kind of axiomatization for granted as the only sensible way to develop set theory.
Already in *1933o (pages 3–4) he said of the systems of axioms of Zermelo, Fraenkel and von Neumann that they resulted from the theory of
types by removing "certain superfluous restrictions" and then general-
izing to the transfinite.

4.3. In the course of preparing this introductory note, I went back
and read some of *Whitehead and Russell 1925*. It struck me that, con-

---

[e]Cf., for example, the excellent discussion of Russell's theory of types and the
axiom of reducibility in Quine's introductory note to *Russell 1908* in *van Heijenoort
1967*.

trary to Gödel, one does not find the sort of *mathematically precise* ramified hierarchy that Gödel describes and attributes to Russell, for example on page 5 of *1939b*.

There seems to me to be a vital distinction between the precise notion of Gödel and the somewhat vaguer discussions of the ramified hierarchy found in Russell's writings. Thus Gödel's well-known comments (cf. page 8 of my introductory note to *Gödel 1938*, these *Works*, Volume II) to the effect that his notion of constructibility may be regarded as a natural extension of Russell's ramified hierarchy into the transfinite now strikes this writer as much too generous.

5. A "Hilbert style" proof of the consistency of *CH*.

On page 1 of *1940a* Gödel writes: "Just recently I have succeeded in giving the proof [of the consistency of the continuum hypothesis] a new shape which makes it somewhat similar to Hilbert's program [as] presented in [*Hilbert 1926*]".

Hilbert, in his *1926*, claimed to sketch a proof of the *truth* of the continuum hypothesis. This attempt was understood by no one at the time, and was generally dismissed (cf. the introductory note to *Hilbert 1926* in *van Heijenoort 1967*, page 368). In light of the later results of Gödel and Cohen that the continuum hypothesis is both relatively consistent with and independent of the usual axioms of set theory, it is hardly surprising that Hilbert's efforts proved unconvincing. Among clearly questionable moves that Hilbert made was his statement that "The solution of the continuum problem can be carried out by means of the theory I have developed [his proof theory], and indeed the first and most important step toward this solution is precisely the demonstration that every mathematical problem can be solved."[f] In addition, Hilbert replaced the question of the cardinality of the continuum by that of the cardinality of the *definable* members of the continuum.

There are two quotes from Gödel that shed light on how he viewed the relationship between his work on the consistency of the continuum hypothesis and the earlier work of Hilbert on the truth of the continuum hypothesis. On page 19 of *1940a* Gödel writes: "The difference between this notion of recursiveness and the one that Hilbert seems to have had in mind is chiefly that I allow quantifiers to occur in the definiens." To my mind, this difference is enormous. Hilbert believed that quantifiers

---

[f]The quote is from p. 384 of Bauer-Mengelberg's translation in *van Heijenoort 1967*.

were essentially eliminable and this is tied to his belief (as indicated in the quote above) that "every mathematical problem can be solved". I find it hard to pin down the writings of Hilbert on this point, but he seems to believe we are living in a world where the halting problem is recursively solvable and every arithmetic set is recursive. Of course, it is because the precise notion of "recursive" had not yet been isolated and understood (at the time when Hilbert wrote "On the infinite") that it is difficult to be sure what Hilbert's words were intended to mean.

Later, in response to a letter of van Heijenoort, Gödel is quite clear about the differences between his work and Hilbert's: "There is a remote analogy between Hilbert's Lemma II and my Theorem 12.2 for $\alpha = 0$ [*1940*, p. 54]. There is, however, this great difference that Hilbert considers only strictly constructive definitions and, moreover, transfinite iterations of the defining operations only up to constructive ordinals, while I admit, not only quantifiers in the definitions, but also iterations of the defining operations up to *any* ordinal number, no matter whether or how it can be defined. The term 'constructible set', in my proof, is justified only in a very weak sense and, in particular, only in the sense of 'relative to ordinal numbers', where the latter are subject to no conditions of constructivity. It was exactly by viewing the situation from this highly transfinite, set-theoretic point of view that in my approach the difficulties were overcome and a *relative* finitary consistency proof was obtained. Of course there is no need in this approach for anything like Hilbert's Lemma I. Hilbert probably hoped to prove it as a special case of a general theorem to the effect that transfinite modes of inference applied to a constructively correct system of axioms lead to no inconsistency."[g] Thus Gödel's description in *1940a* of his proof there of the consistency of CH as being similar in shape to Hilbert's attempt in 1926 is not only excessively generous but downright misleading.

5.1. In the latter part of *1940a* (page 15ff.) Gödel introduces informally a class of relations on an ordinal α which he dubs the "recursive relations of order α". In terms of this notion, he sketches a new version of his proof of the consistency of the continuum hypothesis to obtain a model of type theory where the continuum hypothesis is valid. In this section, we give a precise characterization of Gödel's recursive relations of order α for suitably closed ordinals α in terms of the more usual formulation of constructibility from *Gödel 1939a* (cf. pages 8–9 of the introductory note to that paper in these *Works*, Volume II).

The reader should be warned that if α is ω, then the notion of recursiveness under discussion is true of many more relations on ω than

---

[g]The passage quoted here appears on p. 369 of *van Heijenoort 1967*.

the usual notion of a recursive relation on ω in the sense of Turing. To avoid confusion on this point, I sometimes use the term *Gödel recursive* to emphasize that it is the notion of *1940a* we are dealing with.

I shall not repeat the definition of recursiveness given on pages 15–19 of *1940a*; the reader should review it before reading this section. We give a precise description of the Gödel recursive relations of order α only for ordinals satisfying suitable closure conditions. Since (a) many ordinals satisfy these closure conditions (including any infinite cardinal in any transitive model of set theory) and (b) the notion "recursive of order α" is monotone[h] in α, this will suffice to explicate the general notion.

Let then α be an infinite ordinal which is closed under the usual Gödel pairing function for ordinals[i] as well as ordinal addition and multiplication.[j] Then the recursive relations of order α are characterized as follows:

1. Let $\alpha = \omega$. Then a relation on ω is Gödel recursive of order ω iff it appears in $L_{\omega+\omega}$.[k]

2. Let $\alpha > \omega$. Then a relation on α is Gödel recursive of order α iff it appears in $L_{\alpha\cdot\omega}$.

We only discuss the case $\alpha > \omega$ in what follows. Though there are differences in the details, the case $\alpha = \omega$ involves no new ideas.

Establishing these results is somewhat analogous to grasping which relations on ω are first-order definable in terms of $+$ and $\times$. At first glance, it is not even clear that exponentiation is so definable. However, once one shows that there is a good coding of finite sequences of integers such that the basic functions associated with it are all arithmetical, the full scope of the arithmetical relations follows readily. In a similar fashion, once we establish a coding of the finite sequences of ordinals less than α so that various associated functions and predicates are recursive of order α, the precise description of "recursive of order α" given above will also follow readily.

---

[h]I.e., any relation which is recursive of order α is recursive of order β for any $\beta > \alpha$.

[i]In the sense of *Gödel 1940*, Dfn. 7.9. Cf. these *Works*, Vol. II, p. 62.

[j]It happens that an infinite ordinal is closed under the Gödel pairing function if and only if it is closed under ordinal addition and multiplication. I shall not make use of this fact (which shows that our requirements on α are redundant).

[k]Note that in *1940a* as in *1939a*, Gödel uses the notation M_α for the set that we (following modern usage) denote by L_α.

5.2. We will be focusing attention for a while on the recursive relations of order α restricted to the ordinals less than α. To avoid needless repetition, until further notice we just say "recursive" for "recursive of order α" and "ordinal" for "ordinal less than α".

It is immediate that the collection of recursive relations contains the usual order on α, say $<$, and is closed under first-order definability. Thus it contains such notions as successor ordinal, limit ordinal and the successor function.

The next point to observe is that the graphs of ordinal addition ($\delta = \beta + \gamma$) and ordinal multiplication ($\delta = \beta \cdot \gamma$) can be defined by recursion on δ. One handles first addition and then multiplication in terms of addition. The recursions for addition and multiplication are rather straightforward. As a sample, we give the recursive definition of ordinal addition.

Note that the relation $\beta = \gamma + 1$ is first-order definable in terms of $<$. So we are free to use this special case of addition in giving the recursive definition of addition in general.

Then $\delta = \beta + \gamma$ if and only if:

1. $\gamma = 0$ and $\delta = \beta$

    or

2. $(\exists \gamma' < \gamma)(\exists \delta' < \delta)(\gamma = \gamma' + 1)$ and $\delta' = \beta + \gamma'$ and $\delta = \delta' + 1$

    or

3. γ is a limit ordinal and $(\forall \gamma' < \gamma)(\exists \delta' < \delta)(\delta' = \beta + \gamma')$ and $(\forall \delta' < \delta)(\exists \gamma' < \gamma)(\delta' = \beta + \gamma')$.

Since ordinary addition and multiplication are special cases of their ordinal analogues, one can now see that any arithmetical relation on ω is recursive.

Next let $P : \alpha^2 \to \alpha$ be the usual Gödel pairing function. One shows that the relation $\delta = P(\beta, \gamma)$ is recursive (via a recursion on δ). This ordinal pairing has the inconvenience that there are δ with $P(0, \delta) = \delta$. So let $Q(\beta, \gamma) = P(1 + \beta, 1 + \gamma)$. Then Q maps α^2 injectively into α. It is not onto since 0 is not in the range. Clearly, Q has a recursive graph (since P does). Moreover, if $Q(\beta, \gamma) = \delta$, then we have $\max(\beta, \gamma) < \delta$. We use Q to construct a coding of finite sequences of ordinals by ordinals as follows. The empty sequence is assigned the code 0. If $n \geq 1$, then the n-tuple $\langle \beta_1, \ldots, \beta_n \rangle$ is assigned the code $Q(\beta_1, \delta)$, where δ is the code assigned to the tail $\langle \beta_2, \ldots, \beta_n \rangle$.

It is then easy to verify that the following relations are recursive (via recursions on γ):

1. γ codes a finite sequence of ordinals.
2. γ codes a sequence of length n.
3. γ codes a finite sequence of ordinals, i is less than the length of this sequence and β is the ith component of γ.

5.3. The next step is to set up a ramified language for the structure L_α. In such a language, there will be infinitely many variables of each rank $\leq \alpha$. Moreover there will be a term of the language (in fact many such terms) for each element of L_α. Furthermore, it will be arranged that each term or formula of the language is naturally coded by a finite sequence of ordinals less than α.

Here is one way of implementing the details:

1. The alphabet of our ramified language will include the familiar logical symbols $\wedge$, $\neg$ and $\exists$ as well as the two binary predicates $\in$ and $=$.

2. In order to permit the formation of "abstraction terms" one also has two new symbols, $\{$ and $\}$. (These will be used as follows: If Φ is a formula having only the variable x free, then $\{x\ \Phi\}$ denotes the set of x such that Φ.)[1]

3. There is an infinite stock of *global* variables, v_i $(i \in \omega)$. (In the intended interpretation of our language, these will range over L_α.)

4. For each $\beta < \alpha$, there is an infinite set of variables of rank β, v_i^β $(i \in \omega)$. (In the intended interpretation, these variables will range over L_β.)

The next step is to define the notion of a *ranked formula* of rank β and of an *abstraction term* of rank β. These are defined by a simultaneous induction on β. We remark that all our abstraction terms are *closed*, i.e., do not contain free variables. We use Polish notation. For the definition of "ranked formula of rank β" we give only the clauses for atomic formulas and for the quantifier, and leave the rendition of the clauses involving Boolean connectives to the reader:

1. Let t_1 and t_2 be either abstraction terms of rank less than β or variables of rank β. Let R be one of the relation symbols $\in$ and $=$. Then the concatenation $R\frown t_1 \frown t_2$ is a formula of rank β.

2. Let ϕ be a ranked formula of rank β and v a ranked variable of rank β. Then the concatenation $\exists\frown v \frown \phi$ is a formula of rank β.

Now let x be a ranked variable of rank β and let Φ be a ranked formula of rank β having at most the variable x free. Then by definition the concatenation $\{\frown x \frown \Phi \frown\}$ is an abstraction term of rank β.

---

[1] There is no need for a separating symbol such as a $|$ between the x and the Φ in order to insure "unique readability".

Our heuristic remarks uniquely determine a semantics for the ranked formulas and abstraction terms so that each abstraction term of rank β denotes an element of $L_{\beta+1}$ and every element of $L_{\beta+1}$ is the denotation of some abstraction term.

There is an analogous definition of an unranked formula. (It is essentially a formula of rank α.) The only variables that appear in such an unranked formula (not within the scope of an abstraction term) are the unranked variables. Each unranked *sentence* has a truth value determined by interpreting the abstraction terms as previously indicated and interpreting the variables as ranging over L_α.

5.4. We want to associate to the objects (terms and formulas) of our ramified language finite sequences of ordinals less than α. In fact it is easy to give a "natural" bijection between the alphabet of our language and the ordinals less than α. Associate the eight non-variable symbols to the integers less than 8 in some fixed one-one way. Assign the integer $8 + i$ to v_i. Assign the ordinal $\omega \cdot (1 + \beta) + i$ to the ranked variable v_i^β.

The next step is to define the notion of truth for the limited formulas of our ramified language. This can be done by a recursion of length α. In fact, since $\omega < \alpha$ and α is closed under ordinal multiplication, we have $\omega \cdot \alpha = \alpha$. The definition of truth for limited formulas is done in $\omega \cdot \alpha$ steps, where at stage $\omega \cdot \beta + i$ one is defining the truth of Σ_i formulas of L_β. There is no trouble doing this via a recursion of the sort Gödel allows, though writing out the details is a bit tedious.

Finally, it is easy to define truth for the structure $\langle L_\alpha, \in \rangle$ (in a first order language with constants for all the closed terms of our previous ramified language) by a recursion of length ω.

5.5. One can bootstrap this procedure. By imitating the procedure just outlined, one can successively prove (by induction on $n < \omega$) that the truth predicate for $L_{\alpha \cdot n}$ in a suitable ramified language is recursive.[m] The upshot then is that every subset of α which lies in $L_{\alpha \cdot \omega}$ is Gödel recursive of order α.

The converse is also easy to establish. If the relations $R_1, \ldots, R_m$ on α lie in $L_{\alpha \cdot n+1}$ and S is defined from these by a recursion of length α, then it is not hard to show (by an induction on $\beta < \alpha$) that $S \upharpoonright \beta$ lies in $L_{\alpha \cdot n+1+\beta+1}$. And by a similar argument (once this inductive claim is verified) one shows that S lies in $L_{\alpha \cdot (n+1)+1}$. Thus the recursive relations on α are precisely those relations on α that appear in $L_{\alpha \cdot \omega}$.

---

[m] One still wishes to construe the various syntactic objects as finite sequences of ordinals less than α. This is easy to do given a suitably simple bijection of $\alpha \cdot n$ with α. For that one can take the map that assigns to the ordinal $\alpha \cdot m + \beta$ (where $m < n$ and $\beta < \alpha$) the ordinal $n \cdot \beta + m$.

5.6. Of course, once one has the connections between Gödel's notion of recursive subset of order α and the usual L_α hierarchy, it is trivial to verify the lemmas of *1940a* using the results about the latter proved in *Gödel 1939a*. Admittedly, this is rather inelegant. But the same tools (namely, a coding of the finite sequences of α such that various basic functions are Gödel recursive) can easily be employed for direct proofs of these lemmas.

There is one key technical point which Gödel handles appropriately, but on which he places perhaps too little emphasis. It is possible that ω_1 is strictly greater than the ω_1 of L. Lemma 3 is phrased in a way to deal with this difficulty, since it gives a map of the reals of L into the constructibly countable ordinals. For similar reasons, it may well happen that in the construction of the model of type theory many elements of the $(n+1)$-st layer will determine the same subset of the nth layer, and so must be identified. I think this is what Gödel is alluding to in the sentence on page 29 that reads "In order to have the pure hierarchy of

Vortrag Göttingen
(*1939b*)

Ich möchte hier über ein *Ergebnis* referieren, zu dem ich im Juni letzten Jahres gelangt bin, von dem aber bisher nur ein kurzes Résumé in den *Proceedings of the National Academy of Sciences, U.S.A.* [*Gödel 1939a*] publiziert ist. Dieses Ergebnis betrifft die *Cantorsche Kontinuumsvermutung*, daß die reellen Zahlen den Ordinalzahlen der zweiten Zahlenklasse eineindeutig zugeordnet werden können, oder anders ausgedrückt, daß $2^{\aleph_0} = \aleph_1$, und zwar in ihrer *generalisierten Form* dieser Kontinuumhypothese, $2^{\aleph_\alpha} = \aleph_{\alpha+1}$ für jede Ordinalzahl α. Diese sogenannte verallgemeinerte Kontinuumhypothese heißt nichts anderes als das folgende: Für jede beliebige Menge hat die *Menge aller ihrer Teilmengen* gerade die nächst höhere Mächtigkeit verglichen mit der Mächtigkeit der Menge selbst. Wie Sie wissen, vermutet man seit langem, daß dieser Satz gilt, aber alle Bemühungen um einen *Beweis* sind *bisher erfolglos* geblieben.

Das hatte zur Folge, daß man in seinen Ansprüchen bescheidener wurde und versuchte, *wenigstens die Widerspruchsfreiheit* dieses Satzes zu beweisen: d. h. also zu zeigen, daß dieser Satz nicht widerlegt werden kann. Dieser Beweis ist mir nun tatsächlich gelungen, und das ist das Ergebnis,

types we would have to make some restrictions, which, however, are not very essential."

5.7. The virtue of the alternate approach sketched by Gödel in this lecture is that it gives a very quick definition of the constructible sets of ordinals. For detailed proofs, and for further developments of the theory of L, the usual hierarchy of L_α's and its refinement in *Jensen 1972* to the J_α hierarchy seem much more flexible and useful.

Robert M. Solovay

The text of *1939b* was transcribed from Gabelsberger shorthand by Cheryl Dawson and edited by Cheryl Dawson, Stefan Bauer-Mengelberg and William Craig. The translation was drafted by John Dawson and revised in consultation with William Craig.

Lecture at Göttingen
(*1939b)

Here I would like to report on a *result* which I obtained in June of last year, but of which hitherto only a short resumé has been published, in the *Proceedings of the National Academy of Sciences, U.S.A.*, 1939 [*Gödel 1939a*]. This result concerns *Cantor's continuum conjecture*, that the real numbers can be put in one-to-one correspondence with the ordinal numbers of the second number class, or, otherwise expressed, that $2^{\aleph_0} = \aleph_1$, and indeed, in the *generalized form* of this continuum hypothesis, that $2^{\aleph_\alpha} = \aleph_{\alpha+1}$ for every ordinal number α. This so-called generalized continuum hypothesis means nothing other than the following: For any set whatever, the *set of all its subsets* has, compared with the cardinality of the set itself, just the next higher cardinality. As you know, it has long been conjectured that this proposition holds, but all efforts to *prove* it have *hitherto* remained *unsuccessful*.

This had the consequence that [mathematicians] became more modest in their demands, and attempted to prove *at least the consistency* of this proposition: thus, in other words, to show that this proposition cannot be refuted. I have now in fact succeeded in proving this, and that is the result

über welches ich referieren möchte. Selbstverständlich muß die *Wider-spruchsfreiheit des* zugrunde gelegten *mathematischen Formalismus* dabei vorausgesetzt werden. Das Ergebnis lautet also genauer so: Wenn der übliche mathematische Formalismus widerspruchsfrei ist, so bleibt er es, wenn man die verallgemeinerte Kontinuumhypothese als neues Axiom hinzufügt. Dabei ist nun zu bemerken, daß der *Terminus* "üblicher ma-thematischer Formalismus" nicht ganz eindeutig ist, denn es gibt bekannt-lich viele verschiedene mathematische Formalismen, z. B. den Russellschen,

2 den Hilbertschen, den Formalismus der axiomatischen | Mengenlehre und andere; aber, noch mehr, man weiß heute sogar, daß jeder mathematische Formalismus notwendig[erweise] unvollständig ist, und durch neue evidente Axiome erweitert werden kann. Es gibt also den Formalismus der Mathe-matik streng genommen gar nicht, sondern nur eine unübersehbare Folge von immer umfassenderen Formalismen. Aber der Widerspruchsfreiheits-beweis, über den ich referieren möchte, wird dadurch nicht tangiert. Er ist auf alle bisher aufgestellten Formalismen anwendbar, und man kann zeigen, daß er auch bei den erwähnten Erweiterungen durch neue evidente Axiome ungeändert bestehen bleibt, sodaß also die Widerspruchsfreiheit in einem absoluten Sinn gilt, soweit es heute überhaupt einen Sinn hat von absoluter Widerspruchsfreiheit zu sprechen.

Wie Sie wissen, ist der erste, der ein *Programm* für einen Widerspruchs-freiheitsbeweis der Kontinuumhypothese entworfen hat, *Hilbert*, in seinem Vortrag "Über das Unendliche" aus dem Jahre 1925 [*Hilbert 1926*]. Am Schluß dieses Vortrags setzt Hilbert auseinander, daß die Methode der Beweistheorie auch zu einer Lösung des Cantorschen Kontinuumproblems führt, d. h., zu einer Beantwortung der Frage, ob die reellen Zahlen (oder, was auf dasselbe hinauskommt, die Funktion[en] ganzer Zahlen) den Zah-len der zweiten Zahlenklasse eineindeutig zugeordnet werden können. Ich erinnere Sie daran, daß das Hilbertsche Beweisprogramm für diesen Satz in ganz großem Umriß in folgendem bestand: es wurde zuerst eine gewisse Klasse von Funktionen ganzer Zahlen ausgesondert, nämlich die rekursiv definierten. Über diese rekursiven Funktionen sollten dann zwei Lemmata bewiesen werden, nämlich:

3 | 1. Diese rekursiv definierten Funktionen können den Zahlen der zweiten Zahlenklasse zugeordnet werden.

2. Die anderen in der Mathematik vorkommenden Definitionen, nämlich, die durch Quantifikatoren, d. h. All- und Existentialzeichen, führen nicht über den Bereich der rekursiv definierbaren Funktionen hinaus, oder zumin-dest, man kann widerspruchsfrei annehmen, daß sie nicht über den Bereich der rekursiv definierbaren Funktionen hinausführen.

Dabei sollte der *Beweis des ersten Lemmas* über die Anzahl der rekursiven Funktionen darauf beruhen, daß man bei der rekursiven Definition von Funktionen natürlicher Zahlen *höhere Variablentypen* als solche der zweiten Zahlenklasse vermeiden kann.

on which I would like to report. Needless to say, the *consistency of the* underlying *mathematical formalism* must thereby be assumed. The result thus reads, more precisely: If the usual mathematical formalism is consistent, it remains so if the generalized continuum hypothesis is adjoined as a new axiom. Now here it is to be noted that the *phrase* "usual mathematical formalism" is not entirely clear, since, as is well known, there are many different mathematical formalisms, such as the Russellian, the Hilbertian, the formalism of axiomatic set theory, and others; besides, today in fact we know that every mathematical formalism is necessarily incomplete and can be extended by means of new evident axioms. So, strictly speaking, there is no [one] mathematical formalism at all, but rather only an unsurveyable sequence of ever more comprehensive formalisms. But the consistency proof on which I would like to report is not affected by this. It is applicable to all formalisms hitherto set up, and one can show that it holds unchanged even for the aforementioned extensions by new evident axioms, so that consistency therefore holds in an absolute sense, insofar as it makes any sense at all today to speak of absolute consistency.

As you know, the first to outline a *program* for a consistency proof of the continuum hypothesis was *Hilbert*, in his lecture "On the infinite" from the year 1925 [*Hilbert 1926*]. Toward the end of that lecture, Hilbert outlined how the method of [his] proof theory also leads to a solution of Cantor's continuum problem, that is, to an answer to the question whether the real numbers (or, what comes to the same thing, the functions of integers) can be put in one-to-one correspondence with the numbers of the second number class. Let me remind you that, in very broad outline, Hilbert's program for proving this proposition consisted in the following: first, a certain class of functions of integers was [to be] singled out, namely, those that are defined recursively. About these recursive functions, two lemmas were then to be proved; namely:

1. These recursively defined functions can be put in [one-to-one] correspondence with the numbers of the second number class.

2. The other definitions that occur in mathematics, namely, those involving quantification, that is, universal and existential quantifiers, do not lead outside the domain of the recursively definable functions, or at any rate, one can consistently assume that they do not lead outside the domain of the recursively definable functions.

Thereby the *proof of the first lemma*, about the number of recursive functions, was to rest on the fact that in the recursive definitions of functions of natural numbers one can avoid [the use of] *variables of types higher* than those of the second number class.

Der Beweis, über den ich hier referieren möchte, ist diesem Programm insofern analog, als ebenfalls eine gewisse Klasse von Funktionen oder, was auf dasselbe hinauskommt, von Mengen ausgesondert wird, welche ich konstruierbare Mengen nenne (die genaue Definition [von dem], was ich unter einer konstruierbaren Menge verstehe, verschiebe ich auf später). Von diesen konstruierbaren Mengen wird nun ebenfalls zweierlei gezeigt, nämlich:

1. Die Mächtigkeit der konstruierbaren Mengen natürlicher Zahlen ist höchstens $\aleph_1$, die Mächtigkeit der konstruierbaren Mengen von Mengen natürlicher Zahlen ist höchstens $\aleph_2$, usw., entsprechend der verallgemeinerten Kontinuumhypothese.

2. Die sonst in der Mathematik angewendeten Definitionsmethoden (insbesondere auch die nicht prädikativen) führen nicht über den Bereich der konstruierbaren Mengen hinaus.

Der Beweis des ersten Lemmas über die Anzahl der konstruierbaren Mengen beruht auch hier auf der Vermeidbarkeit allzu hoher Variablentypen in der Definition von konstruierbaren Mengen. Was das zweite Lemma betrifft, so zeigt es sich bei näherem Zusehen, daß diese Aussage nichts anderes bedeutet, als daß die *konstruierbaren Mengen ein Modell für die Mengenlehre* bilden, denn die wesentlichen Axiome der Mengenlehre besagen ja nichts anderes, als daß Mengen, die durch gewisse in der Mathematik übliche Verfahren definiert werden, existieren. | Bei dem Lemma 2 handelt es sich also darum zu beweisen, daß die konstruierbaren Mengen ein Modell für die Mengenlehre bilden und auf Grund von Lemma 1, gelingt es dann zu zeigen, daß in diesem Modell die verallgemeinerte Kontinuumhypothese gilt und damit ist dann ihre Widerspruchsfreiheit nach der Modellmethode bewiesen. Es ist dabei nur folgendes zu bemerken: *Das Modell*, um welches es sich handelt (d. h. das System der konstruierbaren Mengen) *ist durchaus nicht finit*, d. h., es gehen in seine Definition die transfiniten und imprädikativen Verfahren der Mengenlehre wesentlich ein, und das ist der Grund, weshalb man nur zu einem relativen Widerspruchsfreiheitsbeweis gelangt, nämlich, zu dem Ergebnis: Wenn die übliche Mengenlehre widerspruchsfrei ist, so auch die durch die Kontinuumhypothese erweiterte. Dieser Zusatz, "Wenn die übliche Mengenlehre widerspruchsfrei ist", ist notwendig, weil eben die übliche Mengenlehre in der Konstruktion des Modells verwendet wird. Der Sachverhalt ist hier ein *ganz ähnlicher wie* bei dem Widerspruchsfreiheitsbeweis der *nichteuklidischen Geometrie* durch das Kleinsche Modell. Dort wird ja auch nur gezeigt, wenn die euklidische Geometrie widerspruchsfrei ist, so auch die nichteuklidische, weil eben die Axiome der euklidischen Geometrie bei der Konstruktion des Kleinschen Modells verwendet werden.

Bevor ich auf die Konstruktion des Modells näher eingehe, möchte ich nur noch sagen auf *welche Weise man aus dem Lemma 1* schließen kann, daß die verallgemeinerte Kontinuumhypothese im Modell gilt. Das ist nicht

The proof on which I would like to report here is analogous to this program insofar as there is likewise singled out a certain class of functions, or, what comes to the same thing, of sets, which I call constructible sets. (The precise definition of what I understand by a constructible set I postpone until later.) Now likewise two things are shown about these constructible sets, namely:

1. The cardinality of the [[set of all]] constructible sets of natural numbers is at most $\aleph_1$, the cardinality of the [[set of all]] constructible sets of sets of natural numbers is at most $\aleph_2$, and so on, corresponding to the generalized continuum hypothesis.

2. The methods of definition otherwise applied in mathematics (in particular, the impredicative as well) do not lead outside the domain of the constructible sets.

The proof of the first lemma, about the number of constructible sets, here too rests on the avoidability in the definition of constructible sets of variables of types that are too high. As to the second lemma, closer examination reveals that this statement means nothing other than that the *constructible sets* form *a model for set theory*, inasmuch as the basic axioms of set theory assert nothing other than that sets defined by certain procedures commonly used in mathematics exist. As to Lemma 2 therefore it is a matter of proving that the constructible sets form a model for set theory, and on the basis of Lemma 1, one can then show that the generalized continuum hypothesis holds in that model, and thus its consistency is proved by the method of models. Here just the following is to be noted: *the model* in question (that is, the system of constructible sets) *is by no means finitary*; in other words, the transfinite and impredicative procedures of set theory enter into its definition in an essential way, and that is the reason why one obtains only a relative consistency proof, namely the result: If the usual set theory is consistent, then so is its extension by the continuum hypothesis. This added clause, "If the usual set theory is consistent", is necessary, because it is just the usual set theory that is used in the construction of the model. The situation here is *quite similar to* the consistency proof of *non-Euclidean geometry* by means of Klein's model. There too it is shown only that if Euclidean geometry is consistent, so is non-Euclidean, because it is just the axioms of Euclidean geometry that are used in the construction of Klein's model.

Before I discuss the construction of the model in more detail, I should still like to say just *how, from Lemma 1, one* can conclude that the generalized continuum hypothesis holds in the model. Without further argument,

ohne weiteres klar, denn Lemma 1 besagt ja nur: Die Mächtigkeit der
Menge der konstruierbaren Mengen ist höchstens $\aleph_1$; bei dem Übergang zu
einem Modell wird aber die Mächtigkeit relativiert und es kann vorkommen, daß eine Menge, die an sich die Mächtigkeit $\aleph_1$ hat, innerhalb des
Modells eine ganz andere Mächtigkeit hat. Es folgt also aus dem Lemma

5 1 nicht ohne weiteres, daß die | Kontinuumhypothese im Modell gilt. Son-
dern man muß zunächst noch einen weiteren Hilfssatz beweisen, und zwar
geht man am einfachsten folgendermaßen vor: Der Begriff der *konstru-*
ierbaren Menge kann wie jeder mengentheoretische Begriff auch auf beliebige Modelle der Mengenlehre relativiert werden; d. h., ebenso wie man
bei einem beliebigen Modell der Mengenlehre von den in diesem Modell
abzählbaren Mengen sprechen kann, so auch von den in diesem Modell
konstruierbaren Mengen. Dabei brauchen die in diesem Modell konstruierbaren Mengen nicht wirklich konstruierbar zu sein, ebenso wie die in einem
Modell abzählbaren Mengen nicht wirklich abzählbar zu sein brauchen. Der
Begriff der Konstruierbarkeit hat nun aber eine gewisse *Invarianzeigen-*
schaft, welche man als ein drittes Lemma zu 1 und 2 hinzufügen muß,
nämlich, die folgende: Der Begriff konstruierbar ändert sich nicht, wenn
er auf das Modell der konstruierbaren Mengen relativiert wird; d. h., die
im Modell der konstruierbaren Mengen konstruierbaren Mengen sind die
selben wie die konstruierbaren Mengen. Daher gilt in unserem Modell der
Satz: "*Jede Menge ist konstruierbar.*" Denn zunächst ist selbstverständlich jede Menge des Modells konstruierbar nach Definition des Modells.
Wegen der behaupteten Invarianzeigenschaft des Begriffs "konstruierbar"
ist dann aber auch jede Menge des Modells konstruierbar im Modell, und
d. h., es gilt im Modell der Satz: "Jede Menge ist konstruierbar." Daraus folgt aber, daß dieser Satz, "Jede Menge ist konstruierbar", den ich
mit A bezeichne, *widerspruchsfrei* ist, denn wir haben ein Modell gefunden, in dem er gilt. Der Satz A impliziert aber nun offenbar auf Grund
des Lemmas 1 die Kontinuumhypothese. Der Weg des Beweises ist also
der, daß die Widerspruchsfreiheit einer stärkeren Aussage A bewiesen wird
und damit ist dann selbstverständlich die Widerspruchsfreiheit aller Folgerungen von A mitbewiesen. Es ist zu bemerken, daß die Cantorsche

6 Kontinuumhypothese nicht die | einzige interessante Folgerung von A ist.
Eine andere ist, z. B., die Existenz von nichtmeßbaren projektiven Mengen.
Und außerdem hat die Widerspruchsfreiheit des Satzes A an und für sich
ein gewisses Interesse.

Ich nehme an, Sie sind nach alledem schon begierig [[zu wissen]], wie die
Definition der konstruierbaren Mengen eigentlich aussieht. Sie beruht in
der Idee auf der sogenannten *verzweigten Typentheorie*. Diese wiederum
geht zurück auf die Poincaréschen Einwände gegen die sogenannten imprädikativen Definitionen. Man spricht von einer *imprädikativen Definition*,
z. B. einer reellen Zahl, bekanntlich dann wenn eine bestimmte reelle Zahl
definiert wird durch Bezugnahme auf die Gesamtheit aller reellen Zahlen.

that is not clear, since, after all, Lemma 1 merely says [that] the [actual] cardinality of the set of constructible sets [of natural numbers] is at most $\aleph_1$; however in the passage to a model, the cardinality [of a set] is relativized, and it can happen that a set that actually has cardinality $\aleph_1$ has a quite different cardinality within the model. So, without further argument, it does not follow from Lemma 1 that the continuum hypothesis holds in the model. Rather, one must first prove yet another auxiliary proposition, and in fact it is simplest to proceed as follows: the notion of *constructible* set can, like every set-theoretic notion, be relativized to arbitrary models of set theory; that is, just as with regard to an arbitrary model of set theory one can speak of the sets that are countable in that model, so too one can speak of the sets that are constructible in that model. The sets that are constructible in that model need not by that fact actually be constructible, just as the sets that are countable in a model need not actually be countable. The concept of constructibility has, however, a certain *invariance property*, which must be adjoined to 1 and 2 as a third lemma, namely the following: The notion "constructible" does not change when it is relativized to the model of constructible sets; in other words, the sets constructible in the model [that consists] of the constructible sets are the same as the [actually] constructible sets. Hence in our model the proposition "*Every set is constructible*" holds. For, first of all, every set of the model is evidently constructible according to the definition of the model. But then, on account of the asserted invariance property of the notion "constructible", every set of the model is also constructible in the model, and that means that the proposition "Every set is constructible" holds in the model. From that it follows, however, that this proposition, "Every set is constructible", which I denote by A, is *consistent*, for we have found a model in which it holds. But now, by reason of Lemma 1, proposition A obviously implies the continuum hypothesis. So the way the proof goes is that the consistency of a stronger statement A is proved, and thereby, obviously, the consistency of all consequences of A is proved along with it. It is to be noted that Cantor's continuum hypothesis is not the only interesting consequence of A; another, for example, is the existence of nonmeasurable projective sets. And besides, the consistency of A has a certain interest in and of itself.

I assume that after all this you are quite eager [to know] what the definition of the constructible sets really looks like. In its conception it rests on the so-called *ramified type theory*. That, in turn, goes back to Poincaré's objections to the so-called impredicative definitions. As you know, one speaks of an *impredicative definition* of a real number, for example, when a specific real number is defined through reference to the totality of all real numbers.

Sprechen wir statt von reellen Zahlen von Mengen natürlicher Zahlen, was auf dasselbe hinauskommt, so wäre also eine imprädikative Definition für eine Menge a natürlicher Zahlen, z. B., die folgende: Eine natürliche Zahl $x \in a \equiv_{\mathrm{DF}} (m)\varphi(x,m)$. Solche Definitionen, die in der Mathematik häufig vorkommen, sind in einem gewissen Sinn *zirkelhaft*, denn es wird ein bestimmter Gegenstand a definiert durch eine Festsetzung $x \in a$, wenn für alle Mengen natürlicher Zahlen, insbesondere also auch a selbst eine gewisse Aussage φ zutrifft. Das Definiendum a kommt also implizit im Definiens bereits vor, was das Charakteristikum der [imprädikativen Definition ist]. Um solche Zirkel zu vermeiden, verwarfen Russell und Weyl derartige Allzeichen oder Quantifikator[en], die sich auf alle reellen Zahlen beziehen, *und Russell* stellte die sogenannte verzweigte *Typentheorie* auf. In dieser Theorie beschränkt man sich zunächst auf solche reelle Zahlen, in deren Definiens kein Quantifikator für reelle Zahlen vorkommt. Das ist eine gewisse abzählbare Menge M von reellen Zahlen. Dann führt man Quantifikatoren (d. h. All- und Existenzzeichen) ein, die sich auf die Gesamtheit der reellen Zahlen dieser Menge beziehen. Diese Quantifikatoren bezeich-

7   nen also: Für alle Reellen | aus M gilt etwas, bzw., es gibt reelle Zahlen in M für die etwas gilt. Mit Benutzung dieser Quantifikatoren kann man neue reelle Zahlen definieren, die in M noch nicht vorkommen. Sei also die Menge der so definierbaren reellen Zahlen M', dann ist $M \subset M'$. Es hindert nun nichts [daran] Quantifikatoren einzuführen, welche sich auf alle reellen Zahlen aus M' beziehen, und die mittels dieser Quantifikatoren definierbaren reellen Zahlen werden nun eine noch umfassendere Menge M'' bilden, und so kann man weiter gehen und für jede natürliche Zahl [n] eine Menge $M^{(n)}$ bilden. Soweit ist Russell in seiner verzweigten Stufentheorie gegangen. Es hindert nun aber nichts [daran] diesen Prozeß in das *Transfinite fortzusetzen*; d. h., man kann zunächst die Summe aller $M^{(n)}$ bilden und diese Summe M^{ω} nennen ($M^{\omega} = \sum_{i<\omega} M^{i}$), und dann kann man Quantifikatoren einführen, die sich auf alle Elemente dieser Menge M^{ω} beziehen. Die mittels solcher Quantifikatoren definierbaren reellen Zahlen werden dann eine noch umfassendere Menge bilden, die man konsequenterweise mit $M^{\omega+1}$ zu bezeichnen hat, und so kann man fortfahren und jede beliebige Ordinalzahl als Index erreichen.

Dieses *Konstruktionsverfahren* ist selbstverständlich nicht *auf reelle Zahlen beschränkt*, sondern kann ebenso zu der Konstruktion von Mengen reeller Zahlen, und von Funktionen reeller Zahlen usw. angewendet werden; aber noch mehr: Man kann auch z. B. die Konstruktion von reellen Zahlen und von Mengen reeller Zahlen nach diesem Verfahren miteinander kombinieren, indem man in der Definition einer reellen Zahl Quantifikatoren verwendet, die sich auf die Mengen reeller Zahlen beziehen und ähnlich in noch komplizierterer Weise. Was ich nun "konstruierbare Menge" nenne,

8   ist, kurz gesagt, das allgemeinste, was | man auf diesem Wege überhaupt bekommen kann, wobei die Quantifikatoren sich nicht nur auf Mengen

If, instead of real numbers, we speak of sets of natural numbers — which comes to the same thing — then an impredicative definition for a set a of natural numbers would be, for example, the following: a natural number $x \in a \equiv_{DF} (m)\varphi(x,m)$. Such definitions, which occur frequently in mathematics, are in a certain sense *circular*, since a particular object a is defined by means of a stipulation [of the form] "$x \in a$ if for all sets of natural numbers" (and so, in particular, also a itself) "a certain statement φ holds true". The definiendum a thus already occurs implicitly in the definiens, which [is] the characteristic feature [of an impredicative definition]. In order to avoid such circles, Russell and Weyl rejected universal symbols or quantifiers of the kind that refer to all real numbers, *and Russell* set up the so-called ramified *type theory*. In that theory, one restricts oneself, first of all, to such real numbers in whose definiens no quantifier over real numbers occurs. That is a certain countable set M of real numbers. One then introduces quantifiers (that is, universal and existential symbols) that refer to the totality of real numbers of that set. These quantifiers thus denote: For all reals from M, it is the case that ..., [or,] respectively, there are real numbers in M for which it is the case that .... With the use of these quantifiers, new real numbers can be defined that do not yet occur in M. So, if we let M' be the set of the real numbers so definable, then $M \subset M'$. Now nothing prevents the introduction of quantifiers that refer to all real numbers from M', and the real numbers definable by means of these quantifiers will now form a still more comprehensive set M'', and so one can continue and form a set $M^{(n)}$ for each natural number [n]. This is as far as Russell went in his ramified theory of levels. But nothing prevents the *continuation* of this process into the *transfinite*; that is, one can first form the sum of all the $M^{(n)}$ and call that sum $M^\omega (M^\omega = \sum_{i<\omega} M^i)$, and then one can introduce quantifiers that refer to all elements of that set M^ω. The real numbers definable by means of such quantifiers will then form a still more comprehensive set which, to be consistent, one should denote by $M^{\omega+1}$, and so one can proceed to reach any ordinal number whatever as index.

This method of *construction* is *evidently not restricted to real numbers*, but can just as well be applied to the construction of sets of real numbers and of functions of real numbers, and so on; in fact, [one can do] even more: one can also, for example, intermix the construction according to this procedure of real numbers and of sets of real numbers by using in the definition of a real number quantifiers that refer to sets of real numbers, and similarly in still more complicated ways. What I now call [a] "constructible set" is, to put it succinctly, the most general [object] that can at all be obtained in this way, where the quantifiers may refer not only to

reeller Zahlen, sondern auch auf Mengen von Mengen reeller Zahlen und so ad transfinitum beziehen können, und wobei auch die Iterationsindices der Menge M beliebige transfinite Ordinalzahlen (nicht bloß solche der zweiten Klasse) sein können. Man könnte vielleicht glauben, daß man, um diese große Allgemeinheit zu erreichen, die Definition der konstruierbaren Menge unerträglich kompliziert machen muß. Das ist aber durchaus nicht der Fall. Im Gegenteil, die *größte Allgemeinheit ergibt* hier, wie so oft, zugleich die *größte Einfachheit.*

Bevor ich die genaue Definition der konstruierbaren Mengen gebe, muß ich nur noch auf eines aufmerksam machen: schon nach dem bisher Gesagten sehen Sie, daß diese Definition keineswegs finit sein kann, vielmehr wird ja in ihr (abgesehen von Quantifikatoren) auch von dem Begriff der beliebigen transfiniten Ordinalzahlen Gebrauch gemacht in den Iterationsindices. Die folgenden Überlegungen über den Begriff der konstruierbaren Menge vollziehen sich also durchaus im Raum der transfiniten Mengenlehre und zwar zunächst der naiven (nicht axiomatisierten). Es ist dann ein Leichtes festzustellen, daß alle diese Überlegungen auch innerhalb der axiomatischen Mengenlehre durchführbar sind. Diese Feststellung ist für den Widerspruchsfreiheitsbeweis selbstverständlich wesentlich. Ich habe diese Dinge in allen Details in einer Vorlesung in Princeton im Herbst 1938 behandelt, die auch demnächst vervielfältigt erscheinen soll. Aber hier habe ich natürlich nicht die Zeit darauf einzugehen, und bitte Sie daher die Worte Ordinalzahl und Menge im Sinn der naiven Mengenlehre zu verstehen.

Ich erkläre zunächst daran, daß die Objekte, von denen [[in der]] Mengenlehre die Rede ist, in eine transfinite Folge von Russellschen Typen 9 zerfallen. Wir haben | zunächst einen Bereich von Individuen T_0: das sind die Gegenstände von 0-tem Typus, dann haben wir die Menge T_1, sämtliche Mengen von Individuen, das sind die Gegenstände ersten Typus, die Menge T_2, die Menge von Mengen von Individuen, usw. für jede natürliche Zahl n. Diese Folge kann aber auch in das Transfinite fortgesetzt werden. Eine Menge ω-ten Typus wäre, z. B., die Menge aller Mengen endlichen Typus. Allgemein vollzieht sich die Definition der T_α durch die vollständige Induktion in folgender Weise: $T_{\alpha+1} = \text{Pot}(T_\alpha)$, für Limeszahlen β ist $T_\beta = \sum_{\alpha<\beta} T_\alpha$. Mit welchem Bereich T_0 von Individuen man dabei beginnt, ist ziemlich gleichgültig und es ist daher am bequemsten *mit dem leeren Bereich von Individuen* zu beginnen; d. h., wir setzen $T_0 = \Lambda$. Dann ist $T_1 = \{\Lambda\}$, $T_2 = \{\Lambda, \{\Lambda\}\}$, T_ω ist abzählbar und ist, wie leicht zu sehen, der Bereich aller Mengen, die sich mit ausschließlicher Verwendung der Symbole Λ, $\{$, $\}$ darstellen lassen. $T_{\alpha+1} - T_\alpha$ ist die Menge der Mengen α-ten Typus.

Ich habe diese Hierarchie der Russellschen Typen deshalb so ausführlich besprochen, weil die Definition der konstruierbaren Mengen dieser Definition der T_α in einem gewissen Sinn analog ist. Wir definieren auch hier eine gewisse *transfinite Folge von Mengen* M_α, und zwar setzen wir wiederum

sets of real numbers, but also to sets of sets of real numbers and so on, *ad transfinitum*, and where the indices of iteration of the set M can also be arbitrary transfinite ordinal numbers (not merely those of the second class). One might think that in order to achieve this great generality, one would have to make the definition of constructible set unbearably complicated. But that is not at all the case. On the contrary, here the *greatest generality yields*, as it so often does, at the same time the *greatest simplicity*.

Before I give the precise definition of the constructible sets, I have only to call attention to one more thing: from what has already been said, you see that this definition can by no means be finitary; rather, in it (apart from quantifiers) use is also made of the notion of arbitrary transfinite ordinal numbers in the indices of iteration. So the following considerations about the notion of constructible set take place entirely in the ambit of transfinite set theory, and indeed initially of naive (unaxiomatized) [set theory]. It is then an easy matter to ascertain that all these considerations can also be carried out within axiomatic set theory. This ascertainment is evidently essential for the consistency proof. I dealt with these matters in full detail in a lecture in Princeton in the autumn of 1938, which should also soon appear in mimeographed form. But here, of course, I haven't the time to go into that, and therefore I ask you to understand the words "ordinal number" and "set" in the sense of naive set theory.

With regard to that [theory], let me explain first of all, that the objects of which set theory speaks fall into a transfinite sequence of Russellian types. We have, first, a domain T_0 of individuals: they are the objects of the 0th type; then we have the set T_1 of all sets of individuals, which are the objects of the first type; the set T_2, the sets of sets of individuals; and so on for each natural number n. But this sequence can also be continued into the transfinite. A set of the ωth type would, for example, be the set of all sets of finite type. In general, the definition of the T_α is carried out by complete induction in the following way: $T_{\alpha+1}$ = the power set of T_α, [and], for limit numbers β, $T_\beta = \sum_{\alpha < \beta} T_\alpha$. It is rather immaterial which domain T_0 of individuals one begins with, and it is therefore most convenient to begin *with the empty domain of individuals*; that is, we put $T_0 = \Lambda$. Then $T_1 = \{\Lambda\}$, $T_2 = \{\Lambda, \{\Lambda\}\}$, [and] T_ω is countable and, as is easy to see, is the domain of all sets that may be represented by the sole use of the symbols Λ, $\{$ and $\}$. $T_{\alpha+1} - T_\alpha$ is the set of the sets of type α.

I have discussed this hierarchy of Russellian types so thoroughly because the definition of constructible sets is analogous in a certain sense to this definition of the T_α. Here too we define a certain *transfinite sequence of*

$M_0 = \Lambda$, und $M_\beta = \sum_{\alpha<\beta} M_\alpha$ für Limeszahlen β; für ⟦eine⟧ isolierte Zahl $\alpha + 1$ war das $T_{\alpha+1}$ vorhin definiert als Menge aller Teilmengen von T_α.

10   Jetzt setzen wir $M_{\alpha+1} = $ die Menge aller | definierbaren Teilmengen von M_α und zwar definierbar mittels Quantifikatoren, die sich bloß auf die Gesamtheit der schon vorher auftretenden Mengen beziehen, d. h., Quantifikatoren (y), $(\exists y)$, welche bezeichnen: Für alle Elemente y von M_α gilt etwas, bzw., es gibt Elemente y in M_α, für welche etwas gilt. Hier sind *zwei Beispiele solcher* Definitionen angeschrieben:

$$x \in m \equiv_{\mathrm{DF}} (\exists y)\,[y \in x \,.\, y \in a]$$
$$x \in m \equiv_{\mathrm{DF}} (y)[y \in x \supset y \in a]$$

In beiden ist a irgendein gegebenes Element von M_α, m ist das zu definierende Element von $M_{\alpha+1}$. Die erste Definition besagt: x gehört zu m, wenn x mit a ein Element in M_α gemeinsam hat; die zweite: x gehört zu m, wenn alle in M_α liegenden Elemente von x zu a gehören. Man kann auf Grund der logistischen Symbolik ganz *präzis sagen*, wie die zur Konstruktion der Mengen von $M_{\alpha+1}$ zugelassene Definition aussehen muß, nämlich so: Es dürfen im Definiens nur die folgenden Symbole vorkommen:

1. Symbole, die einzelne Elemente aus M_α bezeichnen (in den Beispielen das a);

2. Das $\in$ Symbol, welches die Relation des Elementseins bezeichnet, und das Symbol der Identität;

3. Die Symbole für Verknüpfungen des Aussagenkalküls, $\sim$, $\vee$, ., $\supset$, etc.;

4. Quantifikatoren (y), $(\exists y)$, deren Variablen aber nur die Menge M_α als Variabilitätsbereich haben.

D. h., also, um es nochmals zu wiederholen: $M_{\alpha+1}$ ist per Definitionem die Menge derjenigen Teilmengen von M_α, welche definiert werden können durch Aussagefunktionen $\varphi(x)$, die sich ausschließlich aus den vier angegebenen Arten von Symbolen aufbauen. Solche Aussagefunktionen nenne ich, um einen kurzen Ausdruck zu haben, "Aussagefunktionen über M_α". Die M_α sind also durch drei hier angeschriebene Forderungen definiert:

⟦1.⟧ $M_0 = \Lambda$;

⟦2.⟧ $M_\beta = \sum_{\alpha<\beta} M_\alpha$ (wenn β Limeszahl);

⟦3.⟧ $M_{\alpha+1} = $ Menge der durch Aussagefunktionen über M_α definierbaren Teilmengen von M_α.

11 | Aus dieser Definition ergeben sich unmittelbar einige Eigenschaften der M_α, nämlich,

1. $M_\alpha \subseteq M_{\alpha+1}$, denn jede Element a von M_α kann durch eine Aussagefunktion über M_α definiert werden, nämlich durch $x \in a$, aber auch,

⟦2.⟧ $M_\alpha \in M_{\alpha+1}$, weil die Menge M_α definiert ist durch die Aussagefunktion $x = x$.

sets M_α, and in fact we again put $M_0 = \Lambda$ and $M_\beta = \sum_{\alpha<\beta} M_\alpha$ for limit numbers β. For a successor ordinal $\alpha + 1$, $T_{\alpha+1}$ was defined just now as the set of all subsets of T_α. Now we set $M_{\alpha+1} =$ the set of all definable subsets of M_α — to be precise, definable by means of quantifiers that refer only to the totality of sets already occurring previously, that is, quantifiers (y) and $(\exists y)$ that denote: For all elements y of M_α something holds, [or,] respectively, there are elements y in M_α for which something holds. Here are written *two examples of such* definitions:

$$x \in m \equiv_{\text{DF}} (\exists y)[y \in x \,.\, y \in a]$$
$$x \in m \equiv_{\text{DF}} (y)[y \in x \supset y \in a]$$

In both [of these] a is any given element of M_α [and] m is the element of $M_{\alpha+1}$ to be defined. The first definition says: x belongs to m if x has an element of M_α in common with a, [while] the second says: x belongs to m if all elements of x that lie in M_α belong to a. On the basis of the logical symbolism one can *say* quite *precisely* how the definition permitted in the construction of the sets of $M_{\alpha+1}$ must look: namely, in the definiens only the following symbols may occur:

1. Symbols that denote individual elements of M_α (the a in the examples);

2. The $\in$ symbol, which denotes the relation of being an element, and the symbol for identity;

3. The symbols for propositional connectives, $\sim, \lor, ., \supset$, and so on;

4. Quantifiers, (y) and $(\exists y)$, whose variables, however, have only the set M_α as their domain of variability.

Thus, to repeat once more: $M_{\alpha+1}$ is by definition the set of those subsets of M_α that can be defined by propositional functions $\varphi(x)$ that are built up solely from the four given kinds of symbols. In order to have a short expression [for them], I call such propositional functions "propositional functions over M_α". So the M_α are defined by three requirements, written here:

[1.] $M_0 = \Lambda$;

[2.] $M_\beta = \sum_{\alpha<\beta} M_\alpha$ (if β is a limit number);

[3.] $M_{\alpha+1} =$ the set of the subsets of M_α definable by means of propositional functions over M_α.

From this definition, a few properties of the M_α follow directly, namely:

1. $M_\alpha \subseteq M_{\alpha+1}$, since each element a of M_α can be defined by a propositional function over M_α, namely, by $x \in a$;

but also,

[2.] $M_\alpha \in M_{\alpha+1}$, because the set M_α is defined by the propositional function $x = x$.

Aus diesen beiden Beziehungen folgt $M_\alpha \subset M_{\alpha+1}$ und daher

[3.] $M_\alpha \subset M_\beta$ für $\alpha < \beta$.

Die M_α sind also monoton wachsend; ferner gilt selbstverständlich

[4.] $M_\alpha \subseteq T_\alpha$, und für endliches n gilt sogar $M_n = T_n$, weil jede endliche Teilmenge von M_α durch eine Aussagefunktion über M_α definiert ist. Daher ist auch $M_\omega = T_\omega$.

Für höhere Indices unterscheiden sich aber die M_α und T_α sehr wesentlich, denn nach dem Theorem über die Mächtigkeit der Potenzmenge hat man:

$$\overline{\overline{T_{\alpha+1}}} > \overline{\overline{T_\alpha}}.$$

Dagegen gilt:

[5.] $\overline{\overline{M_{\alpha+1}}} = \overline{\overline{M_\alpha}}$ für unendliches α.

Das ergibt sich da[durch], daß es höchstens soviele Elemente von $M_{\alpha+1}$ gibt, als es Aussagefunktionen über M_α gibt, und es gibt ebensoviele Aussagefunktionen über M_α, als es Elemente von M_α gibt. Denn jede solche Aussagefunktion ist ja eine endliche Kombination von Symbolen, die Elemente von M_α bezeichnen (hier das a) und gewissen anderen Symbolen, z. B., $\in$, $\vee$, etc., die aber nur in endlicher Anzahl vorhanden sind. Nach wohl bekannten Sätzen der Mengenlehre ist die Anzahl dieser Kombinationen ebensogroß als die Anzahl der Elemente, die kombiniert werden. Aus der eben bewiesenen Gleichung folgt weiter, daß allgemein

[6.] $\overline{\overline{M_\alpha}} = \overline{\overline{\alpha}}$ für unendliches α.

Das erste unabzählbare M_α ist also das mit dem Index ω_1 (M_{ω_1}).

Ich nenne nun eine Menge x "konstruierbar", wenn sie irgendwo in dieser Hierarchie der M_α vorkommt, d. h., | wenn es eine Ordinalzahl α gibt, sodaß $x \in M_\alpha$. Das α braucht dabei durchaus nicht eine Ordinalzahl der zweiten Zahlenklasse zu sein, sondern kann beliebig groß sein. Selbstverständlich gibt es für jede konstruierbare Menge x ein kleinstes α, sodaß $x \in M_\alpha$. Dieses α muß, wie sofort zu sehen, eine isolierte Zahl $\beta + 1$ sein, und β nenne ich dann DIE ORDNUNG VON x. M_β ist also die Menge der konstruierbaren Mengen einer Ordnung $< \beta$. Die Menge der konstruierbaren Mengen der Ordnung genau β ist $M_{\beta+1} - M_\beta$. Das sind also, kurz gesagt, die Mengen, welche bei dem β-ten Schritt neu hinzukommen. Diese bei dem β-ten Schritt neu hinzukommenden Mengen haben, wie sofort zu sehen, höchstens den Typus β, weil ja $M \subset T$. Es werden aber darunter auch Mengen von viel kleinerem Typus vorkommen. Darin drückt sich die Tatsache aus, daß es Mengen *niedrigeren Typus* gibt, die sich nur mit Hilfe von *Quantifikatoren für Mengen höheren Typus definieren lassen*.

Betrachten wir zu der näheren Erleuchtung dieses Sachverhalts die *konstruierbaren Mengen natürlicher Zahlen* oder, was auf dasselbe hinausläuft, die konstruierbaren Teilmengen von M_ω. M_ω ist ja, wie vorhin erwähnt, die Menge derjenigen Mengen, welche sich durch endlichen Kombinatio-

From these two relations it follows that $M_\alpha \subset M_{\alpha+1}$, and therefore

[3.] $M_\alpha \subset M_\beta$ for $\alpha < \beta$.

The M_α are thus monotone increasing; furthermore evidently,

[4.] $M_\alpha \subseteq T_\alpha$ holds, and for finite n, even $M_n = T_n$, because every finite subset of M_α is defined by a propositional function over M_α. Therefore $M_\omega = T_\omega$ also.

But for higher indices the M_α and T_α differ quite fundamentally, since, according to the theorem about the cardinality of the power set, one has

$$\overline{\overline{T_{\alpha+1}}} > \overline{\overline{T_\alpha}}.$$

On the contrary, one has

[5.] $\overline{\overline{M_{\alpha+1}}} = \overline{\overline{M_\alpha}}$ for infinite α.

That results from the fact that there are at most as many elements of $M_{\alpha+1}$ as there are propositional functions over M_α, and there are just as many propositional functions over M_α as there are elements of M_α. For, every such propositional function is of course a finite combination of symbols that denote elements of M_α (here the a) and certain other symbols—for example, $\in, \vee$, and so on—of which, however, only a finite number occur. According to well-known theorems of set theory, the number of these combinations is just as large as the number of elements that are combined. From the equation just proved, it follows further that, generally,

[6.] $\overline{\overline{M_\alpha}} = \overline{\overline{\alpha}}$ for infinite α.

The first uncountable M_α is therefore that with the index $\omega_1 (M_{\omega_1})$.

I now call a set x constructible if it occurs somewhere in this hierarchy of the M_α, that is, if there is an ordinal number α such that $x \in M_\alpha$. The α here need not at all be an ordinal number of the second class, but can be arbitrarily large. Obviously, for each constructible set x there is a smallest α such that $x \in M_\alpha$. As can be seen at once, this α must be a successor ordinal $\beta + 1$, and I then call β THE ORDER OF x. So M_β is the set of the constructible sets of order $< \beta$. The set of the constructible sets of order exactly β is $M_{\beta+1} - M_\beta$. These are then, in short, the sets which are newly adjoined at the βth step. As can be seen at once, these sets newly adjoined at the βth step have at most the type β, because $M \subset T$. But among them there will also occur sets of much lower type. This expresses the fact that there are sets *of lower type* that *can* only *be defined* with the help of *quantifiers for sets of higher type.*

For more detailed elucidation of this state of affairs, let us consider the *constructible sets of natural numbers*, or, what comes to the same thing, the constructible subsets of M_ω. M_ω is, after all, as was mentioned a little earlier, the set of those sets that may be expressed by means of

nen der Symbole Λ, $\{$, $\}$ ausdrücken lassen, und diese können offenbar
in sehr einfacher Weise auf die natürlichen Zahlen eineindeutig abgebildet
werden. Wir fragen also: Was sind die konstruierbaren Teilmengen von
M_ω? Zunächst sind offenbar sämtliche Elemente von $M_{\omega+1}$ solche konstruierbare Teilmengen von M_ω, aber das sind bei weitem nicht alle, denn
die Elemente von $M_{\omega+2}$ sind zwar per Definitionem davon Teilmengen von
$M_{\omega+1}$. Aber $M_\omega \subseteq M_{\omega+1}$ und daher werden unter diesen Teilmengen von
13 | $M_{\omega+1}$ insbesondere auch solche vorkommen, welche nur Elemente aus
M_ω enthalten, d. h., also neue Teilmengen von M_ω, und ebenso kann es a
priori für beliebig große Indices α vorkommen, daß in $M_{\alpha+1}$ gewisse neue
Teilmengen von M_ω auftreten, d. h., kurz gesagt, a priori ist es möglich,
daß es konstruierbare Teilmengen von M_ω von beliebig höherer Ordnung
gibt. Hier gilt nun aber ein FUNDAMENTALES THEOREM, welches der Kern
des ganzen Widerspruchsfreiheitsbeweises für die Kontinuumhypothese ist.
Es besagt, daß ω_1 eine Schranke für die Ordnungen der konstruierbaren
Teilmengen von M_ω ist, oder präzis formuliert:

DIE ORDNUNG JEDER KONSTRUIERBAREN TEILMENGE VON
M_ω IST EINE ZAHL DER ZWEITEN KLASSE, d. h. $< \omega_1$.

Das bedeutet also: Wenn die Konstruktion der M_α über M_{ω_1} hinaus fortgesetzt wird, so treten keine neuen Teilmengen von M_ω mehr auf. Man
kann vielleicht sagen, daß erst durch diesen Satz die Definition der konstruierbaren Mengen natürlicher Zahlen einen präzisen Sinn erhält. Denn
man kann jetzt sagen: Eine Teilmenge x von M_ω ist konstruierbar, wenn
es eine Zahl α der zweiten Zahlenklasse gibt, sodaß $x \in M_\alpha$.

Aus dem fundamentalen Theorem folgt sofort, daß es höchstens $\aleph_1$ konstruierbare Mengen natürlicher Zahlen gibt (das war aber das Lemma 1).
Daß es auch *mindestens* $\aleph_1$ konstruierbare Mengen natürlicher Zahlen gibt,
kann ich nicht beweisen. Was ich zeigen kann, ist bloß, daß man das Gegenteil nicht beweisen kann. Denn es ist ja sogar die Annahme, daß alle Mengen natürlicher Zahlen konstruierbar sind, widerspruchsfrei, um so mehr die
Annahme, daß es mindestens $\aleph_1$ konstruierbare Mengen natürlicher Zah-
14 len gibt. Die Frage, ob es $\aleph_1$ oder $\aleph_0$ | konstruierbare Mengen natürlicher
Zahlen gibt, ist allem Anschein nach ein in der Mengenlehre unentscheid-
barer Satz ebenso wie auch der Satz A. Bewiesen ist allerdings bisher nur
die eine Hälfte, nämlich, daß A widerspruchsfrei ist, nicht aber daß auch die
Negation von A widerspruchsfrei ist. Das eben formulierte *fundamentale
Theorem läßt* sich auch *für höhere Typen* beweisen; d. h., es gilt allgemein
für jede Ordinalzahl α: Jede konstruierbare Teilmenge von M_{ω_α} hat eine
Ordnung $< \omega_{\alpha+1}$. Daraus folgt dann sofort, daß es höchstens $\aleph_2$ konstruierbare Mengen von Mengen natürlicher Zahlen gibt, usw. wie es die
verallgemeinerte Kontinuumhypothese verlangt.

Ich möchte noch erwähnen, daß das fundamentale Theorem den berich-
tigten Kern des sogenannten *Russellschen Reduzibilitätsaxioms* darstellt.

finite combinations of the symbols Λ, { and }, and these can obviously be mapped in a very simple way one-to-one onto the natural numbers. So we ask: What are the constructible subsets of M_ω? First of all, all elements of $M_{\omega+1}$ are obviously such constructible subsets of M_ω, but they constitute by no means all of them, since the elements of $M_{\omega+2}$ are in fact, by its definition, subsets of $M_{\omega+1}$. But $M_\omega \subseteq M_{\omega+1}$, and therefore among these subsets of $M_{\omega+1}$ there will, in particular, also occur such that contain only elements from M_ω — in other words, new subsets of M_ω — and similarly, a priori, it can happen that for arbitrarily large indices α, certain new subsets of M_ω appear in $M_{\alpha+1}$; that is, in short, a priori it is possible that there are constructible subsets of M_ω of arbitrarily higher order. But here a FUNDAMENTAL THEOREM holds, which is the crux of the entire consistency proof for the continuum hypothesis. It says that ω_1 is a bound for the orders of the constructible subsets of M_ω, or, precisely formulated:

THE ORDER OF EVERY CONSTRUCTIBLE SUBSET OF M_ω
IS AN ORDINAL OF THE SECOND CLASS, THAT IS, $< \omega_1$.

That means, therefore: When the construction of the M_α is continued beyond M_{ω_1}, no more new subsets of M_ω appear. One can, perhaps, say that only through this theorem does the definition of the constructible sets of natural numbers acquire a precise sense. For one can now say: A subset x of M_ω is constructible if there is a number α of the second number class such that $x \in M_\alpha$.

From the fundamental theorem it follows at once that there are at most $\aleph_1$ constructible sets of natural numbers (and that was *Lemma 1*). I cannot prove that there are also *at least* $\aleph_1$ constructible sets of natural numbers. What I can show is merely that one cannot prove the contrary. For since, after all, the assumption that all sets of natural numbers are constructible is consistent, so, all the more, is the assumption that there are at least $\aleph_1$ constructible sets of natural numbers. The question whether there are $\aleph_1$ or $\aleph_0$ constructible sets of natural numbers is to all appearances a proposition that is undecidable in set theory, as is, likewise, also the proposition A. To be sure, up to now only the one half has been proved, namely, that A is consistent, but not that the negation of A is also consistent. The *fundamental theorem* just formulated *may* also *be* proved *for higher types*; that is, it holds in general for every ordinal number α: Every constructible subset of M_{ω_α} has order $< \omega_{\alpha+1}$. From that it then follows at once that there are at most $\aleph_2$ constructible sets of sets of natural numbers, and so on, as the generalized continuum hypothesis requires.

I should also like to mention that the fundamental theorem constitutes the corrected core of the so-called *Russellian axiom of reducibility*. After all,

Russell hatte ja, wie vorhin erwähnt, eine [[der von]] den M_α ähnliche Konstruktion schon vorher angegeben, sich dabei aber auf endliche Ordnungen beschränkt. Sein Reduzibilitätsaxiom besagt dann, daß die Ordnungen der Mengen jedes Typus durch eine feste endliche Zahl beschränkt sind. Er war selbstverständlich weit davon entfernt das beweisen zu können. Es zeigt sich nun aber, daß wenn man die Konstruktion der Ordnungen ins Transfinite fortsetzt, gewisse transfinite Schranken tatsächlich beweisbar werden. Das ist der Inhalt des fundamentalen Theorems.

Dieser Charakter des fundamentalen Theorems als [[der]] eines Reduzibilitätsaxioms ist auch der Grund dafür, daß *die Axiome der klassischen* Mathematik für das Modell der konstruierbaren Mengen gelten. Denn es sind ja, wie Russell gezeigt hat, das Reduzibilitäts-, das Unendlichkeits-, und das Auswahlaxiom die einzigen Axiome der klassischen Mathematik, die nicht tautologischen Charakter haben. Allerdings muß man | beachten, daß das Reduzibilitätsaxiom in verschiedenen mathematischen Systemen unter verschiedenen Namen und in verschiedenen Formen auftritt, z. B., im Zermeloschen System der Mengenlehre als Aussonderungsaxiom, in den Hilbertschen Systemen in Form von Rekursionsaxiomen usw. Das Reduzibilitätsaxiom gilt nun für die konstruierbaren Mengen auf Grund des fundamentalen Theorems, das Unendlichkeitsaxiom gilt in trivialer Weise, und das Auswahlaxiom gilt ebenfalls, da es eine leichte Folgerung der verallgemeinerten Kontinuumhypothese ist. *So ergibt sich also der Beweis* für *Lemma 2. Lemma 1* ist, wie schon erwähnt, eine unmittelbare Folge des fundamentalen Theorems. Was schließlich *Lemma 3* betrifft, so will ich auf den Beweis nicht näher eingehen, sondern möchte nur erwähnen, daß ein wesentlicher Punkt dabei ist, daß der Begriff der Ordinalzahl absolut ist: d. h., Ordinalzahl im Modell der konstruierbaren Mengen bedeutet dasselbe wie Ordinalzahl selbst. Die Durchführung des Beweises ist etwas umständlich, aber im Grunde handelt es sich um eine ziemlich triviale Sache. Ich möchte daher die restliche Zeit lieber zu einer Skizze des Beweises für das fundamentale Theorem verwenden.

Zunächst einige Vorbereitungen: Wenn m eine konstruierbare Menge der Ordnung α ist, so ist m definiert durch irgend eine Aussagefunktion über M_α, die ich mit $\varphi_\alpha(x)$ bezeichne. Selbstverständlich wird es im Allgemeinen mehrere φ_α geben, durch die m definiert werden kann. Aber es macht keine Schwierigkeit jeder konstruierbaren Menge in eindeutiger Weise *eine bestimmte Aussagefunktion* über dem entsprechenden M_α zuzuordnen, *durch welche m definiert werden kann.*

| Diese eindeutig bestimmte Aussagefunktion nenne ich *die definierende Funktion für m.* Die eindeutige Zuordnung gelingt, weil es offenbar leicht möglich ist eine Wohlordnung sämtlicher konstruierbaren Mengen nach steigender Ordnung zu definieren, und daher auch eine Wohlordnung sämtlicher φ_α. Eine Aussagefunktion φ_α setzt sich nach der Definition aus den oben angegebenen vier Gattungen von Symbolen zusammen. Die Sym-

as was mentioned a while ago, Russell had previously given a construction similar to [that of] the M_α, but had restricted himself to finite orders. His axiom of reducibility then says that the orders of the sets of every type are bounded by a fixed finite number. He was evidently far from being able to prove that. But it now turns out that if the construction of the orders is continued into the transfinite, [the existence of] certain transfinite bounds actually become[s] provable. That is the content of the fundamental theorem.

This character of the fundamental theorem as an axiom of reducibility is also the reason why *the axioms of classical* mathematics hold for the model of the constructible sets. For after all, as Russell showed, the axioms of reducibility, infinity and choice are the only axioms of classical mathematics that do not have [a] tautological character. To be sure, one must observe that the axiom of reducibility appears in different mathematical systems under different names and in different forms, for example, in Zermelo's system of set theory as the axiom of separation, in Hilbert's systems in the form of recursion axioms, and so on. Now the axiom of reducibility holds for the constructible sets on the basis of the fundamental theorem, the axiom of infinity holds in a trivial way, and the axiom of choice likewise holds, since it is an easy consequence of the generalized continuum hypothesis. *This is how the proof of Lemma 2 is obtained. Lemma 1* is, as was already mentioned, a direct consequence of the fundamental theorem. Finally, as concerns *Lemma 3*, I do not want to give details of the proof; rather, I should only like to mention that an essential point in it is that the notion of ordinal number is absolute: that is, ordinal number in the model of the constructible sets means the same as ordinal number itself. To carry out the proof is rather tedious, but basically it is a fairly trivial matter. Therefore I would rather use the remaining time for a sketch of the proof of the fundamental theorem.

First, a few preliminaries: if m is a constructible set of order α, then m is defined by some propositional function over M_α, which I denote by $\varphi_\alpha(x)$. In general, there will of course be several $\varphi_\alpha(x)$ by which m can be defined. But there is no difficulty in assigning to each constructible set, in an unambiguous way, *a definite propositional function* over the corresponding M_α *by which m can be defined.*

I call this unambiguously determined propositional function *the defining function for m.* The unambiguous assignment succeeds because it is obviously easy to define a well-ordering of all the constructible sets according to increasing order, and therefore also a well-ordering of all the φ_α. By definition, a propositional function φ_α consists of the four kinds of symbols mentioned above. The symbols of the first kind, which denote

bole der Gattung 1, welche einzelne Elemente von M_α bezeichnen, nenne ich insbesondere *die Konstanten von* φ_α. Wir nehmen von vornherein an, daß jede φ_α, von dem wir sprechen, auf NORMALFORM gebracht sei, d. h., daß sämtliche Quantifikatoren zu Beginn des Ausdrucks stehen und zwar soll der erste Quantifikator immer ein All-, der letzte ein Existenzzeichen sein. Das ist in trivialer Weise erreichbar. Die einfachste Gestalt, die eine solche Normalform haben kann, sieht also aus: $(u)(\exists v)\mathfrak{R}(u,v,x)$. Dabei ist $\mathfrak{R}$ ein Ausdruck ohne Quantifikatoren. Wenn also x ein Element von M_α ist, so bedeutet in diesem Fall die Aussage $\varphi_\alpha(x)$ soviel wie, zu jedem Element u aus M_α, gibt es ein Element v aus M_α, sodaß $\mathfrak{R}(u,v,x)$, und das wäre offenbar äquivalent mit der Aussage: *Es gibt eine Funktion f*, welche jedem Element u von M_α ein Element $f(u)$ zuordnet, sodaß $\mathfrak{R}(u, f(u), x)$ gilt. Dabei ist also das f eine Funktion, die in [[dem]] ganzen M_α definiert ist und deren Werte ebenfalls Elemente von M_α sind. Eine solche Funk-
17   tion will ich kurz eine *Funktion in* M_α nennen. Die Aussage $\varphi_\alpha(x)$ | ist also in diesem Fall äquivalent mit der Aussage: Es gibt eine Funktion in M_α, welche eine gewisse Bedingung $\mathfrak{R}$ erfüllt. Diese Überlegung läßt sich leicht auch auf Ausdrücke mit komplizierter Normalform verallgemeinern. Immer ist die Aussage $\varphi_\alpha(x)$ äquivalent mit einer Aussage der Form: Es gibt Funktionen in M_α, welche eine gewisse Bedingung $\mathfrak{R}$ erfüllen. (Im allgemeinen werden es Funktionen mit mehreren Argumenten sein.) Eine Funktion in M_α nun, welche die erwähnte Bedingung $\mathfrak{R}$ erfüllt, nenne ich, um kürzer sprechen zu können, SKOLEMFUNKTION FÜR φ_α UND x, weil Skolem diese Funktion zuerst zum Beweis des nach ihm benannten logischen Satzes verwendet hat. Mit Verwendung dieser Terminologie kann man also sagen: Die Aussage $\varphi_\alpha(x)$ ist äquivalent damit, daß es Skolemfunktionen für φ_α und x gibt. Selbstverständlich sind diese Skolemfunktionen durch φ_α und x durchaus nicht eindeutig bestimmt, sondern es wird viele verschiedene geben, aber auf Grund der Wohlordbarkeit der konstruierbaren Mengen kann man jedem φ_α, x eindeutig [[eine]] bestimmte Skolemfunktion zuordnen, welche ich die *ausgezeichnete Skolemfunktion für* φ_α, x nenne. Schließlich brauche ich auch noch den Begriff DES ISOMORPHISMUS zwischen Mengen beliebigen Typus: Wenn a, a' irgend zwei Mengen sind, so nenne ich "a isomorph mit a'", wenn es eine eineindeutige Abbildung der Elemente von a auf die Elemente von a' gibt, welche die Bedingung erfüllt $x' \in y' \equiv x \in y$, also eine Abbildung, welche die $\in$ Relation invariant läßt.

Jetzt bin ich soweit, daß ich mit der Skizze des Beweises für das funda-
18   mentale Theorem beginnen kann, und zwar beschränke ich | mich auf den Fall $\alpha = 1$ (hier in der ersten Zeile) [[$m \subseteq M_\omega \supset m \in M_{\omega_1}$]]. Der allgemeine Fall kann wörtlich ebenso bewiesen werden. Sei also eine konstruierbare Teilmenge m von M_ω gegeben. Es ist zu beweisen, daß die Ordnung von $m < \omega_1$, also abzählbar ist. *Die Idee des Beweises ist die* folgende: Die Menge m ist nach Annahme konstruierbar, d. h., sie tritt irgendwo in der transfiniten Folge der M_β auf, wenn auch vielleicht sehr spät; d. h., die

individual elements of M_α, I call, in particular, *the constants of* φ_α. We assume at the outset that every φ_α of which we speak has been brought into NORMAL FORM, that is, that all the quantifiers stand at the front of the expression, and, in fact, the first quantifier is always supposed to be a universal one and the last an existential one. That can be achieved in a trivial way. So the simplest form that such a normal form can have looks like $(u)(\exists v)\mathfrak{R}(u, v, x)$, where $\mathfrak{R}$ is an expression without quantifiers. Thus if x is an element of M_α, the proposition means, in this case, just that for each element u from M_α there is an element v from M_α such that $\mathfrak{R}(u, v, x)$, and that would obviously be equivalent with the proposition: *There is a function* f which assigns to each element u of M_α an element $f(u)$ such that $\mathfrak{R}(u, f(u), x)$ holds. The f here is thus a function that is defined on all of M_α and whose values are likewise elements of M_α. Such a function I will, for short, call a *function in* M_α. So, in this case, the proposition $\varphi_\alpha(x)$ is equivalent with the proposition: There is a function in M_α which satisfies a certain condition $\mathfrak{R}$. This idea may also easily be generalized to expressions with more complicated normal form. The proposition $\varphi_\alpha(x)$ is always equivalent with a proposition of the form: There are functions in M_α which satisfy a certain condition $\mathfrak{R}$. (In general, they will be functions with several arguments.) Now, to be able to speak more succinctly, I call a function in M_α that satisfies the aforementioned condition a SKOLEM FUNCTION FOR φ_α AND x, because Skolem first used this function in the proof of the theorem of logic named after him. Using this terminology one can thus say: The proposition $\varphi_\alpha(x)$ is equivalent with the proposition that there are Skolem functions for φ_α and x. Evidently, these Skolem functions are not at all unambiguously determined by φ_α and x, but rather, there will be many different ones; but on the basis of the well-orderability of the constructible sets, one can unambiguously assign to each φ_α and x a definite Skolem function that I call the *designated Skolem function for* φ_α *and* x. Finally, I still need also the notion OF ISOMORPHISM between sets of arbitrary type: If a and a' are any two sets, I say a is "isomorphic with" a' if there is a one-to-one mapping of the elements of a onto the elements of a' that satisfies the condition $x' \in y' \equiv x \in y$; in other words, a mapping that leaves the $\in$-relation invariant.

I am now far enough along that I can begin the sketch of the proof of the fundamental theorem; to be precise, I am restricting myself to the case $\alpha = 1$ ([written] here on the first line [of the blackboard]: $[m \subseteq M_\omega \supset m \in M_{\omega_1}]$). The general case can be proved in the same way, word for word. So let a constructible subset m of M_ω be given. It is to be proved that the order of m [is] $< \omega_1$, [and] so is countable. *The idea of the proof is the* following: the set m is by assumption constructible, that is, it appears somewhere in the transfinite sequence of the M_β, even if perhaps very

Konstruktion des m beruht möglicherweise auf einer sehr großen Anzahl
von Vorgängern, welche zuerst definiert werden müssen, bevor m definiert
werden kann. Es gelingt nun aber zu zeigen, daß m auch auf einem kürzeren
Weg erreicht werden kann, in dem man unter den Vorgängern von m eine
gewisse Teilmenge aussondert, welche zu der Konstruktion von m bereits
hinreicht. Diese Teilmenge (ich bezeichne sie mit K) hat die folgenden
Eigenschaften:

1. K ist abzählbar.

2. K ist isomorph in dem vorher definierten Sinn mit einem Anfangsstück
der konstruierbaren Mengen, d. h., isomorph mit einem M_η. [Jede M_η ist
ja zugleich ein Anfangsstück der nach wachsender Ordnung angeordneten
konstruierbaren Mengen.]

Schließlich,

3. $m \in K$, $M_\omega \subseteq K$.

Man sieht leicht, daß aus der Existenz eines K mit den drei genannten
Eigenschaften sofort folgt, daß m eine Ordnung $< \omega_1$ hat, denn: m gehört
ja gemäß 3 zu K, wird also durch den Isomorphismus zwischen K und
M_η auf ein gewisses Element m' von M_η ab[[ge]]bildet. Dieses m' hat nun
sicher eine Ordnung $< \omega_1$; alle Elemente von M_η haben nämlich eine
Ordnung $< \omega_1$, denn M_η | ist abzählbar (weil es mit der abzählbaren
Klasse K isomorph ist). Aus der Abzählbarkeit von M_η folgt aber, daß η
eine Ordinalzahl der zweiten Klasse ist, daraus daher, daß jedes Element
eine Ordnung $< \omega$ hat. m' hat also eine Ordnung $< \omega_1$. Anderseits
aber folgt, daß m und m' dieselben Elemente von M_ω enthalten, d. h.,
$x \in m' \equiv x \in m$ für alle $x \in M_\omega$. Durch den Isomorphismus wird nämlich,
wie sofort zu sehen, [[das]] ganze M_ω elementweise auf sich abgebildet,
d. h., wir haben $x = x'$ für $x \in M_\omega$, und daraus folgt die zu beweisende
Äquivalenz, denn es gilt ja jedenfalls $x \in m \equiv x' \in m'$, weil es sich um
einen Isomorphismus bezüglich der $\in$ Relation handelt. Die eben bewiesene
Äquivalenz besagt aber nichts anderes, als daß die Teilmenge m von M_ω
bereits durch eine Menge m' einer Ordnung $< \omega_1$ definiert werden kann,
was zu beweisen war.

Es kommt also jetzt nur noch darauf an die Existenz einer den obigen drei
Bedingungen genügenden Menge K zu beweisen. Es ist am bequemsten K
simultan mit einer gewissen Menge von Ordinalzahlen O und einer gewissen
Menge F von Skolemfunktionen durch gewisse Abgeschlossenheitseigen-
schaften zu definieren, wie folgt:

1. $m \in K, M_\omega \subseteq K$;

2. $x \in K \supset$ (Ordnung von x) $\in O$;

3. $x \in K \supset$ (Konstante der Definition von x) $\in K$;

4. $a \in O$. (Konstante von φ_α) $\in K \supset$ (das durch φ_α definierte x) $\in K$;

5. $a \in O$. (Konstante von φ_α) $\in K$. $y \in K$. $\supset$ (ausgezeichnete Skolem-
funktion von φ_α, y) $\in F$, falls sie existiert;

6. $h \in F$. $x_1, \ldots, x_n \in K \supset h(x_1, \ldots, x_n) \in K$;

late; that is, the construction of m possibly rests on a very large number of predecessors that must first be defined before m can be defined. But now one can show that m can also be reached in a shorter way, in which, from among the predecessors of m, one separates out a certain subset that already suffices for the construction of m. This subset (I denote it by K) has the following properties:

1. K is countable.

2. K is isomorphic, in the sense defined earlier, with an initial segment of the constructible sets, that is, isomorphic with an M_η. (Every M_η is, after all, at the same time an initial segment of the constructible sets, arranged according to increasing order.)

Finally,

3. $m \in K$ and $M_\omega \subseteq K$.

One easily sees that from the existence of a K with the three stated properties, it follows at once that m has an order $< \omega_1$, for: m belongs to K according to 3, and so, by the isomorphism between K and M_η, is mapped to a certain element m' of M_η. Now this m' certainly has an order $< \omega_1$; namely, all elements of M_η have an order $< \omega_1$, since M_η is countable (because it is isomorphic to the countable class K). But, from the countability of M_η, it follows that η is an ordinal number of the second class, from which, therefore, [it follows further] that every element has an order $< \omega_1$. So m' has an order $< \omega_1$. But, on the other hand, it follows that m and m' contain the same elements of M_ω; that is, $x \in m' \equiv x \in m$, for all $x \in M_\omega$. Namely, as can be seen at once, by the isomorphism all of M_ω is mapped elementwise onto itself, that is, we have $x = x'$ for $x \in M_\omega$, and from that the equivalence to be proved follows, since in any case $x \in m \equiv x' \in m'$, because one is dealing with an isomorphism with respect to the $\in$-relation. But the equivalence just proved says nothing other than that the subset m of M_ω can already be defined by means of a set m' of an order $< \omega_1$, which is [what was] to be proved.

So it comes down now only to proving the existence of a set K satisfying the three conditions above. It is most convenient to define K simultaneously with a certain set O of ordinal numbers and a certain set F of Skolem functions, by means of certain closure properties, as follows:

1. $m \in K$ and $M_\omega \subseteq K$;

2. $x \in K \supset$ (the order of x) $\in O$;

3. $x \in K \supset$ (the constants in the definition of x) $\in K$;

4. $\alpha \in O$. (the constants of φ_α) $\in K \supset$ (the x defined by φ_α) $\in K$;

5. $\alpha \in O$. (the constants of φ_α) $\in K$. $y \in K$. $\supset$ (the designated Skolem functions of φ_α and y) $\in F$, if it exists;

6. $h \in F$. $x_1, \ldots, x_n \in K \supset h(x_1, \ldots, x_n) \in K$;

7. $x, y \in K \, . \, x - y \neq \Lambda \supset$ (erstes Element von $x - y) \in K$;

8. K, O, F sind die kleinsten Mengen, die 1–7 erfüllen.

Wenn wir die Menge $K + O + F$ bilden, so besagt also jede der Postulate 2–7, daß diese Menge abgeschlossen ist gegen eine gewisse Operation. Wenn man die Postulate 1, 8 mitberücksichtigt, so besagen also alle Forderungen 1–8 zusammen, daß $K + O + F$ die abgeschlossene Hülle der Menge $M_\omega + \{m\}$ | gegen die durch 2–7 gegebenen Operationen ist. Diese Operationen *sind aber, wie leicht zu sehen,* alle höchstens abzählbar vielwertig. 2, 6 und 7 sind eindeutig, denn die Ordnung von x ist durch x eindeutig bestimmt. Ebenso ist das erste Element von $x - y$ durch x und y eindeutig bestimmt. Die Operation 3 ist endlich vielwertig, denn in der Definition einer Menge kommen höchstens endlich viele Konstanten vor. Endlich, die Operationen 4 und 5 sind abzählbar vielwertig, denn mit gegebenen Konstanten kann man abzählbar viele Aussagefunktionen über M_α bilden. Nun ist es aber eine leicht zu beweisen[de] Tatsache, daß, wenn man von einer unendlichen Menge die abgeschlossene Hülle gegen endlich viele höchstens abzählbar vielwertige Funktionen bildet, die Mächtigkeit dieser Hülle nicht größer ist als die Mächtigkeit der ursprünglich gegebenen Menge. Daraus folgt also, daß $K + O + F$ abzählbar ist, und daher ist auch K abzählbar. Die Eigenschaft 1 von K ist also bereits bewiesen. Die Eigenschaft 3 gilt nach Forderung 1 der Definition von K. Es bleibt also nur noch übrig die Eigenschaft 2 zu beweisen.

Dazu hat man also einen Isomorphismus zwischen K und einem gewissen M_η zu definieren. Das geschieht durch vollständige Induktion nach der Ordnung der Elemente von K. *Die Menge sämtlicher Ordnungen, die bei Elementen von K überhaupt vorkommen,* ist gemäß Bedingung 2 die Menge O. O ist also eine gewisse Menge von Ordinalzahlen, aber im allgemeinen durchaus | nicht etwa ein Anfangsabschnitt der Reihe der Ordinalzahlen. Denn es kann durchaus vorkommen, daß gewisse Ordinalzahlen unter den Ordnungen der Elemente von K fehlen, und andere, grössere wieder auftreten. Jedenfalls aber können die Elemente von O der Größe nach mittels eines Anfangsabschnitts der Reihe der Ordinalzahlen durchnumeriert werden. Für diese Numerierung werden alle Ordinalzahlen kleiner als eine *gewisse Ordinalzahl η* aufgebraucht werden (und dieses η gibt zugleich den Index des M, mit welchem K sich als Isomorph erweisen wird). Wenn β ein beliebiges Element von O ist, so bezeichne ich mit $\overline{\beta}$ seine Nummer in dieser Durchnumerierung. Der Isomorphismus zwischen K und M_η wird nun so beschaffen sein, daß die Elemente der Ordnung β von K genau auf sämtliche Elemente der Ordnung $\overline{\beta}$ von M_η abgebildet werden. Bei dieser Abbildung wird also die Ordnung der Elemente gewissermaßen ziehharmonika-artig verkleinert. *Die Definition der Abbildung* vollzieht sich durch Induktion nach β, d. h., wir nehmen an, daß für die Elemente von K von der Ordnung $< \beta$ eine Isomorphabbildung auf die Elemente von M_η einer Ordnung $< \overline{\beta}$ bereits definiert ist, und zeigen, wie sich diese Ab-

7. $x, y \in K . x - y \neq \Lambda \supset$ (the first element of $x - y$) $\in K$;

8. K, O and F are the smallest sets that satisfy 1–7.

If we form the set $K + O + F$, then each of the postulates 2–7 says that this set is closed under a certain operation. If one also takes postulates 1 and 8 into account, then all the stipulations 1–8 together mean that $K + O + F$ is the closed hull of the set $M_\omega + \{m\}$ under the operations given by 2–7. *However, as is easily seen,* these operations *are* all at most countably multi-valued. [Operations] *2, 6 and 7* are single-valued, since the order of x is uniquely determined by x. Likewise, the first element of $x - y$ is uniquely determined by x and y. Operation 3 is finitely multi-valued, since at most finitely many constants occur in the definition of a set. Finally, operations 4 and 5 are countably multi-valued, since with given constants one can form [only] countably many propositional functions over M_α. But now it is an easily proven fact that if one forms the closed hull of an infinite set under finitely many at most countably multi-valued functions, the cardinality of this hull is not larger than the cardinality of the set that was originally given. From that it thus follows that $K + O + F$ is countable, and therefore K is also countable. So property 1 of K is already proven. Property 3 holds according to stipulation 1 of the definition of K. So only property 2 still remains to be proved.

For that, an isomorphism has to be defined between K and a certain M_η. That is done by complete induction on the order of the elements of K. *The set of all orders which* occur at all for elements *of K* is, according to condition 2, the set O. O is thus a certain set of ordinal numbers, but in general, by no means an initial segment of the sequence of ordinal numbers. For it may well happen that certain ordinal numbers fail to be among the orders of the elements of K, while other larger ones again appear. But in any case, the elements of O can, in their entirety, be consecutively enumerated by means of an initial segment of the sequence of ordinal numbers. For this enumeration, all the ordinal numbers less than a *certain ordinal number η* will be used (and this η at the same time gives the index of the M with which K will prove to be isomorphic). If β is an arbitrary element of O, I denote by $\overline{\beta}$ its number in this enumeration. Now, the isomorphism between K and M_η will be such that the elements of K of order β are mapped precisely onto all the elements of M_η of order $\overline{\beta}$. With this mapping, the orders of the elements are thus collapsed, so to speak, accordion-like. *The definition of the mapping* is carried out by induction on β, that is, we assume that for the elements of K of order $< \beta$ an isomorphic mapping onto the elements of M_η of order $< \overline{\beta}$ is already defined, and we show how this mapping can be extended to the elements of

bildung auf die Elemente der Ordnung β von K (respective, $\overline{\beta}$ von M_η) erweitern läßt. Die Menge der Elemente einer Ordnung $< \beta$ von K ist offenbar nichts anderes [als] $K \cdot M_\beta$ (denn M_β ist ja die Menge sämtlicher konstruierbaren Mengen einer Ordnung $< \beta$). Ähnlich ist die Menge der Elemente einer Ordnung $< \overline{\beta}$ von M_η nichts anderes als $M_\eta \cdot M_{\overline{\beta}}$, und das ist $= M_{\overline{\beta}}$ wegen des monotonen Wachsens der M_α. Die induktive Annahme

22   besagt also, daß eine eineindeutige und isomorphe Abbildung | von $K \cdot M_\beta$ auf $M_{\overline{\beta}}$ gegeben ist. Wir haben daraus eine Abbildung der Elemente der Ordnung β von K auf Elemente der Ordnung $\overline{\beta}$ von M_η zu definieren. *Sei also ein Element v aus K der Ordnung β gegeben.* v ist dann definiert durch eine gewisse Aussagefunktion φ_β über M_β (d. h., $x \in v \equiv \varphi_\beta(x)$). In dieser definierenden Aussagefunktion φ_β kommen gewisse Konstanten $a_1, \ldots, a_n$ vor, welche selbstverständlich Elemente von M_β sind (da es sich um eine Funktion über M_β handelt). Die $a_1, \ldots, a_n$ sind aber zugleich auch Elemente von K, denn die Forderung 3 der Definition von K besagt ja, daß die Konstanten, welche in der Definition eines Elements von K vorkommen, auch zu K gehören. Die a_i sind demnach Elemente des Durchschnitts $K \cdot M_\beta$. Daher sind sie nach induktiver Annahme auf gewisse Elemente $a'_1, \ldots, a'_n$ von $M_{\overline{\beta}}$ abgebildet, und wir ordnen jetzt der Funktion φ_β diejenige Funktion $\varphi_{\overline{\beta}}$ über $M_{\overline{\beta}}$, zu welche man erhält, indem man in φ_β die Konstanten a_i ersetzt durch a'_i, und indem man weiter sämtliche gebundene Variable in φ_β (deren Variabilitätsbereich M_β ist) ersetzt durch gebundene Variable mit dem Variabilitätsbereich $M_{\overline{\beta}}$, und alles übrige an der Aussagefunktion φ ungeändert läßt. Die so erhaltene Aussagefunktion $\varphi_{\overline{\beta}}$ ist offenbar eine Aussagefunktion über $M_{\overline{\beta}}$ (denn ihre Konstanten gehören zu $M_{\overline{\beta}}$ und ihre gebundenen Variable haben den Bereich $M_{\overline{\beta}}$). $\varphi_{\overline{\beta}}$ definiert daher ein gewisses Element von $M_{\overline{\beta}+1}$, und dieses lassen wir dem gegebenen v entsprechen; d. h., also wir setzen $x \in v' \equiv_{\mathrm{DF}} \varphi_{\overline{\beta}}(x)$,

23   während andererseits $x \in v \equiv \varphi_\beta(x)$ war. Unter Benutzung der anderen | *Postulate für K* kann man nun leicht zeigen, daß auf diese Weise tatsächlich eine eineindeutige Abbildung aller Elemente der Ordnung β von K auf alle Elemente der Ordnung $\overline{\beta}$ erzielt wird. Es bleibt nur noch zu zeigen, daß diese Abbildung ein Isomorphismus ist, d. h., daß $x \in v \equiv x' \in v'$. Auf Grund der Definitionen für v und v' bedeutet das also: $\varphi_\beta(x) \equiv \varphi_{\overline{\beta}}(x')$. Zeigen wir zunächst, daß $\varphi_\beta(x) \supset \varphi_{\overline{\beta}}(x')$. Nehmen wir also an, $\varphi_\beta(x)$ gilt; nach dem vorher erwähnten bedeutet das, daß es Skolemfunktionen für φ_β und x gibt. Das sind Funktionen in M_β, d. h., sie geben angewendet auf Elemente aus M_β wiederum Elemente aus M_β. Wegen Postulat 5 gehören diese Funktionen zu F, und wegen Postulat 6 geben sie angewendet auf die Elemente aus K wiederum Elemente aus K. Daher geben sie angewendet auf die Elemente aus dem Durchschnitt $K \cdot M_\beta$ Elemente aus dem Durchschnitt $K \cdot M_\beta$, d. h., sie sind Funktionen in $K \cdot M_\beta$. Daher werden sie

K of order β (respectively, of M_η of order $\overline{\beta}$). The set of the elements of K of order $< \beta$ is obviously nothing other than $K \cdot M_\beta$ (since M_β is after all the set of all the constructible sets of order $< \beta$). Similarly, the set of the elements of M_η of order $< \overline{\beta}$ is nothing other than $M_\eta \cdot M_{\overline{\beta}}$, and that is equal to $M_{\overline{\beta}}$ on account of the monotone growth of the M_α. So the inductive assumption says that a one-to-one and isomorphic mapping of $K \cdot M_\beta$ onto $M_{\overline{\beta}}$ is given. From that we have to define a mapping of the elements of K of order β onto the elements of M_η of order $\overline{\beta}$. *So let an element v from K of order β be given.* Then v is defined by means of a certain propositional function φ_β over M_β (that is, $x \in v \equiv \varphi_\beta(x)$). In this defining propositional function certain constants $a_1, \ldots, a_n$ occur that are evidently elements of M_β (since one is dealing with a function over M_β). But the $a_1, \ldots, a_n$ are at the same time also elements of K, since stipulation 3 of the definition of K says that the constants that occur in the definition of an element of K also belong to K. The a_i are accordingly elements of the intersection $K \cdot M_\beta$. Therefore, according to the inductive assumption, they are mapped to certain elements $a_1', \ldots, a_n'$ of $M_{\overline{\beta}}$, and we now assign to the function φ_β that function $\varphi_{\overline{\beta}}$ over $M_{\overline{\beta}}$ that is obtained by replacing the constants a_i in φ_β by a_i', and further replacing all the bound variables in φ_β (whose domain of variability is M_β) by bound variables with the domain of variability $M_{\overline{\beta}}$, leaving everything else in the propositional function φ unchanged. The propositional function $\varphi_{\overline{\beta}}$ so obtained is obviously a propositional function over $M_{\overline{\beta}}$ (since its constants belong to $M_{\overline{\beta}}$ and its bound variables have domain $M_{\overline{\beta}}$). Therefore $\varphi_{\overline{\beta}}$ defines a certain element of $M_{\overline{\beta}+1}$, and we let that correspond to the given v; that is, we set $x \in v' \equiv_{\mathrm{DF}} \varphi_{\overline{\beta}}(x)$, while on the other hand we had $x \in v \equiv \varphi_\beta(x)$. By using the other *postulates for K*, one can now easily show that in this way a one-to-one mapping of all elements of K of order β onto all elements of order $\overline{\beta}$ is in fact obtained. It remains only to show that this mapping is an isomorphism, that is, that $x \in v \equiv x' \in v'$. On the basis of the definitions for v and v', this therefore means: $\varphi_\beta(x) \equiv \varphi_{\overline{\beta}}(x')$. First we show that $\varphi_\beta(x) \supset \varphi_{\overline{\beta}}(x')$. So we assume that $\varphi_\beta(x)$ holds; according to what was mentioned earlier, that means that there are Skolem functions for φ_β and x. They are functions in M_β, that is, applied to elements from M_β, they again give elements from M_β. On account of postulate 5, these functions belong to F, and on account of postulate 6, [when] applied to the elements from K, they again give elements from K. Therefore, applied to the elements from the intersection $K \cdot M_\beta$, they give elements from the intersection $K \cdot M_\beta$; that is, they are functions in $K \cdot M_\beta$. Therefore, by the isomorphism between $K \cdot M_\beta$ and

durch den Isomorphismus zwischen $K \cdot M_\beta$ und $M_{\overline{\beta}}$ übergeführt in gewisse Funktionen in $M_{\overline{\beta}}$, und es ist leicht zu sehen, daß diese, weil es sich um einen Isomorphismus handelt, Skolemfunktionen für $\varphi_{\overline{\beta}}$, x' sein müssen. Daher gibt es also Skolemfunktionen für $\varphi_{\overline{\beta}}$, x', und d. h., es gilt $\varphi_{\overline{\beta}}(x')$.

24 So kann also die Implikation $\varphi_\beta(x) \supset \varphi_{\overline{\beta}}(x')$ bewiesen werden und | die Implikation $\sim\varphi_\beta(x) \supset \sim\varphi_{\overline{\beta}}(x')$ wird wörtlich ebenso bewiesen, denn $\sim\varphi_\beta(x)$ ist ja ebenfalls eine Funktion über M_β.

Ich hoffe, daß es mir gelungen ist in dieser kurzen Skizze die Grundgedanken des Beweises für das Haupttheorem klar zu machen, und möchte abschließend nur noch bemerken, daß der skizzierte Beweis für Widerspruchsfreiheit der Kontinuumhypothese zugleich ein Widerspruchsfreiheitsbeweis für das Auswahlaxiom ist, denn es ist leicht zu kontrollieren, daß weder bei der Definition der konstruierbaren Menge noch bei dem Beweis der Lemmata 1–3 das Auswahlaxiom verwendet wird, während anderseits das Auswahlaxiom, wie schon erwähnt, im Modell der konstruierbaren Mengen gilt. Der Sachverhalt ist also der, daß wir innerhalb der Mengenlehre ohne Auswahlaxiom ein Modell für die Mengenlehre mit Auswahlaxiom konstruiert haben, und daraus folgt dann, daß ein Widerspruch aus dem Auswahlaxiom auch zu einem Widerspruch in der Mengenlehre ohne Auswahlaxiom führen müßte, ganz ebenso wie es mit der Kontinuumhypothese oder der nichteuklidischen Geometrie der Fall ist. Ein anderes Nebenresultat hinsichtlich nichtmeßbarer projektiver Mengen habe ich schon vorhin erwähnt. Schließlich ist aber auch die Widerspruchsfreiheit des Satzes A (daß jede Menge konstruierbar ist) an sich interessant besonders deswegen, weil es sehr plausibel ist, daß es sich bei A um einen absolut unentscheidbaren Satz handelt, an dem die Mengenlehre sich in zwei verschiedene Systeme gabelt, ähnlich wie euklidische und nichteuklidische Geometrie. Bewiesen ist, wie gesagt, nur die eine Hälfte, nämlich, daß die Annahme, jede Menge sei konstruierbar, widerspruchsfrei ist. Ich bin vollkommen überzeugt, daß auch die Annahme, daß nichtkonstruierbare Mengen existieren, widerspruchsfrei ist. Ein Beweis dafür würde vielleicht

25 den Schlüssel zu dem Beweis der Un|abhängigkeit der Kontinuumhypothese von den anderen Axiomen der Mengenlehre liefern. Das würde dann erst das abschließende Resultat ergeben, daß man sich tatsächlich mit einem Beweis für die Widerspruchsfreiheit der Kontinuumhypothese begnügen muß, weil dann eben gezeigt wäre, daß ein Beweis für den Satz selbst nicht existiert.

$M_{\overline{\beta}}$, they are carried over into certain functions in $M_{\overline{\beta}}$, and, because one is dealing with an isomorphism, it is easy to see that these must be Skolem functions for $\varphi_{\overline{\beta}}$ and x'. So therefore there are Skolem functions for $\varphi_{\overline{\beta}}$ and x'; that is, $\varphi_{\overline{\beta}}(x')$ holds. Thus the implication $\varphi_{\beta}(x) \supset \varphi_{\overline{\beta}}(x')$ can be proved, and the implication $\sim\varphi_{\beta}(x) \supset \sim\varphi_{\overline{\beta}}(x')$ is proved, word for word, in the same way, since $\sim\varphi_{\beta}(x)$ is likewise a ⟦propositional⟧ function over M_{β}.

I hope that I have succeeded in making clear in this brief sketch the basic ideas of the proof of the main theorem; in conclusion, I would still like only to remark that the proof sketched for the consistency of the continuum hypothesis is at the same time a consistency proof for the axiom of choice, since it is easy to check that neither in the definition of the constructible sets nor in the proof of lemmas 1–3 is the axiom of choice used, while on the other hand, as already mentioned, the axiom of choice holds in the model of the constructible sets. The situation is thus that within set theory without the axiom of choice we have constructed a model for set theory with the axiom of choice, and from that it then follows that a contradiction from the axiom of choice would also have to lead to a contradiction in set theory without the axiom of choice, just as is the case with the continuum hypothesis or non-Euclidean geometry. Earlier I have already mentioned another secondary result regarding nonmeasurable projective sets. Finally, the consistency of the proposition A (that every set is constructible) is also of interest in its own right, especially because it is very plausible that with A one is dealing with an absolutely undecidable proposition, on which set theory bifurcates into two different systems, similar to Euclidean and non-Euclidean geometry. As I said, only the first half is proved, namely, that the assumption that every set is constructible is consistent. I am fully convinced that the assumption that nonconstructible sets exist is also consistent. A proof of that would perhaps furnish the key to the proof of the independence of the continuum hypothesis from the other axioms of set theory. That would then yield the definitive result that one must really be content with a proof of the consistency of the continuum hypothesis, because then what would have been shown is exactly that a proof of the proposition itself does not exist.

Introductory note to *193?*

This article is taken from handwritten notes in English, evidently for a lecture, found in the *Nachlass* in a spiral notebook. Although the date of the piece is not known, some conjectures about this will be discussed later. The lecture was carefully thought out, with blackboard material laid out on facing pages (here incorporated in text between horizontal brackets); nevertheless in some ways it was still a rough draft. For example, there is some ambiguity about whether the "integers" referred to were to be understood as meaning the positive integers or whether 0 was to be included as well.

The manuscript was given no title; that provided by the editors emphasizes a technical result about (polynomial) diophantine equations which would have been new at the time of writing. The result concerns sentences of the form

$$(\forall a_1, \dots, a_m)(\exists x_1, \dots, x_n)F(a_1, \dots, a_m, x_1, \dots, x_n) = 0,$$

where F is a polynomial with integer coefficients, which Gödel calls "propositions of class A". For the purpose of putting Gödel's results in the context of later work on undecidable diophantine problems, it will be useful to speak of "diophantine relations", that is, relations D satisfying conditions of the form:

$$D(a_1, \dots, a_m) \Leftrightarrow (\exists x_1, \dots, x_n)F(a_1, \dots, a_m, x_1, \dots, x_n) = 0,$$

where F is as above. Gödel's class A then consists of all sentences of the form

$$(\forall a_1, \dots, a_m)D(a_1, \dots, a_m)$$

for such D.

Gödel stated his result in two parts:

1. There is no mechanical procedure for determining whether given propositions of class A are true.
2. In any suitable formal system, there is some proposition of class A which is undecidable in that system, that is, which is neither provable nor refutable in the system.

He noted that in fact the proof of these results yields more: they continue to hold if A is replaced by the proper subclass A_N in which the polynomials are restricted to have both their degree and the total number of their variables $(m + n)$ less than N, for some fixed sufficiently large N. The second part of the result is obtained as a corollary of the first, using the now familiar remark that if the result 2 were false, then the enumeration of "all possible proofs in the given formal system" would eventually yield, for each given proposition P of the class A, a proof of P or of its negation, thereby providing "a procedure for deciding every proposition of the class A".[a] Gödel emphasizes that the possibility of stating undecidability results in such generality only became possible with a "mathematically satisfactory definition ... of mechanical procedure ..." and attributes the filling of this "gap" to "Herbrand, Church, and Turing".[b] Of course, it was Gödel himself who added the computational rules that made Herbrand's idea of allowing arbitrary recursion equations appropriate for such a "definition". In line with Gödel's later writing on the subject is the assertion that this Herbrand–Gödel notion being "the correct definition" was "established beyond any doubt by Turing".

Citing *Kleene 1936*, Gödel bases his proof on the existence of a triadic primitive recursive relation R which serves to enumerate all computable sets of integers, in the sense that for every such set W, there exists a number n such that $x \in W \Leftrightarrow (\exists z)R(z, x, n)$. Then the antidiagonal set $K = \{x \mid \neg(\exists z)R(z, x, x)\}$ is not computable. The desired result then follows at once from the following:

KEY LEMMA. *Every primitive recursive relation* $S(u_1, \ldots, u_l)$ *can be written in the form*

$$(\forall a_1, \ldots, a_m)D(a_1, \ldots, a_m, u_1, \ldots, u_l),$$

where D is a diophantine relation.

The basic technique needed for proving this lemma was introduced in *Gödel 1931* in order to show that the undecidable sentence there obtained could be put into arithmetic form. This technique depends on the observation that a recursive definition can be made explicit by using the Chinese remainder theorem to represent an arbitrary finite sequence of integers by a single pair of integers, using the function $\beta(a, b, i)$, defined as the least non-negative remainder of a modulo $1 + b(i + 1)$. In his lectures at the Institute for Advanced Study (*Gödel 1934*), it was

---

[a]Although Gödel doesn't make it explicit, this argument assumes that the given formal system is *correct* with respect to propositions of class A, in the sense that for a proposition P of the class A, if P is provable in the formal system, then P is true, whereas if the negation of P is so provable, then P is false.

[b]This is particularly interesting in the light of some of Gödel's later statements concerning these matters, as well as the question of Gödel's own role in helping to fill the "gap". For discussion of these matters, see *Kleene 1981*, *Davis 1982*, and *Gandy 1988*.

further noted that by putting the undecidable arithmetic sentence into prenex form and using the relations

$$F = 0 \vee G = 0 \Leftrightarrow FG = 0, \quad F = 0 \wedge G = 0 \Leftrightarrow F^2 + G^2 = 0,$$

the sentence could be seen to be making a statement about the solutions of a diophantine equation. The conclusion that the assertion in question had quite an ordinary mathematical content is of course strengthened by noting the particular form of the quantificational prefix obtained in the present paper.[c]

The proof of the key lemma as sketched by Gödel goes essentially as follows: Let s be the primitive recursive function which has the value 0 when the relation S is true and the value 1 otherwise. Let us be given the complete set of recursion equations defining s, beginning with the initial functions and proceeding through a sequence of compositions and primitive recursions. To compute any particular value $v = s(u_1, \ldots, u_l)$, only a finite number of values of each of the functions in these equations is required, and so the Chinese remainder theorem can be used to code the corresponding finite sequences. Thus, using the abbreviation $\mathbf{u_l}$ to represent the l-tuple $u_1, \ldots, u_l$ (and similarly for other letters), we may write

$$v = s(\mathbf{u_l}) \Leftrightarrow (\exists \mathbf{a_r}, \mathbf{b_r})(\forall \mathbf{i_r}) W(\mathbf{u_l}, \mathbf{a_r}, \mathbf{b_r}, \mathbf{i_r}),$$

where the various sequences are coded as

$$\beta(a_j, b_j, i_j), \ i_j = 1, 2, \ldots, k_j; j = 1, 2, \ldots, r,$$

and W is a quantifier-free expression formed with propositional connectives from inequalities and (possibly) nested equations containing β. Therefore,

$$v \neq s(\mathbf{u_l}) \Leftrightarrow (\forall \mathbf{a_r}, \mathbf{b_r})(\exists \mathbf{i_r}) \neg W(\mathbf{u_l}, \mathbf{a_r}, \mathbf{b_r}, \mathbf{i_r}).$$

But it is easy to see that the expression following the universal quantifiers can be transformed into an equivalent expression that defines a diophantine relation: push the $\neg$ through the propositional connectives, expand inequalities into equations in additional unknowns,[d] eliminate nested β's,[e] and use the relations given above for combining conjunctions and disjunctions of equations. Finally, the lemma follows from the equivalence:

---

[c]In his additions for the 1964 revision of the 1934 lectures, Gödel stated, but did not prove, this improved result (cf. *Davis 1965*, p. 73; *Gödel 1986*, pp. 370–371).

[d]$w \neq 0 \Leftrightarrow (\exists z)(w = z + 1)$ and $p \leq q \Leftrightarrow (\exists z)(q = z + p)$.

[e]E.g., $w = \beta(a, b, \beta(c, d, i)) \Leftrightarrow (\exists z)[w = \beta(a, b, z) \wedge z = \beta(c, d, i)]$.

$$S(u_1, \ldots, u_l) \Leftrightarrow 1 \neq s(u_1, \ldots, u_l).$$

The tenth problem in the famous list of *Hilbert 1900* sought a procedure which "in a finite number of steps" could test a given (polynomial) diophantine equation for solvability in integers. It is easy to see that it is equivalent to ask for a test for solvability in natural numbers, and it is in this form that the problem has been studied. Thus, to say that Hilbert's tenth problem is unsolvable is precisely to say that the results of the present paper can be strengthened by replacing the class A by the subclass consisting of those sentences with $m = 0$. Gödel noted that this question was open, but did not refer to it as Hilbert's tenth problem.

It is clear from the basic unsolvability results of recursion theory (in particular from the non-computability of the antidiagonal set K mentioned above) that to prove the unsolvability of Hilbert's tenth problem, it would suffice to show that every recursively enumerable (r.e.) relation is diophantine, or, equivalently, that every primitive recursive relation is diophantine. In my dissertation *Davis 1950*, I attempted to prove this using Gödel's β function, but had to content myself with the much weaker normal form: *Every r.e. relation can be expressed in the form*

$$(\exists y)(\forall k)_{\leq y}(\exists \mathbf{x_n})[F(k, y, \mathbf{x_n}, \mathbf{u_l}) = 0]$$

where F is a polynomial.[f] It is interesting to note that Gödel's results in the present paper follow readily from this theorem.[g] Namely, to obtain the lemma above, put the primitive recursive relation $\neg S$ in this normal form and use the β function once again as follows:

$$\neg S(\mathbf{u_l}) \Leftrightarrow (\exists y)(\forall k)_{\leq y}(\exists \mathbf{x_n})[F(k, y, \mathbf{x_n}, \mathbf{u_l}) = 0]$$
$$\Leftrightarrow (\exists y)(\exists \mathbf{a_n}, \mathbf{b_n})(\forall k)_{\leq y}[F(k, y, \beta(a_1, b_1, k), \ldots,$$
$$\beta(a_n, b_n, k), \mathbf{u_l}) = 0]$$
$$\Leftrightarrow (\exists y)(\exists \mathbf{a_n}, \mathbf{b_n})\neg D(k, y, \mathbf{a_n}, \mathbf{b_n}, \mathbf{u_l}),$$

[f]The relevant result in *Gödel 1931* is that every primitive recursive relation (and therefore every r.e. relation) is equivalent to an arithmetic formula. However, Gödel's proof actually shows that all universal quantifiers in this arithmetic formula can be taken to be *bounded*. Although this fact was noted in my dissertation, I made no use of it, but started instead with Post production systems. At the referee's suggestion, a different proof of my result was given in the published version *Davis 1953*; the published proof did begin with an arithmetic formula with all universal quantifiers bounded. In fact, the methods of the present paper would have sufficed to obtain my result.

[g]To my knowledge, this consequence has not previously been noted in the literature, presumably because interest was focused on the purely existential case.

where D is diophantine. The key lemma above follows on taking negations.

The history of further work leading to the final resolution of Hilbert's tenth problem has been detailed in *Davis 1973* and *Davis, Matiyasevich and Robinson 1976*; it is briefly reviewed here for the reader's convenience. Julia Robinson began her investigations of diophantine relations from the opposite direction, attempting to show that various particular relations are diophantine. She introduced the hypothesis

J.R.: *There is a diophantine function f such that $f(x) = O(x^x)$ but for all natural numbers k, $f(x) \neq O(x^k)$.*

Using the Pell equation $x^2 - (a^2 - 1)y^2 = 1$, she showed (*Robinson 1952*)

J.R. $\Rightarrow$ the exponential relation $z = x^y$ is diophantine.

In 1957 and 1958, Hilary Putnam and I applied the β function to invert the order of quantifiers in the normal form of my dissertation (much as was done above to derive the key lemma) to obtain a small set of "critical" relations with the property that their being diophantine would imply the same for all r.e. relations (*Davis and Putnam 1958*). Finally, combining our methods with those of Julia Robinson, we were able to show that, assuming the (still) unproved hypothesis that there are arbitrarily long arithmetic progressions consisting entirely of prime numbers, *every r.e. relation can be expressed in the form*

$$(\exists x_1, \ldots, x_n) F(u_1, \ldots, u_l, x_1, \ldots, x_n) = 0,$$

where F is a polynomial-like expression in which some of the exponents may be variables. Julia Robinson quickly showed how to eliminate the hypothesis about primes in arithmetic progressions (*Davis, Putnam and Robinson 1961*). At this point the situation was that the desired result that all r.e. relations are diophantine was seen to be equivalent to the hypothesis J.R., since by Robinson's earlier work, J.R. implied that all variable exponents could be eliminated. There the matter rested until Yuri Matiyasevich (*1970*) proved J.R. by providing an ingenious diophantine definition for the function F_{2n}, where F_n is the nth Fibonacci number, which grows in accordance with the requirements of J.R. Thus, Matiyasevich's result combined with the earlier work provided a proof of:

[*]   *A relation is recursively enumerable if and only if it is diophantine.*

As with Gödel's present result, the enumeration theorem of recursion theory (or equivalently, the existence of a universal Turing machine) implies the existence of a fixed bound N on the degree and the number of variables in the diophantine equation given by [*]. Specific and reasonably low bounds can be obtained by considering the degree and number of unknowns separately; for details, see *Matiyasevich and Robinson 1975* and *Jones 1982.*[h]

The theorem [*] asserts the equivalence of two concepts, one logical, the other number-theoretic. As such its importance goes well beyond its role in showing that Hilbert's tenth problem is unsolvable. The survey paper *Davis, Matiyasevich and Robinson 1976* discusses applications and further developments. There continues to be much work on analogues of Hilbert's tenth problem for various kinds of rings. The most important question of this kind is Hilbert's tenth problem for the ring of rational numbers. There has been very little progress on this question, and the prevailing opinion is that it will require much deeper methods than were needed to establish [*]. However, there has been considerable progress in proving the unsolvability of Hilbert's tenth problem for various rings. The technique typically used is to provide a diophantine definition of the rational integers over the ring in question. Some examples of rings for which this has been done are:

1. The ring of integers of any totally real algebraic number field (*Denef and Lipschitz 1978, Denef 1980*).

2. The ring of integers of any algebraic number field admitting exactly one pair of conjugate complex embeddings (*Shlapentokh 1989, Pheidas 1988*).

3. The ring of integers of any algebraic number field whose Galois group over the rationals is abelian (*Shapiro and Shlapentokh 1989*).

4. The field of rational functions in one indeterminate over a field of finite characteristic $\neq 2$ (*Pheidas 1991*).

5. Certain rings of elements of algebraic function fields (*Shlapentokh 1992, Shlapentokh 1992a*).

It is certainly remarkable that Gödel had found a simple undecidable class of statements about diophantine equations a decade before anyone else was giving the matter serious thought. However, the foundational ideas in this paper are in a way even more striking. Gödel begins with Hilbert's belief that, as Gödel puts it, "... for any precisely formulated mathematical question a unique answer can be found ...".[i] Gödel formulates this as follows:

---

[h] *Jones 1981* also contains a thorough study of the status regarding decidability of the classes defined by various quantificational prefixes preceding a diophantine equation.

[i] Statements to this effect can be found in *Hilbert 1900* and in *Hilbert 1926*.

> Given an arbitrary mathematical proposition A there exists a
> proof either for A or for not-A...

If the notion of "proof" is explicated via mathematical logic, then
Hilbert's assertion "becomes amenable to mathematical treatment" and
can in fact be disproved "even in the domain of number theory". However, Gödel continues,

> this negative answer may have two different meanings:
>
> 1. it may mean that the problem in its original formulation has a
>   negative answer, or
> 2. it may mean that through the [[formalization]] something was lost.
>
> It is easily seen that actually the second is the case ...

This is because the number-theoretic questions shown to be undecidable
in a given formalism by the methods Gödel is using are decidable by
principles "exactly as evident as those of the given formalism".

After sketching the proof of his result concerning $\forall\exists$ arithemetic,
Gödel returns to this issue, emphasizing once again that the proof of formal undecidability is "... at the same time a decision of this proposition.[j]
... So the belief in the decidability of every mathematical question is
not shaken by this result." And then Gödel concludes with a bomb-
shell: "Apparently" there are questions "which very likely are really
undecidable" in a sense independent of any particular formalism; if true
this would contradict Hilbert's assertion in the first of the two meanings
suggested by Gödel. These questions have an $\forall\exists$ form much like those
of Gödel's class A except "... only that also variables for real numbers
appear in this polynomial. Questions connected with Cantor's continuum hypothesis lead to problems of this type." Although Gödel had not
yet been able to prove their undecidability, "... there are considerations
which make it highly plausible that they really are undecidable."

This last is one of a small number of statements made by Gödel
during the 1930s which hint at a progression towards the full-blown
Platonism revealed in *Gödel 1944* and *Gödel 1947*. In his *1933o lecture before the Mathematical Association of America, he suggested that
the "kind of Platonism" needed to interpret the axioms of set theory as

---

[j]This is not true of the reductio ad absurdum method of this paper for showing the
existence of formally undecidable sentences; however, one of the side notes outlines
an explicit construction.

"meaningful statements ..., cannot satisfy any critical mind ...". In his *1938* announcement of his consistency proof for AC and GCH, Gödel suggests that the axiom of constructibility "... seems to give a natural completion of the axioms of set theory, in so far as it determines the vague notion of an arbitrary infinite set in a definite way." Even in his *1940*, the word "axiom" is used in connection with the proposition $V = L$, suggesting that Gödel still supported this idea.

The point of view expressed in *Gödel 1947* is that a proof of the independence of the continuum hypothesis from ZFC should in no way be regarded as implying its *absolute* undecidability, that is, its undecidability in the first sense of the present paper. Rather, it should be regarded as a reflection of the inadequacy of our axioms. Since no hint is given as to the precise nature of the Π_2^1 sentence Gödel believed might be demonstrably undecidable in this absolute sense (except that it is related to CH), we can not really determine whether he later changed his mind. It is possible that Gödel came to think that CH, in particular, ought to be decidable from a proper axiomatization of set theory, because of the simplicity of what it asserts. Indeed, in *1947* he rather says as much. On the other hand, the emphasis in the paper under consideration certainly suggests a different philosophical outlook.

Because the present paper does not mention the consistency proof for GCH, it may be that it was written before Gödel was ready to announce it. *Gödel 1938* was "communicated" to the National Academy of Sciences on 9 November 1938, shortly after Gödel had arrived for his third visit to the U.S. That would suggest early 1938 as the date. On the other hand, Gödel's short list of references includes *Kondô 1938*, which only appeared in December 1938, and this suggests a later date. Perhaps Gödel knew Kondô's paper early in 1938 as a preprint. Alternatively, John Dawson has noted that Gödel accepted an invitation to deliver one of the two major addresses on logic at the 1940 International Congress of Mathematicians, which was scheduled to be held September 4–12 in Cambridge, Mass., but, due to the outbreak of World War II, never took place. What Gödel intended to discuss in his address is not known; Dawson suggests that the present manuscript may have been prepared for that occasion.

Martin Davis[k]

Translation of the German in the insertions is by Cheryl Dawson and William Craig. The original German is to be found in the textual notes for **193?*.

[k]In preparing this introduction, I benefited greatly from conversations with John Addison, Solomon Feferman, and Robert Solovay.

[Undecidable diophantine propositions]
(*193?)

Almost everything I am going to say in this lecture was proved and published several years ago, but the publications are scattered in different papers and so it is perhaps not useless to collect all these results together in order to survey this whole domain of questions. The only thing which is new is a certain simplification of the undecidable proposition, which is of no great importance in principle but will perhaps be of interest to number theorists.

1 | You all know Hilbert's famous words that every mathematician is convinced that for any precisely formulated mathematical question a unique answer can be found and that it is exactly this conviction which is the chief stimulus in mathematical research work. Hilbert himself was so firm in this belief that he even thought a mathematical proof could be given for it, at least in the domain of number theory. How can we imagine that such a proof could ever be obtained? In order to find that out we have first to analyze the meaning of the theorem to be proved. For every unprejudiced man it can only mean this: Given an arbitrary mathematical proposition A there exists a proof either for A or for not-A, where by "proof" is meant

2 something which starts | from evident axioms and proceeds by evident inferences. Now formulated in this way the problem is not accessible for mathematical treatment because it involves the non-mathematical notion of evidence. So what is to be done first is to make this notion more explicit by analyzing the actual mathematical proofs. If that is done, and it has been done by mathematical logic and Hilbert's *Beweistheorie*, then our problem becomes amenable to mathematical treatment and the answer turns out to be negative even in the domain of number theory. But it is clear that this negative answer may have two different meanings: (1) it may mean that the problem in its original formulation has a negative answer, or

3 (2) it may mean that through the | transition from evidence to formalism something was lost. It is easily seen that actually the second is the case, since the number-theoretic questions which are undecidable in a given formalism are always decidable by evident inferences not expressible in the given formalism. As to the evidence of these new inferences, they turn out to be exactly as evident as those of the given formalism. So the result is rather that it is not possible to formalise mathematical evidence even in the domain of number theory, but the conviction about which Hilbert speaks remains entirely untouched. Another way of putting the result is this: It is not possible to mechanise mathematical reasoning, i.e., it will never be possible to replace the mathematician by a machine, even if you confine

yourself to number-theoretic | problems. There are of course certain por-   4
tions of mathematics which can be completely mechanised or formalised;
for example, elementary geometry is such a portion but already the theory
of ordinary integers is not. The interest now lies of course in finding, so to
speak, the smallest portion of mathematics which cannot be formalised. I
would like to sketch in this lecture the proof for the sharpest result in this
respect which is known at present.

Consider the well-known Pell equation $x^2 - ay^2 = 1$. You know that it
has a solution in positive integers whenever a is a positive integer and not
a square, so we can also say that the diophantine equation

$$(x^2 - a) \cdot (x^2 - ay^2 - 1) = 0$$

has a solution in positive integers x, y for every positive value of the parame-
ter a. | Now let's generalise this problem. Let a polynomial F in $m+n$ vari-   5
ables $a_1, \ldots, a_m, x_1, \ldots, x_n$ and with integer coefficients be given (in this
example we would have $m = 1$, $n = 2$ and the integer coefficients could be
calculated by carrying out this multiplication) and now let's ask the question: Has the given diophantine equation $F(a_1, \ldots, a_m, x_1, \ldots, x_n) = 0$
solutions for arbitrary integer values of the parameters a_i? There are of
course only enumerably many problems of this type because there are only
enumerably many polynomials with integer coefficients and each of them
gives rise only to a finite number of such problems according to which
variables you consider as parameters and which as the unknown⟦s⟧. Let's
denote the class of these problems, or rather the class of propositions by
which they are expressed, by | A; then it can be shown that already the   6
class A of propositions cannot be completely formalised. That means two
things:

1. There exists no mechanical procedure for deciding every proposition
of the class A.

2. In every formal theory which allows ⟦one⟧ to formulate all propositions
of the class A there exists an undecidable proposition of the class A, i.e.,
there exists a polynomial with integer coefficients and with the variables
divided into two groups, the parameters and unknowns, and it is impossible
to decide in the given formalism whether or not this particular diophantine
equation has solutions for arbitrary values of the parameters.

You see a problem of the class A is a | little more complicated than   7
the question whether a solution for a given diophantine equation exists.
It is asked whether solutions for every value of the parameters exist. If
you want to consider only questions about the existence of solutions of
given diophantine equations, you have a certain subclass A' of A, namely,
those problems of class A for which m (the number of parameters) is 0.
It is an unsolved problem whether for A' instead of A the same theorem
holds; in other words one does not know whether a mechanical procedure

for deciding about the solubility of arbitrary diophantine equations exists. But the result can be strengthened in another direction. Let's denote by A_N
8 the subclass of A which consists of those problems of class A for which the |
degree of the polynomial F is $< N$ and in addition the number of variables $m + n$ is $< N$. Then there exists a certain number N such that Theorems 1 and 2 can be proved already for A_N. I have not actually calculated this number N but it is certainly not a very great number, perhaps on the order of magnitude of 20, and it may be possible to diminish it still further. So the result is that even for polynomials with limited degree and limited number of variables the problems of the type described can [not] be solved by a uniform mechanical procedure, nor can a complete formal theory be given for them.

When I first published my paper about undecidable propositions the
9 result could not be pronounced in this generality, because for | the notions of mechanical procedure and of formal system no mathematically satisfactory definition had been given at that time. This gap has since been filled by Herbrand, Church and Turing. The essential point is to define what a procedure is. Then the notion of a formal system follows easily since a formal theory is given by the following three things:

1. a finite number of primitive symbols whose finite combinations are called expressions;
2. a finite class of expressions (called axioms);
3. certain so-called "rules of inference".

Now a rule of inference is nothing else but a mechanical procedure which allows [one] to determine of any given finite class of expressions whether anything can be inferred from them by means of the rule of inference under consideration, and if so to write down the conclusion. The finite number of
10 | axioms is no restriction since infinitely many axioms can be replaced by rules of inference. We could even confine ourselves to one axiom and one rule of inference. Now you see the essential difficulty of this definition is the notion of mechanical procedure which comes in and which needs further specification. These mechanical procedures are here applied to expressions or rather finite classes of expressions, and the result is either again an expression or the answer Yes or No, and the same holds for the decision-procedures which occur in the first half of the theorem to be proved.

Now expressions and finite classes of expressions can be mapped on integers, and therefore a procedure in the sense we want is nothing else but a function $f(x_1, \ldots, x_r)$ whose arguments as well as its values are integers and which is such that for any given system of integers $n_1, \ldots, n_r$ the
11 value can actually be calculated. | This includes also the case where the answer Yes or No is to be obtained by a procedure, because Yes or No can be replaced by the integers 0 and 1. So the concept we have to specify is the concept of a calculable function of integers. Now the standard way of

defining calculable functions is by recursion. If you define a function f by these two stipulations:

(I)
$$f(1, y) = h(y),$$
$$f(n + 1, y) = g(f(n, y), y),$$

where g and h are supposed to be previously defined calculable functions, then f will likewise be calculable. Another way is substitution:

(II)
$$\varphi(n) = \psi(\chi(n)).$$

If φ is defined by this equation and if ψ and χ are calculable, φ will likewise be so. Now all functions obtained by a finite iteration of these two kinds of definition beginning with the function $+1$ are called "primitive recursive". So all primitive recursive functions are calculable, but it is easy to see that there are other calculable functions. Namely, all | definitions of primitive 12 recursive functions (with one argument) can evidently be enumerated in such a manner that given a number n you can actually write down the nth definition; but then the antidiagonal sequence $f_n(n) + 1$ is evidently calculable but different from all primitive recursive functions.

This diagonal procedure makes it appear hopeless at first sight to arrive at a satisfactory definition of calculable functions, but a closer examination leads to the result that it is by no means so. The diagonal procedure really excludes only the existence of, so to speak, a calculable definition of calculability, but not of a definition at all. One way, essentially Herbrand's, of arriving at such a definition is to generalise this scheme of recursive definition. It can be thought of as a definition by postulates for this function f, where we have to add also the similar postulates | by which g and h are 13 defined, etc. Now all these postulates have the following two characteristics:

I. Each of them is an equation and the sides of these equations are certain expressions which are usually called "terms", where the word "term" is explained by the following two stipulations:

    1.   Every variable $x, y, \ldots$ and every number $1, 2, \ldots$ is a term.

    2.   If f is a letter denoting a function and if $T_1, \ldots, T_r$ are terms, then also $f(T_1, \ldots, T_r)$ is a term.

II. The second characteristic of the recursive postulates is that they allow [one] to calculate the values of the function defined. Now by analyzing in which manner this calculation proceeds you will find that it makes use only of the following two rules:

    R1.   to substitute particular integers such as 3 or 4 for the variables $x, y, \ldots$;

    R2.   to substitute equals for equals, i.e., if | you have arrived in the 14 course of your calculation at an equation $T_1 = T_2$, you are al-

lowed to substitute T_1 for T_2 within any other equation which
you may obtain.

And now it turns out that these two characteristics are exactly those that
give the correct definition of a computable function. Let's call "admissible postulate" for some functions $g_1, \ldots, g_n$ an entirely arbitrary equation
whose sides are terms formed of the g_i. Furthermore let's call any equation
of the form $k = f(n_1, \ldots, n_r)$, where k, $n_1, \ldots, n_r$ are fixed integers, an
"elementary equation" for f. If k is the value of the function f for the
arguments $n_1, \ldots, n_r$ it will be a true elementary equation, if not it will be
a wrong elementary equation. And now we shall call a function of integers
"computable" if there exists a finite number of admissible postulates in f
15   and perhaps | some other auxiliary functions $g_1, \ldots, g_n$ such that every true
elementary equation for f can be derived from these postulates by a finite
number of applications of the rules R1 and R2 and no false elementary
equation for f can be thus derived.

That this really is the correct definition of mechanical computability
was established beyond any doubt by Turing. At first you can easily see
that it is not possible to construct by the diagonal procedure a computable
function not comprised in the definition because, although you can easily
enumerate all possible admissible postulates, you have no procedure for
deciding whether a given system of admissible postulates actually defines a
function, i.e., whether it allows [one] to compute the values of the function.
And for this reason the antidiagonal sequence will not be computable. But
16   Turing has shown | more. He has shown that the computable functions
defined in this way are exactly those for which you can construct a machine
with a finite number of parts which will do the following thing. If you write
down any number $n_1, \ldots, n_r$ on a slip of paper and put the slip into the
machine and turn the crank, then after a finite number of turns the machine
will stop and the value of the function for the argument $n_1, \ldots, n_r$ will be
printed on the paper.

Now by means of the notion of computability we can now formulate
Theorem 1 as follows: Let us number through [enumerate] all problems
of class A by means of integers and let us denote by W the class of those
integers for which the corresponding problem has the answer Yes. Then
Theorem 1 asserts that there is no computable function $f(x)$ whose values
17   are only 0 and 1 and such that $f(x)=0. \equiv . x \in W$. I shall call a class |
for which such a function exists a "computable class"; if in addition this
function f is primitive recursive, I call the class "primitive recursive", and
in a similar way I define "computable relation" and "primitive recursive
relation". It is immediate from this definition that the complement of a
computable class or a primitive recursive class will again be computable,
respectively primitive recursive, and similarly for relations. You have only
to interchange 0 and 1 in this function f in order to see it. So what I
have to prove is that this class W is not computable. Now it is a result

by Kleene that all computable classes can be represented in a uniform manner by means of one triadic primitive [[recursive]] relation R. Namely, for every computable class K there exists an integer n such that $x \in K \equiv (\exists z)R(z, x, n)$; so R is kept fixed and you obtain all computable classes if you let n run over all integers.

The proof is obtained by taking for R the [[following]] relation defined here, which turns out to be primitive recursive. It is assumed that all finite systems of admissible postulates and all finite deductions from them by rules 1 and 2 have been numbered through in some manner. $R(z, x, n) \equiv z$ is the number of a deduction by rules R1 and R2 of the proposition $f(x) = 0$ from the nth system of admissible postulates (R is primitive recursive). Now this representation | here leads very easily to a non-computable class of integers; namely, the class defined by $x \in K \equiv \sim(\exists z)R(z, x, x)$ is evidently non-computable since, if it were [[computable]], there would exist an integer n such that $x \in K \equiv (\exists z)R(z, x, n)$, but then substituting n for x in both   18

Construction of the undecidable sentence:

Assume K is the class of numbers n for which $\sim(\exists x, z)R(x, z, n, n)$ is provable.

$n \in K \equiv (\exists m)B(m, n)$ and N is the number of B by
$B(x, y) \equiv (\exists z)R(z, x, y, N)$.
$y \in K \equiv (\exists x, z)R(z, x, y, N)$.
$N \in K \equiv (\exists x, z)R(z, x, N, N)$
$(= P) = \text{Bew}[\sim(\exists x, z)R(z, x, N, N)]$.
$\sim P \equiv \sim\text{Bew}(\sim P)$.
$\sim P$ has the form: $(n)\varphi(n)$.
$x \in K \equiv \sim(\exists z)R(z, x, x)$.
$x \in K \equiv (\exists z)R(z, x, n)$.

---

these equivalencies you have a contradiction. But that means that Theorem 1 is proved already if you take, instead of A, the class B of all problems of the form $(\exists z)R(z, n, n)$, where n runs over all integers. So for every integer n you have a different problem of this type, namely, the question whether or not there exists an integer z satisfying this condition $R(z, n, n)$. So you can consider n as the number of the problem and the class K is then the class of numbers which correspond to problems with the answer No, and this class is | not computable (as we have just proved). The same   19 holds then for the class of those which have the answer Yes, and that was the class W. And now the proof for Theorem 1 can be accomplished by showing that each problem of class B is equivalent with a certain problem of class A, where the corresponding problem of class A can actually be written down.

As to Theorem 2, it turns out to be equivalent with Theorem 1. Namely, if you have a formal system which for every proposition P of the class A allows [one] to prove either P or not-P, then you have also a procedure for deciding every proposition of the class A. This procedure consists simply in writing down successively all possible proofs in the given formal system until you reach a proof either for P or for not-P, which must happen after a finite number of steps. | That may of course take very long, but we are not speaking of short procedures but of procedures in the most general sense. So what remains to be done is only to set up the correspondence of the problems of class B and class A, and that is done by means of the following lemma: For every primitive recursive relation $S(u_1, \ldots, u_l)$ there exists a polynomial $F(a_1, \ldots, a_m, x_1, \ldots, x_n, u_1, \ldots, u_l)$ with integer coefficients and altogether $m + n + l$ variables such that $S(u_1, \ldots, u_l)$ is equivalent with the fact that the diophantine equation $F(\ldots) = 0$ has solutions in $x_1, \ldots, x_m$ for arbitrary integer values of the parameters $a_1, \ldots, a_m$. Now if you take for S in particular the [primitive] recursive relation $\sim R(z, x, x)$ of the Kleene representation, then you obtain for F a definite polynomial

$$S(u_1, \ldots, u_l)$$
$$\equiv (a_1, \ldots, a_n)(\exists x_1, \ldots, x_m)F(a_1, \ldots, a_n, \ x_n, \ldots, x_m,$$
$$u_1, \ldots, u_l)[\![= 0]\!].$$

$$\sim R(z, x, x)$$
$$\equiv (a_1, \ldots, a_n)(\exists x_1, \ldots, x_m)G(a_1, \ldots, x_i, \ldots, z, x)[\![= 0]\!].$$

which you can actually write down, and this equivalence gives you for each problem of class B an equivalent problem of class A and even of | class A_N with a fixed number N. Namely, you can take for N the maximum of the degree and the number of variables of this polynomial G. So the lemma gives you at once the proof for the stronger theorem I mentioned before.

Now as to the proof of this lemma, it is obtained by a transformation of the recursive definition into an explicit definition due to Dedekind. If f is defined in terms of g and h by means of these two postulates [I and II], then f can be defined explicitly in the following manner: z is to be the value of f for the arguments x, y if and only if there exists a function φ of integers satisfying these two postulates and such that $z = \varphi(x, y)$. This is

$$z = f(x,y) \equiv (\exists \varphi, a)\{x < a \cdot y < a \cdot (u,v)[u,v < a$$
$$\supset \cdot \varphi(0,v) = h(v) \cdot \varphi(u+1,v) =$$
$$g(\varphi(u,v),v)] \cdot \varphi(x,y) = z\}.$$

clear, since we know that there exists one and only one function satisfying these postulates, namely f. But we don't have to assume that φ is really defined for all integers, but only for a certain quadratic ⟦(square)⟧ domain which includes the point x,y; i.e., the statement | $z = f(x,y)$ is   22 also equivalent to the following statement: There exists a function φ and an integer a greater than x,y such that φ satisfies the recursive postulates only for arguments less than a and such that $z = \varphi(x,y)$. Outside of the square with side a, φ need not even be defined. And now the same transformation can be applied if a function f is defined by a chain of recursive definitions and substitutive definitions like this: Let the equations by which these definitions are expressed be called $G_1, \ldots, G_k$ and the functions which they define $f_1, \ldots, f_l$ (where $f_l = f$). Then the statement $z = f(x,y)$ is equivalent to the following statement: There exist functions $\varphi_1, \ldots, \varphi_l$ and numbers $a_1, \ldots, a_l$ such that the equations G_i are satisfied for the φ_i within the domains given by the a_i and such that $z = \varphi_l(x,y)$ and such that $x,y < a_l$[1], and finally such that certain conditions $B_1, \ldots, B_l$

$$z = f(x,y) \equiv (\exists \varphi_1, \ldots, \varphi_l)(\exists a_1, \ldots, a_l)(u_1,u_2)$$
$$[u_1,u_2 < a_1 \supset G_1$$
$$u_1,u_2 < a_2 \supset G_2$$
$$\vdots$$
$$\vdots$$
$$z = \varphi_l(x,y)$$
$$x,y < a_l$$

$$B_1 \qquad \begin{cases} u_1,u_2 < a_j \supset \varphi_k(u_1,u_2) < a_s \\ \cdots \\ B_n \qquad \cdots \end{cases}$$

[1] Corresponding to the analogous condition here.

23 | are satisfied for the a_i. The function of these conditions B_i is to accomplish this, that if in this chain of definitions f_i is defined in terms of, say, f_p and f_q, that then the knowledge of f_p and f_q within their respective domains a_p, a_q is sufficient to compute f_i within its whole domain a_i. I have no time to formulate these conditions B explicitly. The only thing which is essential for our purpose is that they can be expressed in the form specified here: For all numbers less than a certain a_j the value of a certain function φ_k is less than a certain other a_s.

Now if you look more closely at the matrix of this expression [that is the portion which follows after the quantifiers] you will see that it is formed by means of operations of the propositional calculus (namely, implications and
24 conjunctions out of certain | elementary formulas). And these elementary formulas are only equalities and inequalities between terms formed with these functions φ_i and the number variables u_i and a_i. The G_i have this form because they are equalities of this type, and for the other parts of the expression you see it immediately ($x, y < a_l$ of course means $x < a_l . y < a_l$). Now in order to obtain from this expression here the expression which we wish, we have first to replace the functional variables $\varphi_1, \ldots, \varphi_l$ by number variables $b_1, \ldots, b_l$. That this can be done is not surprising, because the φ_i need be defined only in certain finite domains, and the functions defined in finite domains of integers can be numbered through by integers. It is more convenient to replace each φ_i by two integers b_i, c_i
25 (and if it is a function with two | arguments by three integers b_i, c_i, d_i) as follows. I define at first a certain function with three arguments, $W(b, c, x)$, which has the following property: Given an arbitrary function $f(x)$ of integers and any arbitrarily great integer N, you can represent $f(x)$ for arguments $< N$ in the form $W(b, c, x)$ with some fixed integers b, c (and similarly for functions with several arguments). It is easily seen that the function defined here, [the] smallest positive rest [remainder] of b modulo $1 + cx$, will do the job. The main point of the proof is to show that there exist arbitrarily long arithmetic progressions with mutually relatively prime members, which follows from the remark written on the blackboard.

$f(x) = W(b, c, x)$ for $x < N$

$f(x, y) = V(b, c, d, x, y)$ for $x, y < N$

$W(b, c, x) =$ smallest positive rest [remainder] of b modulo
$1 + cx$

$V(b, c, d, x, y) =$ smallest positive rest [remainder]
of b modulo $1 + cx + dy$

$1 + N!x$ for $0 \le x < N$ mutually relatively prime

| Now this function W allows [[one]] to replace functions by pairs of integers   26
as long as the functions are considered only in a finite domain. But this
is the case in this expression, and therefore it will be equivalent to the
following one: There exist integers b_1, c_1, d_1 up to b_l, c_l, d_l and integers
$a_1, \ldots, a_n$ such that for all integers u_1, u_2 the subsequent is true, where the
subsequent is obtained out of [[(from)]] the previous matrix by replacing
$\varphi_i(x, y)$ by $V(b_i, c_i, d_i, x, y)$. So the elementary formulas of the new matrix
are now equalities and inequalities between terms formed exclusively with
this function V and the variables occurring in the prefix. Now it remains

$$(\exists b_1, c_1, d_1, \ldots, b_l, c_l, d_l)(\exists a_1, \ldots, a_n)(u_1, u_2)$$
$$[u_1, u_2 < a_2 \supset G_1$$
$$\vdots$$
$$\vdots$$
$$z = V(b_l, c_l, d_l, x, y)$$
$$x, y < a_l$$
$$u_1, u_2 < a_l \supset V(b_k, c_k, d_k, u_1, u_2) < a_s]$$

only to transform this matrix | into a diophantine equation. I prefer first to   27
put a negation sign to both sides of this equivalence and then shift it across
the quantifiers to the matrix. Then the prefix will have the desired form—
some universal quantifiers followed by some existential quantifiers—and, as
to the matrix, I can bring it to the normal form of the calculus of propositions, i.e., write it as a disjunction of conjunctions of equalities and inequalities between terms formed with this function V. Next I can eliminate nested
occurrences of this function V by this general scheme. Here new existential

$$f(f(x)) < u \equiv (\exists z)[z = f(x) . f(z) < u]$$
$$V(b, c, d, x, y) \equiv (\exists v)[$$

quantifiers are introduced, but they can be shifted in front without changing the form of the prefix which we need. Next we can eliminate also the
unnested occurrences of V by means of the definition of | V. $V(b, c, d, x, y) =$ 28
u means u is the least positive rest [remainder] of b modulo $1 + cx + dy$,
that is, $V(\ldots) = u$ is equivalent with $u < \text{modulo } 1 + cx + dy$ and there

exists an integer v such that $u = b - v(1 + cx + dy)$. The existential quantifier [for] v can again be shifted in front, and now, as you see, the new matrix is built up of equalities and inequalities between polynomials. These equalities and inequalities can be reduced to the two forms $P = 0$, $P > 0$, and, remembering that our variables run over positive integers, we have $P > 0 \equiv (\exists v)P{=}v \equiv (\exists v)[P{-}v{=}0]$, so that finally

$$V(b, c, d, x, y) = u$$
$$\equiv u \le 1 + cx + dy \,.\, (\exists v)[u = b - v(1 + cx + dy)]$$
$$P > 0 \equiv (\exists v)P = v \equiv (\exists v)P - v = 0$$

we obtain a conjunction of disjunctions of equalities of the form $P = 0$, where P is a polynomial with integer coefficients. And now we have these two trivial equivalences | $P = 0\,.\,Q = 0 \equiv P^2 + Q^2 = 0$ and $P = 0 \vee Q = 0 \equiv P \cdot Q = 0$, which allow [us to] transform a disjunction of conjunctions of diophantine equations into one single diophantine equation, so that our expression has now finally taken on the form we wish. Only the left hand side of the equivalence is not quite what we wish. What we want to have is an arbitrary primitive [recursive] relation, and what we have is the expression $\sim[z = f(x, y)]$, where f is an arbitrary primitive recursive function. But by definition of primitive recursive relation any such relation is equivalent to $f(x, y) = 0$, where f is a primitive recursive function which takes on only the values 0 and 1, so this is equivalent with $\sim(f(x, y) = 1)$, and that is what we want.

$$R(x, y) \equiv (f(x, y) = 0) \equiv \sim[f(x, y) = 1]$$

So Theorems 1 and 2 are proved. As to Theorem 2, it says that in every formal | system which allows [one] to express the propositions of class A, there exists an undecidable proposition of class A. If the axioms and rules of inference of this formal system [are given], such an undecidable proposition can actually be written down. But I wish to stress again that this undecidability holds only with respect to this particular system, and still more is true. The proof that the proposition is undecidable is really at the same time a decision of the proposition, but, of course, [is] not expressible in the given system. So the belief in the decidability of every mathematical

question is not shaken by this result. It is true we have also found a problem
which is absolutely unsolvable, namely, to find a mechanical procedure for
deciding every proposition of class *A*. But this is not a problem in the
form of a question with an answer Yes or No, but rather something similar
to | squaring the circle with compass and ruler. As to problems with the   31
answer Yes or No, the conviction that they are always decidable remains
untouched by these results. However, I would not leave it unmentioned
that apparently there do exist questions of a very similar structure which
very likely are really undecidable in the sense which I explained first. The
difference in the structure of these problems is only that also variables for
real numbers appear in this polynomial. Questions connected with Cantor's
continuum hypothesis lead to problems of this type. So far I have not been
able to prove their undecidability, but there are considerations which make
it highly plausible that they really are undecidable.

Lecture [on the] consistency
[of the] continuum hypothesis
(Brown University)
(*1940a)

[The introductory note to *Gödel *1940a* can be found on page 114,
immediately preceding *1939b*.]

If I want to sketch a proof for the consistency of Cantor's continuum
hypothesis, I have a choice between many possibilities. Just recently I
have succeeded in giving the proof a new shape which makes it somewhat
similar to Hilbert's program presented in his lecture *Über das Unendliche*
[*1926*]. I would like to sketch this proof today because it is perhaps the
most perspicuous. But nevertheless I don't want to omit completely my
former proofs, because they contain the heuristic viewpoints. So let me
begin with some heuristic considerations: what we have to do in order to
prove Cantor's continuum hypothesis is | to set up a well-ordering of the   2
real numbers according to the order type ω_1.
   It readily suggests itself to try to effect this well-ordering by reference to
the definitions of the real numbers, i.e., to well-order not the real numbers
themselves but rather all possible definitions of real numbers. But there
arise immediately two objections which seem to destroy this idea at once.

Namely, (1) it seems impossible that a well-ordering of order type ω_1 of the definitions of real numbers exists, because there seem to exist only
3   enumerably | many definitions, and (2) even if such a well-ordering did exist it apparently would not solve Cantor's continuum problem, because we would obtain only a well-ordering of the definable real numbers and we have no proof that every real number is definable; on the contrary, the objection mentioned just before concerning the enumerability of all definitions seems to prove just the opposite. Now, is it possible to overcome these objections and how is it possible? As to the first, concerning the enumerability of
4   definable real numbers, the answer is this: | It turns out that in order to construct all possible definitions of real numbers we have to make use of the ordinal numbers. So it is natural to presuppose the ordinals to be given in advance, i.e., the real numbers which we consider are really not simply the definable ones but rather those definable in terms of ordinals, where each ordinal is to be considered, so to speak, as a primitive undefined notion, so that we have non-enumerably many primitive notions to start with. Of course, for any single real number we need only a finite number of ordinals in order to define it, but altogether we have non-enumerably many.

As to the second objection, that there may exist real numbers not definable even with the help of ordinals, the reply is this: It is of course perfectly possible that such real numbers exist, but we can be sure that nobody will be able to define such real numbers, and hence we can reasonably expect
5   that nobody will | be able to prove their existence; i.e., the assumption that they do not exist will be free from contradiction. But if we have succeeded in well-ordering the definable real numbers by the order type ω_1 and if it can be assumed without contradiction that all real numbers are definable, then it can likewise be assumed without contradiction that all real numbers can be well-ordered by the order type ω_1. So the second objection is correct insofar as it shows that we actually do not obtain a proof for the continuum hypothesis in this way, but only a proof for its consistency.

In the sequel I shall use the term "constructible" instead of "definable", in order to stress that this notion does not completely coincide with the intuitive meaning of definability. And I shall denote by A or A_n the proposition which says that every real number (and more generally) every set
6   is constructible. So one of the main points of the proof | will be to show the consistency of A with the axioms of mathematics. One may at first doubt that this assertion has a meaning at all, because A is apparently a metamathematical statement since it involves the manifestly metamathematical term "definable" or "constructible". But now it has been shown in the last few years how metamathematical statements can be translated into mathematics, and this applies also to the notion of contructibility and the proposition A, so that its consistency with the axioms of mathematics is a meaningful assertion. As to the idea [of] its proof let me remark first that the essential axioms of mathematics, in particular, e.g., the axiom of

reducibility and its set-theoretical equivalents are existential axioms of the following type: Given certain sets $a_1, \ldots, a_n$, a set b having a certain relation to $a_1, \ldots, a_n$ exists. | This set b is always in a certain way defined or   7 constructed out of the sets $a_1, \ldots, a_n$, so that if the notion of constructibility is defined correctly we can expect b will be constructible if $a_1, \ldots, a_n$ are; i.e., the constructible sets will satisfy the axioms of mathematics; i.e., they will form a model for mathematics. And of course it is to be expected that in this model the proposition A will hold. It is to be expected but it is by no means trivial, since the fact that A holds in a model means that every set of the model is constructible in the sense of constructibility which holds in the model; and a priori it may very well happen that a constructible set is not constructible in a given model, just as an enumerable set may be non-enumerable in a given model. So it will be one of the essential steps of the proof to show | that for the notion of constructibility this situation   8 cannot arise, i.e., that it has a certain invariance property with respect to different models.

You see that the central concept of this whole program of a proof is the notion of definable or constructible real number (or more generally of constructible set). How is this concept to be defined in order to have all [the] mentioned properties? There are several different ways of doing this, of which I want to sketch two in these lectures. The first consists in this, that we call those sets "constructible" which can be obtained by means of the ramified hierarchy of types if extended to transfinite orders; | that   9 is to say, we begin with a certain system M_0 of real numbers in whose definitions no quantifiers for sets occur. It is not very essential what we take for M_0 (we can even put $M_0 = \Lambda$). Then we can define certain new sets by means of quantifiers relating to all elements of M_0. Let the totality of all these new sets be called M_1. In a similar manner we obtain M_2 from M_1, and generally M_{n+1} will consist of the sets definable by means of quantifiers relating to all elements of M_n. It is easily seen that in each of the M_n new sets, and in particular new sets of integers (i.e., new real numbers), will actually occur | essentially for the reason that, by means of   10 quantifiers relating to all elements of M_n, you can apply Cantor's diagonal procedure to the real numbers occuring in M_n and thus obtain a new real number. But the real numbers which you obtain for finite indices n are by no means all which you can obtain by this procedure, because nothing prevents you from continuing it into the transfinite as follows: You form at first the sum of all M_n with finite indices and call it M_ω, and then you introduce quantifiers relating to all elements of M_ω and call $M_{\omega+1}$ the totality of sets definable by means of these quantifiers. Then again, for the same reason as before, $M_{\omega+1}$ will contain new real numbers, and similarly for higher transfinite ordinals, and that is why the transfinite ordinals are necessary in order to obtain a satisfactory definition | of definability of   11 a real number. In order to obtain a notion of definability which is so comprehensive that you can assume consistently that every real number is definable, you have to continue this process even for arbitrarily high ordinals; i.e., you have to call a set or a real number "constructible" if it occurs in some M_α, whether this α is definable or not and whether it is enumerable or not. [That's what I meant when I said before that I consider, so to speak, every ordinal as a primitive notion and that I consider the real numbers definable in terms of these.] One may perhaps object to the use of arbitrarily great (even non-enumerable) ordinals in connection with the ramified theory of types. The answer to this is very simple. It depends on the purpose for which you want to use this theory. If you want to use it for

12   giving an unob|jectionable foundation to mathematics our procedure would of course be preposterous, but for proving the consistency of the continuum hypothesis it is perfectly all right, since what we want to prove is of course only a relative consistency of the continuum hypothesis; i.e., we want to prove its consistency under the hypothesis that set theory, including all its transfinite methods, is consistent. Therefore we are justified in using the whole set theory in the consistency proof [because if a contradiction were obtained from the continuum hypothesis and if, on the other hand, we could prove its consistency by means of set-theoretical arguments, then these set-theoretical arguments would be contradictory].

13 | Now let's come back to this transfinite sequence of the M_α. The interesting question in connection with the continuum problem is of course this: At which place will this sequence stop producing new real numbers? It is to be expected that this will happen for $\alpha = \omega_1$ and that is what actually can be proved, i.e., every real number which can be obtained at all in this manner occurs already with an enumerable index α and similar theorems (for higher indices) hold for sets of real numbers, etc. You will see that this theorem is actually nothing else but an axiom of reducibility for transfinite orders, for it says that an arbitrary propositional function with integers as arguments is always formally equivalent with a propositional function of an order $< \omega_1$. So since an axiom of reducibility holds for constructible sets it

14   is not | surprising that the axioms of set theory hold for the constructible sets, because the axiom of reducibility or its equivalents, e.g., Zermelo's Aussonderungsaxiom, is really the only essential axiom of set theory. I would only remark that from the heuristic point of view the matter goes rather the other way around. One constructs first a model for set theory out of the constructible sets and then by applying Skolem's theorem to it one obtains certain enumerable submodels which lead to the theorem about the enumerability of the orders of real numbers. If you wish, I can explain that in more detail in the discussion but now I would rather begin with the second possible definition of constructibility and the scheme of proof,

15   which, as I said, is similar in some respects | to Hilbert's program.

Since we have decided to consider definitions of real numbers in terms of ordinals it is quite natural to apply also the kind of definition most appropriate for ordinals, namely, recursive definition. I am considering quite arbitrary sets of ordinals and relations between ordinals $R, S, \ldots$, about which I only assume that their field is limited, i.e., that there exists an ordinal α greater than all elements of their field. The set of ordinals $< \alpha$ I denote by $\bar{\alpha}$. Furthermore I am considering propositional functions formed of such relations R, S. In particular I want the notion of a propositional function of height α in some relations R_i, where α is an arbitrary ordinal and this notion is defined by the following stipulations:

1. Any expressions of the form $R_i(\ldots, x_j, \ldots, \alpha_j)$, where the | argu-   16
ments of the relations R_i are partly constant ordinals $< \alpha$ and partly variables with range $\bar{\alpha}$, are propositional functions of height α in the R_i.

2. By disjunction, negation and quantification of propositional functions of height α, you obtain again such a propositional function. ⟦Here some examples were apparently given.⟧
The α_i occurring here I call the constants of the expression. So in a propositional function of height α the constants are smaller than α and the quantifiers refer to ordinals $< \alpha$. ⟦Examples were again given.⟧

It is to be noted that if α is non-enumerable then there exist non-enumerably many propositional functions of height α in a given system $R_1, \ldots, R_n$ because there are then non-enumerably many constants which may occur. Now, one can recursively define a relation S in terms of given relations $R_1, \ldots, R_n$ in the following manner: I set

$$S(x_1, \ldots, x_n) \equiv \varphi(S \restriction x_1, R_1, \ldots, R_n),$$

φ a given propositional function in $R_1, \ldots, R_n$ and S confined to x_1, where I mean confinement with respect to the first argument only, i.e., $S \restriction x_1$ is defined by this equi|valence, $S(u_1, \ldots, u_k) . u_1 < x_1$. So I had better   17
say φ is a propositional function in $R_1, \ldots, R_n$ and constituents of this type (such conjunctions). If S occurs in φ in several places then x_1 is everywhere the same, namely, the variable x_1 occurring on the left-hand side, whereas the u_i may vary. You will see that the right-hand side of this defining equivalence depends only on the values of S for $u_1 < x_1$, so that we actually have a recursive definition with respect to the variable x_1, since the value of S for x_1 is defined by means of the values $u_1 < x_1$; the other variables appear only as parameters. Here is an example of such a recursive definition which defines the class G of even ordinals. ⟦An example was presumably again given.⟧ I consider in particular the case where φ is a propositional function of a given height α and where also this equivalence is supposed to hold only for $x_i < \alpha$ (so that the field of the defined relation S is contained in the set $\bar{\alpha}$). In this case—i.e., if S can be defined in this way in terms of $R_1, \ldots, R_n$—I call S "recursive of order α" | with respect   18
to $R_1, \ldots, R_n$ and I call a relation S "recursive of order α" (or of "height" α) if it can be obtained by finitely iterated applications of this method

of definition beginning with the relation $<$ or, to be more exact, S is a recursive relation of height α if there exists a finite sequence of relations $S_1, S_2, \ldots, S_k$ such that:

1. $S_1 = <$,

2. $S_k = S$, and

3. each relation of the sequence is recursive of order α with respect to some of the preceding members of the sequence.

So in particular the field of any recursive relation of order α is contained in the set $\overline{\alpha}$. It is to be noted that the explicit definition is a particular case of the recursive definition, namely, the case where φ does not really contain $S \restriction x$ but only $R_1, \ldots, R_n$. Furthermore it is immediate that a recursive

19 relation of order α is *a fortiori* a recursive relation | of any higher order β, because every propositional function φ of height α is formally equivalent to a propositional function of any given greater height β, since a quantifier $(\exists x)$ with range $< \alpha$ can be replaced by the combination $(\exists x)[x < \alpha]$, where now $\exists x$ is a quantifier with an arbitrarily greater range. The difference between this notion of recursiveness and the one that Hilbert seems to have had in mind is chiefly that I allow quantifiers to occur in the definiens. This makes one of Hilbert's lemmas superfluous and the other demonstrable in a certain modified sense, and this allows [one] by means of some more

20 lemmas to construct a model for set theory in which | Cantor's continuum hypothesis holds. These lemmas I am going to formulate now. At first two trivial remarks: First, the finite sequences of ordinals $\alpha_1, \ldots, \alpha_k$ can be mapped one-to-one on[to] ordinals and in particular in such a manner that the ordinal corresponding to $\alpha_1, \ldots, \alpha_k$, which I denote by $\langle \alpha_1, \ldots, \alpha_k \rangle$, is $< \omega_n$ if the α_i are $< \omega_n$. This mapping allows [one] to define recursiveness also for classes of complexes of ordinals; namely, one may call such a class "recursive of order α" if the corresponding class of ordinals is recursive of order α. The second remark is that the recursive functions of order ω_n can be numbered through by means of the ordinals $< \omega_n$, since each of them is given by a finite number of defining equivalences of this kind and each of these formulas is a finite combination of variables, of logical symbols and of constants. But these constants are ordinals $< \omega_n$ (by definition of [a] propositional function of height ω_n); hence the aforementioned mapping of

21 k-tuples of ordinals on[to] | ordinals allows [one] to define this numbering. Let us denote the recursive relation which corresponds, owing to this numbering, to the ordinal α by R_α, or by $R_\alpha^{\omega_n}$ if it is necessary to remind us that it is the recursive relations of height ω_n which are numbered. Then the lemmas about recursive relations which I need are the following:

1. Let us define a class W^{ω_n} of complexes of ordinals by the stipulation: $\langle \alpha_1, \alpha_2, \ldots, \alpha_r \rangle \in X^{\omega_n}$ if and only if $R_{\alpha_1}^{\omega_n}$ holds between $\alpha_2, \ldots, \alpha_r$ [(i.e.,]$\langle \alpha_1, \ldots, \alpha_r \rangle \in X^{\omega_n} \equiv R_{\alpha_1}^{\omega_n}(\alpha_2, \ldots, \alpha_r)[)]$; then the first lemma says that W^{ω_n} is a recursive class of order ω_{n+1}.

The point is that W is recursive; the order is irrelevant. It is easily seen that the order must be $> \omega_n$ because this W allows [one] to define a kind of antidiagonal sequence of all recursive classes of order ω_n.

The next lemma is similar to Hilbert's Lemma 1. It says:

2. A recursive relation | between integers (i.e., whose field consists only 22 of integers) is always of an order $< \omega_1$;

i.e., if you define a relation between integers by a recursive definition of an arbitrarily high order, it is always possible to define it already by [a] recursive definition of order ω_1. (An analogous theorem holds for recursive relations between ordinals of the second number-class. They have always an order $< \omega_2$, etc.)

Finally, the third lemma which I need is only a more constructive statement of the second. According to the second lemma there exists a mapping of all recursive classes of integers on[to] a subclass of the second number-class. | Now the third lemma says (i) that this mapping itself is recursive 23 and (ii) that it gives for each recursive class of integers not only a definite ordinal $\alpha < \omega_1$ associated with it but even a definite representation of this ordinal α by a well-ordering of integers; to be more exact, lemma 3 says:

3. There exists a recursive relation $R(\alpha, m, n)$ such that if α is the number of a recursive class of integers, then the relation between m, n defined by $R(\alpha, m, n)$ is a well-ordering of integers, and if α, β are numbers of extensionally different recursive classes of integers, then the corresponding well-ordering is of different order type.

The proof | for this third lemma is obtained simply by analyzing the proof 24 of the second, which allows [one] actually to define this relation R. As to the proof of the first lemma, it comes down to defining recursively the metamathematical notion of truth for propositions built up of recursive relations and the other constituents mentioned before in the definition of a propositional function of height α. For this class W is really the class of ordinals associated to the true propositions of this special kind, i.e., of the true atomic propositions of our domain. Now this metamathematical notion of truth, i.e., the class of numbers of true propositions, can be defined by a method similar to the one which Tarski applied for the system of *Principia mathematica*. The point is to well-order all propositions of our domain in such a manner that the truth of each depends in a precisely describable manner on the truth of some | of the preceding; this gives then the 25 desired recursive definition. So, e.g., the propositions P, Q have to precede the proposition $P \vee Q$, and the propositions $S(\alpha_1, \ldots, \alpha_n)$ with smaller α_1 have to precede those with greater α_1, since S is defined recursively, etc. I have no time to go into more details of this but would rather sketch the proof of lemma 2.

So let S be a recursive relation for integers of an arbitrarily high order μ and let this be the corresponding chain of recursive definitions leading up to S:

$$\Sigma \begin{cases} (x_i)[R_1(x_1, \ldots, x_{r_1}) \equiv \varphi_1(<, R_1 \restriction x_1)] \\ (x_i)[R_2(x_1, \ldots, x_{r_2}) \equiv \varphi_2(<, R_1, R_2 \restriction x)] \\ (x_i)[S(x_1, \ldots, x_r) \equiv \varphi_{n+1}(<, R_1, \ldots, R_n, S \restriction x_1)], \\ \text{constants } \alpha_1, \ldots, \alpha_m (< \mu) \end{cases}$$

So each R_i is defined in terms of the preceding R_i by means of these equivalences, the quantifiers in the φ_i refer to the ordinals $< \mu$, and the constant ordinals α_i which may occur in the φ_i are less than μ (by our previous stipulations). Let the system of these $n + 1$ equivalences be called Σ.

Now we can consider these equivalences as axioms for which the relations R_i form a model, and then we can apply Skolem's theorem about the existence of enumerable submodels; then we obtain the result that there exists an enumerable subset E of the set $\bar{\mu}$ such that these $n + 1$ equiva-
26 lences will hold also if | all quantifiers are restricted to E. Let's call this new system of equivalence⟦s⟧ $\Sigma \restriction E$. In particular, E can be chosen in such a manner that it contains all constants $\alpha_1, \ldots, \alpha_n$ and all integers (i.e., all ordinals $< \omega$) simply by beginning Skolem's construction with these objects. But then also the relation $R_i \restriction E$ confined with respect to all arguments will satisfy $\Sigma \restriction E$, because no ordinals outside E occur in the system $\Sigma \restriction E$ as arguments of R_i. Now E is a set of ordinals, hence well-ordered by the "smaller⟦-than⟧" relation, and since it is moreover enumerable, its order type must be an ordinal of the second number-class; call it η. So E can be mapped in an order-preserving way on the ordinals $< \eta$. By this mapping the relations $R_i \restriction E$ will go over into certain relations R_i' in the field of ordinals $< \eta$ and the constants α_i into certain ordinals $< \eta$. In particular, each integer will go over into itself because all integers belong to E by definition of E. So they form an initial segment of E. Hence also
27 S, which is a relation between integers, | will go over into itself, and the "smaller⟦-than⟧" relation, of course, also goes over into itself because the mapping is order-preserving. So we have a complete isomorphism between the objects written down in this ⟦top⟧ line and the corresponding objects below them:

$$\begin{array}{ccccccc} E & < & R_1 \restriction E & R_n \restriction E & S & \alpha_1, \ldots, \alpha_n \\ \bar{\mu} & < & R_1' & R_n' & S & \alpha_1', \ldots, \alpha_n' \end{array}$$

Therefore the system Σ of propositions will also hold for the R_i' and S and α_i' with quantifiers restricted to $\bar{\mu}$ (instead of E). Call this new system Σ'; then evidently Σ' is nothing else but a chain of recursive definitions of order μ leading to S. Hence S is recursive of an order $< \omega_1$, which was to be proved. When you analyze this proof in more detail you will find that it gives an actual construction of this number μ in the form of a well-ordering of the integers by the order type μ, and this remark leads to the proof of lemma 3. So it remains only to | be shown that with the help of lemmas 1–3 it is possible to construct a model for mathematics in which Cantor's continuum hypothesis holds. But that is almost trivial. I choose as the formal system for mathematics the system of *Principia mathematica* with simplified hierarchy of types and consider as individuals the integers, assuming Peano's axioms for them, in addition to the logical axioms. So the primitive objects of this system are individual classes and relations, and, according to the theory of types, these entities split up into an infinite sequence of layers such that a relation of the $(n + 1)$-st layer subsists only between objects of the n-th layer, and similarly for classes, which we can consider as monadic relations. In our model these primitive objects will be [represented by] ordinals, to wit, the individuals by the ordinals $< \omega$ and the relations of the n-th layer by the ordinals of the interval $\omega_{n-1} \leq x < \omega_n$. In order to complete the definition of the model I have to specify | also under which circumstances a relation of the $(n + 1)$-st layer (i.e., an ordinal of this interval) will be said to "subsist" between any objects of the n-th layer (i.e., any ordinals of the preceding interval). For this purpose I remind you that the recursive relations of order ω_n can be numbered through by the ordinals $< \omega_n$, hence also by the ordinals of this interval; and now I shall stipulate, of course, that an ordinal α of this interval "subsists" between some ordinals $\beta_1, \ldots, \beta_k$ of the preceding interval if the recursive relation number α (which we denoted by $R_\alpha^{\omega_n}$) subsists between $\beta_1, \ldots, \beta_k$ or, in other words, if $\langle \alpha, \beta_1, \ldots, \beta_k \rangle \in W^{\omega_n}$ according to our previous definition of W. (In order to have the pure hierarchy of types we would have to make some restrictions, which, however, are not very essential.) So this is the definition of the model.

As to the question whether the axioms of *Principia mathematica* | hold for this model, the only difficulty is the axiom of reducibility (or in set-theoretical formulation the axiom of Aussonderung). It requires that given any propositional function $\varphi(x_1, \ldots, x_r)$ whose free variables belong to the n-th layer there exists a relation R of layer $(n + 1)$ such that

$$(x_1, \ldots, x_r)[R(x_1, \ldots, x_r) \equiv \varphi(x_1, \ldots, x_r)].$$

Now in the model this φ will become a propositional function built up from this relation W^{ω_i} and logical symbols, and its quantifiers will refer to ordinals of certain intervals, since all objects of the model are ordinals. Now, since W is recursive by lemma 1, φ defines a recursive relation between $x_1, \ldots, x_r$, because explicit definition is a special case of recursive definition as remarked before. But the field of this relation is $\bar{\omega}_n$ since the x_i are variables of the n-th layer. Hence this recursive relation is of order ω_{n+1} by lemma 2; call its number α, so $\alpha < \omega_{n+1}$. Then we have

$R_\alpha(x_1, \ldots, x_n) \equiv \varphi(x_1, \ldots, x_n)$, i.e., there exists an element of the model,
31   namely α, which satisfies the requirement of the axioms. | That also Cantor's continuum hypothesis is satisfied for the model is an immediate consequence of lemma 3, which says that this relation R which maps the real numbers on [[to]] ordinals of the recursive number-class is recursive, hence represented by an element of the model. Now before concluding let me make a few remarks about this proof. First, it is easily seen that the recursive relations are exactly the same which I formerly called constructible. Hence also the proposition which says that every real number is recursive is free from contradiction (and similarly for sets of higher type). That can be seen directly by means of the model which I have just constructed. The proposition that every real number is recursively definable will hold in this model. However, I wish to call your attention to the fact that the consistency of A does not mean that for every real number definable, say, in the system of *Principia mathematica* one could prove that it is constructible or recursive. On the contrary, it is possible actually to define certain real numbers which very likely cannot be proved to be constructible, although one can of course assume consistently that they are. That situation arises for impredicative definitions and is due to the fact that by confining ourselves to the ramified types or to recursive definitions we have excluded impredicative definitions.

        The second remark is that this consistency proof for the continuum
32   hypothesis and for the proposition A is in a sense | absolute, i.e., independent of the particular formal system which we choose for mathematics. You know every formal system is incomplete in the sense that it can be enlarged by new axioms which have approximately the same degree of evidence as the original axioms. The most general way of accomplishing these enlargements is by adjoining higher types, e.g., the type ω for the system of *Principia mathematica*. But you will see that my proof goes through for systems of arbitrarily high type. For the system of *Principia mathematica* we had to use the recursive relation of the orders ω_n (with finite index n). For a system including all types less than some transfinite ordinal α, we have to use the recursive relation of orders less than ω_α. So this consistency proof goes through for systems of arbitrarily high type, and you will see that the system in which you can formalize the consistency proof must be
33   approximately of the same type as the system which you want to | prove consistent with A or the continuum hypothesis, since in order to prove the existence of ω_β we need the type β. That has the consequence that for systems of arbitrarily high types we can generally give an intuitionistically admissible proof that if the system is consistent it remains so if A is adjoined as a new axiom. This, so to speak, absolute consistency of A is very interesting from the following point of view: It is to be expected that also $\sim A$ will be consistent with the axioms of mathematics, because an inconsistency of $\sim A$ would imply an inconsistency of the notion of a random

sequence, where by a random sequence I mean one which follows no mathematical law whatsoever, and it seems very unlikely that this notion should
imply a contradiction. Another argument which makes the consistency of
$\sim A$ plausible is that an inconsistency of $\sim A$ would yield a proof for the
axiom of choice, whereas the axiom of choice is generally conjectured to
be independent. So A is very likely a really undecidable proposition (quite
different from the undecidable proposition which I constructed some years
ago and which can always be decided in logics of higher types). This conjectured undecidability of A becomes particularly surprising if you investigate
the structure of A in more | detail. It then turns out that A is equivalent   34
to a proposition of the following form: $(P)[F(x_1, \ldots, x_k, n_1, \ldots, n_l) = 0]$,
where F is a polynomial with given integer coefficients and with two kinds
of variables x_i, n_i, where the x_i are variables for real numbers and the n_i
variables for integers, and where P is a prefix, i.e., a sequence of quantifiers composed of these variables x_i and n_i. I have not yet succeeded in
proving that A, and hence this proposition about this polynomial, really is
undecidable, but what I can prove owing to the results which I presented in
this lecture is of course this: Either this proposition is absolutely undecidable or Cantor's continuum hypothesis is demonstrable (since A implies the
continuum hypothesis). But I have not yet been able to determine which
one of these two possibilities is realized.

| A further remark about the proof is that in the construction of the model   35
we didn't make use of the axiom of choice, but on the other hand the axiom
of choice holds in the model (because the continuum hypothesis is a stronger
proposition); therefore we have obtained as a by-result a consistency proof
for the axiom of choice.

Finally a last remark is this: In the definition of a recursive real number I have allowed that arbitrary ordinals occur as constants, so that we
have non-enumerably many recursive definitions. It is clear that this pro-
cedure | is necessary in order to obtain a notion of recursiveness for which   36
the proposition A is absolutely consistent (i.e., consistent in systems of
arbitrarily high types), because if we don't admit non-enumerably many
undefined constants, we could enumerate all recursive definitions and hence
prove the existence of non-recursive real numbers in some formal system.
If, however, we confine the consistency proof to one formal system (say,
Principia mathematica), then we can always restrict ourselves to constant
ordinals definable in this system. In this way we obtain certain enumerable
models, which are also worthwhile to investigate because they have the following interesting property: Each set occuring in such a model is definable
by a propositional function in this | model. So in these models we have in   37
a certain sense realised the situation that every real number is definable
in a finite number of words, although the totality of real numbers is not
enumerable (i.e., not enumerable by a mathematically defined sequence).

Introductory note to *1941*

In this lecture, delivered at Yale 15 April 1941, Gödel gives a beautifully clear account of what later became known as the "Dialectica Interpretation". The introductory note to *1958* and *1972* in Volume II of these *Works* contains a detailed discussion of Gödel's published account of this interpretation; here we concentrate on differences between the lecture and the published account. Notwithstanding Gödel's assertion in later years that originally he was more interested in the mathematical applications of his interpretation than in its philosophical aspects (see Volume II, pages 217–8), the lecture first and foremost presents the interpretation as a foundational contribution, namely, the replacement of abstract intuitionistic concepts, which Gödel treats very critically in *1938a*, by more strictly constructive ones. In 1941, Gödel also gave a series of lectures at Princeton on intuitionistic logic. For those lectures Gödel made fairly detailed notes, and for some of our comments we have drawn on that material.

We first summarize the contents of the text. For unexplained notations, see the introductory note to *1958* and *1972* in Volume II.

(1) According to Gödel, intuitionists have two objections to classical mathematics: (a) the use of impredicative definitions, and (b) the use of the principle of the excluded middle. Objection (b) is at first sight the more serious one, but this is only apparent, as may be seen from the embedding of classical systems into the corresponding intuitionistic ones (see *Gödel 1933e*).

(2) On the other hand, the strictly constructive character of intuitionistic mathematics is in doubt, since the interpretation of logical operators such as implication and the universal quantifier relies on imprecise abstract notions such as "construction" and "(constructive) proof". (This is a less emphatic restatement of the criticisms in section V of *1938a*.) Gödel remarks that the abstract notion of proof cannot be read as formal proof; see *Gödel 1933*.

(3) The question is then whether these abstract, imprecise notions can be replaced by more strictly constructive ones; Gödel intends to show that this can be done at least for number theory.

(4) Strictly constructive systems should meet the following criteria: (a) the primitive functions and relations are calculable and decidable, respectively, for all arguments; (b) existential quantifiers appear only as abbreviations, for statements with explicit realizations of those quantifiers; propositional operators are not to be applied to universal statements. To meet criterion (b), one considers counterexamples to universal statements instead of negations of universal statements. This results in

a class of meaningful statements in $\exists\forall$-form. These criteria also appear in *1938a*, section II, together with two further criteria. The distinction between two components of finitism (*1958*, page 283) is not found here.

(5) The objects needed for the interpretation are to be primitive recursive functionals of finite type. (In the Princeton notes, Gödel stresses that only equality between numbers is to be taken as a primitive relation. This excludes "intensional" systems of functionals as in the theory **I-HA**$^\omega$, described in Volume II, page 230.) Gödel remarks that it is pretty complicated to see in which manner the primitive recursive functions of finite type are calculable (more detailed comments below). In his *1958*, Gödel treats "calculable functional of finite type" as a primitive notion. Gödel's *1938a*, section IV, contains a very sketchy description of the primitive recursive functionals of finite type.

(6) Intuitionistic arithmetic is reduced to a system Σ which is almost the system **T** of *1958*, except that the statements are not quantifier-free but in $\exists\forall$-form.

(7) The clauses of the Dialectica translation are described and motivated, in particular that for implication. Gödel proceeds to demonstrate some steps in the soundness of intuitionistic predicate logic under this interpretation; the discussion is more leisurely and intuitive than in the published paper.

(8) Four applications are mentioned:

- There is a number-theoretic formula $A(x)$ such that $\neg\forall x(A(x) \vee \neg A(x))$ can be consistently added to intuitionistic arithmetic **HA**. The Princeton notes contain Gödel's proof of this fact; see below.

- If **HA** $\vdash \exists x A(x)$, then for $A^D \equiv \exists\mathbf{x}\forall\mathbf{y}\, A_D(\mathbf{x}, \mathbf{y})$ there is an explicit realization by terms $\mathbf{t}$ such that $\Sigma \vdash A_D(\mathbf{t}, \mathbf{y})$; this does not hold for classical number theory **PA**.

- Combining the negative translation with this new interpretation, the consistency of classical arithmetic can be reduced to Σ as well.

- A version of Markov's rule holds: if **PA** $\vdash \exists x A(x)$ for quantifier-free A (which is equivalent to **HA** $\vdash \neg\neg\exists x A(x)$), it follows that $\Sigma \vdash A(t)$ for suitable t.

(9) The possibility of a generalization to analysis is mentioned, for which an essential extension of Σ will be needed. On the other hand, it is noted that for the consistency of **PA** and **HA** all finite types of Σ are needed.

In his lecture Gödel gives no hint as to the method of proof used for the applications in (8). However, the proof of the first application is sketched in the notes for his lectures at Princeton in the same year. There Gödel describes a (version of) the model *HEO* of the hereditarily effective operations and sketches a proof that *HEO* is a model for his system of functionals. (*HEO* is an extensionalized version of the hereditarily

recursive operations *HRO* defined in Volume II, page 233; in particular, objects of type 2 are given by codes of partial recursive operations
defined on codes of total recursive functions, and respect equality between total recursive functions. For a full definition, see *Troelstra 1973*,
2.4.11.) For the formula $A(x)$ Gödel takes (in modern notation):

$$A(x) := \exists y \forall z \neg Txyz$$

that is, x is the code of a non-total partial recursive function. The Dialectica translation of $\forall x(A(x) \vee \neg A(x))$ is then equivalent to

$$B \equiv \exists v^1 y^1 w^{0 \to (0 \to 0)} \forall x^0 z^0 u^0 C(x, z, u, v, y, w), \quad \text{where}$$
$$C(x, z, u, v, y, w) \equiv [(vx = 0 \wedge \neg T(x, yx, z)) \vee (vx = 1 \wedge T(x, u, wux))].$$

The Dialectica interpretation of $\neg B$ requires functionals X, Z, U of type
$1 \to (1 \to ((0 \to 1) \to 0))$. Now $C(x, z, u, v, y, w)$ is recursively decidable
for any v, y, w of the appropriate types, and for any such v, y, w there
will exist integers x, z, u such that $\neg C(x, z, u, v, w, y)$, since otherwise
v would provide a recursive solution to the totality problem for partial
recursive functions. Thus, given v, y, w we can search for the least triple
$\langle x, z, u \rangle$ (in some standard enumeration) such that $\neg C(x, z, u, v, y, w)$
holds. Since this triple does not depend on the codes for v, y, w, but
only on their extension, this yields suitable X, Z, U of the appropriate
type in the model *HEO*.

As noted above, Gödel considered only equality of type 0 as a primitive notion, and there is no indication that he considered non-extensional
type structures as possible models. (If one permits non-extensional models, an even simpler example of the consistency of intuitionistic arithmetic with a statement of the form $\neg \forall x(A(x) \vee \neg A(x))$ becomes available: one simply takes $A(x) := \exists y Txxy$ and shows that the Dialectica
interpretation of $\neg \forall x(A(x) \vee \neg A(x))$ is satisfied in the model *HRO* of
the hereditarily recursive operations.)

In the Princeton notes Gödel observes that it is not difficult to prove
calculability of the functionals of Σ in the following sense: a functional
term F of type $\sigma_1 \to (\sigma_2 \to (\sigma_3 \to \ldots (\sigma_n \to 0) \ldots))$ is said to be
calculable if for arbitrary calculable $t_1, t_2, \ldots, t_n$ of types $\sigma_1, \sigma_2, \ldots, \sigma_n$,
respectively, $Ft_1t_2 \ldots t_n$ can be proved to be equal to a numeral. Gödel
goes on to say (page 61 of the notes) "I don't want to give this proof
in more detail because it is of no great value for our purpose for the
following reason: if you analyze this proof it turns out that it makes use
of logical axioms, also for expressions containing quantifiers and ... it
is exactly these axioms which we want to deduce from the system Σ."
This is strongly reminiscent of the computability method of Tait (*1967*).

Gödel then continues (page 62): "There exists however another proof. Namely, it is possible instead of making use of the logical operators applied to quantified expressions to use the calculus of the ordinal numbers (to be more exact of the ordinal numbers $< \epsilon_0$). I shall speak about this proof later on." Then follows a description of the idea of an ordinal assignment to terms, such that an evaluation step decreases the ordinal assigned. Page 63′′′ contains another version of these remarks, where Gödel says, "However it seems to be possible to give another proof which makes use of transfinite induction up to certain ordinals (probably up to [the] first ϵ-number would be sufficient)." Since the notes do not contain any further particulars, it is not likely that Gödel had actually carried through such a proof in detail. See also 5.2 of the introductory note to *1958* and *1972* in Volume II.

A. S. Troelstra

In what sense is intuitionistic logic constructive?
(*1941*)

I appreciate very much the opportunity of speaking here in this famous university where there are so many persons interested in mathematical logic. The subject I have chosen is perhaps a little out of fashion now, but it seems to me that this is not quite justified because there are many unsolved problems left, problems which may be of fundamental importance for the foundations of mathematics. I shall be happy if my talk helps to draw attention to these problems.

| The objections of the intuitionists against classical mathematics are directed against two entirely different kinds of inference. First are the so-called impredicative definitions, second is the law of excluded middle and related theorems of the propositional calculus. These two kinds of objection are so entirely independent that one can very well approve one and reject the other, as did the half-intuitionists. At first sight it may seem as if the second objection, concerning the law of excluded middle, were much more serious in its consequences, because it concerns a more fundamental law and also because larger domains of mathematics are involved, since impredicative definitions play a role only in analysis and set theory, not in number theory or algebra. But a closer examination shows that quite the opposite is true. In fact it turns out that nothing at all is lost by

dropping the law of excluded middle, but only the interpretation of the theorems has to be changed. That means, it is true, the intuitionists reject the law of excluded middle $p \lor \neg p$ for their notion of disjunction, but
2 it is possible to define in terms | of their other primitive logical symbols another notion of disjunction for which the law of excluded [middle] does hold also in intuitionistic logic. This is quite trivial; you have only to define $p \lor q = \neg(\neg p . \neg q)$. Then $p \lor \neg p$ becomes $\neg(\neg p . \neg\neg p)$, and this is a case of the law of contradiction, which holds also in intuitionistic logic. But it turns out that this procedure applies not only to the law of excluded middle itself, but also to all its consequences, in particular to the so-called non-constructive existence proofs, provided only that no impredicative definitions are involved. You have only to define the notion of existence by the equivalence $(\exists x)A(x) =_{\mathrm{DF}} \neg(x)\neg A(x)$, and then a non-constructive existence proof rejected by Brouwer as meaningless will become an intuitionistically correct proof for existence in this sense. And it can be shown quite generally that, as long as no impredicative definitions come in, every proof of classical mathematics becomes a correct intuitionistic proof if existence and disjunction are defined in this way. In particular this applies to all purely number-theoretic and algebraic proofs where impredicative definitions never occur.
3 | So intuitionistic logic, as far as the calculus of propositions and of quantification is concerned, turns out to be rather a renaming and reinterpretation than a radical change of classical logic. E.g., negation of universal propositions is rejected as meaningless, but the predicate of absurdity is applied without restriction and the axioms which intuitionists consider as evident about this predicate lead, with suitable definitions of the other terms, to exactly the same calculus as classical negation, provided the other logical notions are suitably defined. Now this state of affairs makes one doubtful whether the intuitionists have really remained faithful to their constructive standpoint in setting up their logic or if not perhaps they have allowed some non-constructive elements to creep unnoticed into their axioms. And actually a closer examination shows at least this: the primitive terms of
4 intuitionistic logic lack the complete | perspicuity and clarity which should be required for the primitive terms of an intuitionistic system. E.g., $P \to Q$ in intuitionistic logic means that Q can be derived from P, and $\neg P$ means a contradiction can be derived from P. But the term "derived" cannot be understood in the sense of "derivation in a definite formal system". (For this notion the axioms of intuitionistic logic would not hold.) So the notion of derivation or of proof must be taken in its intuitive meaning as something directly given by intuition, without any further explanation being necessary. This notion of an intuitionistically correct proof or constructive proof lacks the desirable precision. In fact one may say that it furnishes itself a counterexample against its own admissibility, insofar as it is doubtful whether a proof utilizing this notion of a constructive proof is constructive

or not. For this reason it seems that if one wants to take constructivity in a really strict sense that the primitive | notions of intuitionistic logic cannot be admitted in their usual sense. This however does not exclude the possibility of defining in some way these notions in terms of strictly constructive ones and then proving the logical axioms which are considered as self-evident by the intuitionists. It turns out that this can actually be done in a certain sense, namely, not for intuitionistic logic as a whole but for its applications in definite mathematical theories (e.g., number theory), and this is the result I would like to sketch in this talk. In order to do it I must first explain more fully what seem to me the necessary requirements of a strictly constructive system, namely, the following:

1. All primitive (undefined) functions which one introduces must be calculable for any given arguments and all primitive relations must be decidable for any given arguments.

2. Existential assertions must have a meaning only as abbreviations for actual constructions, i.e., the existential quantifier must not appear as a primitive term but as an abbreviation, and if we write everything without abbreviation we have no existential quantifier at all.

3. Universal propositions can only be negated in the sense that a counterexample exists | in the sense just defined (i.e., in the sense of an abbreviation for an actual construction of a counterexample).

Therefore, leaving out abbreviations, universal propositions cannot be negated at all. But this is only a special case of a more general principle which is its natural consequence, and which says that none of the connectives of the calculus of propositions is to be applied to universal propositions. For if you admit, e.g., $\rightarrow$, you admit negation implicitly, since $\sim P$ can be defined by $P \rightarrow (0 = 1)$. Furthermore you will see that the same reason for not admitting them applies to all operations of the propositional calculus. Namely, they cannot be admitted in the sense of classical logic because there they are defined by truth tables, but truth tables make no sense in a constructive system, if you apply them to universal propositions about whose truth or falsehood you cannot in general decide. But also in the sense of intuitionistic logic they cannot be admitted for the reasons I explained before.

Let me call a system strictly constructive or finitistic if it | satisfies these three requirements (relations and functions decidable, respectively, calculable, no existential quantifiers at all, and no propositional operations applied to universal propositions). I don't know if the name "finitistic" is very well chosen, but there is certainly a close relationship between these systems and what Hilbert called the "finite Einstellung". Of course the three requirements of finitism apply only as long as you don't introduce abbreviations or defined symbols. In a finitistic system, as in every other, you can of course introduce new symbols, either by explicit definition or by definition of use, in an arbitrary manner if only the requirement of every correct definition is satisfied, namely, that the rules by which you introduce these new symbols are such that they allow you to eliminate them in any consideration. This is exactly the manner in which I intend to introduce the notions of intuitionistic logic. Intuitionistic logic itself (as presented, e.g., in Heyting's system) does not satisfy these three requirements. The existential quantifier there is not introduced by definition but appears as a primitive symbol and the propositional connectives are applied without restriction.

8 | But as I said, in its application to definite mathematical theories intuitionistic logic can be reduced to finitistic systems. In order to explain how, I must first investigate the structure of a finitistic system in more detail. What can a proposition containing no existential quantifiers and no propositional connectives applied to universal propositions look like? The

9 answer is very easy, namely this: | The universal quantifiers must all be at the beginning of the expression so that the remaining part of the expression is formed of certain atomic expressions (or prime formulas) by means of the propositional connectives alone (without quantifiers). Now as to the prime formulae, they are composed of certain relations (like, e.g., less than), certain functions (like, e.g., addition and multiplication), certain constant arguments (like, e.g., the numbers 5 or 2), and the variables x, y, z. To be more exact: Each prime formula states that a certain relation subsists between certain terms, where by a term is meant anything obtainable from constants or variables for individuals by simple or iterated application of the functions. Now since by assumption the functions are calculable and the relations decidable, the truth or falsehood of any atomic expression for

10 any | given arguments x, y, ... can be decided. Therefore the application of the propositional operations in this expression M is not problematic. They can be understood in the sense of the truth tables. If understood in this sense they are nothing else but a special kind of these calculable functions, namely, functions which take as arguments and values only the two objects (True and False).

Hence, the meaning of such a proposition (which is the most general form of a meaningful statement in a finitistic system) is the following: If any arbitrary objects x, y, z of a certain domain are given and if you perform certain given calculable operations f, g, $\rightarrow$, etc., in a certain order

11 on them, | then you always obtain the truth-value T as the result, whatever the given objects may have been.

Now as to the axioms and rules of inference of such a finitistic system, they depend on the primitive objects, relations and functions, which may be of various kinds. The only axioms common to all are the axioms of the ordinary two-valued calculus of propositions applied to prime formulae. Existential quantifiers can now be introduced by the following rule: If $A(t)$ is a meaningful expression, i.e., of this structure, and if A contains a constant term t then you can infer $(\exists x)A(x)$, and similarly simultaneously for several terms, say t_1, t_2:

$$A(t_1, t_2) \qquad (\exists x, y)A(x, y),$$

and this is the only rule concerning the existential quantifiers, i.e., an existential assertion can only appear as the last formula of a proof and the last but one formula of the proof must give the corresponding construction. This way of introducing the existential quantifier is of course not an | explicit definition. It has not the form $(\exists x)A(x) =_{DF}$ some expression composed of previously defined symbols, but rather it is a definition of use, which states how propositions containing the new symbol are to be handled in proofs, i.e., from which premises they can be inferred, namely these, and what can be inferred from them, namely nothing. Now such an implicit definition must satisfy the requirement of eliminability. To be more exact: If a proposition not containing the new symbol can be proved with the help of the new symbol, it must be demonstrable without the help of the new symbol (otherwise we would not have to do with a definition but with a new axiom). But this requirement is trivially satisfied by this manner of introducing the existential quantifier.

After introducing existential quantifiers in this manner, every meaningful proposition has, as you see, the following structure: $(\exists x_1, \ldots, x_i)$ $(y_1, \ldots, y_n)M(x_1, \ldots, x_i, y_1, \ldots, y_n)$, where M contains no quantifiers. Needless to say, in intuitionistic logic as formalized by Heyting this is by no means the only rule concerning the existential quantifier. Already the possibility of substituting existential propositions into the schemes of the propositional calculus gives a great number of other rules.

The analysis of a finitistic system I have given so far describes only an abstract scheme and now we have to fill it with concrete objects, i.e., I have | to specify what the primitive objects, the functions and relations are to be. This can be done in various ways and with constructivity (or calculability) in a stricter or looser sense, so that we obtain different layers or levels of finitistic mathematics. The lowest level is recursive number theory, where the primitive objects are the natural numbers and the relations and functions must all be defined by ordinary recursion or complete induction. This system is the simplest and most perspicuous of all finitistic systems, but unfortunately much too weak for the purposes needed in the foundations of mathematics. (You cannot even prove the consistency of classical number theory within this system.) But now one can extend it in various ways, e.g., by allowing recursive definitions of a more general character in the following manner: Instead of ordering the integers by increasing magnitude, you can order (or, to be more exact, well-order) them by some other relation R and then you can allow also such recursive definitions | as define $f(n)$ in terms of some $f(n')$, where now n' precedes n in the ordering

R. Since R is a well-ordering the sequence $n,\ n',\ n'',\dots$ must break off somewhere, and hence $f(n)$ can be calculated in a finite number of steps, but you don't know in advance the number of steps necessary (whereas in the case of ordinary recursive definitions this number is n). By means of such recursive definitions (where the order-type of R is the first ϵ-number) Gentzen succeeded in proving the consistency of classical number theory. The weak point of this kind of extension of recursive number theory lies of course in the question: How does one know that R is a well-ordering without using set-theoretical methods of proof?

But there are other ways of extension, e.g., the following (which I shall use): In recursive number theory we have to do only with integers and
15   functions of integers | (i.e., functions which are applied to integers and yield integers as result), but now we can introduce functions of functions of integers, i.e., functions which are applied to functions and yield functions as a result, say, e.g., the operation Q of squaring a function f, which is defined as follows: $Q(f)$ is the function g which is performed on an argument x by applying f twice.

$$Q(f) = g, \text{ where } g(x) = f(f(x)).$$

Hence

$$Q(f)(x) =_{\text{DF}} f(f(x)).$$

So this squaring is an operation which produces a new function of integers out of a given one, and therefore is called a function of second type. It is clear how still higher types can be defined. In general, if two types t_1, t_2 have been defined already, we can introduce a new type of function which applied to functions of type t_2 gives a function of type t_1 as the result.
16 | Now the chief question for our purpose is: In which way must such functions of higher types be defined in order to be admissible in a finitistic system, i.e., in order to be calculable?

Of course more complicated functions are not defined in one step, but by a chain of definitions where each step defines new functions in terms of previously defined ones. So the question is: What must such a step look like in order to give a calculable function, provided that the previously defined functions are calculable? It is plausible that at least the following two schemes will meet this requirement:

1. Explicit Definition, i.e., a function F is defined by the stipulation that $F(x_1)\dots(x_n)$—i.e., F applied to x_1, the result (which is a function again) applied to x_2, and so on—is equal to a term T composed of the variables $x_1,\dots,x_n$, and of previously defined functions. (The definition of the function Q I gave before is an example.)
17 | 2. We can admit Recursive Definition, i.e., a function G whose arguments are integers and whose values may be functions of any type is defined by

the stipulations that: $G(0) = T_1$; $G(x + 1) = T_2(x, G(x))$, where T_1 and T_2 are terms composed of previously defined functions and the arguments marked here.

So the schemes of definition are formally the same as in recursive number theory, the only difference being that the objects with which we are dealing now are not only numbers but also functions or, in other words, procedures for obtaining numbers out of given numbers (respectively, for obtaining procedures out of given procedures, as this Q here). Accordingly, we have a new primitive operation, namely, the operation of applying the procedure to an object of appropriate type. But this operation is actually calculable since it is contained in the notion of a procedure that it can always be carried through. Therefore also a term composed out of this operation of application and letters denoting special procedures will be calculable, and therefore the construction F introduced by such a definition will again constitute a procedure; and the same holds for this second kind of definition. This is the underlying idea. A closer examination of the question in which manner the functions obtained by these two schemes are really calculable is pretty complicated. But I shall not deal with this question now, but shall rather show how the notions of intuitionistic logic can be defined in terms of these functions or procedures of higher types.

| Let us denote by Σ the formal system in which functions of arbitrarily 18 high type but only these two schemes of definition are admitted. As axioms of Σ, I assume essentially only these:

    1. the ordinary calculus of propositions for decidable expressions,
    2. the rule of complete induction,
    3. the two rules of substitution (namely, substituting equals for equals and substituting special objects in a universal proposition).

So Σ is a finitistic system in the former sense, and we may introduce existential quantifiers in the previously described manner so that every meaningful proposition of Σ will have the form: There exist objects $f_1, \ldots, f_n$ of certain types such that for all objects $g_1, \ldots, g_m$ of certain types $M(f_i, g_j)$ is true. Recursive number theory is a portion of Σ, and now it turns out that intuitionistic logic as applied in number theory can be reduced to this system; and, more generally, if you apply intuitionistic logic in any branch of mathematics you can reduce it to a finitistic system of this kind under the sole hypothesis that the primitive functions and primitive relations of this branch of mathematics are calculable, respectively, decidable. This hypothesis is essential, but it is perhaps a requirement of a sound intuitionistic system that this hypothesis be satisfied. This finitistic system to which intuitionistic logic, applied in the branch of mathematics under consideration, can be reduced is always obtained by introducing functions

19   of | higher types analogous to these, with the only difference that the in-
dividuals upon which the hierarchy of functions is built up are no longer
the integers but the primitive objects of the branch of mathematics under
consideration. Now in what sense can intuitionistic logic be reduced to
these systems?

Let us stick to the example of number theory and the corresponding
system Σ. It is possible to define the meaning of the logical operations
$A \to B$, $\neg A$, $A \vee B$, $(x)A$, $(\exists x)A$, applied to propositions A, B of Σ in such
a manner that

  1. $A \to B$, etc., will again be propositions of Σ and
  2. the axioms and rules of intuitionistic logic become demonstrable theorems of Σ.

It is to be pointed out that in the system Σ we have already the operations of the two-valued calculus of propositions with all their axioms and
rules, but applied only to decidable statements (i.e., such as contain no
quantifiers). So what we are doing by these definitions is to reduce the
problematic case in which A and B contain quantifiers to the unproblematic [[one]] in which they don't.

Now let me expose this definition of the logical operations for the most
important of these, namely, implication. So let A and B be two meaningful
propositions of Σ, A the proposition $(\exists x)(y)M(x,y)$ and B the proposition
20    $(\exists u)(v)N(u,v)$, | where I assume for the sake of simplicity that we have
only one existential quantifier and one universal quantifier in each expression. The general case can be treated exactly the same way; in fact, you
can reduce it to the simpler case by introducing n-tuples of functions or
individuals as new objects. So our problem is to find an expression of Σ (i.e.,
an expression of this structure $[[(\exists f_1, \ldots, \exists f_n)(g_1, \ldots, g_m)M(f_1, \ldots, f_n,$
$g_1, \ldots, g_m)]])$ which has the meaning of this implication: $(\exists x)(y)M(x,y) \to$
$(\exists u)(v)N(u,v)$. But that is not difficult. This implication means: If there
exists an object x satisfying a certain condition then there exists also an
object u satisfying a certain other condition. But in a constructive logic
that can only mean: Given such an x you can construct such a u, i.e.,

$$(\exists f)(x)[(y)M(x,y) \to (v)N(f(x),v)].$$

But this expression still has not the form we want, since implication still
21   is applied to expressions containing quantifiers. But now we have | only
universal quantifiers here, and what can in a constructive logic be the meaning of the assertion: If $(x)F(x)$ then $(y)G(y)$? The simplest interpretation
which suggests itself is this: Given a counterexample for G you can construct a counterexample for F, i.e., the expression in square brackets will
mean

$$(\exists g)(v)[\sim N(f(x),v) \to \sim M(x,g(v))];$$

but here the symbol of implication is applied only to expressions without quantifiers. Therefore, we can replace it by the ordinary implication of two-valued logic. Furthermore, we can apply transposition to this implication, interchanging the order of the two terms, and obtain in this manner for the whole expression:

$$(\exists f)(x)(\exists g)(v)[M(x, g(v)) \supset N(f(x), v)].$$

But now, that for every x there exists such a function g means that g is really a function of two variables x and v, i.e., we finally obtain:

$$(\exists f, g)(x, v)[M(x, g(x, v)) \supset N(f(x), v)],$$

| and now we are through, since this expression has the form we want, i.e.,   22
it is again an expression of Σ. But it is to be noted that this new expression contains functional variables of a higher type than the given expressions A and B. You see, f and g are functions whose arguments and values are of the type of the variables of the given expression; hence they are themselves of a higher type. And this heightening of types produced by the operation $\rightarrow$ explains why it was necessary to adjoin these functions of arbitrarily high type in order to interpret intuitionistic logic. The logical operations can of course be iterated (you can form, e.g., $A \rightarrow (B \rightarrow C)$, etc.), and in this way you can obtain expressions of arbitrarily high type owing to this interpretation.

The definition of the other logical operations is still simpler. $A \& B$, where A and B are these two expressions $[(\exists x)(y)M(x, y)$ and $(\exists u)(v)$ $N(u, v)]$, simply means: $(\exists x, u)(y, v)[M(x, y) . N(u, v)]$, and $A \lor B$ is defined in exactly the same manner, | replacing $\&$ by $\lor$. [The operation] $\neg$ 23
need not be defined separately since it can be defined in terms of implication: $\neg A =_{DF} A \rightarrow (0 = 1)$. So it remains only to explain quantification. I.e., given an expression A of Σ containing some constant t, $A(t)$, what shall we understand by $(\exists z)A(z)$ and by $(z)A(z)$? Now the first is no problem, since $(\exists z)A(z)$ is again an expression of this form (i.e., of Σ) if A is one. But for the universal quantifier it is different, for assume $A(t)$ is this expression $[(\exists x)(y)M(x, y, t)]$. Then $(z)A(z)$ becomes $(z)(\exists x)(y)M(x, y, z)$, which is not an expression of Σ. But here we can again apply the consideration used before, and define this to mean: There exists a function f producing a corresponding x for any given z, i.e., we can define $(z)A(z)$ to mean $(\exists f)(z)(y)M(f(z), y, z)$, and this is again an expression of Σ. So all logical operations are now defined in such a manner as to yield expressions | of   24
Σ if applied to expressions of Σ. Now we have only to prove that for this interpretation of the logical operations the axioms and rules of inference of intuitionistic logic become demonstrable theorems in Σ. Let's consider as examples the axiom $A \rightarrow A$ and the rule of implication: From A and $A \rightarrow B$ we can infer B. What we can show is this: If A is an arbitrary

expression of Σ and "implies" has the meaning just defined, then $A \rightarrow A$ is a demonstrable proposition of Σ, and likewise if A, B are arbitrary propositions of Σ and A, $A \rightarrow B$ are demonstrable in Σ then B is demonstrable in Σ. And exactly the same thing is true for the other axioms and rules of inference. In these two examples the proof is particularly easy. This expression here means $A \rightarrow B$

$$[(\exists f)(\exists g)(x, v)[M(x, g(x, v)) \rightarrow N(f(x), v)]].$$

If A is the same expression as B, then M is N, and we can immediately define explicitly the two functions f, g, namely, $f(x) = x$, $g(x, v) = v$. Then the implicans and implicatum | become the same. Now as to the rule of implication, let us assume that A and $A \rightarrow B$ are demonstrable in Σ. Then owing to the manner in which existential symbols were introduced in Σ, we can define an entity a such that this holds:

$$(y)M(a, y),$$

and two functions φ, ψ such that this holds:

$$(x, v)[M(x, \varphi(x, v)) \supset N(\psi(x), v)].$$

Now these formulas hold for all x, v, y. Hence we can substitute $\frac{a}{x}$ in the second, getting this:

$$(v)[M(a, \varphi(a, v) \supset N(\psi(a), v)].$$

Next we substitute $\varphi(a, v)$ for y in the first, getting this:

$$(v)M(a, \varphi(a, v)).$$

Hence by the rule for the ordinary two-valued implication, this:

$$(v)N(\psi(a), v),$$

and consequently, by the definition of the existential quantifier, this:

$$(\exists z)(v)N(z, v),$$

which is B. The proofs for the other axioms and rules are a little longer but quite as obvious. The only thing you have to do is to write down the corresponding expressions and then everything becomes obvious. A particularly interesting case is the rule of exportation and importation, which says that from the premise | $A \rightarrow (B \rightarrow C)$ you can infer $(A \& B) \rightarrow C$ and vice versa. If you write down the corresponding expressions of Σ you will find that the premise and the conclusion are exactly the same expression.

Almost the same thing happens with the rules of inference concerning the quantifiers. Namely, there it turns out that premise and conclusion differ only in the arrangements of the arguments of these functions f, g.

Now let me give some applications of this interpretation. First of all, if you translate the law of excluded middle $A \vee \neg A$ or of double negation $\neg\neg A \to A$ in it you will find that it cannot be proved in Σ. You can develop this into an independence proof for these axioms, which gives somewhat more than the independence proofs known so far. Namely, Heyting has proved that $A \vee \neg A$ is not demonstrable in intuitionistic logic. But on the other hand, it is not possible to assume without contradiction that $(\exists A)\neg(A \vee \neg A)$, because you can prove in intuitionistic logic the absurdity of the absurdity of the law of excluded middle, i.e., for every proposition $(A)\neg\neg(A \vee \neg A)$. But what you can assume without contradiction is this: $\neg(A)(A \vee \neg A)$. And that this assertion really is consistent in intuitionistic logic can be proved by this interpretation. To be more exact, you can construct a certain number-theoretic propositional function $\varphi(x)$ for which it is free from contradiction to assume in intuitionistic mathematics that $\neg(x)[\varphi(x) \vee \neg\varphi(x)]$.

Another application of this definition of the primitive terms of intuitionistic logic is the following: If you have an arbitrarily complicated expression A composed of formulas of Σ and the symbols $\neg$, $\vee$, $\cdot$, $($, $)$, etc., then its "meaning" (owing to this definition of the logical terms) will be a certain expression A' of Σ, and if A is demonstrable from the axioms of intuitionistic logic, then A' is demonstrable in Σ. And this fact allows [us] to answer the question in which sense intuitionistic logic as applied in number theory, or more generally in any theory with decidable primitive terms, is constructive. Namely, | if you are able to derive in intuitionistic 27 number theory an existential proposition $(\exists x)\varphi(x)$, then the corresponding proposition $(\exists x)\varphi'(x)$ will be demonstrable in Σ, but we know an existence proof in Σ furnishes the construction, i.e., we can find a term t composed of the functions of the system Σ such that $\varphi'(t)$. Hence an intuitionistic existence proof in number theory always allows you to construct an example by means of the functions of the system Σ. For an existence proof of classical number theory this is by no means true. It is quite trivial to construct examples for existence proofs in ordinary number theory where the corresponding construction would give a solution of Fermat's problem. So although these existence proofs a[re] quite trivial in classical logic, nobody is able to give a corresponding construction, and this is not due to the present state of development of mathematics since it can be shown that there exists no procedure | which would allow [us] to construct an example 28 for every existence proof of classical number theory. But for intuitionistic number theory such a procedure exists, as shown by the interpretation I explained. But now, as I remarked in the beginning, classical logic is in a sense contained in intuitionistic logic (provided that no impredicative procedures are involved, which is the case in number theory). Therefore it is to be expected that it will be possible in this way to obtain also some results concerning classical number theory, and this is actually the case. First of all the consistency of classical number theory is reduced to the consistency of Σ, because we know every correct proof of ordinary logic becomes a correct intuitionistic proof if the notions of disjunction and of existence are eliminated by these definitions. Therefore if a contradiction

29 were obtained by means of ordinary [logic] we could obtain | the same contradiction by means of intuitionistic logic; hence we would also have a contradiction in Σ owing to our interpretation. The consistency of number theory has been proved some years ago by Gentzen with the help of the first ϵ-number. Here you have a proof with the help of constructive functions of higher type. Also the by-results which Gentzen obtained can be obtained very easily by this method. The most important one is that even a classical existence proof (for, say, $(\exists x)\varphi(x)$) can be transformed into a construction if only φ is a decidable property of numbers. The proof is immediate. If φ is decidable (i.e., contains no quantifier) then $(\exists x)\varphi(x)$ turns out to be the same formula as $\neg(x)\neg\varphi x$ in this interpretation, but this expression now contains no "or" and no existential quantifier. Therefore if it is classically demonstrable, it is also intuitionistically demonstrable. But for an intu-

30 itionistic | existence proof we know that a corresponding construction in Σ can be found. It is perhaps not altogether hopeless to try to generalize these consistency proofs to analysis by means of functions of still higher (i.e., transfinite) type. Future development will show if that is possible at all and in which sense the system necessary to accomplish this proof will be constructive. By means of the system Σ alone it is certainly not possible to prove the consistency of analysis, because the consistency of Σ can be proved in analysis, nor is it possible to prove the consistency of number theory by a portion of Σ which goes only up to a given finite type. The whole system Σ is actually necessary.

Gödel in front of Firestone Library, Princeton University

Oskar Morgenstern

Introductory note to *1946/9*

1. The texts (see the textual notes for a full description) from which these two versions have been selected for publication here[a] are all preparatory to the article *1949a*, published by Gödel himself, and were destined for the volume *Schilpp 1949* in which that article appeared. As has been remarked in the introductory note to *1949a* (these *Works*, Volume II, pages 199–201), these papers are in content by no means mere preliminary drafts of that article: they are quite extended essays on a subject rather different from that of *1949a*; and, it seems fair to say, they reveal far more of Gödel's philosophic views on the physical world and our knowledge thereof (his views, that is, on the metaphysics and epistemology of natural knowledge). On the other hand, it must be recognized that Gödel's own decision was *not* to publish a work based directly upon these essays, but to begin again and to write the far shorter and more circumscribed *1949a*. In particular, none of the manuscripts here in question was put into final form for publication;[b] there are certain roughnesses, and—of greater moment—certain questionable passages, that one would expect Gödel to have amended in a final revision. (It will be necessary presently to comment on some of these passages.) The papers should be read, therefore, in clear awareness that they are not finished products, but working drafts. As such (as working drafts, that is, on a subject about which Gödel was reflecting deeply over a period of years) they are of a very special interest.

The subject of these essays, in contrast with the published paper *1949a*, may be characterized briefly thus: The very short piece *1949a* is principally concerned to argue that relativity theory supports the doctrine, attributed to "Parmenides, Kant, and the modern idealists"

---

[a]The decision to publish versions *B2* and *C1* was based upon the facts that (1) the manuscripts fall into two distinct groups: *A* and the two labeled *B* on the one hand, the two labeled *C* on the other; (2) within the first group, *B2* is undoubtedly the latest; and (3) of the *C* versions, *C2*, which remains incomplete, has all the earmarks of an attempted but abandoned revision of *C1*. (Cf. also §**2** of this introductory note.)

However, a passage in *A*, pp. 8–11, seemed to the editors to contain remarks of considerable interest not to be found in the later versions. Therefore it is included in this volume as Appendix A.

[b]The title of these texts appears on one of the lists, made by Gödel, with the title *Was ich publizieren könnte* ("What I might [or "could"] publish"); see the preface to this volume. That certainly encourages the belief that Gödel did not mean to disown the contents of these versions when he decided to publish something different. But it is equally certain that the lists in question *cannot* mean that he regarded any of these as in finally acceptable form; some of the detailed points to be discussed below show this very plainly.

(among these last, McTaggart in particular), that *change is illusory.* The present discussion, as its title indicates, singles out Kant among philosophers; it treats in some detail the relationship Gödel finds between Kant's doctrine of the "transcendental ideality" of time and space, on the one hand, and relativity theory on the other. This relationship, in Gödel's view, is significant in both directions: he suggests (a) that Kant's doctrine is in some ways remarkably confirmed by relativity theory, and that to see this is of importance for our understanding of this theory; and (b) that the turn taken in relativity theory requires, if the Kantian principles he discusses are to be maintained seriously, a quite radical reinterpretation and revision of Kantian philosophy itself. But Gödel's extended train of thought defies summary, and no attempt is made here to summarize further his theses or his arguments.

The contents of this introductory note are as follows: §2 deals first with the question of the chronological and substantive relations among the five manuscripts *A, B1, B2, C1, C2.* §3 extends that discussion, with some amplification of a point concerning Gödel's discovery in general relativity. §§4–5 are concerned with questions of mathematical interpretation raised by Gödel's essays, and in part with needed corrections of some of his statements. The remaining §§6–9 deal with philosophical questions: §§6–8 with issues raised by Gödel's interpretation of Kant, §9 with a matter independent of Kant (and previously raised in the introductory note to *1949a*).

It should be noted that the treatment of philosophical issues in §§6–9 is quite independent of what precedes those sections; a reader who wishes to, therefore, can read §§6–9 without (or before) consulting the preceding technical matter.

2. On the question of the chronological placement of these papers, some remarks of David Malament's[c] deserve to be recorded:

> The first thing that struck me after looking through the manuscripts is that they almost certainly were all written *before 1949a*, and that the sequence
>
> $$(A, B1, B2) \rightarrow (C1, C2) \rightarrow 1949a$$
>
> exhibits an interesting line of development. It can be traced in the references Gödel makes to his own technical work in relativity theory. Here are three pieces of evidence.

[c]In a letter, dated 23 August 1986, to Solomon Feferman, from which I quote with the writer's permission.

a. ... there is no mention in B2 that Gödel spacetime exhibits closed timelike curves, even though this fact significantly
strengthens Gödel's argument (for the connection between relativity theory and a Kantian conception of time). It is cited in
C1 and C2, and given prominence in *1949a*. I just cannot imagine that Gödel would have undertaken to write on Kant, time,
and relativity theory after the composition of *1949a* and have
suppressed reference to the fact. Incidentally, it is *not* difficult
for me to imagine that some interval of time lapsed between his
initial discovery of Gödel spacetime and his subsequent realization that it contains closed timelike curves. (I have in mind a
certain technical point here, but shall not bother to elaborate.)
Presumably A, B1, B2 were written sometime in that interval,
and C1, C2 were written after.

b. In a note added to *1949a* on "2 September 1949" Gödel
announced that for *every value* of the "cosmological constant"
there exist rotating solutions that do not admit (any) universal
time functions. (For spacetime models of the sort Gödel was
considering this condition is not only entailed by the existence
of closed timelike curves, but is strictly equivalent to it.) The
discovery was important because it is customary to take the constant to be zero, and Gödel spacetime only qualifies as a solution
to Einstein's equation if one allows it to be non-zero (or rein-
terprets it as a large "unphysical" negative pressure term in the
energy momentum tensor field). The discovery clearly bolstered
Gödel's case. He obviously thought it sufficiently important to
warrant a late change in his manuscript. (Publication of the
Schilpp Einstein volume came in December of 1949.) Yet there
is no mention of the discovery in any of the five manuscripts
here under consideration. This suggests, at least, that they
were written before September 2, 1949.

c. In footnote #53 [in this edition, footnote 15]^d of C1 Gödel
anticipates the objection that the idea of "time travel" leads

---

^{d}A concordance is given in the textual notes between the footnote numbers as
they appear in Gödel's manuscripts and those in the present edition. (It should be
noted that when Gödel interpolated new footnotes, he did not renumber the older
notes, but gave higher numbers to the new ones. Therefore the order of his numbering
does not always follow the order of the text, but does convey information about the
chronological development of his thoughts.)

to absurdity, and hence that Gödel spacetime can be rejected *a priori* as an "unphysical", extraneous solution to Einstein's equation. In response he argues that the objection presupposes the "practical feasibility" of time travel, and that this "*may very well* be precluded by the velocities very close to that of light which would be necessary for it ...". (emphasis added) The statement of the objection is very close in formulation to that in *1949a*. But in the latter Gödel's response is more confident, and is supported by the citation of a calculation. He asserts that "the velocities which would be necessary to complete the [time travel] voyage in a reasonable length of time *are* far beyond everything that can be expected ever to become a practical possibility". (emphasis added) Presumably Gödel did the calculation he cites in the interval between the composition of C1 and *1949a*.

The evidence cited by Malament in (a), as tending to show that the three manuscripts *A*, *B1*, *B2*, were completed before Gödel's discovery of the existence of closed time-like world-lines in the cosmological solutions he calls (in the *C* drafts) "R-worlds", gains still further support from another point of contrast between those versions and that of the *C*s. In each of the former three there occurs the remark—referring to "relativity theory" quite generally—that "for the series of events happening to one material point the 'before' has always an objective meaning", in that "it subsists relative to all observers".[e] In both *C1* and *C2*, on the other hand, we read that "*that* passing of time which is directly experienced has no objective meaning in the R-worlds," because "it is possible in these worlds to travel arbitrarily far into the future or the past and back again". Although the former passage ascribes objective meaning to the "before" of a material point (rather than to that "directly experienced"), the latter stands in essential contradiction to it; for the travel Gödel describes goes (in part of its course) *from* what is "after" for some material point *to* what is "before" for it—thus directly refuting the claim that "it [the 'before'] subsists relative to all observers".

To these remarks there should be added a reference to an illuminating passage in Gödel's lecture *1949b*,[f] in which he comments on the motivation of his own search for rotating solutions to Einstein's field equation. Gödel there notes the equivalence of (a) the vanishing everywhere

[e]See p. 13 of *B2* below, fn. 23 and the text above it.

[f]I wish to thank Malament for calling this passage from *1949b* to my attention at a time when I had not seen that text. Cf. further his introductory note to *1949b* in this volume.

of the angular velocity of matter and (b) the existence of a one-parameter system of three-spaces everywhere orthogonal to the world-lines of matter; and adds (page 12):

> This incidentally also was the way in which I happened to arrive at these rotating solutions. I was working on the relationship ... between Kant and relativistic physics insofar as in both theories the objective existence of a time in the Newtonian sense is denied. On this occasion one is led to observe that in the cosmological solutions known at present there does exist something like an absolute time. This has been pointed out by epistemologists, and it has even been said by the physicist Jeans that this circumstance justifies the retention of the old intuitive concept of an absolute time. So one is led to investigate whether or not this is a necessary property of all possible cosmological solutions.

Thus one sees that it was his reflections on the philosophical bearings of relativity theory—themselves presumably stimulated by the request from Schilpp for a contribution to the Einstein volume—that led Gödel to the discovery of the R-worlds. More particularly, one sees quite explicitly that the starting point of his technical discovery was the realization that (in a matter-filled general relativistic cosmos) the existence of a "natural" cosmic time—defined by slices orthogonal to the world-lines of (cosmic) matter—depends upon the non-rotation of that matter. This is the "technical point" Malament had in mind in his parenthetic remark near the end of (a) in the passage quoted above (written at a time when he had not seen *1949b*): Gödel sought, and found, rotating solutions, because he knew that no "natural" cosmic time can exist in worlds represented by such solutions; only in the further study of the geometry of these rotating solutions did the *stronger* result emerge that these contain closed time-like lines, and therefore *no cosmic time whatever* that is free of anomalies can be defined for them.

3. That perhaps deserves some amplification:

The first, weaker, result—invoked by Gödel, in the passage just quoted, against the contentions of unnamed "epistemologists" and of "the physicist Jeans"—is characterized in the A and B drafts as follows:[g]

---

[g]See p. 10 of *B2* below. This paragraph is carried through with very minor alterations of phrasing from *A* through *B2*.

This view [namely (with attribution to Jeans) that "matter and the curvature of space-time produced by it, if the structure of the world as a whole is taken into account, may enable us to determine some objectively distinguished ordering of all events to which the properties contained in our intuitive idea of time could consistently be attributed"] is supported by the fact that in all known cosmological solutions ... such an "absolute world time" really can be defined. But nevertheless the conclusions drawn above [namely (see Gödel's page 7 *ad finem*), that "what remains of time in relativity theory as an objective reality inherent in the things neither has the structure of a linear ordering nor the character of flowing or allowing of change"] can be maintained because there exist other cosmological solutions for which a definition in terms of physical magnitudes of an absolute world time is demonstrably impossible. If however such a world time were to be introduced in these worlds as a new entity, independent of all observable magnitudes, it would violate the principle of sufficient reason, insofar as one would have to make an arbitrary choice between infinitely many physically completely indistinguishable possibilities, and introduce a perfectly unfounded asymmetry.

Now, the state of affairs asserted here as obtaining in Gödel's new cosmological solutions is—so far as Gödel's explicit statement goes—hardly different from that obtaining in the space-time structure of *special* relativity: that is, here as there, the choice of a particular "world time" is seen as not inherent in the objective structure of the world because it is an *arbitrary* choice among objectively "equivalent" alternatives (introducing "a perfectly unfounded asymmetry"). Indeed, it appears most probable that the train of Gödel's thought in this investigation was one that *did in fact* start from special relativity. For everything Gödel says about relativity theory up to the paragraph running from the end of page 9 to page 10 of *B2* is entirely true of the special theory, and could be understood as referring to that theory alone, except for footnote 4— which is numbered 46 in Gödel's original text (i.e., it is a fairly late addition). And in the paragraph just mentioned, the transition to general relativity is explicitly motivated by the following remark:

> With much more justification [than something just criticized], it is objected [footnote: "See *Jeans 1936*"] that the impossibility of defining any absolute time ... *in the empty space-time scheme of special relativity theory* (upon which the foregoing considerations have been based) does not exclude [an "absolute world time" in the cosmological solutions of general relativity].
> [Emphasis added.]

Thus we see the likelihood of the following remarkable sequence in
Gödel's investigation: (a) philosophical reflections on the status of time
and change in relativity theory, with primary attention to the special
theory, leading him to make his comparisons with Kant; (b) a possible check to this line of thought—namely, the objection (associated
with Jeans) that *general* relativity might warrant a *reinstitution* of time
as "absolute" and change as correspondingly "real"; (c) the technical investigation, motivated by that objection and by the idea that
in a "rotating world" a situation *resembling that of special relativity*
might obtain, so far as the problem of time is concerned—an investigation that was successful in finding examples of such "rotating worlds";
(d) finally, the *entirely unexpected* discovery that in these "rotating
worlds" a very much more thoroughgoing impossibility holds for anything like the conventional notions of time and change.[h]

The difference between what was envisaged by Gödel in (c), and what
was eventually found by him in (d), is this: The principle by which, in
the ordinary cosmological models appealed to by Jeans, an "absolute"
world-time is distinguished, is that events ought to be regarded as "absolutely simultaneous" when they lie in some one three-dimensional (necessarily space-like) submanifold of space-time that is *everywhere orthogonal to the world-lines of "cosmic" matter*. No other "natural" criterion
suggests itself; and in the absence of such a criterion, Gödel says, the
choice of one among the "infinitely many physically ... indistinguishable possibilities" would be arbitrary, and would "introduce a perfectly
unfounded asymmetry". Still, to put the case a little crudely, a determined "absolutist" might argue from the theological postulate that there
is a uniquely distinguished "God's time" (*"Gottes Zeit ist die allerbeste
Zeit!"*)—even if merely physical considerations are inadequate to tell
us which time that is. In the rotating worlds of Gödel, however, the
situation is radically different: there are not "infinitely many physically

---

[h](*Note by the editors.*) Researches of John W. Dawson, Jr., have shown that confirmation of the stages in this sequence is provided by entries in Oskar Morgenstern's
diaries (among his papers at the Perkins Memorial Library of Duke University) and
by Gödel's own statements in letters to his mother. On 23 September 1947 Morgenstern reported that Gödel had found "a world in which simultaneity cannot be
defined", and on 7 November Gödel wrote that his investigations had "led to purely
mathematical results" that he intended to publish separately. The following January
Gödel told Morgenstern that he was eager to return to work in logic, and the next
month declared to him that he expected to have his cosmological results written up
"soon". Instead, however, there followed a three-month hiatus, at the end of which
(on 10 May 1948) Gödel wrote his mother that he had at last settled "a problem" that
had for weeks pushed everything else out of his mind. Two days later Morgenstern
then noted that Gödel was again "making good progress with his cosmological work":
in particular, "Now in his universe one can travel into the past". (The quotations in
this note are Dawson's translations of German originals.)

indistinguishable possibilities" among which to choose; rather, there are *no possibilities at all*: not only are there no three-dimensional slices of space-time *everywhere orthogonal to* the world-lines of cosmic matter, there are simply no three-dimensional space-like slices *whatever* that so much as *intersect* all the world-lines of cosmic matter, each in a single point.

This follows almost immediately from what is surely the most striking characteristic of the rotating worlds: the existence in them of smooth, everywhere time-like, everywhere "future-oriented" world-lines, connecting any point in the space-time to any other point.[i] It is this that leads to the consequence stated by Gödel (version *C1*, page 10, paragraph 2) as follows: "Above all they [the "absolute time-like relations" in the R-worlds] define only a partial ordering, or to be more exact: in whatever way one may introduce an absolute 'before', there always exist either temporally incomparable events or cyclically ordered events."

4. Gödel's statement here is open to some criticism (cf. the warning in the opening paragraph of this note that such criticism would be necessary); it seems likely to me that this is a sign of the haste, and perhaps intellectual tumult, in which these thoughts were penned by him after his unexpected discovery.[j] In the space-time structure of *special* relativity, a "natural" (and in *this* sense "absolute") *partial temporal ordering* does indeed exist: namely, "*a* strictly precedes *b*" just in case *a* and *b* are not identical, and *b* lies within or on the mantle of the "forward" light-cone of *a* (or, equivalently, *a* lies within or on the mantle of the "backward" light-cone of *b*). This temporal ordering satisfies the "natural" demand that whenever "influence", or "a signal", can be propagated from *a* to *b*—and in particular, whenever a *body* can "travel" from *a* to *b*—the relation "*a* precedes *b*" holds.[k] That is also the situation in

[i]The consequence holds in view of elementary facts about the R-worlds: There, each world-line of cosmic matter is an "open line" (i.e., is isometric with the real-number line); and these world-lines have a canonical and mutually coherent time-ordering. A slice of the envisaged kind, then, would divide each matter world-line into a "past" and a "future" part; and at the same time would divide the entire manifold into two components—one past, the other future, to the slice. A time-like line starting from a point on the future side and ending at a point on the past side would have to cross the slice, and at the crossing-point would have to be directed future-to-past; thus no everywhere "future-oriented" line could start at a point on the future side and end at one on the past side of the slice.

[j]To be sure, the statement is repeated verbatim in the revised text *C2*; on the other hand, the fact that this revision was left incomplete lends some independent support to the view that Gödel did not thoroughly reconsider and criticize what he had written.

[k]If *b* belongs to the mantle of the forward cone of *a* (or, equivalently, *a* to the mantle of the backward cone of *b*), a *signal* can be propagated from *a* to *b*; if to the

those general-relativistic *cosmological* solutions that were known before Gödel's work (although the existence *in general* of space-time structures in which this "natural" condition cannot be met had been noticed long before—cf. below). But the state of affairs described in the preceding paragraph tells us *directly* (since a smooth, everywhere time-like, everywhere future-oriented world-line is a *possible* world-line of a *body*) that our "natural" condition in a Gödel R-world demands that every point of space-time strictly precede every other; and this contradicts what one means by a (strict) ordering relation. In short, Gödel's results, taken with the most natural conditions one would place upon a temporal ordering, imply that *not even a partial temporal ordering is possible* in an R-world.

It might, however, be thought that Gödel had something else in mind: namely, not to define the temporal ordering by the possibility of travel or influence, but to define it simply by the light cones themselves. What one finds if one pursues this line exhibits in yet another way the strangeness of the Gödel universes. The situation is most easily described in a restriction to two dimensions of space (with of course one of time). If one starts at a point a of the space-time, and defines as "later than a" all points in the set swept out by the future-oriented time-like or null geodesics starting from a, one indeed arrives at a non-trivial relation; but this relation *is not an ordering*: it is not transitive. The region in question presents the following picture: It starts, cone-like as one would expect, expanding on all sides from a, and admits a succession of space-like two-dimensional slices of increasing radius, until a certain critical radius is reached. These slices, taken in order until the critical radius is attained, can thus be regarded, so far as all this is concerned, as representing "successive epochs" in the "development" of a part of the world; and so we do indeed have (again: "so far") what looks like a bit of an "absolute", "temporal" ordering. But at the critical epoch (or phase), the slice has ceased to be entirely space-like: rather, one tangent line to the slice at each point of its extremity is a null line—and is indeed tangent to the world-line of propagation of a light signal emitted from the initial point a. These bounding null tangents envelop a circle at the critical radius (a circle, thus, of a definite space-like radius, but of circumference zero). The light-lines from a, however, although they are all tangent to this circle, do not merge with it; they pass beyond and *reconverge* to a point b on the "cosmic" geodesic (what Gödel calls the

interior of the cone, then a signal can be propagated, and (even) a *body* (of non-zero rest mass) can travel, from a to b. (I here make *transport of a body* a more special notion than *propagation of a signal* only as a convenient way to record a distinction; in a quite reasonable sense, "particles" of zero rest-mass, which are propagated along the mantle of the cone, can of course also be regarded as "bodies".)

"world-line of matter") through a. If we think of the "points of cosmic matter" as the *stars* of this world, then light from a given star always reconverges to it—in fact, pulsates periodically out to the critical radius and back again. And so do all other geodesics ("ballistic trajectories") from a given star: they all reconverge, with the same "period" as the light signals.[1] The non-transitivity of the relation defined by this procedure is clear, for instance, from the fact that whereas, in the account just given, no point in space-like relation to b is "later than" a, if one takes a point c on the geodesic from a to b and halfway between them—a point, thus, "later than" a—and constructs its future, this will contain a space-like slice through b of the maximum possible radius (the critical radius). So the idea fails: it, too, does not define an ordering.[m]

A corollary of the result just described is worthy of special note. First let it be remarked that in the R-worlds here under discussion—those of Gödel's initial publication *1949*, in contrast with the *expanding* rotating universes of *1952*—the cosmic masses maintain constant relations to one another, in the sense of the space-time geometry, throughout the history

---

[1]In the full four-dimensional case this picture is modified: in the "remaining spatial dimension", omitted from the discussion above, geodesics (whether null or time-like) can be thought of as diverging at a constant rate ("linearly") from their starting point.

[m]One remark should be added to avoid a misunderstanding on a small technical point. I have spoken above of defining a relation "simply by the light cones themselves"; but the construction just described uses time-like as well as null geodesics—and of course that exploits the full metric structure of the space-time, not merely the structure defined by the light cones (the "conformal" structure). This leads to a perspicuous geometrical picture, and its extra details are of interest in themselves. The relation can, however, indeed be characterized entirely with the help of the light cones—that is, of null geodesics alone. For in the construction above (still restricting the account to two dimensions of space), the null geodesics emanating from a have been seen to reconverge to b. In doing so, they sweep out, when one considers just the segments from a to b, a surface (the "mantle" of the forward cone from a). This surface has, from the differential point of view, singularities at a and b; but it is topologically a closed manifold, and in fact a topological *sphere*. It therefore has a well-defined "interior region" in the space-time (which itself has the topology of a three-dimensional Euclidean space). The relation "x is later than a" has in effect been defined to hold, *first*, for all points x other than a itself that belong to that surface or to its interior region. To obtain the full relation defined in the text, one must iterate this construction: Let the relation just defined be expressed as $L(x,a)$; let us set $a_1 = b$; and in general, a_n having been defined, let a_{n+1} be obtained from a_n in the same way b was from a. Then the relation "x is later than a" means: for some n, $L(x, a_n)$. (Adding the suppressed spatial dimension, the construction is slightly more complicated but not essentially different. The mantle of the light cone is now a hypersurface, and topologically a 3-sphere in the now four-dimensional Euclidean space. Instead of the focal point b corresponding to a, there now occurs a focal *space-like line segment*, and the construction has to be iterated from this entire line segment—the successive segments, corresponding to the successive points a_n above, growing longer at each stage. But the successive portions of the mantle continue to be 3-spheres, and everything else is as before.)

of each. The above result means that despite this fact—i.e., despite the
fact that the *bodies are in no sense "moving apart"*—observers on "sufficiently distant" cosmic masses *cannot reach one another at all* with
direct light signals (nor, by the same token, can they reach one another
with direct "ballistic signals"). On the other hand, they *can* (in principle) reach one another with *indirect*, or *relayed*, light signals (or ballistic
signals). This is in effect just a more vivid formulation of the "non-
transitivity" noted above: a light-cone C that originates within, or on
the mantle of, a "preceding" one, A, develops so as to "escape from" A.
It is also noteworthy that this situation contradicts the conception of the
direct light-signal as a "first-signal", or "fastest" signal, in relativistic
physics—cf., e.g., *Reichenbach 1928* (version *1957*), pages 143, 166, 238;
and *Reichenbach 1924, passim.* (To be sure, it is obvious that there can
be no "first-signal" in a world in which *every* space-time point can signal
to *every* other.)

The only remaining possibility there seems to be for an interpretation
that might, *prima facie*, render Gödel's statement correct as it stands
is to take, for the "absolute partial ordering" he refers to, just the relation "a and b are two points on a single world-line of (cosmic) matter,
and a precedes b on this line". Then, indeed, one has a partial ordering; but it is not true that every extension of this relation either leaves
some events temporally incomparable or makes some events "cyclically
ordered". For Gödel himself, in his mathematical treatment of the R-
worlds, introduces a global space-time coordinate system, with a global
time-coordinate that orders "correctly" the events on each world-line of
matter. So the ordering by this time-coordinate extends, to the entire
cosmos, the partial ordering along the world-lines of matter, in contradiction to Gödel's professedly "more exact" statement.

Nevertheless, what one is entitled to regard as the essential intended
content of that statement is true: the ordering by the global time-
coordinate violates the requirement that *events "experienceable" by any
observer be ordered as they are in that observer's experience.*[n] This

---

[n]One incidental remark to this: Gödel refers (*B2*, p. 13) to "the fact that for the
series of events happening to one material point the 'before' has always an objective
meaning"; and of course one will apply the same proposition to the events happening to one observer. But a certain qualification is necessary. In Gödel's R-worlds,
one can conceive a material point—or an observer—as having a world-line with the
topology of a *circle*. In this case, there ceases to be, for the world-line as a whole, a
well-defined ordering of "before" and "after". However, there is still what one may
call an "infinitesimal ordering": a determinate direction "past-to-future" *at each
point* of the line (that is, "at each instant" for the body or the observer); and this
gives rise to a determinate "local ordering ", in the sense that there is induced a determinate ordering on every "sufficiently small" segment of the line. Translated into
terms of the "experience" of such an observer, what one would have is, at each instant

requirement is merely the reformulation in a subjective mode of that earlier stated in terms of the direction of "physical influence"; and since the order of influence, or of experience, can be "from" *any* point of space-time "to" *any* other point (or to *the same* point—hence Gödel's invocation of the possibility of "cyclic order"), the requirement (as we have seen) is incompatible with even a partial ordering.

5. Several further passages present deeper puzzles of mathematical interpretation.

The first of these occurs in *C1* shortly before the statement we have just been concerned with, and, like it, deals specifically with the situation in the R-worlds. Gödel speaks (see pages 8–9) of the procedure of defining time relative to an observer "*by direct measuring with clocks and Einstein's concept of* [*simultaneity*]" (emphasis added) and states in a footnote that this concept, originally designed for special relativity only, "can be extended to general relativity theory, at least under certain conditions which are satisfied in the R-worlds". This claim is very hard to make good sense of. Einstein's criterion of simultaneity was intended to be employed by *a collection of observers* who are in *the same state of uniform motion* (that is, by *inertial* observers who are, mutually, in a state of relative rest). Among such observers, the relation of simultaneity—or, what comes to the same thing, the synchronization of clocks—was to be established by light signals. In the Gödel R-worlds, one does not clearly see what classes of observers are to play anything like that role. The only plausible analogue of *inertial* observers, in general relativity, is *gravitationally free* observers (i.e., ones who move solely under the influence of the gravitational field). But if any class of these latter can be regarded as "at mutual relative rest" in the R-worlds, it would appear to be the class of those observers who move with the cosmic matter (since the geometric relations of such observers remain stable). And yet these certainly *cannot* be treated as a class among whom Einstein's simultaneity criterion is to be employed: indeed, we have already

and over each small enough segment, a definite sense of the "order of becoming"— the "order of changes"; whereas over the totality of experiences of such an observer (a totality of experiences that would be *without end*), one would have an *eternal recurrence* of those experiences. (The question might be raised whether, besides the possibility of such a *closed* world-line, one might also envisage *self-intersecting* world-lines: ones that "circle back" [or "spiral back"] to earlier stages of the line and *cross* those earlier stages. The answer is that such lines are indeed geometrically describable; but if one wants such a line to represent the history of a *body* (in which case it must be "thickened out" into a *continuum* of lines), then the self-intersection would represent a *collision* of *two* bodies that would proceed to *move through* one another. In so far as one speaks of a physics of "bodies", such a process is probably to be regarded as physically impossible.)

seen that between two sufficiently distant observers of this class *no direct light-signals at all* can be sent.[o]

The problem having been noted, it need not be further pursued. Gödel *may* have had some construction in mind that would form a suitable analogue in the R-worlds to the Einstein criterion in special relativity; and if so, one would like to understand what it was; but whether he did, or whether his remark represents an oversight, there is no serious consequence for his general argument. We may rephrase his conclusion with a qualification: "The fundamental temporal relation '*A* before *B* by *t* seconds' obtained in this way, [*if such a relation exists at all,*] is certainly not something inherent in the events ...". The interpolated reservation does not weaken Gödel's assertion.

The perplexities of the next passage are more complicated. In each of the manuscript versions, there occurs a discussion centering on the claim (a *limitation*, according to Gödel, of the parallel he is exploring between relativity theory and Kant's doctrine of space and time) that in general relativity one can define "absolute spatial relations" that "lead to a structure ... not so very different from that of intuitive spatial relations".[p] The detailed description of the generation of this structure is not quite the same in, first, the texts of *A* and *B*; second, footnote 20 (footnote 50 in Gödel's numbering, thus a fairly late addition) of those versions; and, third, the *C* texts. But setting aside the varying details, if we take the latest version to represent Gödel's more considered view, it appears that the alleged space of "absolute spatial relations" is meant to be obtained in the following way:

1. We restrict attention to relativistic worlds in which there is a cosmic "matter-flow" defined everywhere; so there is a distinguished family of time-like world-lines—in fact, geodesics—of which exactly one passes through each space-time point. Each of these distinguished world-lines corresponds to (is the *history* of) one "material point"; and it is the material points—or, in other words, the distinguished world-lines themselves—that are taken to be the "points" of the space we are to describe.[q]

---

[o]It might be suggested that relayed signals be used; but this would altogether defeat the intent of Einstein's criterion, and would lead to no coherent simultaneity-relation.

[p]*B2*, pp. 11-12, with fn. 20; cf. also *C1*, pp. 17-22.1.

[q]It has already been remarked that there are variants in Gödel's discussion. In *B2*, fn. 20, the question is raised of the possibility of defining "'absolute' points of space persisting in time and identifiable at different times" in some way that does *not* simply identify these with the material points. (Gödel's formulation of this question is a little unclear—its phrasing does not directly imply that the "absolute points" are to be

2. The matter-flow (or the vector-field associated with it) is assumed to be *smooth*;[r] and Gödel seems to infer tacitly that this entails the existence of a well-defined differentiable-manifold structure on the set of material points (i.e., on our "space"). That, however, does not follow without some further assumption. It is indeed "ordinarily" the case—i.e., it holds for "reasonable" space-times—and the simplest procedure is just to make this the further assumption: that is, if we let M be the space-time manifold, we assume the existence of a three-dimensional differentiable manifold M_1, and a smooth mapping π of M onto M_1, such that the inverse image under π of any point of M_1 is the entire world-line of a material point. M_1 is, then, the underlying differentiable manifold of the claimed geometry of "absolute spatial relations". (It should be noted that, given the assumption of its existence, the "absolute" character of M_1 itself—that is, its geometrically objective status, grounded in the structure of M together with its matter-flow field—is secure.)

3. It remains for us to define a *metric* structure on the spatial manifold M_1. For each point x of M_1, and each point ξ in $\pi^{-1}(x)$ (that is to say, *each epoch in the history of the material point x*), the map π induces a projection of the tangent-vector space to M at ξ onto the tangent-vector space to M_1 at x. If we consider, in the former tangent space, the subspace *orthogonal to the direction of matter-flow* (that is, to the direction of the world-line x itself at the point ξ), this projection is an *isomorphism* of that (three-dimensional) subspace onto the tangent space to M_1 at x. The three-dimensional subspace in question at ξ has a positive-definite metric induced from the metric of space-time; and through the isomorphism induced by π, this metric can be transferred to the tangent space at x. However, as ξ varies over the world-line x,

---

other than the material points. But the sequel removes that uncertainty, for Gödel goes on to speak of the need for "some requirement" to be imposed on this problem, because "otherwise points of space may simply be identified with material points".) At any rate, having raised the question, Gödel remarks that even its meaning is not quite clear (because the needed further requirement is lacking), and he does not pursue it further, nor does he mention it in the C drafts.

[r] Again variants must be noted. In $B2$, fn. 20, Gödel envisages the possibility of a *discontinuous* velocity-distribution of matter. It is hard to see how a construction like the one he envisages could be made in this case; and, indeed, in $C1$, p. 22, he restricts the described construction to the case where "the velocity vector of matter is continuous". It remains hard to see how the construction could be done without the stronger condition of (a suitable order of) differentiability, and it is perhaps allowable to assume that, writing a principally non-technical paper, Gödel used the word "continuous" vaguely to connote "smooth". In any case, as will be explained presently, yet stronger assumptions seem to be needed.

the transferred metric on the tangent space to $\mathcal{M}_1$ at x will in general not remain constant. We have, therefore, on the manifold $\mathcal{M}_1$, not a Riemannian metric structure, which consists in a metric on the tangent space at each point of the manifold (with suitable smoothness conditions as one goes from point to point), but rather a structure that consists in the specification, on each tangent space, of a *one-parameter family* of metrics (where the parameter is to be thought of as the "varying epoch"—say, the "proper time"—along the world-line of $\mathcal{M}$ that *is* the point in question of $\mathcal{M}_1$).

4. Gödel says that we can use this construction to define on $\mathcal{M}_1$ the structure of "an ordinary three-dimensional Riemannian space changing in time"; but in a way that is *not* canonically (or "absolutely") determined by our data: rather, "This space is different for two observers A, B only insofar as the phases of development of its several parts (in consequence of the different meanings of simultaneousness for A and B) are collected together in different ways into one phase of the whole space" (*B2*, footnote 20; cf. *C1*, page 22). This "collecting together of phases" is, however, not unproblematic: it presupposes a further strengthening of the assumptions we have made. What is required for the construction Gödel indicates is the existence of a "cross-section" of the mapping π: that is, a mapping σ from $\mathcal{M}_1$ to $\mathcal{M}$ such that the composite mapping $\pi\sigma$ is the identity mapping of $\mathcal{M}_1$ onto itself. Thus σ assigns to each material point an epoch in its own history—a phase of its development; and these selected epochs "collected together" may be looked upon as forming a single "phase of the whole space". That being done, it is easy to extend the construction to *each* phase of each particle (e.g., if the matter-flow geodesics are metrically infinite "open" lines, by taking σ to define the zero-point of time on each material world-line, and then extending to a "time-coordinate" by measuring the proper time, starting from zero, along each such world-line). In this way, letting $\mathcal{T}$ represent the global "time" so defined, we have a representation of $\mathcal{M}$ as the Cartesian product $\mathcal{T} \times \mathcal{M}_1$; and for each "phase" or "epoch" in $\mathcal{T}$, the construction in (3) above gives us a well-defined Riemannian metric structure on $\mathcal{M}_1$—thus, in accordance with Gödel's claim, the structure of a Riemannian space changing in time.

That, however, is not the end of our problems with this passage. For although the collecting together of phases can indeed be performed (on the *assumption*—by no means entailed by what went before—of the existence of a cross-section), one has to ask what geometrical or physical significance such a representation of "space changing in time" will have for a general-relativistic cosmos. Gödel appears to imply that the "collecting together of phases of development" is well-defined *relative to an observer*. Indeed, as we have just seen, he refers to the difference in this representation "for two observers A, B"; and he attributes this difference to "the different meanings of simultaneousness for A and B". (Thus the present difficulties are after all connected with the previous one—that is, with Gödel's supposition that there is a "natural" notion of simultaneity "relative to an observer".) However, even if there were such a well-defined concept of "relative simultaneity" in general relativity, it would remain unclear what significance a geometry pieced together in the fashion described would have (for it must be noted that, for instance, the "distances", or "lengths of curves", in the space $\mathcal{M}_1$ at some phase of its development are *not*, in general, space-like distances, or lengths, of anything whatever in the space-time $\mathcal{M}$). And the absence of a well-defined concept of even *relative* simultaneity makes this difficulty all the stronger. At the very least, Gödel's characterization of this as a structure of *absolute* spatial relations would be very hard indeed to defend.

But there is a special point that may have contributed to what seems indisputably a confusion here: namely, the construction described, when it is carried out for a Gödel "R-world", *does indeed* lead to an "absolute" three-dimensional metric geometry (although not one "changing in time"). For in the first place our assumptions are satisfied in these worlds: there is a mapping π of space-time onto a three-dimensional manifold such that the inverse images of points are the "world-lines of matter", and this mapping admits cross-sections. And in the second place, because the R-worlds are dynamically "stationary", *the geometries induced on the tangent spaces to the three-dimensional manifold are constant in time* (that is, constant over the development of proper time along each world-line). It follows not only that the Riemannian structure induced is itself constant in time, but—of far greater importance—that this structure is *independent* of the arbitrariness that affects the "collection into phases" (independent, e.g., of the choice of a cross-section). So for the R-worlds, but *not* for the most general cosmological models, Gödel's distinction, that whereas there is no cosmic structure at all like that of classical time there is an ("absolute") structure "not so very different from that of intuitive spatial relations", does indeed hold.

A further discussion of geometrical relations, this time in connection with a possible vindication of Kant's views on the *a priori* of spatial intuition—a discussion that occurs in all the manuscripts except the incomplete draft *C2*[s]—poses additional problems. It is, in fact, hard to

---

[s]See *B2*, pp. 14–15; *C1*, pp. 23–24.

see this discussion as concerned at all with the situation in the theory
of relativity (special *or* general); for Gödel here speaks of defining "concepts satisfying Euclid's axioms ... also in a non-Euclidean world", by
means that make no reference at all to *time*; so that we seem to be in
a universe of philosophic discourse that altogether antedates Einstein—
roughly, that of Poincaré's *Science and Hypothesis* (see *Poincaré 1905*,
chapters 3–5). Thus the declaration that "we define the 'coordinates' *a*,
b, *c* of any material point *P* by the number of times a measuring rod
must be applied along the directions α, β, γ ... in order to reach *P* from
O" assumes both that these "directions" are well-defined as persisting
through time, and that the result of the measurement is independent of
the kinematic details of the "application" of the measuring rod (thus in
particular of the time taken up by the process).

One point of detail in this passage demands special comment. In *B2*,
footnote 27, Gödel says that his definition of coordinates would, in a
closed (e.g., a Riemannian—that is, metrically spherical) world, not give
unique coordinate values; that it would "have the consequence [either
that] every object is in infinitely many different places simultaneously,
or that for every object there exist infinitely many exactly equal ones".
This is a very inadequate account of the situation. In the first place,
it is only for a manifold that is topologically Euclidean that a global
system of coordinates of the sort Gödel envisages can be defined, assigning "unique coordinate values" to each point. In the second place, the
alternative interpretations envisaged by Gödel, "each object in infinitely
many places" or "each object having infinitely many replicas", appear
to presuppose that the "closed world" in question admits a Euclidean
world as what is called a "covering space": then the covering map represents each point of the covered manifold by infinitely many points,
periodically distributed, of the covering space. But this presupposition
is *false* for spherical spaces: a sphere (of any dimension ≥ 2), because it
is a simply connected space, admits no proper covering spaces whatever.

6. Turning now to issues respecting Gödel's philosophical interpretation
of Kant, it should be noted with emphasis—as Gödel does himself in
his opening footnote to both texts below—that he does not regard himself as a follower of Kant; and one may perhaps fairly conclude that
Gödel is concerned, not primarily with the scholarly authenticity of his
reading, but with what he has found suggestive, in Kant's discussions,
of themes that in his view are illuminating for thought about modern
physics: "Above all", he tells us, "I wanted to show that the questions
arising in such a comparison" (namely, "between relativity theory and
the Kantian doctrine about time and space") "are interesting and perhaps even fruitful for the future development of physics."

It might be best to let the matter rest there, and invite the reader to reflect upon Gödel's set of variations on themes from (or suggested by) Kant, without editorial animadversion. There are, however, a few points on which Gödel's remarks about Kant's philosophy appear either extremely enigmatic or seriously astray; and it has seemed best to call attention to some of these—not least in the hope that other commentators may succeed in clarifying the enigmas. There are also passages where an interpretation of Gödel's that may strike a student of Kant as misguided appears to merit discussion aimed at clarifying the rationale of Gödel's point of view, and the relation—in part one of *deliberate* divergence—that he sees between himself and Kant.

One of the most transparently non-Kantian features of the view Gödel puts forward lies in the fact that for him, in clear and conscious opposition to Kant, the character of our "spatial (and temporal) intuition"—if we have any such thing as an "innate intuition representing geometrical entities [of some kind]"[t]—is not decisive, *either* as guaranteeing, *or* as implying a limitation upon, features of our knowledge of "objects". Thus Gödel tells us (*B2*, page 16; *C1*, page 25) that "What ... Kant did not take into account is that ... space and its properties express themselves also in the sensations, which we know only a posteriori; namely, in the fact that by projecting the sensations in a certain way into a three-dimensional Euclidean space *the laws connecting them can be stated much more easily*" (emphasis added).

To a Kantian certainly, and to a Kant scholar very possibly, this will seem grotesque; for according to Kant, space is "the form of all appearances of outer sense" (A 26/B 42), and as such certainly does "express [itself and its properties] ... in the sensations"[u]—a circumstance Kant cannot seriously be accused of neglecting. But let me suggest that Gödel has here merely expressed himself badly. What I think he claims Kant

---

[t] Whether we do or do not have such an "innate intuition" (I place the phrase in quotes because it is itself un-Kantian) is regarded by Gödel as a still open and interesting question; cf. *B2*, pp. 14–17 and fns. 28–29; *C1*, pp. 23–26 and fns. 20–21.

[u] "Sensation" here must not be understood in the rather special sense attached by Kant to the word *Empfindung*; for the latter is called by Kant (A 167/B 209; emphasis added) "the *matter* of perception" (in contrast with space and time, the *forms* of sensibility), and is accordingly said to have "no extensive magnitude". Rather, Gödel must use the word for what Kant calls (e.g. B 162) "sensory intuition" (*sinnliche Anschauung*); and it is in this sense, accordingly, that it is used in these comments on Gödel's remarks.

On citations of Kant, see the Information for the reader, p. x. In particular, the *Kritik der reinen Vernunft* is cited in the standard way. All translations from the German (that is, from the cited works of Kant and Weyl) in this introductory note, except in note h, are by the present author. For the *Kritik*, the version of Norman Kemp Smith (*Kant 1933*) has been a most helpful guide, but has not been followed on all points.

has neglected is not simply the fact that space expresses itself in the sensations, but something about *how* it does so; in particular, Gödel claims it does so in a way that, in his opinion, we do and *can* know "only a posteriori". What way this is, is what the passage italicized in the preceding quotation from Gödel tells us. But this in turn requires expansion—and contrast with the doctrine of Kant.

For Kant—and this is a point that in my own opinion Gödel has not sufficiently appreciated (the question will be discussed presently)—space does not *merely* "express itself" in the moment-by-moment character of our (pure and empirical) "outer intuition". That it does so express itself is something Kant claims to establish in the Transcendental Aesthetic, which is short and rather simple; but it is only after the famously difficult (Transcendental) Deduction, and Schematism, of the Pure Concepts of the Understanding that he claims to establish what he calls "the principle of the Axioms of Intuition"; and of this he tells us: "This transcendental principle of the mathematics of appearance greatly enlarges our *a priori* knowledge. For it alone can make pure mathematics, in its complete precision, *applicable to objects of experience*" (A 165/B 206; emphasis added). The implications of this claim, for Kant, are momentous and very far-reaching; they can be conveniently summed up here by saying that he believes that the mathematical structures cognized by us through the "forms" of outer and inner intuition—space and time—affect, necessarily, not only our momentary sensations, but the ordering of these sensations, at each moment and successively, in regular patterns that make "experience of objects" (and, so, "objects of experience") possible. And even further: on the basis afforded by the *a priori* forms of intuition, and the bringing to bear, upon what those forms provide, of the "pure concepts of the understanding" (the *categories*), Kant believes one can ground the possibility of "mathematical constructions"—and, in consequence, *a priori* knowledge—not only of spatio-temporal configurations (figures; abstractly conceived motions), but of the objects of "pure natural science".[v] In short, in the view of Kant, the *laws governing the interrelations, and succession, of our sensory perceptions*, in so far as these can *possibly* be regarded as perceptions "*of objects*"—specifically, of *empirical* objects (*Gegenstände der Erfahrung*)—are necessarily conditioned by the forms of intuition.

---

[v]Cf. *Metaphysische Anfangsgründe der Naturwissenschaft (1786)*, Preface: "A rational doctrine of nature (*Naturlehre*) ... deserves the name of a natural science only then, when the natural laws that are fundamental to it are known *a priori* and are not mere laws of experience. One calls a natural cognition of the first kind *pure*; that of the second kind on the other hand will be called *applied* ....

"All *genuine* natural science thus requires a *pure* part, upon which can be grounded the apodictic certainty that reason seeks in [such a science]" (pp. 468–469).

That is an aspect of the Kantian philosophy that Gödel does not take seriously. Indeed, as we shall see presently, he goes to some lengths to attempt to exonerate Kant from having held such views at all deeply. Once more deferring that point, what is quite clear is that Gödel is writing from a perspective that knows, that is steeped in, the existence of possibilities for a science of nature not dreamt of by Kant—most notably, the general theory of relativity, with its possibilities for spatio-temporal structure of such a radically new kind as Gödel himself discovered. Gödel's view of this might be expressed as follows: The development of mathematics since Kant has revealed the possibility of "constructions" of a kind that vastly extend the scope of mathematical structures and mathematical reasoning—so that the latter is seen not to depend upon the rather rigid basis that Kant thought indispensable. This more flexible mathematics has led in turn not only to the bare possibility, but, in conjunction with what we have come to know *a posteriori* about "the sensations" (cf. the quotation from Gödel that we are still discussing), to the *fact* that "the laws connecting them" (the sensations) *cannot* be "stated much more easily" by "projecting [them] into a three-dimensional Euclidean space". What Kant can be said, then, to have "overlooked" is simply the possibility of such a development.

But now Gödel makes an interesting un-Kantian move that can be said to have a genuine Kantian ingredient. According to Kant, the "transcendental" *guarantee* of our knowledge in "pure" mathematics and science is correlative to a transcendental *restriction*: The guarantee is strictly dependent upon the forms of intuition as providing the basis of such knowledge (the possibility of "constructing" the objects of knowledge). But the forms of intuition are (Kant thinks) peculiar to *us*; they do not characterize the "things [in] themselves", nor do they necessarily characterize the way the things would appear to other percipient and thinking beings. Therefore the knowledge we obtain through them, and indeed *any* (theoretical) *knowledge we could possibly have,*[w] is restricted in its applicability to objects *as we experience them*; of the "things [in] themselves" we know and can know nothing. If, however, the whole theory of pure science as grounded in the forms of intuition is wrong, and if we have been able to *infer* (or surmise) *from the sensations* a system of laws that do not rest upon a basis determined by those forms, then we may have some grounds for considering the "objects" to which these laws apply to be "the things [in] themselves". And this is exactly the point of view that Gödel suggests. He does not wish to maintain that physics as it exists today has reached true knowledge of what Kant calls *noumena*;

---

[w]Kant does of course partially exempt our "practical"—that is, ethical—knowledge from this restriction.

but he argues (*B2*, page 13, with footnote 24, and pages 18–21; *C1*, pages 27–30, with the corresponding footnote 27) that it is plausible to consider physics as advancing in steps "beyond the appearances and towards the things".

This intimation that the structure discerned in the general theory of relativity may be *closer to what characterizes the "things in themselves"* than is the more "apparent" structure we encounter in our simpler sensory experience is related—so far as it bears upon Kant's doctrines—to Gödel's contention that, *according to Kant*, although spatio-temporal properties as such are not affections of things in themselves, those properties "correspond" to certain "objective relation[s] of the things to us" (*B2*, page 3; cf. *C1*, pages 6–7). I think it is clear that in this statement Gödel uses the words "objective" and "thing" in what for Kant would be their *noumenal* sense: as referring to "the things in themselves", not to what Kant calls *Gegenstände der Erfahrung*. And it is of some interest that in the incomplete *C2*, presumably the latest of Gödel's drafts, there occurs a slight addition to this discussion, with a further quotation from Kant. The passage in question occurs in what corresponds to the second paragraph under numeral '1' in *B2*, at the beginning of page 4. This whole discussion, with its distinction of "two questions" in numbered sections, is compressed greatly in *C1* (pages 6–7, with no distinction into two questions and no numbered sections), but it is restored in *C2* to something close to its original form. At the end of the paragraph referred to, after the comment that the cited passages seem to imply that "the relations under consideration ... cannot consist solely in the act or disposition of representing", there occurs a new sentence—unfortunately elliptical, because in this version the references to previously cited statements of Kant are left blank: "Finally the passage referred to in _____ seems to indicate that the relations in question are something in the nature of physical influence." As to the quotation meant, it seems possible to identify it with certainty; for the list of relevant Kantian passages (corresponding to (1)–(7) in *B2*, pages 2–3, and to the almost identical paragraph—lacking the identifying numbers—in *C1*, pages 4–5) includes one new item, and it is of obvious pertinence: after quoting Kant (from *Prolegomena* §13, page 286) as saying that appearances are something "whose possibility is based on the relation of certain things unknown in themselves ... to our sensibility", Gödel adds: "and that 'sensual cognition represents ... the way in which the things affect our senses'".[x] To be sure, for Kant, to speak of "physical influence" upon

---

[x]Gödel does not name the locus of this passage; it occurs in *Prolegomena* §13, Anmerkung III, p. 290.

us of things in themselves may appear quite impossible. The notions of the "physical", or "natural", have for Kant a necessary restriction to what is *given* (in some kind of "intuition", for us necessarily *sensible*);[y] these notions cannot, therefore, be applied to noumena. But to say this is only to say that Kant was a Kantian, on precisely the point at which Gödel *states* that he himself is not. Remove the problematic adjective: Gödel is pointing out to us that Kant explicitly speaks of "the things"— presumably the same that he has called "unknown in themselves"—as *affecting* our senses; and in a way that is "represented" by "sensual cognition". His non-Kantian gloss is that when we have learned that our theoretical physics is able to make constructions of a kind unforeseen by Kant, and not related in the way Kant thought necessary to the character of our "sensible intuition", the possibility is opened of our extending the notion of the "physical" after all into the domain Kant thought inescapably closed to it.

7. The foregoing rather lengthy discussion has been concerned with a matter on which Gödel differs critically—and, I have said, consciously and deliberately—with Kant. And yet, as also intimated above, in some respects—in some *places*—Gödel seems to minimize this difference, even to the point of denying it altogether. These must now, briefly, be pointed out.

In *B2*, page 14, one finds the statement that "the Kantian a priori is incompatible with relativity theory only in one minor point (made by Kant in only one rather obscure passage) and ... Kant's thesis of the apriority of Newtonian time and Euclidean space as he meant it does not contradict relativity theory at all." What follows makes it clear that the sense in which Gödel takes Kant to have "meant" his thesis is one that restricts its application to the structure of what is given at each moment in sense perception; thus, that his reading of Kant here is the one that has already been criticized above: that space "expresses itself" *only* in the moment-by-moment character of our "outer intuition". But what in the world can Gödel mean by the "one minor point" in which Kant's view *is* incompatible with relativity theory; and in what "rather obscure passage" does he say Kant makes that point?

On page 18 we find: "A real contradiction between relativity theory and Kantian philosophy seems to me to exist only in one point, namely, as to Kant's opinion that *natural science in the description it gives of the world must necessarily retain the forms of our sense perception and*

---

[y] Cf. A 845/B 873: "... *nature*, that is, the totality of *given* objects (whether they are given to the senses, or, if one will, to another sort of intuition)".

can do nothing else but set up relations between appearances within this frame" (emphasis added). That is admirably put, and seems exactly correct. But if this is the single point at which a contradiction occurs between Kant and relativity theory, it must be the point referred to on page 14. How can this point be deemed "minor"? It is not only a point of extremely far-reaching consequences for the philosophy, and the practice, of natural science; it is also absolutely central to Kant's critical philosophy: it is the heart of his doctrine of "transcendental idealism". It does, however, seem to follow from the two passages just cited that this is the point Gödel has called minor. As to the "obscure passage" that he refers to as the only place in which the point is made, Gödel has not told us which place he means, and I am completely unable to guess.

But this is certainly a matter in which we must bear in mind what has been emphasized in §1 of this introduction—that we do not have from Gödel's hand a final considered version of this essay: indeed, it should be noted that the passage cited from *B2*, page 14, has been revised in *C1* (page 23) so as to *eliminate* reference to the "minor point" and the "obscure passage".

There is another place where version *B2* as here published contains a statement about Kant that is both drastically incorrect and, in my opinion, most likely *not* representative of Gödel's own considered reading of Kant. On pages 16–17, we read:

> This insight [namely, that projection of our sensations into a three-dimensional Euclidean space might be "of no use for stating the laws governing them"] does not necessitate any weakening of Kant's thesis of the subjectivity of space, since *the sensations also* are something subjective (based on the actions of the things on us) and therefore any of their properties may as well be due to subjective as to objective conditions, i.e., to properties of the senses as well as to properties of the objects. Nor does it necessitate a weakening of his thesis of the apriority of space, since Kant did not assert that the "adequacy of our representation of space to the relation of the objects to our sensibility" can be known a priori. [That some particular relations of the things to us are adequately represented by Euclidean geometry in every world ... certainly does not mean yet "a complete adequacy to *the* relation of the things to our sensibility".]

The history of this passage through the successive drafts is rather complex, and need not be discussed here in detail. But it is crucial to remark that in *B1*—which in this substantially agrees with *A*—the second sentence of the quoted paragraph reads:

It does however necessitate a weakening of his thesis of the a priority of space, in that [*sic*] sense that the "adequacy of our representation of space to the relation of the objects to our sensibility" can only be known by experience, provided a reasonable meaning is given to this phrase.

The sentence thus says the opposite of what it says in the version printed below (and cited above). Now, what one finds in the *B2* typescript is, first, the *same* text as in *B1*: nothing in the sentence as just quoted from *B1* is canceled. *Then*, however, there are written in, above the *uncanceled* text, the following things: above the word 'It', the word 'Nor'; above 'however', 'it'; above the sequence 'sense that the "adequacy', the phrase 'since Kant did not assert'; and after the sequence 'to our sensibility" '—above a parenthesis that had already been canceled in *B1*—the phrase 'can be known a priori' (with no mark of punctuation— no period to stop the sentence). The remainder of the original sentence, from 'can only be known' to the end, is left on the following lines with no modification.

It will be seen from this that Gödel wrote in what can only be regarded as a *tentative* revision (since he did not cancel the original, or indicate at all which words the new ones were to replace). To be sure, the substance of the new words and phrases and the grammatical structure of the sentence determine uniquely how the revised sentence must read; but the manuscript does not give a decisive sign that the revision was judged favorably by Gödel in the end. The corresponding passage in *C1* (page 25) agrees with *B1* (and *A*).

It should be added that, on the substantive issue of Kantian interpretation, the passage as revised in *B2* is quite indefensible. For instance, in the *Prolegomena*, shortly after the passage cited by Gödel on the complete adequacy of our representation of space to the relation which our sensibility has to the objects, Kant says (page 292; emphasis added):

> Thus it is so far from true that my doctrine of the ideality of space and time makes the whole sensible world a mere illusion, that it is rather the only means of rendering secure the application of one of the most important [kinds of] knowledge—namely, that which mathematics propounds *a priori*—to actual objects, ... because without this observation it would be entirely impossible to tell whether the intuitions of space and time ... were not mere phantoms of the brain, made by ourselves, to which there corresponds no object whatever, *at least not adequately*, so that geometry itself should be a bare illusion ....

In Kant's own opinion, therefore, the only possible grounds of assurance of the adequacy of our spatial representation are the grounds provided by his transcendental idealism; and these, he says, are *a priori* grounds.

It seems fair to conclude that the same impulse to minimize Kant's differences with him—i.e., his with Kant—that we have already seen operating, has here led Gödel *at one moment* to venture an interpretation that is really impossible, and that he himself may in all likelihood have come to think better of.

8. There is another issue of Kantian interpretation that should be noted, although it is of no importance for Gödel's general considerations. Certain remarks of Gödel's assume (and appear to take for granted as uncontroversial and well-known characteristics of Kant's doctrine) that what Kant calls "sensibility" is divided into two faculties: the "faculty of sensation" and the "faculty of representation"; and that the latter of these is "something intellectual".[z] But this is quite inauthentic; and the question arises, just what Gödel himself means here and to what in Kant this distinction of Gödel's may be taken to correspond.

Two possibilities suggest themselves. The first is that he is thinking of Kant's distinction of the *matter* of intuition and its *form*. This reading is supported by the considerations (a) that Gödel calls the "first part" the faculty of *sensation*, and Kant says that "in the appearance I call that which corresponds to sensation [*Empfindung*] its *matter*" (A 20/B 34); and (b) that space and time are indeed, of course, according to Kant, *forms* of appearances, or of intuition. But it seems very strained, on this interpretation, that Gödel calls these two parts two "faculties"; that he calls the second part—the *form* of intuition—"the faculty of *representation*"; and that he characterizes this as "something *intellectual*". The alternative appears to be to identify Gödel's "faculty of representation" with the *productive imagination* of Kant. This is at least, in a certain sense, on the way towards being "intellectual"—i.e., a function of the *understanding*. Thus Kant says (B 151–2): "The imagination ... belongs to *sensibility*; yet to the extent that its synthesis is an exercise of spontaneity, which is determining and not merely like the sense determinable, ... to this extent imagination is ... a faculty to determine the sensibility *a priori*; and its synthesis of intuitions, conformable with the categories, must be the transcendental synthesis of *imagination*, which is an action of the understanding on the sensibility ...." The supposition that what Gödel meant by the "second part of sensibility" was the (productive) imagination has the advantage of making it clearer how he can speak of it as a distinct faculty, and as something intellectual. But if this *is* what Gödel meant, it seems all the stranger that he offered no

---

[z]See *B2*, p. 4b; *C1*, p. 17.

fuller discussion and no textual references whatever on what is surely a rather subtle part of Kant's doctrine.[aa]

This second way of taking Gödel perhaps gains some support from the only passage that can be seen as *introducing* the distinction to which he later refers as if it were already understood. This is, however, not a passage in Gödel's text, but only in a note; and is itself rather cryptic as to there being "two parts" to the faculty: " 'Sensibility' is, according to Kant, the faculty of having sensations under the influence of objects and happenings outside or inside ourselves, *and forming images* of outer objects and psychic processes out of these sensations." (*B2*, page 1, footnote 3; slightly revised version in *C1*, page 2, footnote 8; emphasis added.)

9. Several matters of interest present themselves for comment concerning the philosophical position taken by Gödel himself in these essays, independently of its relation to Kant; but the length to which this introduction has already grown suggests restraint at this point. One such matter, however, must be briefly noted, because it involves a bit of textual evidence that deserves to be put on record.

In the introductory note to *Gödel 1949a*, a question is raised as to why Gödel should have thought that it was necessary to defend the physical non-absurdity of his R-worlds by establishing that the "time travel" *physically* possible in them would not be *practically* possible (see these *Works*, Volume II, page 199, footnote a). The argument given by Gödel in *1949a*, page 561, appears in nearly the same words (but in a weaker form—cf. remark (c) of Malament quoted in §2 above) in *C1*, page 12, footnote 15. However, Malament has informed me of an earlier version of that note, in which Gödel himself presents exactly the case that is made in the cited introductory note. (This is in one respect very reassuring to the commentator; in another, disconcerting: why did Gödel change his mind?) I quote Malament's minute of this text (I have made one emendation, in brackets, of a typographical error; the initial and final passages in brackets are Malament's own, bracketed in his typescript):

[aa]Since this section was written, a new piece of evidence has been found that lends support to the second alternative discussed, and provides a possible source for the phrase used by Gödel. *Kowalewski 1924* (a volume that Gödel could well have seen) contains the notes on Kant's lectures during a sequence of three semesters (Winter 1791/92 – Winter 1792/93) by one of his auditors. On p. 106 of that volume, corresponding to p. 40 of the manuscript notes on Kant's course "Anthropologie", one finds the boldface heading "Von der Einbildungskraft und Phantasie" ("On the imagination and phantasy"); and immediately following that heading, the sentence (within brackets, indicating—*ibid.*, p. 52—that in the manuscript it appears as a marginal note): "Sie ist das 2te Stück der Sinnlichkeit" ("It is the second part of sensibility").

[The original version of the footnote includes a passage that was subsequently crossed out. It is here restored and typed in boldface. The footnote addresses the standard objection that "time travel" is absurd: if it were possible, one could go back in time and undo the past.]

This state of affairs seems to imply an absurdity. For it enables one e.g. to travel into the near past of those places where he has himself lived. There he would encounter a person which would be himself so and so many years ago. Now he could do something to this person which he knows by his own memory has not happened to him. This and similar contradictions, however, **presuppose, not only the practical feasibility of the trip into the past (velocities very close to that of light would be necessary for it) but also certain decision[s] on the part of the traveler; whose possibility one concludes only from vague conviction of the freedom of the will. Practically the same inconsistencies (again by neglecting certain "practical" difficulties) can be derived from the assumption of strict causality and the freedom of the will in the sense just indicated. Hence, as far as the paradoxical situation under consideration is concerned, an R-world is not any more absurd than any world subject to strict causality.**

[Gödel's response (in boldface) seems exactly right. Why did he cross it out and emphasize instead the "practical difficulties of time travel"?]

It seems fitting to conclude with another quotation, bearing on the same issue, containing an early recognition of the possibility of such results as Gödel's, written by another great mathematician and philosopher (and colleague of Gödel and Einstein). The following occurs in *Weyl 1918a*, page 220:

From each world-point there proceeds the double cone of active future and passive past. Whereas in the special theory of relativity these are separated by an intermediate region, it is here [in the general theory] intrinsically quite possible that the cone of the active future encroaches upon that of the passive past; it can therefore happen, in principle, that I now experience events which arise, in part, only as effects of my future decisions and actions. Neither is it excluded that a world-line, though it has time-like direction at each point (in particular, the world-line of my body), returns to the neighborhood of a world-point that it has already once passed. From this there would result a more radical sort of Doppelgängerdom than ever an E. T. A. Hoffmann has thought up. In point of fact, such considerable variations of the g_{ik} as would be required for these effects do not occur in the world-region in which we live; yet to think through these possibilities, as they bear upon the philosophical problem of the relation of cosmic and phenomenal time, has a certain interest. However paradoxical what thus comes to light, a genuine contradiction to the facts given immediately in our experience nowhere arises.

Howard Stein[bb]

Kant's *Critique of Pure Reason* is cited below in the second edition (*Kant 1787*) in the standard way, i.e., the page number preceded by "B". (Gödel does not cite or refer to the first edition *1781*.) Where convenient, these page references are inserted into the text. Page references to Kant's *Prolegomena* (*Kant 1783*) have been supplied by the editors and refer to volume IV of the Academy edition of Kant's works (*Kant 1902–*).

All translations of quotations from Kant are Gödel's own.

The editors wish to acknowledge Howard Stein's considerable contribution to the editing of these texts, in addition to his writing the introductory note. We also relied on an earlier examination and classification of Gödel's drafts by David Malament and Stein.

[bb]I am indebted to Charles Parsons for helpful discussion both of textual points and issues of Kantian interpretation.

David Malament's contribution to this commentary is indicated only in part by the references to him in the foregoing. Numerous conversations with him concerning mathematically obscure points in Gödel's essay have been an invaluable aid to my own attempts to clarify them; I here record my gratitude.

Some observations about the relationship
between theory of relativity and Kantian philosophy
(*1946/9–B2)

I

It is a remarkable fact, to which however very little attention is being paid in current philosophical discussions, that at least in one point relativity theory has furnished a very striking confirmation of Kantian doctrines.[1] In fact it is one of the most surprising and counterintuitive tenets of Kant for which this is the case, namely, for that part of his doctrine about time which says that time is neither "something existing in itself" (i.e., a separate entity besides the objects in it), nor "a characteristic or ordering inherent in the objects",[2] but only exists in a relative sense. That entity relative to which it exists, *according to Kant*, is the perceiving subject or, more precisely, its "sensibility";[3] *according to relativity theory*, it is certain more general and abstract things, such as material points, world lines, and coordinate systems, which, however, likewise can be conceived most conveniently as characteristics of, or as belonging to, a possible observer.[4]

---

[1] In order to prevent any misunderstanding, I wish to say right in the beginning that I am not an adherent of Kantian philosophy in general. The subsequent considerations only try to show that a surprising similarity subsists in some respects between relativity theory and the Kantian doctrine about time and space and that contradictions between them, as far as they occur, are by far not so fundamental as is widely believed. Above all, however, I wanted to show that the questions arising in such a comparison are interesting and perhaps even fruitful for the future development of physics.

[2] Cf. B 49, paragraph a).

[3] "Sensibility" is, according to Kant, the faculty of having sensations under the influence of objects and happenings outside or inside ourselves, and forming images of outer objects and psychic processes out of these sensations. "Faculty of sense perception" would be a more complete, but somewhat cumbersome, expression of what is meant. Sensibility is divided by Kant into outer and inner sense.

[4] Only in [the] case of *special* relativity theory and a straight world line is it unambiguous how a time valid for the whole world is to be defined relative to an observer traveling along this world line. Generalizations to other cases exist, but it is not so clear what the "correct" generalization is, or whether a uniquely determined simplest (or most convenient) generalization at all exists. For the physicist such questions are of little interest, since the measuring results can be predicted also directly from the objective state of affairs. If one takes the viewpoint that time in general relativity theory is the time-like[a] coordinate of any Gaussian coordinate system which is so chosen that it has one everywhere time-like and three everywhere space-like coordinates, the dependence of time on the perceiving subject becomes still more manifest, since then anything (within certain limits) which the perceiving subject chooses to consider as time is time.

---

[a] Handwritten in the margin of the manuscript are the words "too general!" and an arrow pointing to the typewritten line beginning with the word "time-like". Thus the phrase may as well refer to the entire passage as to the specific line or word.

II

There exists, also according to Kant, an objective correlate in the things of our representation of time; e.g., he speaks of "those modifications" (evidently of the things in themselves) "which we represent to ourselves as changes" (B 54). Therefore his thesis of the | "*non-existence of time except* 2 *relative to the perceiving subject*", must be understood to refer ONLY to something which is supposed to have the essential characteristics imagined in *our subjective idea* of time[5] (again in complete agreement with relativity theory).

That furthermore Kant, insofar as he attributes any reality to time *such as we perceive it*, really means[b] that temporal properties are certain relations of the things to the perceiving subject appears from many passages of his writings. E.g., in his explanation of the difference between "Schein" and "Erscheinung" he says quite clearly that the properties of the appearances, although they are not properties of the things in themselves,[6] still are not mere illusions, because they represent relations of the things to the subject. Kant's own words, in literal translation, are: (1) "the predicates of the appearance can be attributed to the object itself in relation to our sense" (B 69 n.), and: (2) "what is not to be found in the object in itself but always in its relation to the subject ... is appearance" (B 69 n.). Again in the same discussion Kant says that the properties which we attribute to the objects in the appearance (3) "depend ... on the mode of intuition of the subject *in the relation* of the given object to it" (B 69 [emphasis Gödel's]). In other passages he says that the outer as well as the inner sense (4) "can contain in its representation only the relation of an object to the subject" (B 67), that[7] time and space are (5) "characteristics not inherent in the things in themselves, but only in their relation to our sensibility", that[8] the appearances are something (6) "whose possibility is based on the relation of certain things unknown in themselves ... to our sensibility". Finally Kant

---

[5] In particular, e.g., [that it] defines a linear ordering of all events. Only for such an entity is the word "time" without further specification used in the sequel.

[6] The term "things in themselves" reduced to the plain meaning which it has in Kant's theory of knowledge simply means "the things", as distinguished from the images they produce in us and the relations they have to our perceptive faculties.

[7] Cf. *Prolegomena*, §11, [p. 284].

[8] See *Prolegomena*, §13, [p. 286]. To this passage, as well as that numbered (2), it is to be noted that for Kant "appearance" does not mean the fact that something appears, but rather an object of thought which has all the properties which are attributed to the object appearing (cf. his definition at B 34). Also the context in which these two passages occur shows clearly that "appearance" is meant in this sense.

---

[b] The phrase "by it" has been deleted following "means" by the editors, under the assumption that Gödel had in mind the German "dabei". As such it was judged to be implied by the remainder of the sentence.

also says[9] that (7) "our representation of space *is completely adequate* to
the relation | which our sensibility has to the objects" (and this evidently
is meant to apply to time as well).

But does perhaps the relation to sensibility of which Kant speaks in
these passages consist solely in the fact that we represent the things in
time and space and as such and such appearances, or that we have the
permanent disposition to do so? In that case time would have no more
reality than a constant illusion, and every analogy with physics, as to the
positive half of Kant's doctrine (as formulated above) would disappear.
Here two questions are to be distinguished:

1. As to the spatial and temporal properties and relations of *objects in
time and space*, it is quite clear that, by Kant's doctrine, they certainly
do not mean only that we imagine things to have these properties and
relations, but on the contrary to each of them there must correspond some
objective relation of the things to us which subsists (independently of our
representations) whenever we ascribe that spatial or temporal property or
relation to something. This follows from the fact that we determine the
spatial and temporal positions and properties of objects by certain well-
defined rules[10] on the basis of our sensations and the sensations are caused
by actions of things upon us,[11] which have nothing to do with our cognitive
activity. Hence, generally, to differences in spatial or temporal properties
differences in the objective state of affairs must correspond.

| That this actually is Kant's meaning not only follows indirectly from
the general considerations just made, but also directly from the wording
of some of the passages quoted, in particular the one[12] speaking about
"those modifications which we represent to ourselves as changes" and that
numbered (3) above, where certain relations of the things to us appear
clearly as existing *besides* our "mode of intuition". Moreover one may
allege the passages (2), (4), (6), which (if one takes account of what is said
in footnote 8) seem to imply that the relations under consideration are in
some sense the object of our representations [and] hence cannot consist
solely in the act or disposition of representing.

On the other hand, Kant's subjectivity of temporal relations cannot be
interpreted to refer solely to their qualitative character, so that to each temporal relation an isomorphic objective relation between the things would

---

[9]See *Prolegomena*, §13, end of note II, [pp. 289–290—emphasis Gödel's].

[10]Almost the whole chapter on the "Second analogy of experience" treats of the rule
by which we determine temporal ordering in the appearances. That the formulation
of this rule given by Kant can hardly be considered as satisfactory does not change
anything in the principle in question.

[11]Cf. B 34: "The effect of an object on the faculty of representation insofar as we
are affected by it is sensation".

[12]Cf. the passage from B 54 cited above.

correspond. For in that case at least the "ordering" given by time would be something "inherent in the objects", which is explicitly denied by Kant (see the formulation given in the beginning). So the state of affairs as conceived by Kant, as to the point in question, *is very similar* to that subsisting in relativity theory.

2. As to the properties of *space and time themselves*, such as the one-dimensionality of time, the Euclidicity of space, etc., the whole tenor of the Kantian exposition, as well as many particular passages, seem to imply that, by Kant's doctrine, they really exist only insofar as we represent things in (i.e., project our sensations on) these schemes, and that, as far as the reality independent of our representations is concerned, we might quite as well project them on other schemes. E.g., Kant says about time: "So it has subjective reality with regard to my inner experience, i.e., I really have the | representation of time and my modifications in it. Hence it is to be   4a considered to be real not as an object, but only as the mode of intuition of myself as an object." (B 53) The theory of the apriority of time and space in particular seems to require this conception.

On the other hand, however, some of the passages quoted before seem to point in the opposite direction and to attribute some kind of objectivity also to time and space themselves, although, of course, only in the sense of | relations of the things to us (but relations subsisting independently of   4b any *cognitive* facts). Especially the passage (7) can hardly be interpreted to mean that space is adequate to the relation which things have to us owing to our representing them in space. Still less could such an adequacy be used (as Kant does use it) as a defense against the objection that his theory transforms the whole world of sense perception into a mere illusion. For it would be a very poor defense to say that this is not so, because it is an adequate representation of this illusion.

So in this passage a view as to the nature of space and time, slightly different from that usually ascribed to Kant, seems to be implied, which however is not incompatible with the latter insofar as, corresponding to the two parts of sensibility (the faculty of sensation and of representation), both kinds of relations of the things to sensibility may subsist beside each other. The entire neglect of the second kind in the discussion of the epistemological character of geometry seems to me to be the reason why Kant failed to see that geometry in one sense is an empirical science (cf. page 16).

| III   5

Having now discussed the historical correctness of the formulation of Kant's thesis given in the beginning, and having thrown some light upon its meaning, let us see in what sense and to what extent this doctrine of Kant has been verified in relativity theory, and let us begin with the negative part of Kant's thesis. Here the agreement is most striking; for the fundamental temporal relation between two events A and B ("A before B by t seconds") is always quantitatively different for suitably chosen different observers and may even be inverse in direction for two different observers. Hence it is, such as it is directly observed, certainly not something inherent in the events.

6 | Of course there might nonetheless exist some objective reality independent of the observer and the frame of reference, which has the character of time such as we imagine it, but which does not adequately express itself in the observations. In relativity theory however this is not the case either, since what remains of time as an objective characteristic inherent in the events themselves, namely, the relation "A is before B for all observers" and its quantitative specification,[13] is quite different from what we imagine by temporal sequence.

First of all it determines only a partial ordering, i.e., some of the pairs of events have no absolute temporal relation whatsoever to each other, insofar[e] as the relation "neither before nor after" cannot be interpreted as "simultaneousness". (If it were, two events simultaneous with a third one would in general not be simultaneous with each other.) Still more

---

[13]I.e., the relation "A before B in general by at least t seconds (but not more than t seconds) for every observer".[c] This magnitude is exactly the main invariant of the space-time scheme of special relativity theory. This is all that remains in relativity theory of time as an objective characteristic inherent in the events. [Or, to be more exact, every generally applicable definition of an absolute "before" leads to a relation which in some possible worlds has the above-explained properties distinguishing it from the intuitive "before". (See the paragraph about the definition of an absolute world time below.[d])]

---

[c]There is an obvious slip here, for "by at least t seconds (but not more than t seconds) for every observer" just means *by exactly t seconds for every observer*—which is an assertion that time is "absolute"! Gödel's original note, in A and $B1$, does not contain the parenthetical phrase; this is inserted, in longhand, in $B2$. But without that phrase the statement is still wrong. For two events, A and B, in Minkowski space-time ("the space-time scheme of special relativity theory"), with A earlier than B, the proper time-lapse from A to B along any time-like path is *less than or equal to* the Minkowski "interval" t between them (the maximum, t, is attained along the "straight" path); the greatest *lower* bound of this time-lapse is always zero.

What seems to have happened is this: First, Gödel overlooked the point just made: he thought of the straight path as the *geodesic* from A to B (which is quite right) and overlooked for the nonce that in Minkowski space-time a *time-like* geodesic does not minimize, but rather *maximizes*, length between its extremities. His intent, therefore, was to characterize the Minkowski interval as the greatest lower bound of the proper time-lapses. Gödel then noticed this logical slip; and (somewhat hastily) thought to correct it by the addition of the parenthetical phrase—which fails to accomplish its purpose. To say what he *meant*, he should have written: "the relation 't is the largest number such that A is before B by at least t seconds for every observer'". To say what he *should* have meant, he should have written: "the relation 't is the *smallest* number such that A is before B by at *most* t seconds for every observer'". A comedy of errors!

[d]Gödel evidently meant to refer to the full paragraph on p. 10 below.

[e]Surely the word intended here is "inasmuch".

different from our idea of temporal sequence is the absolute "before" in its quantitative properties. For it is not additive (i.e., except for special cases, $AB + BC \neq AC$ for the temporal distances of three events A, B, C) and cannot be represented by a line, but only by a four-dimensional space; in short, objective "time" is not one-, but four-dimensional. Furthermore for some of the pairs of events no absolute temporal distance at all exists (for such events any temporal distance whatsoever is valid for some possible observer).

But not only as to the structure, but also as to the specific content of our idea of time—namely, the character of the "flowing" of time and of "change" of things in time—relativity theory gives the unambiguous answer that the objective correlate of our representation of time does *not* have this character. For that time elapses and change exists means, or at least implies, that at any moment of our existence only a certain portion of the totality of facts of which the world | is composed exists (and different 7 portions at different moments). Moreover, if there is to be real change in the things, not only apparent change for the observer, the word "exists" in the preceding sentence must mean existence of the facts in themselves (not just for me), i.e., the facts existing now must be the same no matter which observer, at the present moment, judges about the question. But for the "now" of relativity theory this is not the case; and a generally applicable objective concept of "now", which would have this property, cannot be introduced at all within the framework of relativity theory (see the paragraph below about the definition of an absolute world time[f]). Hence, if relativity theory gives a correct description of reality, the assumption that at any moment of time only a certain portion of the facts composing the world exists objectively is wrong, i.e., there exists no objective change and no objective lapse of time.[14]

This state of affairs matches surprisingly well with a certain passage of the *Critique of pure reason*, where Kant says that for beings with other forms of cognition "those modifications which we represent to ourselves as changes would give rise to a perception in which the idea of time, and therefore also of change, would not occur at all" (B 54) (such as, e.g., the perception of an inclination of the world-lines in a certain place of Minkowski's four-dimensional world, one is almost tempted to add).

---

[14]The unreality of change has been asserted frequently, also quite independently of the theory of relativity (already in antiquity), simply on the ground that the idea of change is self-contradictory (cf., e.g., *McTaggart 1908*). Indeed if change is real, "the present" seems to be the same thing as "the existing" (or at least "the existing" in the proper sense of the word, whereas past and future have only some weakened, shadowy form of existence). But this entails that we can, at different times, rightly assert about the same thing that it exists and that it does not exist.

---

[f]See the paragraph cited in fn. 13.

Summarizing the foregoing considerations we can say that what remains of time in relativity theory as an objective reality inherent in the things neither has the structure of a linear ordering nor the character of flowing or allowing of change. Something of this kind, however, can hardly be called time. So Kant's view about time *has* been confirmed by relativity theory
8 at least | as to its negative part.

As to its positive part the similarity to relativity theory is less [[pronounced]]^g in the details, but still there exists a close affinity in principle. Time in the ordinary sense of the word (i.e., the time obtained from that immediately experienced by a simple process of idealization and extrapolation outside ourselves) is in both theories a relation to the observer, but in relativity theory it is a very simple relation to the sense organs of the observer (to be more precise it depends only on the place and state of motion of his sense organs or other observational apparatus); for Kant it is a relation to his "sensibility" (which presumably means that it depends on the special structure of his organs of sense and representation). Considering what has been said above on pages 3–4 (cf. in particular (7)) this difference between Kant and relativity theory is not so fundamental as it might at first seem because also for Kant the relation in question is to a large extent factual (e.g., physical), not merely cognitive. But still it doubtless means that the relativity of time is asserted in a much stronger sense by Kant than in relativity theory. A similar relationship exists between the two theories also in other respects (see what is said about space on page 13) so that it may be said in general that (as far as the nature of space and time is concerned) Kant and relativity theory go in exactly the same direction, but relativity theory goes only one step in the direction indicated by Kant.

But also as to the positive part of Kant's doctrine there really exists a deep-rooted affinity (not only a connection made up artificially). The linear ordering of all events by the time immediately experienced, e.g., really expresses a "relation of the things to our sensibility". For it comes about only by projecting the events on the world line of our body, i.e., on the line of immediate sensorial contact we have with the reality outside ourselves.

9 | IV

The agreement described in the preceding pages between certain consequences of modern physics and a doctrine which Kant set up 150 years ago, in contradiction both to common sense and to the physicists and philosophers of his times, is greatly surprising, and it is hard to understand why so

---

^g See the textual notes.

little attention is being paid to it in philosophical discussions of relativity theory. Quite on the contrary, as far as any relationship is at all noted, Kant is usually stamped (together with Newton) as an arch-champion of absolute time and space, and then either his doctrine or relativity theory is rejected, or some "reconciliation" is attempted. If however someone sees in relativity theory a confirmation of Kant, he nearly always means it in some entirely different sense, or he passes with a few words over the questions considered above.

Moreover, disregarding the relations to Kant, there is, with very few exceptions, a tendency to minimize the subjectivistic consequences of relativity theory. Much store is laid by the existence of the absolute relation of "before" mentioned above, but it is overlooked that this relation has a structure entirely different from that of the intuitive "before" and in particular does not permit of any objective lapse of time or objective change.

With much more justification, it is objected[15] that the impossibility of defining any absolute time (among or besides the various relative ones) in the empty space-time scheme of special relativity theory (upon which the foregoing considerations have been based) does not exclude that matter and the curvature of space-time produced by it, | if the structure of the world as   10 a whole is taken into account, may enable us to determine some objectively distinguished ordering of all events to which the properties contained in our intuitive idea of time could consistently be attributed, and compared to which the various observed times would appear as something like systematic errors due to motion of the observers.

This view is supported by the fact that in all known cosmological solutions (i.e., relativistically possible structures in the large of non-empty worlds) such an "absolute world time" really can be defined.[16] But nevertheless the conclusions drawn above can be maintained because there exist other cosmological solutions[17] for which a definition in terms of physical magnitudes of an absolute world time is demonstrably impossible. If, however, such a world time were to be introduced in these worlds as a new entity, independent of all observable magnitudes, it would violate the principle of sufficient reason, insofar as one would have to make an arbitrary choice between infinitely many physically completely indistinguishable possibilities, and introduce a perfectly unfounded asymmetry. Therefrom it

---

[15]See *Jeans 1936*, pp. 22–23.

[16]Prima facie only for a world in which matter is strictly homogeneously distributed, but this difficulty could very likely be overcome, e.g., by considering the homogeneous solution with the least total deviation from the given one.

[17]Not yet published result.[h]

---

[h]Published in *Gödel 1949*, with supplementary results in *1952*; cf. also *1949b* in this volume.

follows that the possibility of a determination of an absolute world time, where it exists at all, is certainly not due to the laws of nature (which are satisfied in all cosmological solutions), but only to the special distribution and motion which matter has in those instances. A lapse of time, however, would have to be founded, one should think, in the laws of nature, i.e., it could hardly be maintained that whether or not an objective lapse of time exists depends on the special manner in which matter and its impulse are distributed in the world.

11 |
V

One reason for the neglect in contemporary philosophical literature of the relations set forth between Kant and relativity theory may lie in the fact that space and time occur on the same footing in Kant, whereas the foregoing considerations about time can be transferred to space only in a limited measure.

It is true that something like the space of Newtonian physics, which is supposed to exist beside and in complete independence of matter, likewise has no reality in relativity theory (its non-existence follows even from prerelativistic physics and the principle of the objective equality of states which cannot be distinguished by observations[18]). Kant however denied the objective existence not only of space and time, but also of spatial and temporal relations (B 42 a), and therein exactly consists the novelty of Kant's view. That time and space have no existence independent of and beside the things was asserted already by Leibniz.

Now as to time this view of Kant's has been verified to a large extent by relativity theory insofar as, in general, for some of the pairs of events no absolute temporal relation whatsoever exists in relativity theory, and for the remaining ones [there exist] relations of an entirely different nature and structure. But for space the situation is different. The spatial relations directly observed and measured, it is true, also have no absolute meaning, since they are different for different observers. But, in contradistinction to
12   time, there also exist for any two material | objects (i.e., things persisting in time, not events) absolute spatial relations[19] and they lead to a structure

---

[18] If there existed an absolute space, a state in which the whole world moves in a certain direction with respect to this space would be objectively different but experimentally indistinguishable from a state in which it is at rest.

[19] Of course every property or relation changing in time is, strictly speaking, not "inherent in the things", because it subsists only relative to a certain moment of time. But nevertheless the spatial relations under consideration are absolute as far as anything changing in time can be absolute, i.e., they are the same for different observers at moments of their respective times which correspond to each other in a certain way.[i]

---

[i] Written beside this footnote in shorthand was the word "falsch" (false).

(see below) not so very different from that of intuitive spatial relations. Nor do the latter include anything in their qualitative character which would contradict relativity theory (as does the idea of the flow of time).

In particular a "true spatial distance" of two points of the same or different material objects at a given moment of time (as opposed to the various apparent distances relative to various observers) can be defined (e.g., in case of not too large distances, by their distance in a coordinate system in which the centre of the line segment connecting them is at rest).[20] Moreover, in addition to the system of absolute spatial distances of the actually existing physical objects, there also exists (independent of the observer and the frame of reference) the schema of these distances for all *possible* movements of material points or (which is essentially the same) the schema of all possible movements of rigid bodies.[21] These movements can be described by the world lines of all points of these bodies, and these world lines in their turn are described by their coincidences with or relative positions to the material points of the actually existing bodies. This schema is what may be called the objective physical geometry of the world, of which the various geometries existing for various observers are only different aspects.[22]

[20] In order to obtain a structure not too different from intuitive space, one may confine this definition to infinitely small distances. This leads to a Riemannian space changing in time, possibly with discontinuous g_{ik}, if the velocity distribution is discontinuous. This space is different for two observers A, B only insofar as the phases of development of its several parts (in consequence of the different meanings of simultaneousness for A and B) are collected together in different ways into one phase of the whole space. The asymmetry between time and space which shows up in the situation described is not due to the space-time scheme as such, but only to its content, namely, the fact that there exist world lines of material points and that they always go in time-like directions, whereas in general there do not exist any physically distinguished space-like three-dimensional hypersurfaces. It is worth mentioning that the extent to which it is possible to speak of objective spatial relations in relativity theory (namely, always for objects persisting in time, but in general not for events) is essentially the same as in classical physics supplemented by the principle mentioned above. The difference is only that in classical physics the "true distance" and the apparent distance relative to an observer always agree with each other. Whether perhaps even "absolute" points of space persisting in time and identifiable at different times, can always be introduced by reference to the world as a whole and the matter in it is not known. This question (which corresponds exactly to that about time discussed on pp. 9–10) is complicated by the circumstance that not even its meaning is quite clear. For, evidently, in order to obtain something not too far apart from the intuitive idea of "points of space", some requirement about an invariability (at least in some respect) of this absolute space is necessary. Otherwise points of space may simply be identified with material points.

[21] Strictly speaking only infinitely small bodies (i.e., in practice, bodies small with respect to the stars).

[22] When, e.g., a man is included in a rotating box, the geometrical behaviour of rigid bodies changes for him only insofar as bodies at rest relative to him behave like bodies moving in a certain way in the former state and vice versa, and in general moving bodies behave like bodies moving in another way in the former state.

These absolute spatial relations, it is true, are not strictly isomorphic with the corresponding ones of intuitive space (insofar as, e.g., they are non-Euclidean), but these differences are insubstantial except for one point: namely, the finiteness of Riemannian space and of the straight lines in it

13   disagrees *completely* with our spatial intuition. | [To speak of an ideality (i.e., subjectivity) of space, however, is hardly justified, unless the conditions prevailing among the things are fundamentally, not only slightly, different.] Note that even for temporal relations there exist in relativity theory certain marked similarities between the objective state of affairs and that subsisting relative to an observer, so that a still higher degree of subjectivity is conceivable. This applies in particular to the fact that for the series of events happening to one material point the "before" has always an objective meaning.[23]

In the present imperfect state of physics, however, it cannot be maintained with any reasonable degree of certainty that the space-time scheme of relativity theory really describes the objective structure of the material world. Perhaps it is to be considered as only one step beyond the appearances and towards the things (i.e., as one "level of objectivation", to be followed by others[24]). Quantum physics in particular seems to indicate that physical reality is something still more different from the appearances than even the four-dimensional Einstein-Minkowski world.[25] T. Kaluza's fifth dimension points in the same direction.[26]

---

[23]I.e., it subsists relative to all observers. Kant, incidentally, would not acknowledge this as a sufficient criterion for objectivity. Cf. B 72.

[24]Cf. in this connection *Bollert 1921*, where one may find a description in more detail of these steps or "levels of objectivation", each of which is obtained from the preceding one by the elimination of certain subjective elements. The "natural" world picture, i.e., Kant's world of appearances itself, also must of course be considered as one such level, in which a great many subjective elements of the "world of sensations" are already eliminated. Unfortunately, whenever this fruitful viewpoint of a *distinction* between subjective and objective elements in our knowledge (which is so impressively suggested by Kant's comparison with the Copernican system; see below p. 20) appears in epistemology, there is at once a tendency to exaggerate it into a boundless subjectivism, whereby its effect is annulled. Kant's thesis of the unknowability of the things in themselves is one example, and another one is the prejudice that the positivistic interpretation of quantum mechanics, the only one known at present, must necessarily be the final stage of the theory.

[25]Quantum physics has not yet succeeded in giving a satisfactory description of a physical reality which would make the success of its rules of computation understandable. Cf. *Einstein, Podolsky and Rosen 1935*, where it is proved that the wave function cannot represent the whole reality, unless an action at distance not diminishing with the distance is assumed. Unfortunately, this condition, which is necessary for the proof given, is not stated explicitly in the result.

[26]See *Kaluza 1921*.

VI

14

There exists, in addition to the situation just discussed, a second circumstance which tends to conceal the close relationship which exists in some respects between relativity theory and Kant, namely, the fact that Kant, in addition to the subjectivity of time and space, also asserted the apriority of all our knowledge concerning them, and that he even based his main proof for their subjectivity on their apriority. Their apriority, however, seems flagrantly to contradict relativity theory, since the newly discovered properties of time and space (such as the relativity of simultaneousness, the non-Euclidicity of space, the Lorentz contraction, etc.), far from being a priori recognizable, even flatly contradict our supposed a priori intuition.

Closer examination however shows that the Kantian a priori is incompatible with relativity theory only in one minor point (made by Kant in only one rather obscure passage) and that Kant's thesis of the apriority of Newtonian time and Euclidean space as he meant it does not contradict relativity theory at all.

In the case of geometry, e.g., the fact that the physical bodies surrounding us move by the laws of a non-Euclidean geometry does not exclude in the least that we should have a Euclidean "form of sense perception", i.e., that we should possess an a priori representation of Euclidean space and be able to form images of outer objects only by projecting our sensations on this representation of space, so that, even if we were born in some strongly non-Euclidean world, we would nevertheless invariably imagine space to be Euclidean, but material objects to change their size and shape in a certain regular manner, when they move with respect to us or we with respect to them. | Nor does the non-Euclidicity of physical geometry (defined by the behaviour of rigid bodies) mean that this Euclidean pure intuition, if it exists, is simply wrong. For geometrical concepts satisfying Euclid's axioms can be defined also in a non-Euclidean world, although (*in surprising conformity with Kant's views*) in general only as *relations* to our sense organs (or some other arbitrarily chosen objects).

Such definitions can be obtained, e.g., in the following way: We select some physical object (e.g., the earth or our body), a point O on it, and three (preferably orthogonal) directions α, β, γ at this point. Then we define the "coordinates" a, b, c of any material point P by the number of times a measuring rod must be applied along the directions α, β, γ (respectively, the directions obtained by suitable parallel displacements) in order to reach P from O.[27] If then straight lines are defined by linear

---

15

[27] In a closed (e.g., a Riemannian) world this definition of course would not give unique values for the coordinates, i.e., it would have the consequence that either every object is in infinitely many different places simultaneously, or that for every object there exist infinitely many exactly equal ones. But this undesirable consequence could be

equations in these coordinates, the Euclidean axioms for them are true irrespective of the physical properties of the world. The same, of course, is true for other concepts and axioms of Euclidean geometry.

Hence, if we should possess an innate intuition representing geometrical entities of this kind,[28] the axioms of this intuitive geometry, although referring to reality (namely, to measuring processes) would be a priori true.[29]

16   Nevertheless, | they would not be analytic from Kant's standpoint, since arithmetic, to which geometry is reduced in this way, was not considered as analytic by Kant.[30] What, however, Kant did not take into account is that (irrespective of whether we have such an a priori intuition) space and its properties express themselves also in the sensations, which we know only a posteriori; namely, in the fact that by projecting the sensations in a certain way into a three-dimensional Euclidean space the laws connecting them can be stated much more easily. That this is not a necessary property of our sensations follows from the possibility of a physical non-Euclidean geometry and other considerations along these lines [which show that our sensations might be such that to project them into a three-dimensional Euclidean space, though possible, would be of no use for stating the laws governing them].

This insight does not necessitate any weakening of Kant's thesis of the subjectivity of space, since *the sensations also* are something subjective (based on the actions of the things on us) and therefore any of their properties may as well be due to subjective as to objective conditions, i.e., to properties of the senses as well as to properties of the objects. Nor does it

avoided by choosing other definitions. Only if we were to project our world, by the procedure described above, on a space of a higher dimensionality (e.g., four) would consequences of this kind become inevitable.

[28]Of course I do not want to say that pure intuition according to Kant refers exactly to those geometrical entities which are defined above. Rather I only wanted to show that, surprisingly, an innate Euclidean geometrical intuition which is a priori valid and whose concepts have a well-determined physical meaning is possible and compatible with the existence and physical reality of non-Euclidean geometry and with relativity theory.

[29]This seems to contradict Einstein's well-known aperçu about the relation between geometry and experience (cf. *Einstein 1921*), but the contradiction is rather one of words than of meaning, since the phrase "sich auf die Wirklichkeit beziehen" is evidently meant by Einstein in the sense of "say something about reality", whereas the geometrical axioms as interpreted above, although they refer to physical reality, do not say anything about it. Also, according to Kant, the geometrical propositions are synthetic, not because they say anything about that which is given (i.e., the sensations), or about the things in themselves, but because they say something about that which we, or rather our intellectual faculties, imagine (assume, construct) in the process of mental assimilation of the sensations.

[30]Whether or not Kant was justified in this view depends on what is meant by analyticity which, on the basis of Kant's own definition, is by no means so unambiguous as it looks at first sight.

necessitate a weakening of his thesis of the apriority of space, since Kant did not assert that the "adequacy of our representation of space to the relation of the objects to our sensibility" can be known a priori.[j] [That some particular relations of the things to us are adequately represented by Euclidean geometry in every world (cf. the definitions given above) certainly does not mean yet "a complete adequacy | to *the* relation of the things to   17 our sensibility".]

In point of fact the Kantian assertion about adequacy is factually (although not a priori) true in this sense, that the deviations from Euclidean geometry in the small are far below everything directly observable. [It *might* turn out to be true also in the more interesting sense that some Euclidean geometry defined relative to an observer (i.e., a world line) is in some way physically distinguished—in contradistinction to the Euclidean geometry defined above which is of no physical interest[31] (very unlikely).]

One might moreover say that our supposed a priori intuition (irrespective of the question of its adequacy) would certainly be wrong in a strongly non-Euclidean world, at least insofar as it would represent Euclidean space, not as *our* scheme of arraying things (on the basis of relations of the things to *us*), but as a physical reality entirely independent of our existence. But it may be replied that this objectivation of space is a misrepresentation also by Kant's own theory, no matter whether physical geometry is Euclidean or not; that Kant however avoided this consequence [that pure intuition is in this respect wrong] by laying the blame solely on our interpretation of the data of pure intuition.[32]

I do not think the question whether, in accordance with Kant's view, we have an *innate* (and therefore a priori) intuition of Euclidean space (i.e., whether we would develop the same intuition also in a strongly non-Euclidean world) has yet been decided; nor the related question whether we are able (in our world) to learn to imagine[33] a non-Euclidean space. [For

[31]This question is closely connected with that discussed in fn. 4. It is worth noting that for time the answer to the corresponding question is trivially in the affirmative, since the reasonably defined time relative to an observer is of course isomorphic with Newtonian time.

[32]See B 69 and *Prolegomena*, §13, note III.

[33]Usually it is not realized how greatly different a non-Euclidean spatial intuition would be from ours in case of a sufficiently large curvature of space. E.g., in a certain Lobachevskian space (which stretches into infinity like ours) the circumference of a circle of radius 1 inch would be 1 yard.

[j]Whether this passage, as here rendered, genuinely reflects Gödel's considered opinion is very doubtful, in the light of the details of his manuscript text at this point and of its history. For a discussion, see the introductory note above, §7, especially pp. 224–226.

we can imagine this also in terms of regular changes of size and shape due
to motion in a Euclidean space.[34]] [For we would certainly not identify (in
18   our intuitive world picture) the behaviour of | physically rigid bodies with
geometry in every possible world, e.g., not in a world which is Euclidean
for the most part, but has certain non-Euclidean spots.] But whatever the
answers to these questions may be, there can certainly never result from
them any incompatibility between Kant and relativity theory, but at most
between Kant and psychology (or phenomenology) of sense perception.

VII

A real contradiction between relativity theory and Kantian philosophy
seems to me to exist only in one point, namely, as to Kant's opinion that
natural science in the description it gives of the world must necessarily
retain the forms of our sense perception and can do nothing else but set up
relations between appearances within this frame.

This view of Kant has doubtless its source in his conviction of the unknowability (at least by theoretical reason) of the things in themselves,[35]
and at this point, it seems to me, Kant should be modified, if one wants
to establish agreement between his doctrines and modern physics; i.e., it
should be assumed that it is possible for scientific knowledge, at least partially and step by step, to go beyond the appearances and approach the
world of things.[36]

The abandoning of that "natural" picture of the world which Kant calls
the world of "appearance" is exactly the main characteristic distinguishing
modern physics from Newtonian physics. Newtonian physics, except for the
elimination of secondary qualities (which in principle was known already
to Democritus), is only a refinement, but not a correction, of this picture
of the world; modern physics however has an entirely different character.
This is seen most clearly from the distinction which has developed between
19   "laboratory | language" and the theory, whereas Newtonian physics can be
completely expressed in a refined laboratory language.

[34]Cf., e.g., *Einstein 1921*.

[35]There is of course no strict equivalence between these two tenets. There is, however,
a certain affinity between them, insofar as the abandoning of our ordinary forms of
thinking is at least a first step towards a knowledge of the things in themselves.

[36]Cf. fn. 24.

VIII

So the trend of modern physics is in one respect opposed to Kantian philosophy. On the other hand, however, it should not be overlooked that the very refutation of Kant's assertion concerning the impossibility for theoretical science of stepping outside the limits of our natural conception of the world has furnished in several points a most striking confirmation of his main doctrine concerning this natural world picture, namely, its largely subjectivistic character, even as to those concepts which seem to constitute the very backbone of reality.

Moreover, it is to be noted that the possibility of a knowledge of things beyond the appearances is by no means so strictly opposed to the views of Kant himself as it is to those of many of his followers. For (1) Kant held the concept of things in themselves to be meaningful and emphasized repeatedly that their existence must be assumed, (2) the impossibility of a knowledge concerning them, in Kant's view, is by no means a necessary consequence of the nature of all knowledge, and perhaps does not subsist even for human knowledge in *every* respect. One may compare ⟦with respect⟧ to this point, e.g., what he says in the preface to the second edition of the *Critique of pure reason* (B xxi, Bxxii n. and Bxxvi n.) about the positive use of the *Critique*, not only for religious belief, but possibly also for knowledge (B xxvi n.). In the second of the passages cited above Kant even goes so far as to compare his subjectivistic theory of knowledge, as to its possible positive use, with Copernicus's explanation of the apparent motions of the planets by the motion of the observer, pointing out that this new viewpoint of Copernicus led to the discovery of the "hidden power connecting the structure of the universe, the Newtonian attraction".

Kant, it is true, wanted to base such knowledge on ethics, but, even as far as theoretical reason is concerned, he evidently did not want to say that nothing whatsoever can be asserted about the things in themselves. For Kant himself asserted, | e.g., that they exist, affect our sensibility, and do not exist in time and space, but that the ideas of time and space are completely adequate to their relationship to our sensibility.

Doubtless Kant would least of all have held an overstepping of the world of appearances to be possible, on the basis of conclusions drawn from experiments (already his thesis of the empirical reality of time and space seems to exclude this). But ⟦with respect⟧ to this point it is to be noted that, in perfect conformity with Kant, the experiments by themselves really do not force us to abandon Newtonian time and space as objective realities,[37] but

---

[37] In special relativity theory, e.g., some arbitrarily chosen inertial system can be considered as representing absolute space and time (with all properties Newton attributed to them) and the varying observational results of observers moving with different velocities can be explained by the effect which motion relative to this absolute space has on the

only the experiments together with certain general principles, in particular the principle that two states of affairs which cannot be distinguished by observations are also objectively equal. Generally speaking, it can be said that relativity theory (especially general relativity theory) owes its origin, perhaps more than any other physical theory, to the consistent application of certain very general principles, and was only subsequently verified in its consequences by experience.

physical bodies and processes, in particular on the measuring instruments (effects which incidentally follow without any ad hoc hypothesis from the empirically verifiable electro-magnetic equations of Maxwell and the very natural assumption that the constitution of physical bodies is based on electromagnetic forces or forces of a similar nature). It remains however entirely arbitrary which coordinate system is in this way distinguished as the absolute space-time scheme and this, by the principle formulated above, entails that absolute time and space do not exist.

Some observations about the relationship between theory of relativity and Kantian philosophy (*1946/9–C1)

It is an interesting fact, to which very little attention is being paid in current philosophical discussions, that at least in one point relativity theory has furnished a very striking illustration, in some sense even a verification, of Kantian doctrines.[1] It is a certain part of Kant's doctrine about time for which this is the case, namely, for his assertion that time is neither "something existing in itself" (i.e., a separate entity besides the objects in it) nor "a characteristic or ordering inherent in the objects"[2] but only | a characteristic inherent in the relation of the objects to something else.[3]

This view of Kant's, which constitutes the negative part and a fraction of the positive part of his doctrine about time, is indeed literally true in certain relativistic worlds,[4] provided [that] by "time" is understood what everybody understood by it before relativity theory existed. For this idea of time no doubt includes as its most essential characteristic that time consists of a one-dimensional system of points, isomorphic with a straight line, in which every event happening in the world has a definite place. But in the worlds under consideration an absolute[5] time of this kind does not exist.

---

[1] I wish to say right in the beginning that I am not an adherent of Kantian philosophy in general. The subsequent considerations only try to show that a surprising similarity subsists in some respects between relativity theory and the Kantian doctrine about time and space and that contradictions between them, as far as they occur, are by far not so fundamental as is frequently maintained.

[2] See B 49, paragraph a).

[3] Cf. *Prolegomena*, §11, [p. 284].

[4] I.e., worlds possible by the laws of nature which hold in relativity theory.

[5] "Absolute", in the sense of being independent of the frame or an object of reference, corresponds pretty closely to Kant's phrase "inherent in the object". For the concept "relative" appears in Kant as the opposite. Cf. the passage referred to in fn. [17].[a]

---

[a] In the manuscript, this footnote ended with the phrase "Cnf. the passage referred to in footnote ", with a blank space where the reference number should be. This raised the question whether there is a basis for identifying the passage meant. Each of the quotations from Kant given by Gödel on p. 4 involves the contrast between *properties of the "objects"* (or *"the things"*) and *relations of the objects* to our sensibility; therefore each of these is a possible candidate here (the references to the source, now given in the text, were given by Gödel in distinct footnotes for the distinct passages). This, however, provides no reasonable criterion for choosing *one* of those passages as the one meant here. Furthermore, none of these passages seems quite to fit Gödel's words in the present footnote. But the passage (B42, paragraph a) referred to in Gödel's problematic footnote 63 (fn. [17] in the present version—see editorial note f for discussion of the latter) contains the following statement:

2

There exist, it is true, also in these worlds certain absolute time-like
relations between events, e.g., the time-like distance given by the line-
element ds, in case $ds^2 > 0$; but this does not contradict Kant's doctrine
about time[6] because the structure of these absolute relations lacks the
essential characteristic of time just mentioned; and Kant by no means
2′ denied the existence of some correlate, inherent in the things, of our | idea
of time with properties different from intuitive time. On the contrary he
speaks explicitly of "those affections" (evidently of the things in themselves)
"which we represent to ourselves as changes" and which "in beings with
other forms of cognition would give rise to a perception in which the idea
of time ... would not occur at all" (B 54).

Essential differences between Kant's view about time and the conditions
prevailing in the relativistic worlds mentioned[7] exist, it is true, as to the
positive part of Kant's doctrine, in particular as to the question relative to
what a time in the usual sense does exist. But even in this respect there is
a distinct similarity of principle. A time satisfying the requirement stated
above (i.e., a one-dimensional temporal ordering of the events) exists for

[6]It does contradict certain other parts of Kant's system, namely, Kant's view that
the things in themselves (and therefore evidently also the objective correlate of the idea
of time) are unknowable, not only by sensual imagination, but also by abstract thinking.
Of course it need not be maintained that relativity theory gives a complete knowledge of
the objective correlate of the idea of time, but only that it goes one step in this direction
(cf. fn. [16]), but even this would seem to disagree with Kant (cf. p. [27]).

[7]In the sequel, I shall refer to these worlds as "R-worlds" ("R" being an abbreviation
for: "rotating, rotation-symmetric, stationary"). For their mathematical description cf.
[1949]. The compass of inertia in these worlds rotates everywhere relative to matter,
which in our world would mean that it rotates relative to the totality of galactic systems.
If the current explanation of the red-shift of distant nebulae is correct, our world is not
an R-world. It may, however, be an expanding rotating world.[b] In that case, some of the
subsequent considerations about the non-objectivity of time (in particular those about
travelling into the past and future, cf. below p. 12) might remain applicable.

────────────────

... space does not represent any determination that attaches to the objects
themselves and which remains even when abstraction has been made of
all the subjective conditions of intuition. For no determinations, *whether
absolute or relative*, can be intuited prior to the existence of the things to
which they belong, and none, therefore, can be intuited *a priori*. [Emphasis
added.]

Here we have the explicit contrast of "absolute" and "relative"—a contrast that Gödel
may have been referring to when he wrote "For the concept 'relative' appears in Kant
as the opposite." (As the opposite of what?—of *absolute*.)

[b]For a brief reference to these, see *Gödel *1949b*; for Gödel's sketch of a proof of the
existence of such solutions to Einstein's equation, see *Gödel 1952*.

Kant, relative to the perceiving subject or more precisely its "sensibility";[8] in the R-worlds | it exists relative to certain more general and abstract 3 entities such as material points, world lines, and coordinate systems, which however likewise can be conceived most conveniently as characteristics of, or as belonging to, a possible observer.

| That Kant, insofar as he attributes any objective reality to time, really 4 means[c] that temporal properties represent certain *relations* of the things to the perceiving subject appears from many passages of his writings. E.g., in his explanation of the difference between "Schein" and "Erscheinung" he says quite clearly that the properties of the appearances, although they are not properties of the things in themselves, still are not mere illusions, because they represent relations of the things to the subject. Kant's own words, in literal translation, are: "the predicates of the appearance can be attributed to the object itself in relation to our sense", and: "what is not to be found in the object in itself but always in its relation to the subject ... is appearance"(B 69 n.). Again in the same discussion Kant says that the properties which we attribute to the objects in the appearance "depend ... on the mode of intuition of the subject *in the relation* of the given object to it"(B 69 [emphasis Gödel's]). In other passages he says that the outer as well as the inner sense "can contain in its representation only the relation of an object to the subject" (B 67), that[9] time and space are "characteristics not inherent in the things in themselves, but only in their relation to our sensibility", that[10] the appearances are something "whose possibility is based on the relation of certain things unknown in themselves ... to our sensibility". Finally Kant also says[11] that "our representation of space is completely adequate to the relation | which our sensibility has to 5 the objects" (and this evidently is meant to apply to time as well).

---

[8]"Sensibility" is, according to Kant, the faculty of having sensations under the influence of things and happenings and of forming images of outer objects and of oneself and his affections out of these sensations. Sensibility is divided by Kant into outer and inner sense.

[9]Cf. *Prolegomena*, §11, [p. 284].

[10]See *Prolegomena* §13, [p. 286]. To this passage, as well as the second one from B 69n. referred to above), it is to be noted that for Kant "appearance" does not mean the fact that something appears, but rather an object of thought which has all the properties which are attributed to the object appearing (cf. his definition at B 34). Also the context in which these two passages occur shows clearly that "appearance" is meant in this sense.

[11]See: *Prolegomena* §13, end of note II, [pp. 289–290].

---

[c]The phrase "by it" has been deleted following "means" by the editors, under the assumption that Gödel had in mind the German "dabei". As such it was judged to be implied by the rest of the sentence. In version *C2*, Gödel had replaced "means by it that temporal properties represent" by "is viewing temporal properties as representing".

6 | This Kantian relativity of time certainly was not meant in so weak a sense as to refer solely to the qualitative character of temporal properties. For in that case at least the ordering given by time would be something "inherent in the objects", which is explicitly denied by Kant (see the formulation quoted in the beginning of this paper). On the other hand, Kant's relativity of time cannot be interpreted in so strong a sense that temporal properties would not mean anything else for the things except that we imagine them to have these properties. For this would clearly contradict some of the passages quoted above, which imply that our ideas of temporal conditions depend on the one hand on certain objective "modifications" (cf. the passage [from B 54] referred to on page 2′ above), on the other
7 hand on | certain "relations" of the objects to us (cf. the passage [from B 69] referred to on page 4 above).

So Kant's relativity of time is to some extent similar to that of relativity theory. One essential difference subsists, however, insofar as Kant hardly meant that temporal properties of things could be different for different observers, but rather that (except for the various positions in time of various individuals) they are the same for all human beings, and could be different (or non-existent) only for beings of an entirely different nature.

8 | Let us now compare in more detail Kant's theory of time with the conditions prevailing in the R-worlds and let us begin with the negative part of Kant's doctrine. Here the first thing we note is that if time is defined by direct measuring with clocks and Einstein's concept of simultaneousness,[12] Kant's assertion as formulated in the beginning is an immediate consequence of a simple relativistic fact, even irrespective of the particular structure of the world. For the fundamental temporal relation "A before B by t
9 seconds" obtained in this way is, in relativity theory, | always quantitatively different for suitably chosen different observers and may even be inverse in direction for two different observers. Hence it is certainly not something inherent in the events, nor [does it] allow time to be a substratum in which the events are lying.

10 | This, however, does not yet exclude the existence of some physical structure independent of the observer or any other object of reference which has the properties of Newtonian time, or at least the most essential one mentioned on page [2], but is not connected with the clock readings (or at least not with the clock readings of any arbitrary observer) in the simple and direct way indicated above. Actually such an "absolute world time" can be defined[13] in all known cosmological solutions (i.e., relativistically possi-

---

[12]This concept was originally designed for special relativity theory only, but it is clear that it can be extended to general relativity theory, at least under certain conditions which are satisfied in the R-worlds.

[13]Immediately only for worlds in which matter is strictly homogeneously distributed,

ble worlds).[14] In the R-worlds, however, such a definition is demonstrably impossible; i.e., none of the various systems of "points of time" which can be introduced is objectively distinguishable from the others, but any one of them can be singled out only by reference to individual objects, such as our earth.

As was noted above there exist also in the R-worlds certain absolute time-like relations, but they are quite different from "time" in the usual sense of the term. Above all they define only a partial ordering, or to be more exact: in whatever way one may introduce an absolute "before", there always exist either | temporally incomparable events or cyclically ordered   11 events. It follows that every definable absolute temporal distance lacks the property of additivity (i.e., $AB + BC \neq AC$ except for special cases) and therefore cannot be represented by a line, but only by a more than one-dimensional space. These structural differences further imply that an objective lapse of time, such as is contained in the intuitive idea of time, is impossible in the R-worlds. For by the lapse of time we imagine that reality consists of an infinity of layers which come into existence successively. But the R-worlds cannot be split up in such layers except relative to an observer or other object of | reference on which the layers then will depend. That   12 at least *that* passing of time which is directly experienced has no objective meaning in the R-worlds also follows from the fact that it is possible in these worlds to travel arbitrarily far into the future or the past and back again exactly as it is possible in other worlds to travel to distant parts of space.[15]

Hence summarizing the foregoing considerations one may say that what remains of time in the R-worlds as an objective reality, independent of an observer or other objects of reference, neither defines a linear ordering of the events, nor consists of a one-dimensional system of points, nor can have the character of "passing by". | Something of this kind, however, can hardly   13 be called time and is certainly quite different from what Kant understood

---

but this difficulty can be overcome; e.g., by considering the homogeneous solution with (in some sense) the least total deviation from the given one.

[14]On the ground of the practical success of these solutions for the explanation of astronomical facts, J. Jeans claimed a return to the old idea of an absolute time passing objectively to be justified (cf. *Jeans 1936*, pp. 22–23). However, it can hardly be maintained with certainty that our world really is described by one of these solutions. Moreover cf. what is said below on p. [[17]].

[15]This state of affairs seems to imply an absurdity. For it enables one, e.g., to travel into the near past of those places where he has himself lived. There he would encounter a person which would be himself so and so many years ago. Now he could do something to this person which he knows by his memory has not happened to him. This and similar contradictions, however, in order to prove the impossibility of the R-worlds, presuppose at least the practical feasibility of the trip into the past, which may very well be precluded by the velocities very close to that of light which would be necessary for it, or by other circumstances.

by time. So at least the negative part of Kant's theory of time is doubtless true in the R-worlds.

14

As to its positive part, the agreement is less [pronounced],[d] but still there exists an undeniable affinity in principle. | Not only does[e] a time satisfying the requirement stated on page [2] exist in the R-[worlds] only as a relation to something, but in particular there exists for each possible observer a certain time which may be called *his* time, because it is obtained by the simplest extrapolation of the time directly experienced or measured by him to the whole world. This time (as in Kant) expresses a relation

15/6

of the things to this observer. | There is, however, this difference that in relativity theory it is a purely physical relation to the body of the observer (i.e., it exists relative to him insofar as he is a sensual being in the most general sense); for Kant it is a relation to his sensibility (whereby Kant evidently meant that it exists only relative to the special structure of his organs of sense and imagination). This doubtless means that the only-relative existence of time is asserted in a much stronger sense by Kant than by relativity theory, even for the R-worlds.

A similar relationship subsists between the two theories also in other respects, in particular insofar as Kant doubtless held the difference between intuitive time and its objective correlate to be far greater than it is in relativity theory, even in the R-worlds; in fact, so great that the latter cannot be described at all in concepts understandable for human beings. So it may be said in general that, as far as the nature of time (and also of space, see below) is concerned, Kant and relativity theory go in exactly the same direction, but relativity theory goes only one step in the direction indicated by Kant.[16]

But nevertheless there really exists, also as to the positive part of Kant's doctrine, a deep-rooted affinity, not only a connection made up artificially. For the time valid for a given observer really expresses a "relation of the things to his sensibility", although in a more general sense than for Kant. For it comes about by projecting the events on the world line of his body,

[16]The present state of physics, however, seems to indicate that the future development will continue along these lines. The fifth dimension introduced by Th. Kaluza (cf. his *1921*), e.g., points in this direction. Still more so quantum mechanics, which has so far not been able to give a satisfactory description of an objective reality which would make the success of its rules of computation understandable. [Cf. *Einstein, Podolsky and Rosen 1935*, where it is proved that the wave function cannot represent the whole reality, unless an action at distance with very strange properties is assumed.] If the future development of physics really goes in the direction just indicated, relativity theory would have to be interpreted as only one "level of objectivation" (cf. fn. [26]) to be followed by others.

[d]See the textual notes.
[e]See the textual notes.

i.e., on the line of immediate sensorial contact he has with the reality outside himself.

| It is perfectly true that the term "sensibility" here does not have exactly   17
the same meaning as in Kant. For (1) Kant speaks (in the case of time) of inner sensibility [and] (2) Kant means in the first place the second part of sensibility, the faculty of representation (which is something intellectual) and not the faculty of sensation. But nevertheless the mere fact that the same general terms are appropriate to describe the situation in both cases proves an affinity of principle although not an agreement in detail.

It is important to note that the foregoing considerations are not entirely confined to that special kind of *possible* universes which were called R-worlds above, but that in a somewhat weakened form they also apply to those worlds in which an absolute time can be introduced. For this absolute time, although it agrees with intuitive time as to the type of ordering it creates for the events, still is in other respects quite different from it. Above all it does not exist owing to the laws of nature, but only owing to the particular structure of these worlds, and therefore it hardly conforms to reason to ascribe a passing to it. For this would imply that whether or not an objective passing of time exists depends on the particular way in which matter and its motion are arranged in the world. Hence if one includes in the concept of time certain characteristics of intuitive time in addition to its order type, then time has at most a relative existence also in the cosmological solutions known at present. All the more this would be true of the rotating and expanding universes (cf. footnote [14]).

One may wonder why in the foregoing pages only time was spoken of. Owing to the symmetry which subsists in relativity theory between time and space one should expect that the same would apply to space as well. As a matter of fact, however, this expectation verifies itself only in part and there are some essential points where the analogy does not go through.

| This is due to the fact that (1) the passing of time has no analogue   20
for space and that (2) the symmetry between space and time in relativity theory is not complete, if the existence of matter is taken into account. For matter entails the existence of a three-parametric system of physically distinguished one-dimensional time-like subspaces (namely, the world lines of matter), whereas there exists in general no one-parametric system of physically | distinguished three-dimensional space-like subspaces. For the   21
space-elements orthogonal to the world lines of matter can in general not be fitted together into three-dimensional spaces. This has the consequence that there exist no possible worlds in which | the structure of absolute   22
space-like relations would be so greatly different from intuitive space as is the case for time in the R-worlds. In particular, if the velocity vector of matter is continuous, the space of absolute distances between neighboring material points (defined by the orthogonal distances of their world lines) is an ordinary three-dimensional Riemannian space changing in time. That,

unlike [such space] in classical physics, its dependence on time can be represented in various ways (owing to the relativity of simultaneousness) is due to the different character which time (not space) has in relativity theory, as is seen from the fact that the same applies to every physical property which changes in time. Owing to the vector of mean velocity of matter (the mean being taken over suitably chosen regions of astronomical 22.1   dimen|sions) it is even possible to introduce in some sense "absolute" points of space, but they differ in several respects essentially from the absolute points of Newtonian and intuitive space, in particular insofar as their mutual distances need not remain constant in time (i.e., this space in general contracts, expands or is being deformed.)

23 | There exists one circumstance which tends to conceal the close relationship which subsists in some respects between relativity theory and Kant, namely, [17] the fact that Kant, in addition to the subjectivity of time and space, also asserted the apriority of all our knowledge concerning them, and that he even based his main proof for their subjectivity on their apriority. Their apriority, however, seems flagrantly to contradict relativity theory, since the newly discovered properties of time and space (such as the relativity of simultaneousness, the non-Euclidicity of space, the Lorentz contraction, etc.), far from being a priori recognizable, even flatly contradict our supposed a priori intuition.

Closer examination, however, shows that these properties not recognizable a priori concern rather the objective correlate of our ideas of time and space (i.e., in Kant's terminology, those affections which we represent to ourselves as time and space) and therefore are compatible with the apriority of some subjective space-time scheme which is founded in the intellectual equipment of man, but is likewise applicable to the description of physical reality, namely, as a kind of relative time and space (coordinate system). The apriority of Newtonian time and Euclidean space, understood in this sense, does not contradict relativity theory at all, but constitutes a question of its own.

[17] It is to be noted that Kant explicitly denied the objectivity of space not only if conceived as an entity besides the things, but also if conceived as a system of relations between the things. (Cf. B 42, paragraph a.)[f]

---

[f] Because Gödel had not marked this footnote for deletion from his list of footnotes, and because one of the footnote cross-references left blank in the text—namely, fn. 5 (our numbering)—probably refers to this footnote, we have elected to place this footnote here. (For a concordance of our footnote numbers with the original numbers, see the textual notes.) Further evidence that he did not intend to delete this footnote, but simply forgot to mark a location, consists in the occurrence of this footnote's (original) number in a concordance made by Gödel himself. There the number appears between those for the footnotes that became our fns. 16 and 18. (Note also that there were some pages deleted in that region.) The location we have chosen is in agreement with that evidence.

In the case of geometry, e.g., the fact that the physical bodies surrounding us move by the laws of a non-Euclidean geometry does not exclude in the least that we should have a Euclidean "form of sense perception", i.e., that we should possess an a priori representation of Euclidean space and be able to form images of outer objects only by projecting our sensations on this representation of space, so that, even if we were born in some strongly non-Euclidean world, we would nevertheless invariably imagine space to be Euclidean, but material objects to change their size and shape in a certain regular manner when they move with respect to us or we with respect to them. | Nor does the non-Euclidicity of physical geometry (defined by the behaviour of rigid bodies) mean that this Euclidean pure intuition, if it exists, is simply wrong. For geometrical concepts satisfying Euclid's axioms can be defined also in a non-Euclidean world, although (in conformity with Kant's views) not as objective[18] properties of the things but only as *relations* of the things to our sense organs (or, so one may add, to some other arbitrarily chosen objects).

Such definitions can be obtained, e.g., in the following way: we select some physical object (e.g., the earth or our body), a point O on it and three (preferably orthogonal) directions α, β, γ at this point. Then we define the "coordinates" a, b, c of any material point P by the number of times a measuring rod must be applied along the directions α, β, γ (respectively, the directions obtained by suitable parallel displacements) in order to reach P from O.[19] If then straight lines are defined by linear equations in these coordinates, the Euclidean axioms for them are true irrespective of the physical properties of the world. The same, of course, can be done for other concepts and axioms of geometry.

Hence, if we should possess an innate intuition representing to us geometrical entities of this kind,[20] the axioms of this intuitive geometry although referring to reality (namely, to measuring processes) would be a

---

[18]At least if certain axioms about "objectivity", such as the interchangeability of two different observers, are postulated.

[19]In a closed (e.g., a Riemannian) world this definition would not give unique values for the coordinates, i.e., it would have the consequence that either one object can be in many different places simultaneously, or that for every object there exist infinitely many exactly equal ones; or, if the definition is slightly changed, space would consist of a finite portion of Euclidean space only.

[20]I.e., representing those propositions as evident, which hold for these concepts. Of course I do not want to say that pure intuition, according to Kant, refers exactly to those geometrical concepts which are defined above, or that an intuition of this kind actually exists. I only wanted to show that an innate Euclidean geometrical intuition which refers to reality and is a priori valid is logically possible and compatible with the existence of non-Euclidean geometry and with relativity theory. Such an intuition might also be so constituted that only the representation of space itself and the general idea of its relatedness to reality are inborn, while the particular way in which we project our sensations on it depends on experience.

25   priori true.[21] Nevertheless, | they would not be analytic by Kant's concep-
      tion of analyticity, since arithmetic, to which geometry is reduced in this
      way, was not considered to be analytic by Kant.[22] What, however, Kant
      did not take into account is that (irrespective of whether we have such an
      a priori intuition) space and its properties express themselves also in the
      sensations, which we know only a posteriori; namely, in the fact that by
      projecting the sensations in a certain way on a three-dimensional Euclidean
      space the laws connecting them can be expressed in a certain simple form.
      That this is not a necessary property of our sensations follows from the
      possibility of a physical non-Euclidean geometry and from other considera-
      tions along these lines, which show that our sensations might be such that
      to project them on a three-dimensional Euclidean space, though possible,
      would be of no use at all.
          This insight does not imply anything about the objectivity of space.
      For the sensations also are something subjective and therefore any of their
      properties may as well be due to properties of the things as to the manner
      of interaction between the things and ourselves[23] (the decision depending
      on the kind of reality which must be assumed in order to explain the ap-
      pearances satisfactorily). It does, however, necessitate (even in case we
      have an innate intuition of Euclidean space) a weakening of Kant's theory
      of the apriority of space, at least to this extent, that the "adequacy of our
      representation of space to the relation of the objects to our sensibility" can
      only be known by experience, provided a reasonable meaning[24] is given to

          [21]This seems to contradict Einstein's well-known aperçu about the relation between
      geometry and experience (cf. *Einstein 1921*), but the contradiction is rather one of words
      than of meaning, since the phrase "sich auf die Wirklichkeit beziehen" is evidently meant
      by Einstein in the sense of "say something about reality", whereas the geometrical axioms
      as interpreted above, although they refer to physical reality, do not say anything about
      it. Also according to Kant, the geometrical propositions are synthetic, not because they
      say anything about that which is given from outside (i.e., the sensations), or about the
      things in themselves, but because they say something about that which we, or rather
      our intellectual faculties, imagine or construct in the process of mental assimilation of
      the sensations. These subjective additions, however, do not form a part of objective
      reality in the proper sense of the term.

          [22]It is to be noted that the term "synthetic" today is frequently used in a sense dif-
      ferent from Kant's, namely, for a proposition which expresses a state of affairs subsisting
      independently of the activity of our intellect apprehending it. In this sense, "synthetic"
      and "a posteriori" would doubtless coincide also for Kant. But Kant would deny that
      the activity of our intellect consists solely in explicit definitions and their application.

          [23]The subjectivity here in question is of course of a different kind from Kant's, as
      is seen from the fact that it does not imply apriority. Nevertheless space could very
      appropriately be called "a form of sensibility" also if it were subjective in this second
      sense.

          [24]I.e., by "relation of the things to sensibility" must be meant the way in which outer
      objects produce sensations, not the fact that we represent things to ourselves in space,
      for in the latter case, the adequacy in question would mean nothing at all.[g]

      ───

          [g]In the left-hand margin beside this footnote, Gödel had a question mark.

this phrase. That some particular relations of the things to us are adequately represented by Euclidean geometry in every world (cf., e.g., the definitions given above) certainly does not mean yet "a complete adequacy | to *the* relation of the things to our sensibility".   26

Nevertheless, Kant may be right in the sense explained above. I do not think the question whether we have an innate intuition of Euclidean space, or what representation of space we would actually develop in a strongly non-Euclidean world, has yet been decided; nor the related question whether we are able (in our world) to learn to imagine[25] a non-Euclidean space. That we are able to imagine the behaviour of physically rigid bodies in a non-Euclidean world[26] does not prove it. For we would certainly not identify (in our intuitive world picture) the behaviour of | physically rigid   27 bodies with the properties of space in every possible world, e.g., not in a world which is Euclidean for the most part, but has certain non-Euclidean spots. In a homogeneous non-Euclidean world, it is true, it seems more likely that we *would* develop an intuition of non-Euclidean geometry; there is, however, still room for reasonable doubt. But whatever the answers to these questions may be, there can certainly never result from them any incompatibility between Kant and relativity theory, but at most between Kant and psychology (or phenomenology) of sense perception.

A real contradiction between relativity theory and Kantian philosophy seems to me to exist only in one point, namely, as to Kant's opinion that natural science, in the description it gives of the world, must necessarily retain the forms of our sense perception and can do nothing else but set up relations between appearances within this frame.

This view of Kant has doubtless its source in his conviction of the unknowability (at least by theoretical reason) of the things in themselves, and at this point, it seems to me, Kant should be modified, if one wants to establish agreement between his doctrines and modern physics; i.e., it should be assumed that it is possible for scientific knowledge, at least partially and step by step, to go beyond the appearances and approach the things in themselves.[27]

---

[25]Usually it is not realized how greatly different a non-Euclidean spatial intuition would be from ours in case of a sufficiently large curvature of space. E.g., in a certain Lobachevskian space (which stretches into infinity like ours) the circumference of a circle of radius 1 inch would be 1 yard.

[26]Cf., e.g., *Einstein 1921*.

[27]Cf. in this connection *Bollert 1921*, where one may find a description in more detail of these steps or levels of objectivation, each of which is obtained from the preceding one by the elimination of certain subjective elements. The "natural" world picture, i.e., Kant's world of appearances itself, also must of course be considered as one such level, in which a great many subjective elements of the "world of sensations" have already been eliminated. Unfortunately, whenever this fruitful viewpoint of a distinction between subjective and objective elements in our knowledge (which is so impressively

The abandoning of that "natural" picture of the world which Kant calls the world of "appearance" is exactly the main characteristic distinguishing modern physics from Newtonian physics. Newtonian physics, except for the elimination of secondary qualities (which, in principle, was known already to Democritus), is only a refinement, but not a correction, of this picture of the world; modern physics, however, has an entirely different character. This is seen most clearly from the distinction which has developed between "laboratory | language" and the theory, whereas Newtonian physics can be completely expressed in a refined laboratory language.

| So the trend of modern physics is in one respect opposed to Kantian philosophy. On the other hand, however, it should not be overlooked that the very refutation of Kant's view concerning the impossibility for theoretical science of stepping outside the limits of our natural conception of the world has furnished in several points a verification of one of his main doctrines concerning this natural world picture, namely, its largely subjectivistic[28] character, even as to those concepts which seem to constitute the very backbone of reality.

Moreover, it is to be noted that the possibility of a knowledge of things beyond the appearances is by no means so strictly opposed to the views of Kant himself as it is to those of some of his followers. For (1) Kant held the concept of things in themselves to be meaningful and emphasized repeatedly that their existence must be assumed, [and] (2) the impossibility of a knowledge concerning them is for Kant by no means a necessary consequence of the nature of all knowledge, nor subsists even for human knowledge in *every* respect. One may compare [with respect] to this point, e.g., what he says at B xxi, B xxi n. and B xxvi n. about the positive use of the *Critique*, not only for religious belief, but possibly also for knowledge (B xxvi n.). In the second of the passages cited above, Kant even goes so far as to compare his subjectivistic theory of knowledge as to its possible positive use with Copernicus's explanation of the apparent motions of the planets by the motion of the observer, pointing out that this new view-

---

suggested by Kant's comparison with the Copernican system, see below, p. [29]) appears in the history of science, there is at once a tendency to exaggerate it into a boundless subjectivism, whereby its effect is annulled. Kant's thesis of the unknowability of the things in themselves is one example, another one is the prejudice that the positivistic interpretation of quantum mechanics must necessarily be the final stage of the theory.

[28]One may say that (as also seems to appear from the contents of this paper) relativity theory has proved only the relativity of time, not its subjectivity (i.e., relativity to our imaginative faculty). But it is to be noted that (1) the absoluteness (or uniqueness) forms itself an essential characteristic of intuitive time, and that (2) a relative passing of time (if any sense can at all be given to this phrase) would be something radically different from the passing as described above on p. [11]; for existence by its nature is something absolute. Hence, an entity corresponding really in all essentials to the intuitive idea of time, also in relativity theory, exists only in our imagination.

point of Copernicus led to the discovery of the "hidden power connecting the structure of the universe, the Newtonian attraction".

Kant, it is true, wanted to base such knowledge on ethics, but, even as far as theoretical reason is concerned, he evidently did not want to say that nothing whatsoever can be asserted about the things in themselves. For Kant himself asserted, | e.g., that they exist, affect our sensibility, and do not exist in time and space, but that the ideas of time and space are completely adequate to their relationship to our sensibility.

Doubtless Kant would least of all have held an overstepping of the world of appearances to be possible, on the basis of conclusions drawn from experiments. But [with respect] to this point it is to be noted that, in perfect conformity with Kant, the observational results by themselves really do not force us to abandon Newtonian time and space as objective realities,[29] but only the observational results together with certain general principles, e.g., the principle that two states of affairs which cannot be distinguished by observations are also objectively equal. Generally speaking, it can be said that relativity theory (especially general relativity theory) owes its origin perhaps more than any other physical theory to the application of certain very general principles and ideas, and, for the most part, was only subsequently verified by experience.

[29] In special relativity theory, e.g., some arbitrarily chosen inertial system can be considered as representing absolute space and time (with all properties Newton attributed to them) and the varying observational results of observers moving with different velocities can be explained by the effect which motion relative to this absolute space has on the physical bodies and processes, in particular on the measuring instruments (effects which incidentally follow without any ad hoc hypothesis from the empirically verifiable electromagnetic equations of Maxwell and the very natural assumption that the constitution of physical bodies is based on electromagnetic forces or other forces with similar mathematical properties). It remains, however, entirely arbitrary which coordinate system is in this way distinguished as the absolute space-time scheme and this, by the principle formulated above, entails that absolute time and space do not exist.

Gödel and Einstein, 13 May 1947

Oskar Morgenstern

Introductory note to *1949b*

This item is the slightly edited text of a lecture that Gödel presented at the Institute for Advanced Study on 7 May 1949. It has been prepared from a handwritten manuscript (in English) in the Gödel *Nachlass*.

1. Introductory remarks

In his lecture, Gödel exhibits and discusses the properties of his new exact solution to Einstein's equation. The solution represents a possible universe, compatible with the constraints of relativity theory, in which aggregate matter (on a cosmological scale) is in a state of uniform, rigid rotation. Gödel's presentation here covers much the same ground as does the published account (*1949*) that appeared two months later. But the tone throughout is more relaxed and expansive, and there are a number of specific additions. They include the following:

(a) At the very beginning of the lecture, Gödel exhibits a Newtonian cosmological model in which, as in his own model, aggregate matter is in a state of uniform rigid rotation. In so doing, he partially anticipates later work by Heckmann and Schücking (*1955*) on Newtonian analogues of Gödel space-time. (See also *Lathrop and Teglas 1978*.)

(b) He discusses what it can mean to say that the universe in its entirety is rotating and notes the incompatibility of general relativity with (at least some versions of) "Mach's principle".

(c) He explains a generic connection in general relativity between rotation and temporal structure, and notes that it was this connection that first led him to look for "rotating solutions".

(d) Gödel gives an account of what he calls the "geometric meaning" of his solution that is perhaps more accessible, because less compressed, than the version in *1949*. (For an elegant redescription see *Chakrabarti, Geroch, and Liang 1983*.)

(e) He makes several remarks that help one visualize the configuration of light-cones in Gödel space-time and visualize its time-like and null geodesics. (As represented in a standard space-time diagram, the latter are helices of bounded radius that cyclically intersect the [vertical] worldlines of the major mass points. Thus free particles and light rays exhibit a kind of boomerang effect. See *Kundt 1956*, *Chandrasekhar and Wright 1961*, and *Lathrop and Teglas 1978*.)

(f) He asserts in passing the existence of yet other solutions to Einstein's equation representing possible universes in which (i) aggregate matter is rotating, but no closed timelike curves are present;
(ii) aggregate matter is expanding as well as rotating. (Gödel
sketches an existence proof in *1952*, but he does not exhibit any
solutions explicitly.)

Gödel's remarks (c) concerning how he first came to look for "rotating solutions" are of particular biographical interest. He makes clear
that his technical investigations were driven, at least initially, by antecedent philosophical interests. In what follows, I'll make a number
of non-technical explanatory remarks about the generic connection between rotation and temporal structure that is at issue here, and then
add a few details to Gödel's account (d) of the "geometrical meaning"
of his solution.

2. Cosmic rotation and "objective" time

Gödel came to look for solutions to Einstein's equation representing
rotating universes in the course of trying to bolster an argument that
relativity theory supports a particular conception of time, one that he
identified with Kant and certain species of "idealist philosophy". It is
that argument, in its fully bolstered form, that one finds in *1949a*.

What links Kant and relativistic physics, according to Gödel, is a
common denial of the "objective existence of ... time in the Newtonian
sense". In the case of relativistic physics, of course, denial is supposed
to follow from the relativity of simultaneity (and of temporal relations
more generally). But Gödel himself raises a possible difficulty with this
position.

To be sure, relativity theory teaches us that it does not make sense
to speak of simultaneity until one relativizes consideration to particular
individuals. (It would be more appropriate to speak of relativization to
time-like curves or worldlines. It is not essential that they be "traversed"
by conscious agents, or any objects for that matter.) But in the context
of cosmology, at least, one is given at the outset a distinguished class of
worldlines, namely, those of the major mass points of the universe (stars
or galaxies). And so it is natural to understand all attributions of temporal structure as relativized to that class. This is in fact done all the
time. It is done, for example, by cosmologists when they talk about the
"first three minutes" of the universe. In this context, therefore, one does
recover an "objective", in the sense of uniquely distinguished, temporal
structure.

Gödel's search for "rotating solutions" was prompted by a desire to
counter this objection. It turns out that in relativity theory it is a

highly contingent matter whether one can talk about simultaneity in any natural sense *even after relativizing consideration to a particular family of worldlines.* Gödel apparently understood this early on, and undertook to find a cosmological model, unlike all others previously discovered, in which the worldlines of the major mass points would not support a natural notion of relative simultaneity. But this was tantamount to looking for a rotating universe, because, and this is the crucial point, *in a cosmological model, non-rotation of the major mass points is precisely the necessary and sufficient condition for there to exist a natural notion of simultaneity relative to their worldlines.* This is the generic connection between rotation and temporal structure that Gödel discusses in his lecture.

What does one mean by a "natural" notion of relative simultaneity? Consider a congruence of worldlines, and a spatial slice (i.e., a three-dimensional space-like hypersurface) that intersects the congruence. Standardly, at least, one construes the latter as a "simultaneity slice" relative to the former if and only if the slice is everywhere orthogonal to the individual worldlines of the congruence (with respect to the background space-time metric). This identification of relative simultaneity with orthogonality is taken for granted in standard presentations of "special relativity" and in discussions of cosmology. Moreover, one can easily show that the identification is forced if one requires of any candidate notion of relative simultaneity that it respect certain weak symmetry conditions.

In any case, this is the identification Gödel has in mind when he speaks of "a very natural definition of simultaneity independent of coordinates". And the italicized assertion above can be captured, more precisely, this way: in a cosmological model, the congruence of worldlines of the major mass points is twist-free (i.e., has everywhere vanishing rotation) if and only if the congruence admits an orthogonal foliation (i.e., if there exists a one-parameter family of slices everywhere orthogonal to the worldlines).

The assertion is a special case of Frobenius' theorem. (See *Wald 1984*, page 434.) One can make it at least intuitively plausible by considering an analogue. Think about an ordinary rope. In its natural twisted state, the rope cannot be sliced in such a way that the slice is orthogonal to all fibers. But if the rope is first untwisted, such a slicing is possible. Thus orthogonal sliceability is equivalent to fiber untwistedness. The assertion above merely extends this intuitive equivalence to the four-dimensional "space-time ropes" (i.e., congruences of worldlines) determined by the major mass points of the universe.

By finding the first cosmological model in relativity theory in which the major mass points are in a state of rotation, Gödel found the first one in which there is no natural, distinguished notion of temporal structure, not even one determined by relativization to the worldlines of the major mass points. He could make no claim that our universe, in fact, has this property. But he could hold that out as a possibility; our universe just might exhibit some very small net rotation. And even if it does not, he could claim that, at least so far as relativity theory is concerned, it is an entirely accidental matter that it does not.

3. The "geometrical meaning" of Gödel space-time

Gödel's account of the "geometrical meaning" of his solution is not so terse as that in *1949*, but it does still leave out many details. Here we add just a few, and make explicit the connection between Gödel's abstract characterization of his metric and his two representations involving particular coordinate systems.

First, consider the coordinate-free characterization. Quite generally, a relativistic space-time may be taken to be a structure (M, g_{ik}), where M is a connected, smooth, four-dimensional manifold, and g_{ik} is a smooth semi-Riemannian metric on M of signature $(+, -, -, -)$. In Gödel space-time M is $\mathbb{R}^4$, and (M, g_{ik}) can be decomposed as a metric product of $\mathbb{R}$ (with its usual positive definite metric) and a structure $(\mathbb{R}^3, h_{ik})$, where h_{ik} has signature $(+, -, -)$. Any discussion of the "geometrical meaning" of Gödel's solution concerns the latter. The metric h_{ik} is of form

$$h_{ik} = h'_{ik} + t_i t_k$$

where

1. h'_{ik} is a complete metric on $\mathbb{R}^3$ of signature $(+, -, -)$ and constant positive curvature; and
2. $t^i = h'^{ik} t_k$ is a unit time-like Killing field with respect to h'_{ik}.

(In Gödel's presentation of his metric, he carries a free scale parameter a. The corresponding constant curvature of h'_{ik} comes out to be $(1/4a^2)$.)

The transition from h'_{ik} to h_{ik} is what Gödel describes as "stretching the metric in the ratio $\sqrt{2}$ in a direction of Clifford parallels". The "Clifford parallels" here are the integral curves of the Killing field t^i. Notice that (i) h'_{ik} and h_{ik} agree in their determinations of whether vectors are orthogonal to t^i; (ii) they agree in the squared lengths they assign to vectors orthogonal to t^i; and (iii) whereas t^i has length 1 with respect to h'_{ik}, it has length $\sqrt{2}$ with respect to h_{ik}.

We can easily recover this abstract characterization of the Gödel metric starting from either of the two coordinate-dependent expressions he exhibits in the lecture. Consider the first:

$$a^2[dx_0^2 - dx_1^2 + (1/2)e^{2x_1}dx_2^2 - dx_3^2 + 2e^{x_1}dx_0 dx_2].$$

Here the coordinates $x_0, \ldots, x_3$ range over all of $\mathbb{R}$. We arrive at the structure $(\mathbb{R}^3, h_{ik})$ simply by dropping the term $-dx_3^2$ and restricting the reduced metric to any hyperplane of constant x_3 value. This reduced metric can be cast in the form

$$a^2[(1/2)dx_0^2 + e^{x_1}dx_0 dx_2 - dx_1^2] + (a^2/2)[dx_0 + e^{x_1}dx_2]^2.$$

Let h'_{ik} be the metric determined by the first term

$$a^2[(1/2)dx_0^2 + e^{x_1}dx_0 dx_2 - dx_1^2],$$

and let t_i be the field $(a/\sqrt{2})(dx_0 + e^{x_1}dx_2)$ determined by the second. Then the inverse h'^{ik} is given by

$$\left(\frac{1}{a^2}\right)[-2e^{-2x_1}\left(\frac{\partial}{\partial x_2}\right)^2 + 4e^{-x_1}\left(\frac{\partial}{\partial x_0}\right)\left(\frac{\partial}{\partial x_2}\right) - \left(\frac{\partial}{\partial x_1}\right)^2],$$

and $t^i = h'^{ik}t_k$ comes out to be, simply, $(\sqrt{2}/a)(\partial/\partial x_0)$. Clearly t^i has unit length with respect to h'_{ik} and is a Killing field with respect to that metric (since x_0 does not appear in any of the coefficients of h'_{ik}).

It remains to show that $(\mathbb{R}^3, h'_{ik})$ is a complete manifold with constant curvature $(1/4a^2)$. To do so we define a map

$$(x_0, x_1, x_2) \mapsto (u_0, u_1, u_2, u_3)$$

from $\mathbb{R}^3$ into $\mathbb{R}^4$ by setting

$$u_0 = 2a[\cos(x_0/2\sqrt{2})\cosh(x_1/2) - (1/2\sqrt{2})x_2\, e^{x_1/2}\sin(x_0/2\sqrt{2})]$$
$$u_1 = 2a[\sin(x_0/2\sqrt{2})\cosh(x_1/2) + (1/2\sqrt{2})x_2\, e^{x_1/2}\cos(x_0/2\sqrt{2})]$$
$$u_2 = 2a[-\sin(x_0/2\sqrt{2})\sinh(x_1/2) + (1/2\sqrt{2})x_2\, e^{x_1/2}\cos(x_0/2\sqrt{2})]$$
$$u_3 = 2a[\cos(x_0/2\sqrt{2})\sinh(x_1/2) + (1/2\sqrt{2})x_2\, e^{x_1/2}\sin(x_0/2\sqrt{2})].$$

A straightforward computation determines that

$$u_0^2 + u_1^2 - u_2^2 - u_3^2 = 4a^2$$
$$du_0^2 + du_1^2 - du_2^2 - du_3^2 = a^2[(1/2)dx_0^2 + e^{x_1}dx_0 dx_2 - dx_1^2]$$

at every point in the image of the map. Moreover, one can verify that the map is a diffeomorphism if one restricts x_0, say, to the interval $[-\pi, \pi)$. Indeed, under that restriction one can explicitly solve for the x coordinates in terms of the u coordinates; e.g.,

$$x_0 = 2\sqrt{2}\, \text{arc}\cos[(u_0 + u_3)/((u_0 + u_3)^2 + (u_2 - u_1)^2)^{1/2}].$$

Thus we see that $(\mathbb{R}^3, h'_{ik})$ is an isometric covering space of the manifold

$$H = \{(u_0, \ldots, u_3) \in \mathbb{R}^4 \mid u_0^2 + u_1^2 - u_2^2 - u_3^2 = 4a^2\}$$

with respect to the metric induced on H by the background flat metric on $\mathbb{R}^4$ of signature $(+, +, -, -)$. It is a standard result that the latter is a complete manifold of constant curvature $(1/4a^2)$. (See, for example, *O'Neill 1983*, page 113.)

We can execute very much the same computational argument starting from Gödel's second representation of his metric (in cylindrical coordinates):

$$4a^2[dt^2 - dr^2 - dy^2 + (sh^4 r - sh^2 r)d\varphi^2 + 2\sqrt{2}\, sh^2 r\, d\varphi\, dt].$$

If we drop the term $-dy^2$ and regroup the other terms, we arrive at the expression

$$4a^2[(1/2)dt^2 - dr^2 - \sinh^2 r\, d\varphi^2 + \sqrt{2}\, \sinh^2 r\, d\varphi\, dt]$$
$$+ 2a^2[dt + \sqrt{2}\, \sinh^2 r\, d\varphi]^2$$

for h_{ik}. If we use the two terms to define, respectively, the metric h'_{ik} and the field t_i, then again a simple computation establishes that $t^i = h'^{ik} t_k$ is a unit time-like Killing field with respect to h'_{ik}. (It comes out to be $(1/a\sqrt{2})(\partial/\partial t)$.) To map $(\mathbb{R}^3, h'_{ik})$ isometrically onto H (with its induced metric) we set

$$u_0 = 2a\, \cos(t/\sqrt{2})\, \cosh r$$
$$u_1 = 2a\, \sin(t/\sqrt{2})\, \cosh r$$
$$u_2 = 2a\, \sinh r\, \sin(\varphi - (t/\sqrt{2}))$$
$$u_3 = 2a\, \sinh r\, \cos(\varphi - (t/\sqrt{2})).$$

To gain further insight into the two displayed maps of $(\mathbb{R}^3, h'_{ik})$ onto H, we can recast them, making use of the fact that H has a natural Lie group structure.

Consider the four-dimensional (associative, distributive) algebra of what Gödel calls the "hyperbolic quaternions". Its elements can be construed as vectors

$$\varphi = w_0 + w_1 j_1 + w_2 j_2 + w_3 j_3$$

with real coefficients $w_0, \ldots, w_3$. Multiplication is defined by the requirement that 1 serve as an identity element, and by the table

$$j_1^2 = -1 \quad j_2^2 = j_3^2 = 1$$
$$j_1 \cdot j_2 = -j_2 \cdot j_1 = j_3$$
$$j_2 \cdot j_3 = -j_3 \cdot j_2 = -j_1$$
$$j_3 \cdot j_1 = -j_1 \cdot j_3 = j_2.$$

If we define the conjugate and norm of φ by setting

$$\overline{\varphi} = w_0 - w_1 j_1 - w_2 j_2 - w_3 j_3$$
$$\text{norm}(\varphi) = \varphi \cdot \overline{\varphi} = w_0^2 + w_1^2 - w_2^2 - w_3^2,$$

then it follows easily that for all φ and ψ, $\overline{\varphi \cdot \psi} = \overline{\psi} \cdot \overline{\varphi}$ and hence

$$\text{norm}(\varphi \cdot \psi) = \text{norm}(\varphi)\,\text{norm}(\psi).$$

To simplify notation now, let us identify the hyperbolic quaternion $w_0 + w_1 j_1 + w_2 j_2 + w_3 j_3$ with the corresponding element $(w_0, \ldots, w_3)$ in $\mathbb{R}^4$. Then H is identified with the set of hyperbolic quaternions of norm $4a^2$, and it acquires a natural Lie group structure: given two elements u and u' in H, we simply take their product to be $(1/4a^2)u \cdot u'$. The norm-product condition above guarantees that the operation is well-defined. The element u has $\overline{u}$ for an inverse.

Notice that for all real numbers x_0, x_1, x_2, the quadruples

$$(\cos x_0, \sin x_0, 0, 0)$$
$$(\cosh x_1, 0, 0, \sinh x_1)$$
$$(1, x_2, x_2, 0)$$

all have norm one. So their dot product has norm 1. Straightforward multiplication establishes that the associated map

$$(x_0, x_1, x_2) \mapsto 2a(\cos x_0, \sin x_0, 0, 0) \cdot (\cosh x_1, 0, 0, \sinh x_1) \cdot (1, x_2, x_2, 0)$$

is essentially just the first of the two maps from $(\mathbb{R}^3, h'_{ik})$ onto H displayed above. This is where it "comes from". [To match Gödel's coefficients exactly one has to take the dot product of

$$(\cos(x_0/2\sqrt{2}),\ \sin(x_0/2\sqrt{2}), 0, 0)$$
$$(\cosh(x_1/2), 0, 0,\ \sinh(x_1/2))$$
$$(1, x_2/2, x_2/2, 0).]$$

Notice also that for every fixed element w of norm 1, the map from H to H defined by

$$u \mapsto w \cdot u$$

is an isometry. (Maps of this form are norm preserving. Hence they preserve the flat metric on $\mathbb{R}^4$ and the induced metric on H.) Moreover, elements of the first type above—$(\cos x_0, \sin x_0, 0, 0)$—form a one-parameter group. So the family of maps $\{\varphi_{x_0}\}_{x_0 \in \mathbb{R}}$ defined by

$$\varphi_{x_0} : u \mapsto (\cos x_0, \sin x_0, 0, 0) \cdot u$$

forms a one-parameter group of isometries on H. The infinitesimal generator of this group is just the Killing field $(\partial/\partial x_0)$.

With these ingredients, we can work backwards from the abstract characterization of Gödel's metric to recover the first coordinate expression. First we compute $du_0^2 + du_1^2 - du_2^2 - du_3^2$ in terms of the coordinates x_0, x_1, x_2. This gives us an expression for h'_{ik}. Then we find a multiple of $(\partial/\partial x_0)$ that has length 1 with respect to h'_{ik}. This gives us an expression for t^i. Then we simply add—$h^{ik} = h'^{ik} + t^i t^k$—to gain the final expression for h^{ik}.

To recover the expression for the Gödel metric in cylindrical coordinates one proceeds in much the same way, but one starts with the coordinatization

$$(t, r, \varphi) \mapsto 2a(\cos(t/\sqrt{2}), \sin(t/\sqrt{2}), 0, 0) \cdot$$
$$(\cosh r, 0, \sinh r \, \sin \varphi, \sinh r \, \cos \varphi)$$

of H.

Consider again the abstract characterization of Gödel's metric. Why should one be interested in a metric of this sort in the first place? As Gödel explains, the answer is that one knows "in advance" that it will provide a solution to Einstein's equation for the case in which the energy-momentum source is a cosmic dust field (or a perfect fluid, if one does not want to allow a non-zero "cosmological constant"). One can show, quite generally, that if h'_{ik} is a three-dimensional metric of constant curvature α (not necessarily positive), t^i is a unit time-like Killing field with respect to h'_{ik}, and h_{ik} is of form

$$h_{ik} = h'_{ik} + \beta t_i t_k,$$

then the Ricci tensors corresponding to h'_{ik} and h_{ik} have forms

$$R'_{ik} = 2\alpha h'_{ik}$$
$$R_{ik} = 2(1 - \beta)\alpha h'_{ik} + (2\beta^2 \alpha + 6\beta\alpha)t_i t_k.$$

So if one takes $\beta = 1$, the latter reduces to $R_{ik} = 8\alpha t_i t_k$, or to $R_{ik} = 4\alpha t_i t_k$ if one rescales t^i to be of unit length with respect to h_{ik}. Therefore

(working with the rescaled t^i)

$$R_{ik} - (1/2)h_{ik}R = 4\alpha t_i t_k - 2\alpha h_{ik}.$$

To cast the right hand side in the desired form

$$8\pi\rho\kappa t_i t_k + \lambda h_{ik}$$

one need only set $\rho = \alpha/2\pi\kappa$ and $\lambda = -2\alpha$. Clearly, the construction works only if the initial constant curvature α is positive (since the gravitational constant κ and the mass density ρ are positive), though the magnitude of α is not constrained.

David B. Malament[a]

[a]This note was written while I was a Fellow at the Center for Advanced Study in the Behavioral Sciences. I am grateful for the financial support provided by the National Science Foundation (#BN587-00864) and the University of Chicago. I am also grateful to Howard Stein for helpful comments on an earlier version.

Lecture on rotating universes
(*1949b)

A few years ago, in a note in *Nature*, Gamow [*1946*] suggested that the whole universe might be in a state of uniform rotation and that this rotation might explain the observed rotation of the galactic systems. Indeed, if the primordial matter out of which the galaxies were formed by condensation was in a state of rotation, the galaxies themselves will possess a much faster rotation. For in consequence of the law of conservation of angular momentum, their angular velocity will increase as the square of the ratio of contraction. Therefore they will exhibit a rotation even in the coordinate system in which the primordial matter was at rest, that is to say, a rotation in the frame of reference defined by the totality of galaxies. In exactly the same way, the rotation of the galaxies in its turn can be used to explain the rotation of the fixed stars and planetary systems and therewith essentially all rotations occurring in astronomy. Of course, there arises at once the objection that by this theory the axes of rotation of all astro-

nomical systems would have to be parallel with each other, which seems to
2   contradict observation. But this difficulty is perhaps | not insurmountable.
For a closer statistical examination of the material furnished by observation might show that a certain degree of regularity in the directions of
the axes of rotation is actually there, and the lack of complete regularity
might be explainable by various disturbing influences. In particular, if the
axes of rotation of matter in different places in space are not parallel with
each other, the irregular motion of the galaxies will mix up the different
directions of the axes.

Let us now consider the question of a rotating universe from the standpoint of theory and first of all from the standpoint of Newtonian physics.
It is well-known that Newtonian physics gives a surprisingly good approximation for the expanding universes; so one might expect that the same
will be true here; and this expectation is actually verified. Of course, if you
want to apply the Newtonian law of attraction to cosmology, you have to
formulate it as a local action principle

$$\Delta V = 4\pi\kappa\varrho$$

(because the integrals to which the distant action principle leads would in
general not converge). According to this law, V is not uniquely determined
3 | by ϱ, because we can require no boundary conditions. But one may
make the tentative assumption that any values of V and ϱ which satisfy
this equation represent a physically possible state of affairs. Under this
assumption,
$$V = \pi\kappa\varrho(x^2 + y^2) = \pi\kappa\varrho r^2$$
is a possible gravitational potential for a homogeneous distribution of ϱ. It
evidently has rotational symmetry around the z-axis. But this potential (or
rather the field of force produced by it) is exactly in equilibrium with the
centrifugal force of a uniform rotation around the z-axis with an angular
velocity of $\sqrt{2\pi\kappa\varrho}$. This Newtonian universe is the exact analogue of the
relativistic one I am going to define shortly. The degree of approximation
again is surprisingly good. The angular velocity for the relativistic solution
differs only by a factor $\sqrt{2}$ from this one here. The privileged role which the
z-axis in this Newtonian world plays is only apparent. Actually, there would
exist no experiment by which you could distinguish one place in the world
from another. For if you describe this universe in a coordinate system which
rotates together with matter, the only observable apparent forces would be
the Coriolis forces, because the centrifugal forces are exactly annulled by
the gravitational forces. But the Coriolis forces are independent of the
place. So what is observable of the axis of rotation is only its direction,
but not the place where it is located. This circumstance makes the analogy
with the corresponding relativistic solution very close.
4 | If we now consider the problem from the standpoint of relativity theory,

the first question which arises is in what sense we can speak of a rotation of the whole universe in relativity theory, where we have no absolute space to which we could refer it. The answer, of course, is that in relativity theory, as a substitute for absolute space, we have a certain inertial field which determines the motion of bodies upon which no forces act. In particular, this inertial field determines the behaviour of the axis of a completely free gyroscope or of the plane of a pendulum, and it is with respect to the spatial directions defined in this way (by a free gyroscope or the plane of a pendulum) that matter will have to rotate. In the usual terminology, this means that it will rotate relative to the compass of inertia. You see that this kind of rotation does not involve the idea of an axis around which the whole world rotates. The world may be perfectly homogeneous and still rotate locally in every place, as is actually the case in the example I am going to give. Nevertheless, the world may be said to rotate as a whole (like a | rigid body) because the mutual distances of any two material particles 5 (measured by the orthogonal distance of their world lines) remain constant during all times. Of course, it is also possible and even more suggestive to think of this world as a rigid body at rest and of the compass of inertia as rotating everywhere relative to this body. Evidently this state of affairs shows that the inertial field is to a large extent independent of the state of motion of matter. This contradicts Mach's principle but it does not contradict relativity theory.

Let us now investigate quantitatively what a rotation of matter relative to the compass of inertia means in terms of the g_{ik} and the velocity vector v of matter. For this purpose, it is best to use the concept of a local inertial system. An inertial system in special relativity theory, as you know, is one in which the g_{ik} have values

$$g_{ik} = \eta_{ik} = \begin{bmatrix} 1 & 0 & 0 & 0 \\ 0 & -1 & 0 & 0 \\ 0 & 0 & -1 & 0 \\ 0 & 0 & 0 & -1 \end{bmatrix}, \qquad [\![(*)]\!]$$

which implies that the motion of a free particle is uniform and rectilinear in these coordinates. In general relativity theory, of course, such a coordinate system for the whole space does not exist, but it exists locally, i.e., the coordinates can be so chosen that at the point considered, the g_{ik} | have 6 these values, and moreover all derivatives $\partial g_{ik}/\partial x_l$ vanish, so that also in the neighbourhood of the point considered, the g_{ik} will have these values up to magnitudes of the second order. It is a well-known theorem of differential geometry that at any point of a Riemannian space there exist coordinate systems satisfying these requirements. In the sequel I shall sometimes have to consider coordinate systems which at the point considered only satisfy the first condition. I shall call them "orthonormal", which I am told is a term in use for the corresponding concept in the theory of orthogonal systems of functions.

Now the angular velocity we are interested in is the angular velocity relative to the directions of space defined by axes of gyroscopes moving along with matter. For this is the angular velocity which the astronomer who, with his measuring apparatus, moves along with matter will observe. Therefore we have to introduce an inertial system which moves along with matter, i.e., one at whose origin matter is at rest at the moment considered, or, in other terms, | one at whose origin the velocity vector

$$v^i = \frac{dx_i}{ds}$$

has values

$$v^1 = v^2 = v^3 = 0, \quad v^0 = 1.$$

Then, because directions defined by gyroscopes remain fixed in an inertial system, the angular velocity we are looking for will be equal to the angular velocity at the origin of this coordinate system computed as in classical mechanics. Therefore we obtain the formulae

$$\omega_1 = \frac{1}{2}\left[\frac{\partial(v^2/v^0)}{\partial x_3} - \frac{\partial(v^3/v^0)}{\partial x_2}\right] = \frac{1}{2}\left(\frac{\partial v^2}{\partial x_3} - \frac{\partial v^3}{\partial x_2}\right)$$

$$\omega_2 = \ldots$$

$$\omega_3 = \ldots.$$

(This equality evidently holds in consequence of the values of v^i at the origin.) Of course, instead of a coordinate system in which the compass of inertia is at rest and matter rotates, you may for the computation as well take a coordinate system in which matter is at rest and the compass of inertia rotates. I mean, of course, a coordinate system in which matter is at rest everywhere (not only at the origin), so that v^1, v^2, v^3 vanish identically. Then, under the further assumption that the coordinate system is orthonormal at the origin, | you obtain the formulae

$$\omega_1 = \frac{1}{2}(-\Gamma^2{}_{30} + \Gamma^3{}_{20}) = \frac{1}{2}(\Gamma_{230} - \Gamma_{320}) = \frac{1}{2}\left(\frac{\partial g_{02}}{\partial x_3} - \frac{\partial g_{03}}{\partial x_2}\right)$$

$$\omega_2 = \ldots$$

$$\omega_3 = \ldots$$

for the angular velocity at which the compass of inertia rotates in this system, which is the same as the angular velocity at which it rotates relative to matter. These formulae follow immediately from the fact that the motion of the axis of a free gyroscope in four-space is a parallel displacement. But a parallel displacement of a vector u along the x_0-axis is expressed by the formulae

$$\frac{du^i}{ds} = -\Gamma^i{}_{k0}u^k.$$

So the matrix $-\Gamma^i{}_{k0}$ expresses the apparent deformation and rotation which the vector space undergoes by parallel displacement, and it is known from elasticity theory how the rotation is separated from the deformation. Since the coordinate system is orthonormal, the difference between the two kinds of Christoffel symbols disappears (or rather is reduced to a difference in sign), and we obtain the formulae displayed.

But these two formulae only apply in the very special coordinate systems defined, and | it is of course very desirable to have a covariant expression 9 for the angular velocity, i.e., one which applies in every coordinate system. This covariant expression is

$$a_{ikl} = v_i\left(\frac{\partial v_k}{\partial x_l} - \frac{\partial v_l}{\partial x_k}\right) + v_l\left(\frac{\partial v_i}{\partial x_k} - \frac{\partial v_k}{\partial x_i}\right) + v_k\left(\frac{\partial v_l}{\partial x_i} - \frac{\partial v_i}{\partial x_l}\right).$$

$$\omega^j = \frac{\epsilon^{iklj} a_{ikl}}{12\sqrt{g}}.$$

As you can verify immediately, a_{ikl} is a skew-symmetric tensor; ω is the corresponding vector obtained by inner multiplication with the ϵ-tensor. Using the skew symmetry of the ϵ-tensor, you can verify immediately that ω is always orthogonal to v. Moreover, you can verify by direct computation in a few lines that, for these two special coordinate systems, ω^i for $i = 1, 2, 3$ becomes equal to the expression for ω^i obtained before, and ω^0 vanishes. Of course, strictly speaking, only these relations here justify the preceding definitions, i.e., show that they are independent of the particular inertial [coordinate] system chosen, and similarly for the second definition. Moreover, this covariant representation of the angular velocity is of interest for the following reason: If v is an arbitrary vector field in an n-dimensional space, then the identical vanishing of this skew-symmetric tensor a is the necessary and sufficient condition for the existence of a one-parametric system of $(n-1)$-spaces | which are everywhere orthogonal to the vectors of 10 the field. This is a straightforward generalization of the well-known condition which a vector field [[v]] in Euclidean three-space must satisfy if there is to exist a system of surfaces everywhere orthogonal to the vectors of the field. This condition requires that

$$v \cdot (\text{rot } v) = v_1\left(\frac{\partial v_2}{\partial x_3} - \frac{\partial v_3}{\partial x_2}\right) + \cdots = 0.$$

The true reason why this is the necessary and sufficient condition in both cases is that (1) this expression vanishes if the rot of v [[i.e., the curl of v]] vanishes; (2) if it vanishes for v, it also vanishes for ϱv, where ϱ is an arbitrary scalar field, because this expression is simply multiplied by ϱ^2 if v is multiplied by ϱ. The terms containing derivatives of ϱ cancel out, as you can verify immediately. This, of course, must be so in view of the fact that the existence (or nonexistence) of surfaces orthogonal to the vectors v depends only on the directions, not on the absolute values, of the vectors. So in view of the just explained geometrical meaning of the vanishing of the skew-symmetric tensor a, we arrive at the conclusion that the angular velocity of matter relative to the compass of inertia is everywhere 0 then and only then if there exists a one-parametric system

11 of three-spaces everywhere orthogonal | to the world lines of matter. Now the existence of such a system of three-spaces means, in a certain sense, the existence of an absolute time coordinate. For these three-spaces yield a very natural definition of simultaneity independent of the coordinate system. So there is a very simple connection between the existence of a rotation of matter and the existence of an objectively distinguished world time.

12 | This incidentally also was the way in which I happened to arrive at these rotating solutions. I was working on the relationship between Kant and relativity theory and, more particularly, on the similarity which subsists between Kant and relativistic physics insofar as in both theories the objective existence of a time in the Newtonian sense is denied. On this occasion one is led to observe that in the cosmological solutions known at present there does exist something like an absolute time. This has been pointed out by epistemologists, and it has even been said by the physicist Jeans that this circumstance justifies the retention of the old intuitive concept of an absolute time. So one is led to investigate whether or not this is a necessary property of all possible cosmological solutions.

13 | The rotating cosmological solution I have found on this occasion has certain shortcomings from the physical viewpoint. In particular, it yields no red-shift for distant objects, and it contains closed time-like lines. It has, however, a certain interest in principle insofar as it doubtless is the simplest rotating solution and, moreover, the only homogeneous solution which rotates like a rigid body. The linear element is very simple indeed. It is given by the expression

$$ds^2 = a^2[dx_0^2 - dx_1^2 + (1/2)e^{2x_1}dx_2^2 - dx_3^2 + 2e^{x_1}dx_0dx_2],$$

where a is a constant. Written as a sum of squares it reads like this:

$$ds^2 = a^2[(dx_0 + e^{x_1}dx_2)^2 - dx_1^2 - (1/2)e^{2x_1}dx_2^2 - dx_3^2].$$

The x_0-lines in this coordinate system are the world lines of matter. This linear element is obtained from the underlined three-dimensional linear element simply by adding the term $-dx_3^2$. As you see, x_3 does not occur in the preceding terms. So the four-space with this ds consists of a one-parametric system of copies | of the three-space with the underlined metric. If we introduce cylindrical coordinates in these three-spaces, with the old x_0-axis as rotational axis of the coordinate system [and set $x_3 = 2y$], we obtain the expression

$$ds^2 = 4a^2[dt^2 - dr^2 - dy^2 + (\text{sh}^4 r - \text{sh}^2 r)d\varphi^2 + 2\sqrt{2}\,\text{sh}^2 r\, d\varphi\, dt]$$

for the linear element, which gives a clearer picture of the structure of this three-space. It shows that it has rotational symmetry around the t-axis. Moreover, the non-vanishing of the term $g_{t\varphi}$ shows that the t-lines (i.e., the world lines of matter) are not orthogonal to the r, φ plane; they are orthogonal to the radius vector from the origin (since g_{rt} vanishes), but they are inclined to the r, φ plane and, more specifically, the inclination is always in the direction of increasing φ. So the world lines of matter are helices of a sort, as they must be in a rotating universe.

I shall give the geometrical meaning for this linear element later, but first I would like to show how the field equations can be directly verified very quickly, or at | least describe in a few words the way in which it is most easily done. You can first show that this four-space is homogeneous. Namely, the four transformations

1. $x_0 = x_0' + a_1$
   [$x_i = x_i'$ for $i \neq 0$]

2. $x_2 = x_2' + a_2$
   [$x_i = x_i'$ for $i \neq 2$]

3. $x_3 = x_3' + a_3$
   [$x_i = x_i'$ for $i \neq 3$]

4. $x_1 = x_1' + a_4$
   $x_2 = x_2' e^{-a_4}$
   [$x_0 = x_0'$]
   [$x_3 = x_3'$]

(where a_1 to a_4 are real numbers) carry this linear element into itself. But by suitable choice of the a_i, you can evidently take the origin of the coordinate system into any given point. So it is sufficient to verify the field equations at one point, say the origin. Now at one point you can always introduce orthonormal coordinates by a linear transformation. If you do this by the linear transformation

$$x_0 = x_0' - \sqrt{2}\, x_2'$$
$$x_2 = \sqrt{2}\, x_2',$$

you can without much calculation obtain the components of the Riemann tensor by means of the formula

$$R_{ijkl} = \frac{1}{2}\left[\frac{\partial^2 g_{il}}{\partial x_j \partial x_k} + \frac{\partial^2 g_{jk}}{\partial x_i \partial x_l} - \frac{\partial^2 g_{ik}}{\partial x_j \partial x_l} - \frac{\partial^2 g_{jl}}{\partial x_i \partial x_k}\right]$$
$$+ g^{nn}\Gamma_{n\,il}\Gamma_{n\,jk} - g^{nn}\Gamma_{n\,ik}\Gamma_{n\,jl}.$$

Since almost all first and second derivatives of the g_{ik} vanish, the compu-
16   tation is very easy. | The result is that only the R_{iklm} with two different indices [other than 3] do not vanish and that they all have the same value:

$$R_{0101} = g^{22}(\Gamma_{201})^2 = -1/2$$
$$R_{0202} = g^{11}(\Gamma_{102})^2 = -1/2$$
$$R_{1212} = g^{00}(\Gamma_{012})^2 + g^{22}(\Gamma_{212})^2 = (1/2) - 1 = -1/2.$$

Hence, for the components of the contracted Riemann tensor, you obtain that only R_{00} is not 0, and it is equal to 1. So, if you denote by v the unit vector in the direction of the x_0-axis, you have the equality

$$R_{ik} = v_i v_k.$$

Or rather, this holds in case the constant a is 1. In general, therefore, you have the equality

$$R_{ik} = (1/a^2)v_i v_k.$$

17 | But whenever R_{ik} can be represented in this form and, in addition, R is a constant (as it is in our case, because the space is homogeneous), the relativistic field equations can be satisfied in a very simple manner. We only have to set the cosmological constant λ equal to $-R/2$ and then multiply this equality by g_{ik} and subtract the result from the equality above:

$$R_{ik} - (1/2)g_{ik}R = (1/a^2)v_i v_k + \lambda g_{ik}.$$

The resulting equation is identical with the field equations

$$R_{ik} - (1/2)g_{ik}R = 8\pi\varrho\kappa\, v_i v_k + \lambda g_{ik}$$

if we determine the constant a by this equation:

$$(1/a^2) = 8\pi\kappa\varrho.$$

If we compute the angular velocity by means of the formulae given before, we obtain the value

$$\omega = 2\sqrt{\pi\kappa\varrho}.$$

For the density ϱ observed in our world, i.e. 10^{-30} g/cm^3, this leads to a period of rotation of $2 \cdot 10^{11}$ years. This way of verifying the field equations by direct computation is of course not very interesting, but it [is] surprisingly short. In less than one printed page you can give in this way a complete proof including all details for the existence of rotating solutions.
| Before entering into a discussion of the properties of this solution, I   18
would like to say something about the heuristic argument by which this [solution] is most easily obtained and about the geometrical meaning of this linear element. This will also show the true reason why it satisfies the field equations. If we want to find the relativistic analogue for the Newtonian universe mentioned before, we must of course look for a stationary solution. Moreover, since the Newtonian universe rotates like a rigid body, we have to require that the mutual distances of any two material particles should remain constant in time, i.e., that any two world lines of matter should be equidistant—that is to say that a perpendicular drawn from a point of one line to another line must have the same length no matter what point of the first line you choose. So the world lines of matter in the four-space we are looking for must be a system of pairwise equidistant geodesic lines, and for each point of the space there must exist exactly one line of the system passing through it. However, this system of lines must differ from a system of parallels in Euclidean space insofar as a translation | of the whole space   19
along these lines must not be a parallel displacement of the space. For if it were, the compass of inertia (whose behavior is described by a parallel displacement) would not rotate relative to matter.

Now this description of the system of world lines reminds one at once of a system of Clifford parallels in a three-space of constant positive curvature. It is a well-known fact (for which, moreover, I am going to give a very simple proof later on) that in a three-space of constant positive curvature, there exist in every direction systems of straight lines satisfying all requirements I enumerated. The difference from what we are looking for is only (1) that we are looking for such a system of lines in a four-space, [and] (2) that the metric of relativity theory is not positive definite but has the signature $+---$. But these two defects can easily be remedied. There exist, of course, also three-spaces of constant curvature and signature $+--$ and, as I am going to prove shortly, these three-spaces (or to be more exact those of positive curvature) also contain systems of Clifford parallels. | Now from   20
such a three-space we can very easily obtain a four-space and a system of lines in it which has the desired properties. Namely, we simply add to the linear element a term $-dx_3^2$ [assuming we have started with coordinates x_0 to x_2], that is to say, we construct in the simplest possible way a four-space out of a one-parametric system of congruent three-spaces of the kind described. Then we define in one of these three-spaces a system of Clifford parallels and enlarge it by adding the lines obtained by the translation: $x_3' = x_3 + a$. It can easily be verified that then also any two lines of the

enlarged system will be equidistant. The only thing which remains to be seen is how we can ensure that the field equations of relativity theory will be satisfied. For this purpose, let us see what the contracted Riemann tensor will look like in the four-space we just constructed. Of course this space is homogeneous. So it is sufficient to compute the Riemann tensor at one point P. We may furthermore assume that the coordinate system is orthonormal at this point, [and] that, moreover, the first three basis vectors lie in one of the three-spaces of the one-parametric system of three-
21   spaces used. The fourth basis vector will then lie in the direction | of an x_3-line. Moreover, the first basis vector can be so chosen as to lie in the direction of a line of the chosen system of Clifford parallels.

Now, about the values of the R_{ik} in such a coordinate system, we can say the following: All components R_{i3} must vanish, because the g_{ik} do not depend on x_3 and, moreover, the g_{i3} are constants. From these properties of the g_{ik} you can derive immediately that all $\partial g_{ik}/\partial x_l$ vanish if at least one of the three indices is 3. Next you can derive the same thing for the Γ_{ikl}, namely, that they vanish if at least one index is 3, and finally, the same holds for the R_{iklm} owing to the formula [above expressing R_{iklm} in terms of $\partial g_{ik}/\partial x_l$ and Γ_{ikl}] and therefore also for the R_{ik}. So it follows that the matrix of the R_{ik} has only zeros in the last row and the last column. The numerical value for the remaining three-dimensional matrix follows from this formula for three-spaces of constant curvature, where ϱ is the curvature:

$$R_{ik} = 2\varrho g_{ik}.$$

Since the coordinate system was assumed to be orthonormal, the g_{ik} have
22 | the values [in the matrix (∗)], so that we obtain for the matrix of the R_{ik} this form:

$$\begin{bmatrix} 2\varrho & 0 & 0 & 0 \\ 0 & -2\varrho & 0 & 0 \\ 0 & 0 & -2\varrho & 0 \\ 0 & 0 & 0 & 0 \end{bmatrix}.$$

It is easily seen that an R_{ik} of this form can never satisfy the field equations, whatever the values of ϱ and λ and the unit vector v may be. For if we write the field equations in the form

$$R_{ik} - g_{ik}((R/2) + \lambda) = 8\pi\kappa\varrho v_i v_k,$$

we see that they mean that by subtracting a multiple of g_{ik} from R_{ik}, we must obtain a tensor which is the product of a vector with itself, and this vector moreover must be time-like. But a diagonal tensor which is the product of a vector with itself can only have one component different from zero, and because the vector is to be time-like, this component must be the first one; but it is evidently impossible to make the last three vanish in this manner.

But if we now remember why we started the construction with three-spaces of constant curvature, we see that we did it only because they contained a system of equidistant geodesics. So if we were able to modify the metric in our three-spaces in such a | way that the chosen system of Clifford   23
parallels remained a system of equidistant geodesics, the new three-spaces would be as good for our purposes as the old ones, and we might be able in this way to introduce a new arbitrary parameter which could make it possible to satisfy the field equations. Now this is actually possible and in a very simple manner. We only have to stretch the metric in an arbitrary ratio μ in the direction of the lines of the chosen system of Clifford parallels. By stretching the metric of a space (let's say the Euclidean plane) in a certain direction α I mean the following thing: We decompose the linear element $[\![PQ]\!]$ into a component PR parallel to the direction α and a component RQ orthogonal to it and multiply the first component with the factor μ, while we leave the second one unchanged, so that we obtain the expression

$$PQ'^2 = QR^2 + \mu^2 PR^2$$
$$= QR^2 + PR^2 + (\mu^2 - 1)PR^2$$
$$= PQ^2 + (\mu^2 - 1)PR^2$$

for the new distance. Owing to the equalities here, this operation means the same as adding to the linear element a certain multiple of its projection on α. The same operation can be performed in any Riemannian space if, instead of the direction α, there is given a system of lines simply | covering   24
the whole space. If we denote by v the vector field which consists of all tangent vectors of unit length to the lines of the given system, we obtain the expression

$$ds'^2 = ds^2 + (\mu^2 - 1)(v_i dx^i)^2$$

for the new linear element. For $v_i dx^i$ is the orthogonal projection of the linear element on the lines of the system, and we have to add that to the old linear element multiplied by a factor $(\mu^2 - 1)$. It is easily seen that if the system of lines along which we stretch the metric is a system of equidistant lines, it will be a system of equidistant lines also in the new metric (simply because distances orthogonal to the lines of the system are not changed at all by the operation of stretching). Moreover, if the system of lines consisted of geodesics, the lines will be geodesics also in the new metric. This is particularly easy to see in our case, because the lines of a system of Clifford parallels are axes of rotational symmetry both for the space and for the system of lines. Hence they will be axes of rotational symmetry also for the new metric; but any axis of rotational symmetry is a geodesic line.

| The rotational symmetry of the new metric also allows one to determine   25
very easily the form of the contracted Riemann tensor of the new metric in

the coordinate system defined before. A symmetric tensor R_{ik} in a three-space determines an ellipsoid (or some other surface of the second degree) in the vector space of each point. But if an ellipsoid in a three-space is to have rotational symmetry, it must be an ellipsoid of revolution, one of whose principal axes has the direction of the axis of rotational symmetry and the other two [of which] are orthogonal to it; and the same holds for other surfaces of the second degree. Hence in the coordinate system chosen before (whose first axis is the axis of rotational symmetry), the R_{ik} must have diagonal form, and the second and third entries on the diagonal must be equal to each other. So the whole four-dimensional tensor R_{ik} must have the form

$$\begin{bmatrix} a & 0 & 0 & 0 \\ 0 & b & 0 & 0 \\ 0 & 0 & b & 0 \\ 0 & 0 & 0 & 0 \end{bmatrix},$$

where a and b evidently will depend on the ratio of stretching μ. Hence we may hope that by a suitable choice of μ we may be able to make $b = 0$. But

26   in that [case] R_{ik} will be | equal to $ae_i e_k$ if e is the first unit vector of the coordinate system. If, however, the R_{ik} once have this form, I have shown before how the field equations can be satisfied, provided a is positive. For it must be equal to a multiple of the density of matter. Now it turns out that $\mu = \sqrt{2}$, and only this value, actually gives $b = 0$ and $a > 0$, and this is the space I defined before.

As to the derivation of the two forms of the linear element from the geometrical interpretation just given, there is not much to be said about the second form. In order to obtain it, you only have to introduce cylindrical coordinates with the lines of the system of Clifford parallels as z-lines of the coordinate system. In order to obtain the first form, a closer examination of the group of transformations of our space is necessary. For this purpose let us first consider the simplest representation of the group of an ordinary

27   three-space of constant positive curvature. We can think of such | a space as the surface of a sphere, or rather, hypersphere, around the origin in a Euclidean four-space. Such a sphere is given by the equation

$$x_0^2 + x_1^2 + x_2^2 + x_3^2 = 1.$$

Now we can associate with each point of our space, i.e. the sphere, a quaternion

$$\varphi = x_0 + i_1 x_1 + i_2 x_2 + i_3 x_3,$$

which, owing to the preceding equation, will have the norm 1. For the norm by definition is the product of φ with its complex conjugate, which gives the sum of squares

$$\mathrm{norm}(\varphi) = \varphi \cdot \overline{\varphi} = x_0^2 + x_1^2 + x_2^2 + x_3^2.$$

Then, owing to the fact that the norm of a product is the product of the norms, a multiplication of φ with a quaternion of norm 1 will carry this sphere into itself. This will be true no matter whether you multiply from the left or from the right, hence also if you multiply simultaneously from both sides. So the equation

$$\varphi' = p\varphi q$$

represents a motion of the sphere into itself, and, vice versa, every motion of the sphere into itself can be represented in this way. Now it is easily proved that if you apply the powers of one quaternion p^α, where α is a real number, to one fixed point φ of the sphere, the resulting one-dimensional system of points will form a geodesic on this sphere. In order to prove this, it is sufficient to remark that $p^{-1} \cdot \varphi$, $1 \cdot \varphi$, $p \cdot \varphi$ lie on a geodesic. But this follows because (a) $p^{-1} = \bar{p}$, (b) there subsists a linear relation among p, $\bar{p}$ and 1, namely, $p + \bar{p} = \alpha \cdot 1$ (here α is a real number), and (c) the geodesics on the sphere are given by linear equations. If you now perform this operation $p^\alpha \cdot \varphi$ *with the same quaternion p* on all points of the sphere, | you will obtain a system of geodesics covering the whole space, and it is   28 clear that if any two lines of the system have a point in common they will be identical with each other, i.e., any two different lines of the system will be non-intersecting. Moreover, there exist transformations of the sphere which carry each line of the system into itself, i.e., which translate the whole sphere along these lines; namely, the transformation

$$\varphi' = p^\beta \cdot \varphi,$$

for an arbitrary number β, has that property. But from the existence of such transformations, it evidently follows that any two lines of the system are equidistant. So any such system of lines is a system of Clifford parallels.

Now literally the same considerations can be made if, instead of a four-dimensional Euclidean space, you start out with a space in which the square of the distance between two points is defined by the expression

$$r^2 = (x_0 - x_0')^2 + (x_1 - x_1')^2 - (x_2 - x_2')^2 - (x_3 - x_3')^2$$

(with two negative squares) and if you con|sider, instead of a sphere, the   29 hypersurface of second degree defined by

$$x_0^2 + x_1^2 - x_2^2 - x_3^2 = 1;$$

only instead of the quaternions you now have to take another four-parametric algebra in which the norm [of the element]

$$\varphi = x_0 + j_1 x_1 + j_2 x_2 + j_3 x_3$$

is given by

$$\text{norm}(\varphi) = \varphi \cdot \overline{\varphi} = x_0^2 + x_1^2 - x_2^2 - x_3^2.$$

Such an algebra can best be defined in the following way: You take the complex quaternions, i.e., those for which the coordinates x_0, x_1, x_2, x_3 are arbitrary complex numbers $a + bi$, and then consider that subalgebra of the complex quaternions in which the first two coordinates are real and the last two coordinates x_2, x_3 are pure imaginary. So the unities of this algebra are

$$1 \quad i_1 \quad i \cdot i_2 \quad i \cdot i_3.$$

If you denote [the latter three] by j_1 to j_3, you obtain the rules of multiplication

$$j_1^2 = -1, \qquad j_2^2 = j_3^2 = 1,$$
$$j_1 \cdot j_2 = \quad j_3 [\![= -j_2 \cdot j_1]\!],$$
$$j_2 \cdot j_3 = -j_1 [\![= -j_3 \cdot j_2]\!],$$
$$j_3 \cdot j_1 = \quad j_2 [\![= -j_1 \cdot j_3]\!],$$

which show that this number system (with real coefficients x_0 to x_3) actually is an algebra, i.e., the product of any two numbers in the system is again a number of the system, and, moreover, the associative law holds, because it holds for the complex quaternions. This algebra evidently should
30   be called | the algebra of hyperbolic quaternions. Strangely enough, that name occurs in the literature in a different meaning, but anyhow I shall use this name in the sequel. (Or does there perhaps exist a name in use for this algebra?) Now you can repeat with the hyperbolic quaternions all considerations made before. The fact that the norm of a hyperbolic quaternion may vanish also for non-zero quaternions is of no consequence, since we have to do only with quaternions of norm 1. There is, however, still another difference between the ordinary and the hyperbolic quaternions, namely, that the latter contain three-parametric subalgebras. For example, the subsystem defined by $x_1 = x_2$ is such a subalgebra. As the unities of this subsystem you evidently can take the three quaternions

$$1 \quad\quad j_1 + j_2 \quad\quad j_3.$$

But they reproduce each other by multiplication according to the rules

$$(j_1 + j_2)^2 = 0$$
$$(j_1 + j_2) \cdot j_3 = -(j_1 + j_2)$$
$$j_3^2 = 1.$$

Now it is to be expected that you will obtain a particularly simple form for the linear element if you take the points corresponding to this subalgebra
31   as one of the coordinate planes. | This is actually so—namely, you then

obtain the first form of the linear element

$$ds^2 = a^2[(dx_0 + e^{x_1}\, dx_2)^2 - dx_1^2 - (1/2)e^{2x_1}\, dx_2^2 - dx_3^2].$$

[By means of the hyperbolic quaternions, you also obtain a very simple representation of the group of this space (or rather of the group of the underlined three-space; the extension to the four-space is then trivial). Namely, the group of this space will evidently consist of all those transformations of the space of constant curvature (from which we started) which also carry the given system of Clifford parallels into itself (not each line separately, of course, but the whole system). But these transformations are evidently those of the form

$$\varphi' = p^\alpha \cdot \varphi \cdot q,$$

where p is the quaternion defining the system of Clifford parallels in the manner explained above. As you see, this group has four real parameters, namely α and the three parameters in q (since q is a hyperbolic quaternion of absolute value 1). Therefore the group | of transformations of this four-space will be five-parametric, because also the transformations $x_3' = x_3 + k$ evidently carry the space into itself. Of course, this group of transformations must be five-parametric because (1) the space is homogeneous, which requires four parameters, and (2) it has rotational symmetry at every point, which gives you one more parameter.]

| Now I would like to say a few words about the physical properties of this solution. First of all, you can verify immediately that there exists no red shift for distant objects in this solution. This is true for every solution in which the world lines of matter are equidistant. For any such solution admits a one-parametric group of translations of the space along the world lines of matter [or], to be more exact, transformations expressible as

$$x_0' = x_0 + a_0,$$

if the coordinate system is so chosen that the x_0-lines are the world lines of matter and the x_0-coordinate is the "Eigenzeit" measured along these world lines. Now if you apply this transformation to a light ray leading from one particle of matter to another one, you will again obtain a light ray between the same two particles, but one which leaves the first particle a_0 seconds later and also arrives at the second particle a_0 seconds later. But this means that light signals sent from one particle of matter to | another arrive with the same time intervals in which they are being sent. Hence also, the time intervals between two succeeding crests of a light wave will be the same at the source and at the reception of light, i.e., there will be no red shift. This consideration shows quite generally not only that an expansion implies a red shift but also vice versa, [[that]] a red shift implies an expansion, no matter whether the universe rotates or not, i.e., a rotation alone can never produce a red shift.

The next question to be considered is the value of the angular velocity of the galaxies to be expected in a rotating world of this type. We have obtained before a value for the angular velocity of the universe. In order to derive from it the angular velocity of the galaxies, we have to divide it by the square of the ratio of contraction by which the galaxies were formed from the homogeneous primordial matter, provided, of course, we assume that the galaxies are homogeneous spheres and rotate like rigid

35   bodies. But for a rough esti|mate of the order of magnitude, such an assumption may be made. (Moreover, it turns out that a concentration of mass toward the center of the galaxies changes surprisingly little the value obtained, provided in the rotation the centrifugal and gravitational forces are in equilibrium, which is assumed by astronomers to be true.) Now the ratio of contraction can be estimated from the observed average ratio between distance and diameter of galaxies. This ratio, according to Hubble, is about 200:1. So we have to divide this period of rotation by 200^2, which leads to a value of about $5 \cdot 10^6$ years. As I understand, the periods of rotation of galaxies have been determined only in a few cases. For these few cases the observed value is somewhat larger, but not very much larger, than this number. It is, on an average, about $3 \cdot 10^7$ years. But of course this agreement or disagreement is of very little significance, because in consequence of the lack of an expansion, this solution can hardly be the real universe anyway.

36 | The next property of our universe which is of physical interest is the aforementioned existence of closed time-like lines, which I would like to discuss now. If you examine the second form of the linear element,

$$ds^2 = 4a^2[dt^2 - dr^2 - dy^2 + (\mathrm{sh}^4 r - \mathrm{sh}^2 r)d\varphi^2 + 2\sqrt{2}\,\mathrm{sh}^2 r\,d\varphi\,dt],$$

you will find that the coefficient of $d\varphi^2$ changes its sign for sufficiently great values of r. Namely, for $\mathrm{sh}\,r < 1$, $\mathrm{sh}^4 r < \mathrm{sh}^2 r$, i.e., the coefficient of $d\varphi^2$ is less than zero; for $\mathrm{sh}\,r > 1$, however, we get the opposite inequality, and therefore the coefficient of $d\varphi^2$ is greater than zero. If you choose R so that $\mathrm{sh}\,R = 1$, the coefficient of $d\varphi^2$ becomes equal to zero, i.e., the circle with radius R around the origin in the r, φ plane is a null line. Now a null line is the path of a possible light signal. It is not necessarily the path of a light ray in vacuum because it need not be a geodesic line, and actually it isn't a geodesic line in our case. But by means of a sufficient number of mirrors, you can force a light ray to go along any null line. To be more exact, you

37   can force | it to go along any polygon of geodesic null lines approximating the given null line. If, in particular, you send a light signal along a null line

running back into itself, this means that the light signal will come back at exactly the same moment at which it is sent.

Now the reason for this circle being a null line, of course, is that, exactly as the world lines of matter, so also the light cones surrounding them are getting more and more inclined to the r, φ plane as you move away from the origin of the coordinate system. Hence finally their inclination is getting so great that they touch the r, φ plane. Therefore if you increase r still further, the light cones will even intersect the r, φ plane, i.e., they will in part be situated below the r, φ plane. But this implies that there also exist null lines which, starting from some point in the r, φ plane, circle around the origin but at the same | time are getting farther and farther 38 below the r, φ plane as you go along them, so that they will return to the world line of matter from which they started at some point below the r, φ plane; that is to say, a light signal sent along such a null line will return earlier than it was sent. So it is possible to send light signals into the past and moreover, since a light path can be approximated as closely as you wish by a path of a material particle, you can even travel into the past on a rocket ship of sufficiently high velocity. Of course, these considerations seem to prove the physical impossibility of these worlds. But since these closed lines of particles and of light have an immense radius (it is of the same order of magnitude as the world radius of the Einstein static universe), these experiments are many orders of magnitude above what could | actually be done, and therefore it is questionable whether anything 39 follows from these considerations. [For there may be some impossibility in principle involved which is not known at present.] At any rate, however, the problem of whether our universe rotates is not necessarily connected with this question, because there certainly also exist rotating solutions without closed time-like lines.

In connection with the temporal structure of this universe, I would like to point out one more fact. It may seem that in consequence of the possibility of travelling into the past it becomes impossible to distinguish consistently between the positive and negative direction of time. But this is not at all true. It does remain possible to divide the light cones into two classes, the positive and negative ones, in such a way that a limit of positive light cones is again a positive light cone—this continuity of course is a necessary requirement if the division is to make sense physically—since a division can evidently be effected by defining the positive light cones | to be those 40 which contain the positive direction of the t-lines in this coordinate system. This has two consequences: (1) that after such a definition has been made, each possible world line has a well-defined orientation which determines for any two neighbouring points on it which is earlier and which is later; and (2) that in whatever way you may travel around in space-time, your own positive direction of time will always agree with the positive direction of time subsisting in the places where you arrive. In particular, a closed time-like line will always look like that in figure 1,

Figure 1

and never like that in figure 2,

Figure 2

which would mean that on returning you would find your surroundings look like a film running in the wrong direction.

But in spite of the existence for the whole world of a uniform positive direction of time, there exists, on the other hand, no uniform time for the whole world. This assertion can be given a quite precise meaning. Namely, one may say that a coordinate x_0 is a time coordinate if x_0 always increases if one moves along any possible world line of matter in its positive direction. In special relativity theory and in the non-rotating universes, such time coordinates evidently exist for the whole four-dimensional space. But the existence of closed time-like lines implies that such a time coordinate for the whole four-dimensional space cannot exist. For if you move along a closed time-like line in its positive direction, the | time coordinate x_0 for the initial point must be the same as for the end point (because initial and end point coincide), but then x_0 cannot have been increasing along the whole line. But the existence of a time coordinate in this sense is not absolutely necessary in order that the solution may have a physical meaning. The existence of a well-defined positive direction of time, however, is necessary. For only the direction of time determines, for example, what it means to send a light signal, since the light is to be found only in the positive half of the double light cone issuing from the space-time point from which the signal is sent. The positive direction of time also allows you to formulate the second law of thermodynamics, at least for closed systems; whether it could be applied to the whole universe I do not know.

| I would like to mention one more physical property of our universe—   42
namely, that the light rays in empty space orthogonal to the axis of rotation are circles, i.e., they revert to their origin, arriving, however, always at a later moment than they were sent. This effect has the consequence that looking in a direction orthogonal to the axis of rotation you can see only up to a certain finite distance, and this again has the consequence that the apparent density of galaxies would have to decrease more and more if you look in such a direction. But for the distances covered by the telescopes of today, this effect would be pretty small. That the light rays orthogonal to the axis of rotation are circles again shows the close agreement with the Newtonian approximation. For | if you introduce in the Newtonian ap-   43
proximation a coordinate system in which matter is at rest (i.e., a rotating coordinate system), then also the paths of free particles will be circles in consequence of the Coriolis forces.

There are two more facts of some interest concerning this rotating universe. Namely, (1) by making certain identifications of the points, you can construct out of it finite rotating universes. But they also have a time of finite extension. (2) This universe is essentially the only rotating universe which is spatially homogeneous and has equidistant world lines of matter, where "essentially" means up to the question of connectivity in the large. So all other rotating solutions have either a contraction or an expansion or both at the same time, namely, a contraction in one direction of space and an expansion in another direction. And there actually exist rotating and expanding solutions, but I have not yet represented them in analytic form.

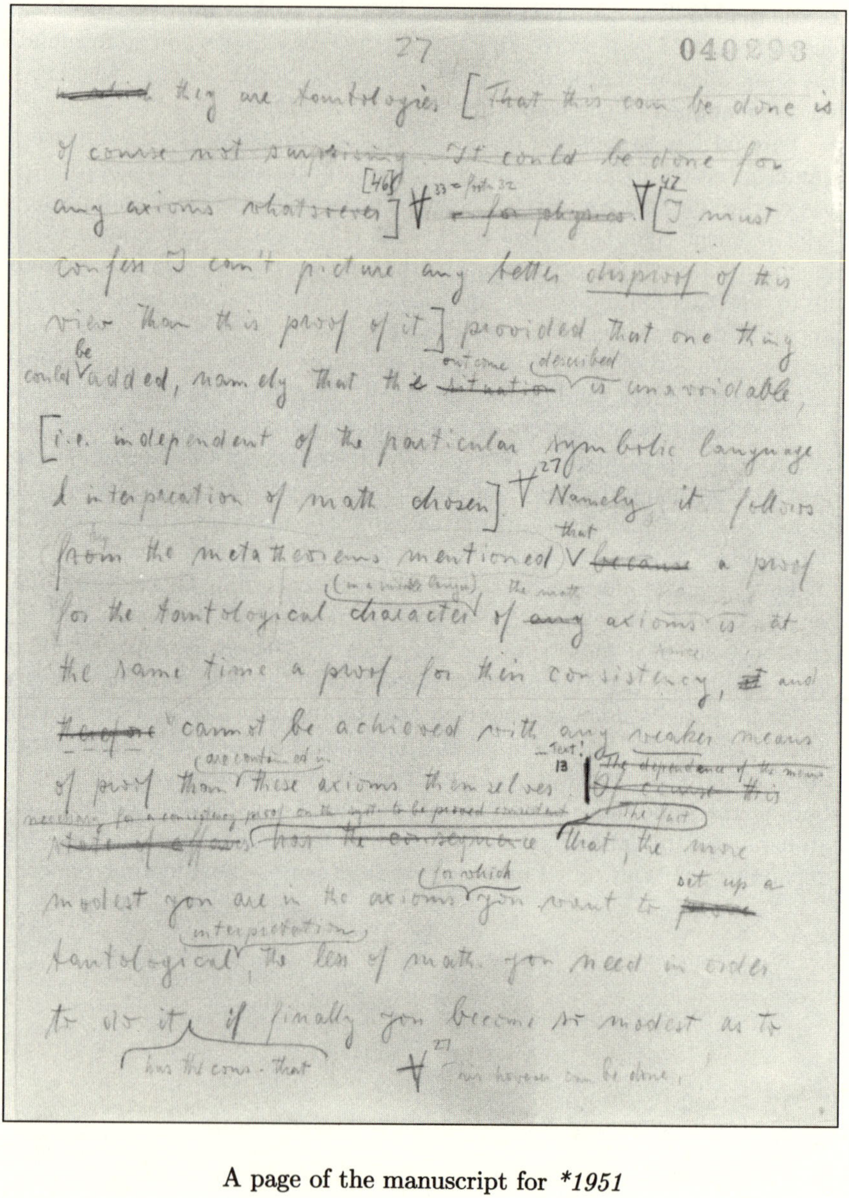

A page of the manuscript for *1951*

↑45 ~~Some of the undesirable consequences of this standpoint~~
are the following: This theorem is a prop. of exactly the same
logical character as 2+2=4 ✱

↑46 results from the fact that symbols can be mapped on the integers & therefore finitistic (& a fortiori classical) one theory
contains all ~~the consideration~~ proofs based solely upon them.
~~The evidence for is far in particular~~ for this fact in [?] is
not absolutely conclusive because ~~it is not sufficiently~~
~~well defined what frontier~~ the ✱ evident axioms referring
to the non-abstract concept under consideration have not been
investigated thoroughly enough. ~~the those~~ However ~~some~~
~~few formalists have~~ the fact itself is acknowledged
even by leading formalists esp. [?] Foote ↑ 8

↑47 Now of course ~~it~~ is clear that the ~~see not pursue the~~ elaboration of this
proon. will does not satisfy the requirement set up on
p. because not ~~only~~ the syst. rules but all of math. in ad[?]
is used in the derivations. But moreover this elab. of
nominalism ~~would~~ yields an outright disproof of it

✱ Some of the untenable consequences of this standpoint are
the following.

Insertions for the manuscript page shown on page 288

Introductory note to *1951

1. Historical information and overview

On 26 December 1951, at a meeting of the American Mathematical Society at Brown University, Gödel delivered the twenty-fifth Josiah Willard Gibbs Lecture, "Some basic theorems on the foundations of mathematics and their implications". It is not known when he received the invitation to give this lecture. It is probable, as Wang suggests (*1987*, pages 117–18), that the lecture was the main project Gödel worked on in the fall of 1951. In letters to Rita Dickstein (21 March 1953) and Yehoshua Bar-Hillel (7 January 1954), preserved in Gödel's *Nachlass*, he expressed his intention to publish the lecture in the *Bulletin of the American Mathematical Society*. These letters lend some support to the conjecture that he continued to work on the text after 1951. The lecture was included on a list Gödel made up bearing the title "Was ich publizieren könnte" (see the preface to this volume) and also preserved in the *Nachlass*. No correspondence with the editors of the *Bulletin* is known, however, and the only text we have is handwritten (and of a rather intricate structure; see the textual notes). Since other papers of Gödel survive in typescripts—in the cases of *1946/9* and *1953/9* in several versions—it may also be conjectured that he did not come close to sending it off for publication.

Gödel's lecture may be divided into two parts, the first of which is an exposition of certain logical results and of philosophical views that he regards as direct consequences of those results. In this part of the lecture Gödel tries to establish that the results show mathematics to be "incompletable" or "inexhaustible", and that one of them demonstrates that *"either ... the human mind (even within the realm of pure mathematics) infinitely surpasses the powers of any finite machine, or else there exist absolutely unsolvable diophantine problems"* (page 13). (It will be explained below what Gödel understands by a "diophantine problem".) By an "absolutely undecidable" problem, Gödel means one that is undecidable, "not just within some particular axiomatic system, but by *any* mathematical proof the human mind can conceive" (page 13).

In the second, more avowedly philosophical, part of the lecture, Gödel's main concern is to adduce a number of considerations favoring the standpoint called realism or Platonism, which can be defined, in Gödel's own words, as the view that mathematical objects and "concepts form an objective reality of their own, which we cannot create or change, but only perceive and describe" (page 30).

2. Set theory and the incompletability of mathematics

The attempt to axiomatize set theory is the first of two illustrations Gödel provides of what he means by the inexhaustibility of mathematics. Gödel claims that in order to avoid the paradoxes "without bringing in something entirely extraneous[a] to actual mathematical procedure, the concept of set must be axiomatized in a stepwise manner" (page 3). He then proceeds to lay out the "iterative" or "cumulative" hierarchy of sets: we begin with the integers and iterate the power-set operation through the finite ordinals. This iteration is an instance of a general procedure for obtaining sets from a set A and well-ordering R: starting with A, iterate the power-set operation through all ordinals less than the order type of R (taking unions at limit ordinals). Specializing R to a well-ordering of A (perhaps one whose ordinal is the cardinality of A) yields a new operation whose value at any set A is the set of all sets obtained from A at some stage of this procedure, a set far larger than the power set of A. We can require that this new operation, and indeed *any* set-theoretic operation, can be so iterated, and that there should also always exist a set closed under our iterative procedure when applied to any such operation.

Axioms can be formulated to describe the sets formed at various stages of this process. But as there is no end to the sequence of operations to which this iterative procedure can be applied, there is none to the formation of axioms. "... nor can there ever be an end to *this* procedure of forming the axioms, because the very formulation of the axioms up to a certain stage gives rise to the next axiom" (page 5).

The elaboration of Gödel's views on the iterative concept of set found in *Wang 1974* makes it clear that the axioms we thus formulate will imply all those of ZF, including the axioms of replacement. An interesting conclusion is immediate: on Gödel's view, the iterative concept of set is only *partially* embodied in the theory ZF.

Gödel seems never to have wavered from the view that ZF only partially characterizes the concept of set. In *1933o* he speaks of "... an infinity of systems, and whichever system you choose out of this infinity, there is one more comprehensive, i.e., one whose axioms are stronger" (page 10). And as late as *1964*, footnote 20, he states that Mahlo's axioms, which assert the existence of Mahlo cardinals but which cannot be proved in ZF, are "implied by the general concept of set".

Gödel observes that higher-level set-theoretic axioms will entail the solution of certain diophantine problems of level 0 left undecided by the

---

[a]It is conceivable that he may have had in mind Quine's set theories NF and ML, in which whether a formula counts as an axiom depends on whether it satisfies a somewhat artificial syntactical restriction.

preceding axioms; the problems, moreover, take a particularly simple form, viz., to determine the truth or falsity of sentences $\forall \mathbf{x} \exists \mathbf{y} P(\mathbf{x}, \mathbf{y}) = 0$, where $\mathbf{x}$ and $\mathbf{y}$ are sequences of integer variables and $P(\mathbf{x}, \mathbf{y})$ is a polynomial with integer coefficients. Let us call this class of sentences "class A". (For Gödel's proof that undecidable sentences can be taken to be in class A, see *193?*).

3. The incompleteness theorems and incompletability

Not surprisingly, Gödel's own incompleteness theorems provide his second illustration of the incompletability of mathematics. Invoking the notion of a Turing machine, he states that the first theorem "is equivalent to the fact that there exists no finite procedure for the systematic decision of all diophantine problems of the type specified" (page 9); little further mention is then made of the first theorem, since it is the *second* theorem (page 10) that he thinks makes the incompletability of mathematics particularly evident:

> *For any well-defined system of axioms and rules ... the proposition stating their consistency (or rather the equivalent number-theoretical proposition) is undemonstrable from these axioms and rules, provided these axioms and rules are consistent and suffice to derive a certain portion of the finitistic arithmetic of integers.*

Gödel's argument that his second theorem shows the incompletability of mathematics runs as follows: No one can set up a formal system and consistently state about it that he perceives (with mathematical certitude) that its axioms and rules are correct and that he believes that they contain all of mathematics, for anyone who claims to perceive the correctness of the axioms and rules must also claim to perceive their consistency; but since the consistency of the axioms is not provable in the system, the person is claiming to perceive the truth of something that cannot be proved in the system, and is therefore obliged to abandon the claim that the system contains all of mathematics.

Gödel moves immediately to prevent a possible misunderstanding. He distinguishes the system of all true mathematical propositions from that of all demonstrable mathematical propositions, calling these mathematics in the objective and subjective senses, respectively, and claims that it is only objective mathematics that no axiom system can fully comprise. He adds that we could not, however, know of any finite rule that might happen to produce all of subjective mathematics that it is correct. The ground for both claims is the indemonstrability of the assertion of consistency. To be sure, we could successively come to recognize, of each proposition produced by subjective mathematics, that that proposition is correct; but we could not know the general proposition that they are *all* correct.

Were there to be such a rule, Gödel says, the mind would be "equivalent to a finite machine that, however, is unable to understand completely its own functioning" (page 12), again on the ground that the insight that the brain produces only "correct (or only consistent) results would surpass the powers of human reason" (footnote 14). Gödel supposes that if a (consistent) machine "completely understands" its own functioning, then it can recognize its own consistency.

Gödel also holds that if the human mind is "equivalent to a finite machine" (page 12), then there is a finite rule producing all the evident axioms of demonstrable mathematics. Since the assertion of consistency can be recast as a sentence in class A, he takes it that it follows that either the human mind surpasses the powers of a finite machine or there exist simple problems about the natural numbers not decidable by any proof the human mind can conceive. He calls his conclusion a "mathematically established fact" (page 13) that seems to him of great philosophical interest.

There is a gap between the proposition that no finite machine meeting certain weak conditions can print a certain formal sentence (which will depend on the machine) and the statement that if the human mind is a finite machine, there exist truths that cannot be established by any proof the human mind can conceive. It is not that no proposition about the "human mind" or human beings or brains can ever be validly inferred from a mathematical proposition. (On the contrary: since 91 is composite, no human being will ever come to know that 91 is prime.) What may be found problematic in Gödel's judgment that his conclusion is of philosophical interest is that it is certainly not obvious what it means to say that the human mind, or even the mind of some one human being, *is* a finite machine, e.g., a Turing machine. And to say that the mind (at least in its theorem-proving aspect), or *a* mind, may be represented by a Turing machine is to leave entirely open just *how* it is so represented. Nevertheless, the following statement about minds, replete with vagueness though it may be, would indeed seem to be a consequence of the second theorem: If there is a Turing machine whose output is the set of sentences expressing just those propositions that can be proved by a mind capable of understanding all propositions expressed by a sentence in class A, then there is a true proposition expressed by a sentence in class A that cannot be proved by that mind.

Apart from the difficulties involved in deriving from the second incompleteness theorem the disjunctive claim that either the mind is not a finite machine or there exist absolutely undecidable mathematical propositions, a further problem for Gödel's view is that the supposition that the second alternative holds does not seem particularly surprising or remarkable at present. (Of course, it may well be that the existence of propositions whose truth we could never recognize is unremarkable precisely *because* we have come to understand the incompleteness theorems so well.) Why, we may wonder, should there *not* be mathematical truths that cannot be given any proof that human minds can comprehend? It may be noted that there are many persons who, influenced by the picture of the mind as a Turing machine, find the falsity of the first and the truth of the second alternative a pair of propositions they are quite willing to maintain. Others, while reserving judgment on the question whether (the mathematical abilities of) a mind can be (represented by) a Turing machine, simply find it extremely plausible that there are mathematical truths unprovable by any humanly comprehensible proof.[b]

According to Wang (*1974*, pages 324–326), Gödel believed that Hilbert was right to reject the second alternative. Otherwise, by asking unanswerable questions while asserting that only reason can answer them, reason would be irrational. (This view may derive from Kant's opinion that "there are sciences the very nature of which requires that every question arising within their domain should be completely answerable in terms of what is known, inasmuch as the answer must issue from the same sources from which the question proceeds" [A 476/B 504, translation from *Kant 1933*]. Kant cites pure mathematics as one such science [A 480/B 508].[c]) Not only did Gödel reject the second alternative, he appears to have thought (at least late in his life) that there were independent reasons for accepting the first as well: Remark 3 of *1972a* is an argument against Turing's view that "mental procedures cannot go beyond mechanical procedures" (page 306).[d]

Gödel's disjunctive conclusion concerning the significance of his incompleteness theorems stands in contrast with the conclusion drawn by writers such as Ernest Nagel and James R. Newman (*1958*), J.R. Lucas (*1961*), and Roger Penrose (*1989*) to the effect that the theorems show outright that the mind is not a Turing machine, since, as they suppose, the mind can see with mathematical certainty that any Turing machine that it might be alleged to be (or be represented by) is actually consistent, and can therefore prove a proposition not provable by that machine. The classic reply to these views was given by Hilary Putnam

---

[b] In their introductory note to Remark 2 of *1972a*, Feferman and Solovay suggest one possible example. Cf. these *Works*, Vol. II, p. 292.

[c] I am grateful to Carl Posy and Sally Sedgwick for calling these passages in the *Critique of pure reason* to my attention.

[d] A critical assessment of Gödel's argumentation is given in unpublished work of Warren Goldfarb.

(*1960*): Merely to find from a given machine M a statement S for which it can be proved that M, if consistent, cannot prove S is not to *prove* S—even if M *is* consistent. It is fair to say that the arguments of these writers have as yet obtained little credence.

Before we turn to the more philosophical part of Gödel's lecture, let us mention some questions that his discussion suggests. Do the impossibility of axiomatizing the concept of set and that of axiomatizing the whole of mathematics bear any interesting relation to each other? Indeed, is there a significant general phenomenon of inexhaustibility or incompleteness of which they are both examples (and if so, what is it)? Is there even a third instance of the incompletability or inexhaustibility of mathematics to be cited?

4. Realism, or Platonism

Gödel remarks that if either the mind is not a finite machine or there exist absolutely undecidable propositions, then the philosophical conclusions to be drawn are "very decidedly opposed to materialistic philosophy" (page 15). If the first alternative holds and the mind's operations cannot be reduced to those of the brain, which is made out of a finite number of neurons and their connections, then vitalism, he states, would seem to be inescapable. Gödel claims that this alternative is not known to be false and that some of the "leading men in brain and nerve physiology" (page 17) deny the possibility of a purely mechanistic explanation of mental processes.

The second alternative, which, he says, "seems to disprove the view that mathematics is only our own creation" (page 15), appears to imply some version of realism or Platonism about the objects of mathematics and gives Gödel considerably more to say.

A creator, he says, "necessarily knows all properties of his creatures, because they can't have any others except those he has given to them" (page 16). Gödel considers poor the objection that the constructor need not know *every* property of what he constructs, that, e.g., we cannot predict the complete behavior of machines we make (or, one might now add, of software we write). His reply to this objection is to argue that if it were correct, it would provide further support for Platonism in mathematics, because we build machines "... out of some given material. If the situation were similar in mathematics, then this material or basis for our constructions ... would force some realistic viewpoint upon us even if certain other ingredients of mathematics were our own creation" (page 18).

Gödel's claim that a creator must know all properties of the things he creates, since they can have no others except those the creator gives them, may strike the reader as a far-fetched defense of the quite plausible claim that mathematics cannot be only (i.e., entirely) our own creation, at least not if our capacity for proving facts about the natural numbers can be adequately represented by a Turing machine. For how, one might wonder, could it have been *we* who brought about the truth of any true proposition in the absence of a proof of that proposition that we could produce? It might be said that the truth of the proposition is a consequence of stipulations we have made concerning the natural numbers. For this reply to be explanatory, however, "consequence" must mean "deductive consequence" and not (say) "higher-order semantic consequence"; but that is precisely what is *not* the case with regard to an undecidable proposition. In any case, the incompleteness theorems suggest that it is doubtful that the view that mathematics is entirely our own creation can be successfully elaborated. (Gödel does not discuss the objection to the other half of his claim, that objects might in fact acquire properties not bestowed upon them by their creator, for example, as a result of being perceived by others.)

To the objection that the meaning of a proposition about all integers can consist only in the existence of a proof of it, and therefore that neither an undecidable proposition nor its negation is true, Gödel makes a particularly interesting response. He suggests that the abhorrence mathematicians display towards inductive methods in mathematics may be "due to the very prejudice that mathematical objects somehow have no real existence. If mathematics describes an objective world ... there is no reason why inductive methods should not be applied in mathematics" (page 20). Thus his second alternative, that there exist absolutely undecidable propositions, favors the standpoint of empiricism in one respect.

As to what such empirical methods might look like, Gödel offers no concrete suggestion; but, in a footnote, he gives an example of a proposition where probabilities, he says, can be estimated even now: The probability that for each n there is at least one digit $\neq 0$ between the n-th and the n^2-th digits of the decimal expansion of π converges toward 1 as one goes on verifying it for greater and greater n. One may, however, be uncertain whether it makes sense to ask what the probability is of that general statement, given that it has not been falsified below $n = 1,000,000$, or to ask for which n the probability would exceed .999.

Gödel then gives three arguments supporting the view he calls conceptual realism (or Platonism) and directed against the view that mathematics is our own creation.

According to the first of these, the attainment of great clarity in the foundations of mathematics has helped us little in the solution of mathematical problems; but this, says Gödel, would be impossible were mathematics our "free creation", for then mathematical ignorance could be due only to failure to understand what we have created (or to computational complexity), and would have to disappear once we attained "perfect clearness".

But, it might be replied, there is no reason to suppose that perfect *clarity* about one of our creations should yield perfect knowledge of it. What is it about creation that guarantees that once we know exactly what a creation of ours is, we must know everything about it? Gödel seems to identify progress in understanding the foundations of mathematics with the attainment of ever greater clarity about mathematics; but, one might think, mathematics might be our own creation and we might have attained perfect clarity about the fundamental properties of what we have created, but we might nevertheless be rather ignorant about non-fundamental properties. There is no reason to suppose that even perfect clarity with respect to all the fundamental properties of our creations must yield *complete* knowledge of those creations.

Gödel's second argument against the view that mathematics is our own creation is that mathematicians cannot create the validity of theorems at will. "If anything like creation exists at all in mathematics, then what any theorem does is exactly to restrict the freedom of creation" (page 22). This consideration is often thought to be a powerful argument on behalf of a realist view of mathematics of the type Gödel wishes to espouse. It is perhaps presented most forcefully as a claim to the effect that the contrary position is confused or incredible: that once it has been made clear exactly *which* objects (including operations, properties and relations) are *in question*, i.e., being talked about, which, all may concede, may well be a matter for choice or decision, the suggestion that there is still room for a decision whether or not those objects have those properties, stand in those relations, etc. cannot be believed to be true. (One might think: Once it is certain that it is 9, 4, 36, multiplication, and equality that are under consideration, how could it possibly be *up to us* whether or not the product of 9 and 4 is 36?) If the creation could not have turned out otherwise, Gödel is arguing, in what sense is there *creation* at all?

Gödel's third argument is that in order to demonstrate certain propositions about the integers, we must employ the concept of a set of integers; but the creation of integers does not "necessitate" that of sets of integers. Thus we appear to be in the "very strange situation indeed" (page 23) of having to make a further creation in order to determine what properties we have given to the integers, which were supposed to be our creation.

This consideration may perhaps best be taken as a "plausibility" argument: Confronted with these facts about integers, sets of integers and our knowledge of the properties of integers, how can we find even slightly plausible the suggestion that mathematics is our own creation?

Whether or not it follows from the view that mathematics is not our own creation that the objects of mathematics have an objective existence that is independent of us will of course depend on how the concepts "objective existence" and "independence" are to be understood: it may be argued that we lack an interpretation of the key terms in this putative consequence under which it is true but not trivially true.

5. Against conventionalism

Conceding that "free creation" is a vague term, Gödel then undertakes to give a more specific refutation of what he takes to be the most precise articulation of that suggestion, the view usually called mathematical conventionalism (though Gödel often refers to it as nominalism), according to which mathematical propositions express only certain aspects of linguistic conventions, "that is, they simply repeat parts of these conventions" (page 23). His discussion is intricate and, in view of the six drafts[e] he made of a projected paper on the philosophy of Rudolf Carnap (at one time the pre-eminent advocate of conventionalism in mathematics), it is highly probable that Gödel was never able to formulate his objections to Carnap's view to his own complete satisfaction. Annotations to the manuscript strongly suggest that he did not intend to read this section of the lecture to his audience in Providence.

He begins by quickly disposing of what he takes to be the simplest form of conventionalism: the view that the truth of mathematical propositions is due solely to the definitions of the terms they contain. Gödel understands this to mean that there is a mechanical method for converting any mathematical truth (and no mathematical falsehood) to an explicit tautology of the form $a = a$ by systematically replacing terms by their definitions. Since any such conversion method would yield a decision procedure for arithmetical truth, this simplest version of conventionalism fails: there is no such decision procedure.

Refined versions, he claims, fare no better. He then attempts to refute the claim that "every demonstrable[f] mathematical proposition can be deduced from the rules about the truth and falsehood of sentences alone (that is, without using or knowing anything else except these rules)" (page 25).

---

[e]See the introductory note to *1953/9* in this volume.

[f]Although "demonstrable" here might be thought to be a slip for "true," "dem." has been inserted and "true" crossed out at this point in the manuscript. (However, subsequent occurrences are not similarly changed, and the view under attack concerns mathematical truth.)

Gödel's argument is that in order to derive the truth of the axioms of mathematics from rules about the truth and falsity of sentences (as, for example, the truth of $p \vee \sim p$ *is* derivable from the usual rules for truth and falsity of disjunctions and negations), one must apply mathematical and logical concepts and axioms to symbols, sets of symbols, sets of sets of symbols, etc. Thus, one who wants to explain mathematical truth as a species of tautology will find that the explanation cannot proceed without the aid of the axioms of mathematics themselves. Mathematical induction provides the central illustration of Gödel's point: any proof that all instances of mathematical induction are true will appeal, in some way or other, to a form of the principle of mathematical induction itself, or to even stronger set-theoretical principles that cannot plausibly be regarded as rules about the truth and falsity of sentences.

He writes, "while the original idea of this viewpoint was to make the truth of the mathematical axioms understandable by showing that they are tautologies, it ends up with just the opposite, that is, the truth of the axioms must *first* be assumed and *then* it can be shown that, in a suitably chosen language, they are tautologies" (pages 26–7).

By "tautology", it should be noted, Gödel does not mean "truth-functionally valid sentence", but rather something like "sentence whose truth can be deduced from rules stipulating the conditions under which sentences are true and false". Gödel's point is thus that the conventionalists' claim that the truth of true mathematical statements can be deduced from such rules is of no interest if true, since strong mathematical axioms, which can in no way be regarded as "syntactical", will have to be assumed in any valid deduction that shows those statements true.

Gödel argues that any attempt to prove the tautological character of the axioms of mathematics would be a proof of their consistency, which, by his second theorem, cannot be achieved with means weaker than the axioms themselves. It may well be, he notes, that not all of the axioms are needed for the proof of consistency, but it is, he claims, a "practical certainty" that to prove consistency some "abstract concepts", such as "set" or "function of integers", together with the axioms governing these notions, will have to be employed in the proof. Since these notions cannot be considered to be syntactical, it follows, he claims, that syntax cannot rationally warrant our "precritical" beliefs concerning the consistency of classical mathematics.

Although some portions of the theory of abstract concepts can be nominalistically based, and fragments of arithmetic, concerning, e.g., numbers less than 1000, reduced to truth-functionally valid statements, a syntactical justification of mathematical induction is unavailable, "since this axiom itself has to be used in the syntactical considerations" (page 27). Thus the well-known reducibility of arithmetical identities like "$5 + 7 = 12$" to explicit tautologies is misleading, Gödel says, not only because this statement is contained in a tiny fragment of mathematics whose reducibility to tautology tells us nothing about the rest of mathematics—which includes statements that can be established only with the aid of induction—but also because either "+" is defined so as to refer only to numbers in some finite domain (in which case it does not refer to ordinary addition), or the concept of set, along with axioms about sets, will have to be used in the definitions and proofs.

Gödel then sums up the previous discussion: The essence of the nominalist-conventionalist view is that propositions which we believe express mathematical facts do not do so, and are true simply because of "an idle running of language", i.e., because the rules which determine when propositions are true or false determine that these propositions are true "no matter what the facts are" (page 29). To this view Gödel raises two objections, of which the first summarizes the main point of the foregoing discussion: in any putative proof that mathematics is tautologous or true solely by virtue of some such rules, one would have to use mathematics that is at least as complicated as that being asserted to be tautologous or thus true.

The second objection is that no justification can be given for regarding certain mathematical statements, such as complete induction, as "void of content", for one can easily construct systems in which certain empirical statements are taken as axioms. (For the notion of "content" which Gödel has in mind, see *Carnap 1937*, pages 42 and 120.) As it would clearly be unjustifiable to classify those empirical statements as therefore lacking in content, so, Gödel claims, it would be no more justifiable to regard those mathematical statements as actually *void of content*. Thus, according to Gödel, no ground has been given for thinking that there are no such things as mathematical facts, a claim Gödel calls "the essence of this view" (page 29).

6. Realism and analyticity

Gödel is prepared to acknowledge a grain of truth in the nominalist position. "A mathematical proposition says nothing about the physical or psychical reality existing in space and time, because it is true already owing to the meaning of the terms occurring in it, irrespectively of the world of real things" (page 30). It is an error to think that the meanings of the terms are man-made or that they consist in semantical conventions. Meanings are concepts, which "form an objective reality of their own, which we cannot create or change, but only perceive and describe" (page 30).[g]

---

[g]Cf. *1944* and Parsons' introductory note thereto.

Philosophers of mathematics and other metaphysicians dispute whether the supposition that mathematical objects or concepts "form an objective reality of their own" is surrogate theology (if not outright craziness), is trivially correct, or is in profound need of philosophical clarification. The matter will not be resolved here. Gödel elaborates, "... a mathematical proposition, although it does not say anything about space-time reality, still may have a very sound objective content, insofar as it says something about relations of concepts" (pages 30–1). But the elaboration helps not at all to settle the dispute. To complicate matters further, it should be noted that the term "world", as in the phrase "world of real things", belongs to the same family as "reality" and "objective", and thus the assertion that mathematical truths say something "about the world" would seem to enjoy the same status (insane, trivial, or unclear) as the claim that they describe an "objective reality".

Gödel's realism takes a strong form: relations between the concepts are not "tautological", because among the axioms that govern the concepts entering into those relations, some must be assumed which are not tautological, but which "follow from the meaning of the primitive terms under consideration" (page 31). Gödel's thought is that statements, such as instances of the comprehension schema in analysis (second-order arithmetic), even those containing quantifiers ranging over all sets of integers, are valid "owing to the meaning of the term 'set'—one might even say they express the very meaning of the term 'set'" (page 32). Gödel distinguishes between truths he calls "analytic" (those true in virtue of the meanings of the terms expressing them or "owing to the nature of the concepts occurring therein") and "tautological" truths (those "devoid of content" or "true owing to our definitions"). It may be emphasized that Gödel does not restrict the term "analytic" to statements of the "oculists are eye-doctors" or "actresses are female" variety. The analytic truths about sets, Gödel states, cannot be proved without appeal to the concept of set itself; and some analytic propositions might well be undecidable, since "our knowledge of the world of concepts may be as limited and incomplete as that of [the] world of things" (page 34). Gödel also discusses the notion of analyticity near the end of *1944*.

Quine's influential attack (*1951*) on the concept of analyticity appeared three months before Gödel delivered his lecture. Gödel's claim that the axioms of set theory are analytic—"true owing to the meanings of the terms they contain or the nature of the concepts those terms express"—is troubling for at least three sorts of reasons that do not entirely depend on Quine's claim that the phrase "true by virtue of meanings" has not been shown to isolate a significant class of truths.

In the first place, there is a difficulty about the truth of the axioms: a number of thoughtful writers believe that the axioms of set theory do not describe anything real, despite Gödel's later assertion (*1964*, page 271) that they force themselves upon us as being true. It is certainly a sensible view to hold both that Cantor's theory of transfinite numbers is a fantasy and that the standard theorems of elementary number theory and analysis are unquestionably true. In any case, the axioms of set theory lack the kind of obviousness one would have expected *axioms* characterized as "analytic" to enjoy.

Secondly, the axiom of extensionality would seem to be the only axiom of ZF that can be properly said to be true in virtue of the *meaning* of the word "set"; indeed, the axiom is often justified on the ground that the criterion of identity of sets it gives, viz., having the same members, is just part of what is meant by "set" (as opposed, say, to "property") and it is the only one that can be thus defended by an appeal to what "set" means. But since Gödel understands "true in virtue of meanings" as so much wider than "true owing to definitions" that it encompasses all axioms of set theory, Quine's questions re-arise: How is the notion of meaning that Gödel is using to be understood? When the axioms of set theory are said to be true in virtue of the meanings of their constituent terms, what more is said beyond that they are true? What is it for them to be true *in virtue of* the meanings of the terms they contain? A possible rejoinder to the effect that it is not the meaning of "set" but the nature of the concept of set that is of primary importance for Gödel is open to the reply that the last two questions remain unanswered under the replacement of "meanings of terms" by "natures of concepts".

Gödel's view raises worrisome questions of a third sort, suggested in part by later writing (*1964*) of his own: Could not the axioms of set theory be true, not in virtue of the concept of set or the meaning of "set", but simply because sets just happen to be as the axioms have it? Why, one might ask, must our knowledge of sets be mediated solely through our understanding of the *concept* of set; could we not know how matters stand with sets by "something like a perception" of them—to quote from the supplement to *1964*—that is as direct as our perception of the *concept* of set? Even lacking such a perception, might we not acquire quasi-empirical evidence, of a sort that Gödel himself has acknowledged may exist, that certain set-theoretic matters happen to stand one way rather than another? One wonders why a *conceptual* realism should be found any more plausible than an "objectual" realism.[h]

Since "our knowledge of the world of concepts may be as limited and incomplete as that of [the] world of things", Gödel holds that the paradoxes of set theory pose no more threat to his Platonism than the illusion of the stick in water poses to the view that there is an "outer world". The interesting implied suggestion is that we are taken in by

---

[h] *Parsons 199?* contains further discussion of Gödel's use of the notion of analyticity.

something like an optical *illusion* when we accept the principles that
lead to set-theoretic contradiction; perhaps we ought to wonder what
we might learn about our mental faculties from a study of these principles.

7. Conclusion

Gödel concludes by claiming that although he has disproved the nominalist standpoint and adduced strong arguments against the more general view that mathematics is our own creation, he could not claim to
have proved the realist viewpoint he favors, for to do so would require
a survey of the alternatives, a proof that the survey was exhaustive,
and a refutation of all the alternatives except realism. Among the alternatives to be refuted are Aristotelian realism, which he characterizes
as the view that concepts are aspects or parts of things, and psychologism, which holds that mathematics is nothing but the psychological
laws by which thoughts, presumably concerning calculation, etc., occur
in us. About Aristotelian realism, Gödel says only that he does not
think it tenable. His principal charge against psychologism, reminiscent
of Frege's objections, is briefly given: if psychologism were correct, there
would be no *mathematical* knowledge, but only knowledge that our mind
is so constituted as to consider certain statements of mathematics true.
His discussion is admittedly cursory, however, and Gödel gives psychologism, whatever its merits, much less attention than nominalism.

The suggestion with which he closes the lecture may seem utterly
strange: that *after sufficient clarification* of the concepts in question, it
will be possible to conduct the discussion of these matters "with mathematical rigor", at which time the result will be that the Platonistic view
is the only one tenable. (Here he characterizes the position somewhat
differently, as "the view that mathematics describes a non-sensual reality, which exists independently both of the acts and [[of]] the dispositions
of the human mind and is only perceived, and probably perceived very
incompletely, by the human mind" [page 38].) What is surprising here
is not the commitment to Platonism, but the suggestion, which recalls
Leibniz's project for a universal characteristic,[i] that there could be a
mathematically rigorous discussion of these matters, of which the correctness of any such view could be a "result". Gödel calls Platonism
rather unpopular among mathematicians; it is probably rather more
popular among them now, forty years after he gave his lecture, in some

[i]In his introductory note to *1944*, Parsons calls Gödel's view, given in the last
paragraph of *1944*, that Leibniz did not regard the *Characteristica universalis* as a
utopian project "one of his most striking and enigmatic utterances".

measure because of his advocacy of it, but perhaps more importantly because every other leading view seems to suffer from serious mathematical or philosophical defects. Gödel's idea that we shall one day achieve sufficient clarity about the concepts involved in *philosophical* discussion of mathematics to be able to prove, mathematically, the truth of some position in the philosophy of mathematics, however, appears significantly less credible at present than his Platonism.

George Boolos[j]

The translation of the quotation from Hermite at the end of *1951* is by Solomon Feferman, with the assistance of Marguerite Frank.

[j]I am grateful to Cheryl Dawson, John Dawson, Solomon Feferman, Warren Goldfarb, and Charles Parsons for much editorial and philosophical advice.

Some basic theorems on the foundations of mathematics and their implications
(*1951)

Research in the foundations of mathematics during the past few decades has produced some results which seem to me of interest, not only in themselves, but also with regard to their implications for the traditional philosophical problems about the nature of mathematics. The results themselves, I believe, are fairly widely known, but nevertheless I think it will be useful to present them in outline once again, especially in view of the fact that, due to the work of various mathematicians, they have taken on a much more satisfactory form than they had had originally. The greatest improvement was made possible through the precise definition of the concept of finite procedure,[1] which plays a decisive role in these results. There are several different ways of arriving at such a definition, which, however, all lead to exactly the same concept. The most satisfactory way, in my opinion, is that of reducing the concept of finite procedure to that of a

---

[1]This concept, for the applications to be considered in this lecture, is equivalent to the concept of a "computable function of integers" (that is, one whose definition makes it possible actually to compute $f(n)$ for each integer n). The procedures to be considered do not operate on integers but on formulas, but because of the enumeration of the formulas in question, they can always be reduced to procedures operating on integers.

machine with a finite number of parts, as has been done by the British mathematician Turing. | As to the philosophical consequences of the re- 1 sults under consideration, I don't think they have ever been adequately discussed, or [have] only [just been] taken notice of.

The metamathematical results I have in mind are all centered around, or, one may even say, are only different aspects of one basic fact, which might be called the incompletability or inexhaustibility of mathematics. This fact is encountered in its simplest form when the axiomatic method is applied, not to some hypothetico-deductive system such as geometry (where the mathematician can assert only the conditional truth of the theorems), but to mathematics proper, that is, to the body of those mathematical propositions which hold in an absolute sense, without any further hypothesis. There must exist propositions of this kind, because otherwise there could not exist any hypothetical theorems | either. For example, *some* implica- 2 tions of the form:

> If such and such axioms are assumed, then such and such a theorem holds,

must necessarily be true in an absolute sense. Similarly, any theorem of finitistic number theory, such as $2 + 2 = 4$, is, no doubt, of this kind. Of course, the task of axiomatizing mathematics proper differs from the usual conception of axiomatics insofar as the axioms are not arbitrary, but must be correct mathematical propositions, and moreover, evident without proof. There is no escaping the necessity of assuming some axioms or rules of inference as evident without proof, because the proofs must have some starting point. However, there are widely divergent views as to the extension of mathematics proper, as I defined it. The intuitionists and finitists, for example, reject some of its axioms and concepts, which others acknowledge, such as the law of excluded middle or the general concept of set.

The phenomenon of the inexhaustibility of mathematics,[2] however, always is present in some form, no matter what standpoint is taken. So I might as well explain it for the simplest and most natural standpoint, which takes mathematics as it is, without curtailing it by any criticism. From this standpoint all of mathematics is reducible to abstract set theory. For example, the statement that the axioms of projective geometry imply a certain theorem means that if a set M of elements called points and a set N of subsets of M called straight lines satisfy the axioms, then the theorem

---

[2]The term "mathematics", here and in the sequel, is always supposed to mean "mathematics proper" (which of course includes formal logic as far as it is acknowledged to be correct by the particular standpoint taken).

3   holds for N, M. Or, to mention | another example, a theorem of number
theory can be interpreted to be an assertion about finite sets. So the problem at stake is that of axiomatizing set theory. Now, if one attacks this
problem, the result is quite different from what one would have expected.
Instead of ending up with a finite number of axioms, as in geometry, one
is faced with an infinite series of axioms, which can be extended further
and further, without any end being visible and, apparently, without any
possibility of comprising all these axioms in a finite rule producing them.[3]
This comes about through the circumstance that, if one wants to avoid the
paradoxes of set theory without bringing in something entirely extraneous
to actual mathematical procedure, the concept of set must be axiomatized
in a stepwise manner.[4] If, for example, we begin with the integers, that is,
the finite sets of a special kind, we have at first the sets of integers and the
axioms referring to them (axioms of the first level), then the sets of sets
of integers with their axioms (axioms of the second level), and so on for
any finite iteration of the operation "set of".[5] Next we have the set of all
4   these | sets of finite order. But now we can deal with this set in exactly the
same manner as we dealt with the set of integers before, that is, consider
the subsets of it (that is, the sets of order ω) and formulate axioms about
their existence. Evidently this procedure can be iterated beyond ω, in fact
up to any transfinite ordinal number. So it may be required as the next
axiom that the iteration is possible for *any* ordinal, that is, for any order
type belonging to some well-ordered set. But are we at an end now? By no
means. For we have now a new operation of forming sets, namely, forming
a set out of some initial set A and some well-ordered set B by applying the
operation "set of" to A as many times as the well-ordered set B indicates.[6]
And, setting B equal to some well-ordering of A, now we can iterate this
new operation, and again iterate it into the transfinite. This will give rise
to a new operation again, which we can treat in the same way, and so on.
So the next step will be to require that *any* operation producing sets out
5   of sets can be iterated up to | any ordinal number (that is, order type of a
well-ordered set). But are we at an end now? No, because we can require

---

[3]In the axiomatizations of non-mathematical disciplines such as physical geometry,
mathematics proper is presupposed; and the axiomatization refers to the content of the
discipline under consideration only insofar as it goes beyond mathematics proper. This
content, at least in the examples which have been encountered so far, can be expressed
by a finite number of axioms.

[4]This circumstance, in the usual presentation of the axioms, is not directly apparent,
but shows itself on closer examination of the meaning of the axioms.

[5]The operation "set of" is substantially the same as the operation "power set", where
the power set of M is by definition the set of all subsets of M.

[6]In order to carry out the iteration one may put $A = B$ and assume that a special well-
ordering has been assigned to any set. For ordinals of the second kind [limit ordinals],
the set of the previously obtained sets is always to be formed.

not only that the procedure just described can be carried out with any operation, but that moreover there should exist a set closed with respect to it, that is, one which has the property that, if this procedure (with any operation) is applied to elements of this set, it again yields elements of this set. You will realize, I think, that we are still not at an end, nor can there ever be an end to *this* procedure of forming the axioms, because the very formulation of the axioms up to a certain stage gives rise to the next axiom. It is true that in the mathematics of today the higher levels of this hierarchy are practically never used. It is safe to say that 99.9% of present-day mathematics is contained in the first three levels of this hierarchy. So for all practical purposes, all of mathematics *can* be reduced to a finite number of axioms. However, | this is a mere historical accident, which is of no impor- 6 tance for questions of principle. Moreover it is not altogether unlikely that this character of present-day mathematics may have something to do with another character of it, namely, its inability to prove certain fundamental theorems, such as, for example, Riemann's hypothesis, in spite of many years of effort. For it can be shown that the axioms for sets of high levels, in their relevance, are by no means confined to these sets, but, on the contrary, have consequences even for the 0-level, that is, the theory of integers. To be more exact, each of these set-theoretical axioms entails the solution of certain diophantine problems which had been undecidable on the basis of the preceding axioms.[7] *The diophantine problems in question are of the following type: Let $P(x_1, \ldots, x_n, y_1, \ldots, y_m)$ be a polynomial with given integral coefficients and $n + m$ variables,* | $x_1, \ldots x_n, y_1, \ldots, y_m$, *and consider* 7 *the variables x_i as the unknowns and the variables y_i as parameters; then the problem is: Has the equation $P = 0$ integral solutions for any integral values of the parameters, or are there integral values of the parameters for which this equation has no integral solutions? To each of the set-theoretical axioms a certain polynomial P can be assigned, for which the problem just formulated becomes decidable owing to this axiom. It even can always be achieved that the degree of P is not higher than 4.* [The] mathematics of today has not yet learned to make use of the set-theoretical axioms for the solution of number-theoretical problems, except for the axioms of the first level. These are actually used in analytic number theory. But for mastering number theory this is demonstrably insufficient. Some kind of

---

[7]This theorem, in order to hold also if the intuitionistic or finitistic standpoint is assumed, requires as a hypothesis the consistency of the axioms of set theory, which of course is self-evident (and therefore can be dropped as a hypothesis) if set theory is considered to be mathematics proper. However, for finitistic mathematics a similar theorem holds, without any questionable hypothesis of consistency; namely, the introduction of recursive functions of higher and higher order leads to the solution of more and more number-theoretical problems of the specified kind. In intuitionistic mathematics there doubtless holds a similar theorem for the introduction (by new axioms) of greater and greater ordinals of the second number class.

8   set-theoretical number theory, still to be dis|covered, would certainly reach
much farther.

I have tried so far to explain the fact I call ⟦the⟧ incompletability of
mathematics for one particular approach to the foundations of mathematics, namely axiomatics of set theory. That, however, this fact is entirely
independent of the particular approach and standpoint chosen appears from
certain very general theorems. The first of these theorems simply states
that, *whatever well-defined system of axioms and rules of inference may
be chosen, there always exist diophantine problems of the type described
which are undecidable by these axioms and rules, provided only that no false
propositions of this type are derivable.*[8] If I speak of a well-defined system
of axioms and rules here, this only means that it must be possible actually
to write the axioms down in some precise formalism or, if their number is
infinite, a finite procedure for writing them down one after the other must
be given. Likewise the rules of inference are to be such that, given any
premises, either the conclusion (by any one of the rules of inference) can
9   be written | down, or it can be ascertained that there exists no immediate
conclusion by the rule of inference under consideration. This requirement
for the rules and axioms is equivalent to the requirement that it should be
possible to build a finite machine, in the precise sense of a "Turing machine", which will write down all the consequences of the axioms one after
the other. For this reason, the theorem under consideration is equivalent
to the fact that there exists no finite procedure for the systematic decision
of all diophantine problems of the type specified.

The second theorem has to do with the concept of freedom from contradiction. For a well-defined system of axioms and rules the question of
their consistency is, of course, itself a well-defined mathematical question.
Moreover, since the symbols and propositions of ⟦any⟧ one formalism are
always at most enumerable, everything can be mapped on⟦to⟧ the integers,
and it is plausible and in fact demonstrable that the question of consistency
can always be transformed into a number-theoretical question (to be more
10 exact, into one of the type described above). Now | the theorem says that
*for any well-defined system of axioms and rules, in particular, the proposi-
tion stating their consistency*[9] *(or rather the equivalent number-theoretical
proposition) is undemonstrable from these axioms and rules, provided these
axioms and rules are consistent and suffice to derive a certain portion*[10] *of*

---

[8]This hypothesis can be replaced by consistency (as shown by Rosser in ⟦his *1936*⟧),
but the undecidable propositions then have a slightly more complicated structure. Moreover, the hypothesis must be added that the axioms imply the primitive properties of
addition, multiplication and <.

[9]It is one of the propositions which are undecidable, provided that no false number-
theoretical ⟦propositions⟧ are derivable (see the preceding theorem).

[10]Namely, Peano's axioms and the rule of definition by ordinary induction, with a
logic satisfying the strictest finitistic requirements.

the finitistic arithmetic of integers. It is *this* theorem which makes the incompletability of mathematics particularly evident. For, *it makes it impossible that someone should set up a certain well-defined system of axioms and rules and consistently make the following assertion about it: All of these axioms and rules I perceive (with mathematical certitude) to be correct, and moreover I believe that they contain all of mathematics.* If someone makes such a statement he contradicts himself.[11] For if he perceives the axioms under consideration to be correct, he also perceives (with the same certainty) that they are consistent. Hence he has a mathematical | insight not derivable from his axioms. However, one has to be careful in order to understand clearly the meaning of this state of affairs. Does it mean that no well-defined system of correct axioms can contain all of mathematics proper? It does, if by mathematics proper is understood the system of all true mathematical propositions; it does not, however, if one understands by it the system of all demonstrable mathematical propositions. I shall distinguish these two meanings of mathematics as mathematics in the objective and in the subjective sense: Evidently no well-defined system of correct axioms can comprise all [[of]] objective mathematics, since the proposition which states the consistency of the system is true, but not demonstrable in the system. However, as to subjective mathematics, it is not precluded that there should exist a finite rule producing all its evident axioms. However, if such a rule exists, we with our human understanding could certainly never know it to be such, that is, we could never know with mathematical | certainty that all propositions it produces are correct;[12] or in other terms, we could perceive to be true only one proposition after the other, for any finite number of them. The assertion, however, that they are all true could at most be known with empirical certainty, on the basis of a sufficient number of instances or by other inductive inferences.[13] If it were so, this would mean that the human mind (in the realm of pure

11

12

[11]If he only says "I believe I shall be able to perceive one after the other to be true" (where their number is supposed to be infinite), he does not contradict himself. (See below.)

[12]For this (or the consequence concerning the consistency of the axioms) would constitute a mathematical insight not derivable from the axioms [[and]] rules under consideration, contrary to the assumption.

[13]For example, it is conceivable (although far outside the limits of present-day science) that brain physiology would advance so far that it would be known with empirical certainty
1. that the brain suffices for the explanation of all mental phenomena and is a machine in the sense of Turing;
2. that such and such is the precise anatomical structure and physiological functioning of the part of the brain which performs mathematical thinking.
Furthermore, in case the finitistic (or intuitionistic) standpoint is taken, such an inductive inference might be based on a (more or less empirical) belief that non-finitistic (or non-intuitionistic) mathematics is consistent.

mathematics) *is* equivalent to a finite machine that, however, is unable to understand completely[14] its own functioning. This inability [of man] to understand himself would then wrongly appear to him as its [(the mind's)] boundlessness or inexhaustibility. But, please, note that if it were so, this would in no way derogate from the incompletability of objective mathematics. On the contrary, it would only make it particularly striking. For if the human mind were equivalent to a finite machine, then objective mathematics not only would be incompletable in the sense of not being contained in

13   any well-defined axiomatic system, but moreover there would exist | *absolutely* unsolvable diophantine problems of the type described above, where the epithet "absolutely" means that they would be undecidable, not just within some particular axiomatic system, but by *any* mathematical proof the human mind can conceive. So the following disjunctive conclusion is inevitable: *Either mathematics is incompletable in this sense, that its evident axioms can never be comprised in a finite rule, that is to say, the human mind (even within the realm of pure mathematics) infinitely surpasses the powers of any finite machine, or else there exist absolutely unsolvable diophantine problems of the type specified* (where the case that both terms of the disjunction are true is not excluded, so that there are, strictly speaking, three alternatives). It is this mathematically established fact which seems to me of great philosophical interest. Of course, in this connection it is of great importance that at least this fact is entirely independent of the special standpoint taken toward the foundations of mathematics.[15]

14 | There is, however, one restriction to this independence, namely, the standpoint taken must be liberal enough to admit propositions about all integers as meaningful. If someone were so strict a finitist that he would maintain that only particular propositions of the type $2 + 2 = 4$ belong to mathematics proper,[16] then the incompletability theorem would not

---

[14]Of course, the physical working of the thinking mechanism could very well be completely understandable; the insight, however, that this particular mechanism must always lead to correct (or only consistent) results would surpass the powers of human reason.

[15]For intuitionists and finitists the theorem holds as an implication (instead of a disjunction). It is to be noted that intuitionists have always asserted the first term of the disjunction (and negated the second term, in the sense that no demonstrably undecidable propositions can exist). [See above, p. [?][a].] But this means nothing for the question which alternative applies to intuitionistic mathematics, if the terms occurring in it are understood in the objective sense (rejected as meaningless by the intuitionists). As for finitism, it seems very likely that the first disjunctive term is false.

[16]K. Menger's "implicationistic standpoint" (see *Menger 1930a*, p. 323), if taken in the strictest sense, would lead to such an attitude, since according to it, the only meaningful mathematical propositions (that is, in my terminology, the only ones belonging to mathematics proper) would be those that assert that such and such a conclusion can

---

[a]We are unable to locate a place in the text to which Gödel would be referring here.

apply—at least not *this* incompletability theorem. But I don't think that such an attitude could be maintained consistently, because it is by exactly the same kind of evidence that we judge that $2 + 2 = 4$ and that $a + b = b + a$ for any two integers a, b. Moreover, this standpoint, in order to be consistent, would have to exclude also *concepts* that refer to *all* integers, such as "+" (or to all formulas, such as "correct proof by such and such rules") and replace them with others that apply only within some finite domain of integers (or formulas). It is to be noted, however, that although the truth of the disjunctive theorem is independent of the standpoint taken, the question as to which alternative holds need not be independent of it. (See footnote [15].)

| I think I now have explained sufficiently the mathematical aspect of the situation and can turn to the philosophical implications. Of course, in consequence of the undeveloped state of philosophy in our days, you must not expect these inferences to be drawn with mathematical rigour.   15

Corresponding to the disjunctive form of the main theorem about the incompletability of mathematics, the philosophical implications *prima facie* will be disjunctive too; however, under either alternative they are very decidedly opposed to materialistic philosophy. Namely, if the first alternative holds, this seems to imply that the working of the human mind cannot be reduced to the working of the brain, which to all appearances is a finite machine with a finite number of parts, namely, the neurons and their connections. So apparently one is driven to take some vitalistic viewpoint. On the other hand, the second alternative, where there exist absolutely undecidable mathematical propositions, seems to disprove the view that mathematics is only our own creation; | for the creator necessarily knows   16 all properties of his creatures, because they can't have any others except those he has given to them. So this alternative seems to imply that mathematical objects and facts (or at least *something* in them) exist objectively and independently of our mental acts and decisions, that is to say, [it seems to imply] some form or other of Platonism or "realism" as to the mathe-

be drawn from such and such axioms and rules of inference in such and such [a] manner. This, however, is a proposition of exactly the same logical character as $2+2 = 4$. Some of the untenable consequences of this standpoint are the following: A negative proposition to the effect that the conclusion B cannot be drawn from the axioms and rules A would not belong to mathematics proper; hence nothing could be known about it except perhaps that it follows from certain other axioms and rules. However, a proof that it does so follow (since these other axioms and rules again are arbitrary) would in no way exclude the possibility that (in spite of the formal proof to the contrary) a derivation of B from A might some day be accomplished. For the same reason also, the usual inductive proof for $a + b = b + a$ would not exclude the possibility of discovering two integers not satisfying this equation.

matical objects.[17] For, the empirical interpretation of mathematics,[18] that
is, the view that mathematical facts are a special kind of physical or psy-
chological facts, is too absurd to be seriously maintained (see below). |
It is not known whether the first alternative holds, but at any rate it is
in good agreement with the opinions of some of the leading men in brain
and nerve physiology, who very decidedly deny the possibility | of a purely
mechanistic explanation of psychical and nervous processes.

As far as the second alternative is concerned, one might object that the
constructor need not necessarily know *every* property of what he constructs.
For example, we build machines and still cannot predict their behaviour
in every detail. But this objection is very poor. For we don't create the
machines out of nothing, but build them out of some given material. If the
situation were similar in mathematics, then this material or basis for our
constructions would be something objective and would force some realistic
viewpoint upon us even if certain other ingredients of mathematics were
our own creation. The same would be true if in our creations we were to
use some instrument in us but different from our ego (such as "reason"
interpreted as something like a thinking machine). For mathematical facts
would then (at least in part) express properties of this instrument, which
would have an objective existence.

One may thirdly object that the meaning of a proposition about all
integers, since it is impossible to verify it for all integers one by one, can
consist only in the existence of a general proof. Therefore, in the case of an

---

[17]There exists no term of sufficient generality to express exactly the conclusion drawn
here, which only says that the objects and theorems of mathematics are as objective
and independent of our free choice and our creative acts as is the physical world. It
determines, however, in no way what these objective entities are—in particular, whether
they are located in nature or in the human mind or in neither of the two. These three
views about the nature of mathematics correspond exactly to the three views about the
nature of concepts, which traditionally go by the names of psychologism, Aristotelian
conceptualism and Platonism.

[18]That is, the view that mathematical objects and the way in which we know them
are not essentially different from physical or psychical objects and laws of nature. The
true situation, on the contrary, is that if the objectivity of mathematics is assumed, it
follows at once that its objects must be totally different from sensual objects because

1. Mathematical propositions, if properly analyzed, turn out to assert nothing about
   the actualities of the space-time world. This is particularly clear in applied propositions such as: Either it has or it has not rained yesterday. The existence of purely
   conceptual knowledge (besides mathematics) satisfying these requirements is not
   excluded by this remark.
2. The mathematical objects are known precisely, and general laws can be recognized
   with certainty, that is, by deductive, not inductive, inference.
3. They can be known (in principle) without using the senses (that is, by means of
   reason alone) for this very reason, that they don't concern actualities about which
   the senses (the inner sense included) inform us, but possibilities and impossibili-
   ties.

undecidable proposition about all integers, neither itself nor its negation is true. Hence neither expresses an objectively existing but unknown property of the integers. | I am not in a position now to discuss the epistemological question as to whether this opinion is at all consistent. It certainly looks as if one must *first* understand the meaning of a proposition *before* he can understand a proof of it, so that the meaning of "all" could not be defined in terms of the meaning of "proof". But independently of this epistemological investigation, I wish to point out that one may conjecture the truth of a universal proposition (for example, that I shall be able to verify a certain property for *any* integer given to me) and at the same time conjecture that no general proof for this fact exists. It is easy to imagine situations in which both these conjectures would be very well founded. For the first half of it, this would, for example, be the case if the proposition in question were some equation $F(n) = G(n)$ of two number-theoretical functions which could be verified up to *very* great numbers n.[19] Moreover, exactly as in the natural sciences, this *inductio per enumerationem simplicem* is by no means the only inductive method conceivable in mathematics. I admit that every mathematician has an inborn abhorrence to giving more than heuristic | significance to such inductive arguments. I think, however, that this is due to the very prejudice that mathematical objects somehow have no real existence. If mathematics describes an objective world just like physics, there is no reason why inductive methods should not be applied in mathematics just the same as in physics. The fact is that in mathematics we still have the same attitude today that in former times one had toward all science, namely, we try to derive everything by cogent proofs from the definitions (that is, in ontological terminology, from the essences of things). Perhaps this method, if it claims monopoly, is as wrong in mathematics as it was in physics.

This whole consideration incidentally shows that the philosophical implications of the mathematical facts explained do not lie entirely on the side of rationalistic or idealistic philosophy, but that in one respect they favor the empiricist viewpoint.[20] | It is true that only the second alternative points in this direction. However, *and this is the item I would like to*

19

20

21

---

[19]Such a verification of an *equality* (not an inequality) between two number-theoretical functions of not too complicated or artificial structure would certainly give a great probability to their complete equality, although its numerical value could not be estimated in the present state of science. However, it is easy to give examples of general propositions about integers where the probability can be estimated even now. For example, the probability of the proposition which states that for each n there is at least one digit $\neq 0$ between the n-th and n^2-th digits of the decimal expansion of π converges toward 1 as one goes on verifying it for greater and greater n. A similar situation also prevails for Goldbach's and Fermat's theorems [*sic*].

[20]To be more precise, it suggests that the situation in mathematics is not so very different from that in the natural sciences. As to whether, in the last analysis, apriorism or empiricism is correct is a different question.

discuss now, it seems to me that the philosophical conclusions drawn under the second alternative, in particular, conceptual realism (Platonism), are supported by modern developments in the foundations of mathematics also, irrespectively of which alternative holds. The main arguments pointing in this direction seem to me [to be] the following. First of all, if mathematics were our free creation, ignorance as to the objects we created, it is true, might still occur, but only through lack of a clear realization as to what we really have created (or, perhaps, due to the practical difficulty of too complicated computations). Therefore it would have to disappear (at least in principle, although perhaps not in practice[21]) as soon as we attain perfect clearness. However, modern developments in the foundations of mathematics have accomplished an insurmountable degree of exactness, but this has helped practically nothing for the solution of mathematical problems.

22 | Secondly, the activity of the mathematician shows very little of the freedom a creator should enjoy. Even if, for example, the axioms about integers were a free invention, still it must be admitted that the mathematician, after he has imagined the first few properties of his objects, is at an end with his creative ability, and he is not in a position also to create the validity of the theorems at his will. If anything like creation exists at all in mathematics, then what any theorem does is exactly to restrict the freedom of creation. That, however, which restricts it must evidently exist independently of the creation.[22]

Thirdly, if mathematical objects are our creations, then evidently integers and sets of integers will have to be two different creations, the first of which does not necessitate the second. However, in order to prove certain propositions about integers, the concept of set of integers is necessary. So 23 here, in order to find out what properties *we* have | given to certain objects of our imagination, [we] must first create certain other objects—a very strange situation indeed!

[21]That is, every problem would have to be reducible to some finite computation.

[22]It is of no avail to say that these restrictions are brought about by the requirement of consistency, which itself is our free choice, because one might choose to bring about consistency *and* certain theorems. Nor does it help to say that the theorems only repeat (wholly or in part) the properties first invented, because then the exact realization of what was first assumed would have to be sufficient for deciding any question of the theory, which is disproved by the first [argument (above)] and the third argument [(below)]. As to the question of whether undecidable propositions can be decided arbitrarily by a new act of creation, see fn. [?][b].

---

[b]No footnote in the manuscript deals with this question. However, the shorthand annotation to p. 29′ (see editorial note g below and the textual notes) does contain the phrase "contin[uous] creation". That could have been a note of Gödel to himself to write something on the question.

What I [have] said so far has been formulated in terms of the rather vague concept of "free creation" or "free invention". There exist attempts to give a more precise meaning to this term. However, this only has the consequence that also the disproof of the standpoint in question is becoming more precise and cogent. I would like to show this in detail for the most precise, and at the same time most radical, formulation that has been given so far. It is that which[c] || interprets mathematical propositions as expressing solely certain aspects of syntactical (or linguistic)[23] conventions, that is,

[23]That is, the conventions must not refer to any extralinguistic objects (as does a demonstrat[ion]-def[inition] [d]), but must state rules about the meaning or truth of symbolic expressions solely on the basis of their outward structure. Moreover, of course these rules must be such that they do not imply the truth or falsehood of any factual propositions (since in that case they could certainly not be called void of content nor syntactical). This, however, entails their consistency, because an inconsistency (in classical logic, which is under consideration here) would imply every factual proposition. It is to be noted that if the term "syntactical rule" is understood in this generality, the view under consideration includes, as a special elaboration of it, the formalistic foundation of mathematics, since according to the latter, mathematics is based solely on certain syntactical rules of the form: Propositions of such and such structure are true [the axioms], and if propositions of ... structure are true, then such and such other propositions are also true; and moreover, as can easily be seen, the consistency proof gives the assurance that these rules are void of content insofar as they imply no factual propositions. On the other hand, also, vice-versa, it will turn out below that the feasibility of the nominalistic program implies the feasibility of the formalistic program. (For very lucid expositions of the philosophical aspects of this nominalistic view, see *Hahn 1935* or *Carnap 1935a, 1935b*.) It might be doubted whether this (nominalistic) view should at all be subsumed under the view that considers mathematics to be a free creation of the mind, because it denies altogether the existence of mathematical objects. Moreover, the relationship between the two is extremely close, since also under the other view the so-called existence of mathematical objects consists solely in their being constructed in thought, and nominalists would not deny that we actually imagine (non-existent) objects behind the mathematical symbols and that these subjective ideas might even furnish the guiding principle in the choice of the syntactical rules.

---

[c]The double vertical lines indicate material marked in the manuscript "Omit from here to p. 29". Since this material was not crossed out, a plausible conjecture is that it was to be omitted only from his oral presentation. But other conjectures are possible, for example, that he came at a later time to think it duplicative of or superseded by discussions in *1953/9.* See also the textual notes.

[d]Gödel writes "demonstrat.-def.", in some other places without the hyphen. It might with almost equal plausibility be read as "demonstrative definition" (which would be stylistically more attractive).

What indication there is as to what he has in mind is given by the following passage:

Of course it is to be noted that a demonstrat[ion]-def[inition] does not mean pointing the finger to the object for which a name is introduced (which in most cases is not possible even for physical concepts), but that it rather means explaining the meaning of a word by means of the situations in which it is used.

(From Gödel's footnote 58. This note is flagged in the alternate version of the text printed from p. 29′ [see the textual notes] and also in Gödel's footnote 26, to which we have found no reference in the text.)

they simply repeat parts of these conventions. According to this view, mathematical propositions, duly analyzed, must turn out to be as void of content as, for example, the statement "All stallions are horses". Everybody will agree that this proposition does not express any zoological or other objective fact, but [rather,] its truth is due solely to the circumstance that we chose to use the term "stallion" as an abbreviation for "male horse". | Now by far the most common type of symbolic conventions are definitions (either explicit or contextual, where the latter however must be such as to make it possible to eliminate the term defined in any context [where] it occurs). Therefore the simplest version of the view in question would consist in the assertion that mathematical propositions are true solely owing to the definitions of the terms occurring in them, that is, that by successively replacing all terms by their definientia, any theorem can be reduced to an explicit tautology, $a = a$. (Note that $a = a$ must be admitted as true if definitions are admitted, for one may define b by $b = a$ and then, owing to this definition, replace b by a in this equality.) But now it follows directly from the theorems mentioned before that such a reduction to explicit tautologies is impossible. For it would immediately yield a mechanical procedure for deciding about the truth or falsehood of every mathematical proposition. Such a procedure, however, cannot exist, not even for number theory. This disproof, it is true, refers only to the simplest | version of this (nominalistic) standpoint. But the more refined ones do not fare any better. The weakest statement that at least would have to be demonstrable, in order that this view concerning the tautological character of mathematics be tenable, is the following: Every demonstrable mathematical proposition can be deduced from the rules about the truth and falsehood of sentences alone (that is, without using or knowing anything else except these rules) and the negations of demonstrable mathematical propositions cannot be so derived.[24] In precisely formulated languages, such rules (that is, rules which stipulate under which conditions a given sentence is true) occur as a means for determining the meaning of sentences. Moreover in all known languages there *are* propositions which seem to be true owing to these rules alone. For example, if disjunction and negation are introduced by those rules:

   1) $p \vee q$ is true if at least one of its terms is true, and
   2) $\sim p$ is true if p is not true,

---

[24] As to the requirement of consistency, see fn. [23?]$^\text{e}$.

$^\text{e}$It is also possible that Gödel intended to write a new note on this subject. In the manuscript, the text as we give it is above some crossed-out text in which something is said about the "requirement of consistency", which, however, he may have thought repeated points in (our) note 23.

then it clearly follows from these rules | that $p \vee \sim p$ is always true what- 26
ever p may be. (Propositions so derivable are called tautologies.) Now it
is actually so, that for the symbolisms of mathematical logic, with suitably
chosen semantical rules, the truth of the mathematical axioms *is* derivable
from these rules;[25] however (and this is the great stumbling block), in this
derivation the mathematical and logical concepts and axioms themselves
must be used in a special application, namely, as referring to symbols, combinations of symbols, sets of such combinations, etc. Hence this theory, if
it wants to prove the tautological character of the mathematical axioms,
must first assume these axioms to be true. So while the original idea of
this viewpoint was to make the truth of the mathematical axioms understandable by showing that they are tautologies, it ends up with just the
opposite, that is, the truth of the axioms must *first* be assumed and *then*
it can be shown that, in a suitably chosen language, | they are tautologies. 27
Moreover, a similar statement holds good for the mathematical concepts,
that is, instead of being able to define their meaning by means of symbolic
conventions, one must first know their meaning in order to understand the
syntactical conventions in question or the proof that they imply the mathematical axioms but not their negations. Now, of course, it is clear that this
elaboration of the nominalistic view does not satisfy the requirement set up
on page [25?], because not the syntactic rules alone, but all of mathematics
in addition is used in the derivations. But moreover, this elaboration of
nominalism would yield an outright *disproof* of it (I must confess I can't
picture any better disproof of this view than this proof of it), provided
that one thing could be added, namely, that the outcome described is un-
avoidable (that is, independent of the particular symbolic language and
interpretation of mathematics chosen). Now it is not exactly this that can
be proved, but something so close to it that it also suffices to disprove
the view in question. Namely, it follows by the metatheorems mentioned
that a proof for the tautological character (in a suitable language) of the
mathematical axioms is at the same time a proof for their consistency, and
cannot be achieved with any *weaker* means of proof than are contained in
these axioms themselves. This does not mean that *all* the axioms of a given
system must be used in its consistency proof. On the contrary, usually the
axioms lying outside the system which are necessary make it possible to

---

[25]See *Ramsey 1926*, pp. 368 and 382, and *Carnap 1937*, pp. 39 and 110. It is worth
mentioning that Ramsey even succeeds in reducing them to explicit tautologies $a = a$
by means of explicit definitions (see p. [24?] above), but at the expense of admitting
propositions of infinite (and even transfinite) length, which of course entails the necessity
of presupposing transfinite set theory in order to be able [to] deal with these infinite
entities. Carnap confines himself to propositions of finite length, but instead has to
consider infinite sets, sets of sets, etc., of these finite propositions.

dispense with some of the axioms of the system (although they do not imply the latter).[26] However, what follows with practical certainty is this: In order to prove the consistency of classical number theory (and *a fortiori* of all stronger systems) certain *abstract* concepts (and the directly evident axioms referring to them) must be used, where "abstract" means concepts which do not refer to sense objects,[27] of which symbols are a special kind. These abstract concepts, however, are certainly not syntactical [but rather those whose justification by syntactical considerations should be the main task of nominalism]. Hence it follows that *there exists no rational justification of our precritical beliefs concerning the applicability and consistency of classical mathematics (nor even its undermost level, number theory) on the basis of a syntactical interpretation.* It is true that this statement does not apply to certain subsystems of classical mathematics, which may even contain some *part* of the theory of the abstract concepts referred to. In this sense, nominalism can point to some partial successes. For it *is* actually possible to base the axioms of these systems on purely syntactical considerations. In this manner, for example, the use of the concepts of "all" and "there is" referring to integers can be justified (that is, proved consistent) by means of syntactical considerations. However, for the most essential number-theoretic axiom, complete induction, such a syntactical foundation, even within the limits in which it is possible, gives no justification of our precritical belief in it, since this axiom itself has to be used

[26]For example, any axiom system S for set theory belonging to the series explained in the beginning of this lecture, the axiom of choice included, can be proved consistent by means of the axiom of the next order (or by means of the axiom that S is consistent) without the axiom of choice. Similarly, it is not impossible that the axioms of the lower levels of this hierarchy could be proved consistent by means of axioms of higher levels, with such restrictions, however, as would make them acceptable to intuitionists.

[27]Examples of such abstract concepts are, for example, "set", "function of integers", "demonstrable" (the latter in the non-formalistic sense of "knowable to be true"), "derivable", etc., or finally "there is", referring to all *possible* combinations of symbols. The necessity of such concepts for the consistency proof of classical mathematics results from the fact that symbols can be mapped on[to] the integers, and therefore finitistic (and *a fortiori*, classical) number theory contains all proofs based solely upon them. The evidence for this fact so far is not absolutely conclusive because the evident axioms referring to the non-abstract concept under consideration have not been investigated thoroughly enough. However, the fact itself is acknowledged even by leading formalists; see [*Bernays 1941a*, pp. 144, 147; *1935*, pp. 68, 69; *1935b*, p. 94; *1954*, p. 2; also *Gentzen 1937*, p. 203][f].

[f]These references are supplied from *1953/9-III*, fn. 24, which is attached to essentially the same remark (§24 of that text). But cf. also *1958*, fn. 1.

in the syntactical considerations.[28] The fact that the more modest you are in the axioms for which you want to set up a tautological interpretation, the less of mathematics you need in order to do it, has the consequence that if finally you become so modest as to | confine yourself to some finite 28 domain, for example, to the integers up to 1000, then the mathematical propositions valid in this field can be so interpreted as to be tautological even in the strictest sense, that is, reducible to explicit tautologies by means of the explicit definitions of the terms. No wonder, because the section of mathematics necessary for the proof of the consistency of this finite mathematics is contained already in the theory of the finite combinatorial processes which are necessary in order to reduce a formula to an explicit tautology by substitutions. This explains the well-known, but misleading, fact that formulas like $5 + 7 = 12$ can, by means of certain definitions, | be 29 reduced to explicit tautologies. This fact, incidentally, is misleading also for the reason that in these reductions (if they are to be interpreted as simple substitutions of the definiens for the definiendum on the basis of explicit definitions), the $+$ is not identical with the ordinary $+$, because it can be defined only for a finite number of arguments (by an enumeration of this finite number of cases). If, on the other hand, $+$ is defined contextually, then one has to use the concept of finite manifold already in the proof of $2 + 2 = 4$. A similar circularity also occurs in the proof that $p \vee \sim p$ is a tautology, because disjunction and negation, in their intuitive meanings, evidently occur in it. ||

The essence of this view is that there exists no such thing as a mathematical fact, that the truth of propositions which we believe express mathematical facts only means that (due to the rather complicated rules which define the meaning of propositions, that is, which determine under what circumstances a given proposition is true) an idle running of language occurs in these propositions, in that the said rules make them true no matter what the facts are. Such propositions can rightly be called void of content. Now it [is] actually possible to build up a language in which mathematical propositions are void of content in this sense. The only trouble is

1. that one has to use the very same mathematical facts[g] (or equally

---

[28]The objection raised here against a syntactical foundation of number theory is substantially the same [as the one] which Poincaré leveled against both Frege's and Hilbert's foundation of number theory. However, this objection is not justified against Frege, because the logical concepts and axioms he has to presuppose do not explicitly contain the concept of a "finite manifold" with its axioms, while the grammatical concepts and considerations necessary to set up the syntactical rules and establish their tautological character do.

---

[g]An unnumbered remark cited at this point appears at the bottom of page 29' of Gödel's manuscript text. Neither a true footnote nor a textual insertion, it is rather a shorthand annotation. For a transcription and translation of its contents, see the textual notes.

complicated other mathematical facts) in order to show that they
don't exist;

2. that by this method, if a division of the empirical facts in[to] two
parts, A and B, is given such that B implies nothing in A, a language
can be constructed in which the propositions expressing B would be
void of content. And if your opponent were to say: "You are arbitrarily disregarding certain observable facts B", one may answer: "You
are doing the same thing, for example with the law of complete induction, which I perceive to be true on the basis of my understanding
(that is, perception) of the concept of integer."

30 | However, it seems to me that nevertheless one ingredient of this wrong
theory of mathematical truth is perfectly correct and really discloses the
true nature of mathematics. Namely, it is correct that a mathematical
proposition says nothing about the physical or psychical reality existing
in space and time, because it is true already owing to the meaning of the
terms occurring in it, irrespectively of the world of real things. What is
wrong, however, is that the meaning of the terms (that is, the concepts
they denote) is asserted to be something man-made and consisting merely
in semantical conventions. The truth, I believe, is that these concepts form
an objective reality of their own, which we cannot create or change, but
only perceive and describe.[29]

Therefore a mathematical proposition, although it does not say anything
about space-time reality, still may have a very sound objective content, in-
31 sofar as it | says something about relations of concepts. The existence of
non-"tautological" relations between the concepts of mathematics appears

---

[29]This holds good also for those parts of mathematics which *can* be reduced to
syntactic rules (see above). For these rules are based on the idea of a finite manifold
(namely, of a finite sequence of symbols), and this idea and its properties are entirely
independent of our free choice. In fact, its theory is equivalent to the theory of [the]
integers. The possibility of so constructing a language that this theory is incorporated
into it in the form of syntactic rules proves nothing. See fn. [?][h].

---

[h]A conjecture as to what Gödel is referring to is that it is his footnote 35, to which
there is no reference in the text. It reads as follows:

To be more exact the true situation as opposed to the view criticized is the following:

1. The meanings of mathematical terms are not reducible to the linguistic rules about
   their use except for a very restricted domain of mathematics (cf. [pp. 25–27?]).
2. Even where such a reduction is possible the linguistic rules cannot be considered
   to be something man-made and propositions about them to be lacking objective
   content because these rules are based on the idea of a finite manifold (in the form
   of finite sequences of symbols) and this idea (with all its properties) is entirely
   independent of any convention and free choice (hence *is* something objective). In
   fact, its theory is equivalent to arithmetic.

It could be, however, that this note was superseded by our note 29 (Gödel's 49).

above all in the circumstance that for the primitive terms of mathematics, axioms must be assumed, which are by no means tautologies (in the sense of being in any way reducible to $a = a$), but still do follow from the meaning of the primitive terms under consideration. For example, the basic axiom, or rather, axiom schema, for the concept of set of integers says that, given a well-defined property of integers (that is, a propositional expression $\varphi(n)$ with an integer variable n), there exists the set M of those integers which have the property φ. Now, considering the circumstance that φ may itself contain the term "set of integers", we have here a series of rather involved axioms about the concept of set. | Nevertheless, these axioms (as the aforementioned results show) cannot be reduced to anything substantially simpler, let alone to explicit tautologies. It is true that these axioms are valid owing to the meaning of the term "set"—one might even say they express the very meaning of the term "set"—and therefore they might fittingly be called analytic; however, the term "tautological", that is, devoid of content, for them is entirely out of place, because even the assertion of the existence of a concept of set satisfying these axioms (or of the consistency of these axioms) is so far from being empty that it cannot be proved without again using the concept of set itself, or some other abstract concept of [a] similar nature.

Of course, this particular argument is addressed only to mathematicians who admit the general concept of set in mathematics proper. For finitists, however, literally the same argument could be alleged for the concept of integer and the axiom of complete induction. For, if the general concept of set is *not* admitted in mathematics proper, then complete induction | must be assumed as an axiom.

I wish to repeat that "analytic" here does not mean "true owing to our definitions", but rather "true owing to the nature of the concepts occurring [therein]", in contradistinction to "true owing to the properties and the behaviour of things". This concept of analytic is so far from meaning "void of content" that it is perfectly possible that an analytic proposition might be undecidable (or decidable only with [a certain] probability). | For, our knowledge of the world of concepts may be as limited and incomplete as that of [the] world of things. It is certainly undeniable that this knowledge, in certain cases, not only is incomplete, but even indistinct. This occurs in the paradoxes of set theory, which are frequently alleged as a disproof of Platonism, but, I think, quite unjustly. Our visual perceptions sometimes contradict our tactile perceptions, for example, in the case of a rod immersed in water, but nobody in his right mind will conclude from this fact that the outer world does not exist.

I have purposely spoken of two separate worlds (the world of things and of concepts), because I do not think that Aristotelian realism (according to which concepts are parts or aspects of things) is tenable.

| Of course I do not claim that the foregoing considerations amount to

a real proof of this view about the nature of mathematics. The most I could assert would be to have disproved the nominalistic view, which considers mathematics to consist solely in syntactical conventions and their consequences. Moreover, I have adduced some strong arguments against the more general view that mathematics is our own creation. There are, however, other alternatives to Platonism, in particular psychologism and Aristotelian realism. In order to establish Platonistic realism, these theories would have to be disproved one after the other, and then it would have to be shown that they exhaust all possibilities. I am not in a position to do this now; however, I would like to give some indications along these lines. One possible form of psychologism admits that mathematics investigates relations of concepts and that concepts cannot be created at our will, but

36   are given to us as a reality, which | we cannot change; however, it contends that these concepts are only psychological dispositions, that is, that they are nothing but, so to speak, the wheels of our thinking machine. To be more exact, a concept would consist in the disposition

1. to have a certain mental experience when we think of it

and

2. to pass certain judgements (or have certain experiences of direct knowledge) about its relations to other concepts and to empirical objects.

The essence of this psychologistic view is that the object of mathematics is nothing but the psychological laws by which thoughts, convictions, and so on occur in us, in the same sense as the object of another part of psychology is the laws by which emotions occur in us. The chief objection to this view I can see at the present moment is that if it were correct, we would have no mathematical knowledge whatsoever. We would not know, for example, that $2 + 2 = 4$, but only that our mind is so constituted as to hold this to be true, and there would then be no reason whatsoever why, by some

37   other train of thought, we should not arrive at the opposite | conclusion with the same degree of certainty. Hence, whoever assumes that there is some domain, however small, of *mathematical* propositions which we *know* to be true, cannot accept this view.[i]

38 | I am under the impression that after sufficient clarification of the concepts in question it will be possible to conduct these discussions with mathematical rigour and that the result then will be that (under certain assumptions which can hardly be denied [in particular the assumption that there exists at all something like mathematical knowledge]) the Platonistic view

---

[i]The rest of this page and part of the next are crossed out in Gödel's manuscript.

is the only one tenable. Thereby I mean the view that mathematics describes a non-sensual reality, which exists independently both of the acts and [of] the dispositions of the human mind and is only perceived, and probably perceived very incompletely, by the human mind. This view is rather unpopular among mathematicians; there exist, however, some great mathematicians who have adhered to it. For example, Hermite once wrote the following sentence:

| Il existe, si je ne me trompe, tout un monde qui est l'ensemble   39
des vérités mathématiques, dans lequel nous n'avons accès que
par l'intelligence, comme existe le monde des réalités physiques;
l'un et l'autre indépendants de nous, tous deux de création
divine.[30] [There exists, unless I am mistaken, an entire world
consisting of the totality of mathematical truths, which is accessible to us only through our intelligence, just as there exists the
world of physical realities; each one is independent of us, both
of them divinely created.]

---

[30]See *Darboux 1912*[, p. 142]. The passage quoted continues as follows:

qui ne semblent distincts qu'à cause de la faiblesse de notre esprit, qui ne
sont pour une pensée plus puissante qu'une seule et même chose, et dont
la synthèse se révèle partiellement dans cette merveilleuse correspondance
entre les Mathématiques abstraites d'une part, l'Astronomie et toutes les
branches de la Physique de l'autre. [and appear different only because
of the weakness of our mind; but, for a more powerful intelligence, they
are one and the same thing, whose synthesis is partially revealed in that
marvelous correspondence between abstract mathematics on the one hand
and astronomy and all branches of physics on the other.]

So here Hermite seems to turn toward Aristotelian realism. However, he does so only figuratively, since Platonism remains the only conception understandable for the human mind.

Introductory note to *1953/9

In 1953 Gödel accepted the invitation of Paul Arthur Schilpp, the editor of the series *The Library of Living Philosophers*, to write a paper for Schilpp's projected volume on Rudolf Carnap. This would have been Gödel's third contribution to the series, after *1944* on Russell and *1949a* on Einstein. Gödel continued working on the paper over the next six years. He wrote Schilpp in 1954 that the paper was essentially done, but he wanted to add some paragraphs on Carnap's more recent work, in 1955 that it would be done soon, in 1956 that it would be done "within two weeks", and in 1957 that, as he was reducing the size of the paper to one-third or one-quarter, he should be able to get it done "very soon". In the end, in a letter of 3 February 1959 answering a final request, Gödel told Schilpp that he did not wish to publish it. It was too late to allow Carnap to frame a reply, and Gödel felt it would be neither fair nor "conducive to an elucidation of the situation" to publish his paper without one. Gödel went on to express deeper reservations that prevented him from submitting the paper earlier:

> The fact is that I have completed several different versions, but none of them satisfies me. It is easy to allege very weighty and striking arguments in favor of my views, but a complete elucidation of the situation turned out to be more difficult than I had anticipated, doubtless in consequence of the fact that the subject matter is closely related to, and in part identical with, one of the basic problems of philosophy, namely the question of the objective reality of concepts and their relations. On the other hand, in view of widely held prejudices, it may do more harm than good to publish half done work.

Here Gödel is displaying his characteristic caution (see *Feferman 1984a*), as well as his overestimation of the extent to which positivist dogmas— what he means by "widely held prejudices"—remained orthodoxy by 1959 (cf. these *Works*, Volume II, page 166).

Despite his reservations, Gödel preserved six versions of the paper, labelling them Fassung I–Fassung VI, and included the paper on one of the lists "Was ich publizieren könnte" (see the preface to this volume). It may have been to his mind work half done, but there is no reason to think he would have repudiated any of what he said.

The relations of the versions are reasonably straightforward. I is a manuscript. II is a typed copy of that manuscript, with many revisions and additions in Gödel's handwriting and some additional typed sheets. III is a typescript of the revision, with further revisions and additions in handwriting and additional typed sheets. As III is the richest in content of the six versions, it is published below. In IV, which is handwritten and only 60 percent as long, Gödel pares and to some extent reorganizes the material of III. V and VI are yet shorter descendants of IV, each about one-quarter the length of III, containing concise formulations of the central argument of the earlier versions. Both are typescripts—in part the same typescript, in part different ones—with handwritten revisions. There is evidence that Gödel's numbering is mistaken, and that V is later than VI: some revisions handwritten on VI are part of the typescript underlying V, while nothing of the reverse is true. (However, there are also handwritten revisions in VI that are not typed in V, so clearly V was not obtained *from* VI.) The content of the two versions overlaps almost completely, but in many places the language or argument in V is clearer than in VI. V is published below.

Throughout all versions, there is no substantive variation in Gödel's position. The only differences are ones of formulation and of the inclusion of additional arguments, attention to various side issues, considerations of objections, and the like.

1

Gödel's aim in all versions of "Is Mathematics Syntax of Language?" is to give a direct, comprehensive, and precise criticism of the "linguistic" accounts of the foundations of mathematics developed by the logical positivists of the Vienna Circle. The criticism, Gödel thinks, is conclusive; in VI he goes so far as to state that the theses of the linguistic conception "are refutable, as far as any philosophical assertion can be refutable in the present state of philosophy" (page 1). In framing parts of his argument, in particular to characterize what is wrong with the positivist conception, Gödel invokes aspects of his own contrasting view of the foundations of mathematics. As we have seen, it was his dissatisfaction with these positive remarks that led him to demur from publication.

Gödel's sharp opposition to logical positivism is not readily apparent in the works he published during his lifetime. To be sure, the mathematical realism he espouses in *1944*, *1947*, and, most forcefully, *1964* is not the view of a positivist. But in these writings it is never explicitly set against positivism. His formulation of realism in *1944*, that classes and mathematical concepts are "real objects ... existing independently of our definitions and constructions" (page 137), seems more directed at constructivist than positivist views. Indeed, the assertion might even be accepted by a positivist, under a suitable interpretation of the notion of existence (see, for example, *Carnap 1956*, page 21). The position of

1964 is more clearly inconsonant with a positivist approach, particularly the claim that there exists a faculty of mathematical intuition much like perception. Here too, though, no direct criticism of positivist views is made.

This opposition becomes explicit, however, in three papers in the *Nachlass*. First, in **1951*, Gödel criticizes the general view that "mathematics is only our own creation" (page 15). He then moves to a consideration of what he calls "the most precise, and at the same time most radical, formulation" of such a view, namely, that "which interprets mathematical propositions as expressing solely certain aspects of syntactical (or linguistic) conventions", and cites Hans Hahn and Rudolf Carnap (page 23). The criticisms that follow overlap to a considerable extent with those in **1953/9*, although they are not as fully developed. Finally, in **1961/?*, Gödel expresses his profound disagreement with positivistic ways of thinking about the world generally, not just in foundations of mathematics.

The positivists' conception of the nature of mathematics stems from ideas broached in Wittgenstein's *Tractatus* (*1921*). The Vienna Circle adopted Wittgenstein's distinction between factual truths and "truths" that are artifacts of the representational system—the latter identified with the sort of truth that holds come what may, no matter how the facts lie, and that consequently has no content. Moreover, they extended to all of mathematics Wittgenstein's idea that logical truths were of this sort.

> If I have succeeded in clarifying somewhat the role of logic, I may now be quite brief about the role of *mathematics*. The propositions of mathematics are of exactly the same kind as the propositions of logic: they are tautologous, they say nothing at all about the objects we want to talk about, but concern only the manner in which we want to speak of them. [*Hahn 1933*; translation from *Ayer 1959*, page 158]

It is Carnap who most fully developed a view of this sort, in his *Logical Syntax of Language* (*1934a, 1937*). (Gödel was familiar with this work; indeed, he read a draft of it in 1932 and gave Carnap many technical suggestions. See *Carnap 1937*, page xvi, and the Gödel–Carnap correspondence of 1932.) Carnap's central notion is that of a language, or, in his later terminology, a linguistic framework. A linguistic framework provides the logical relations of consequence and contradiction among propositions. The fixing of these logical relations is a precondition for rational inquiry and discourse. There are many alternative frameworks, many different logics of inference and inquiry. There can be no question of justifying one over another, since justification is an intra-framework notion; justification can proceed only given the logical relations that a framework provides. This pluralistic standpoint is expressed in Carnap's Principle of Tolerance:

> *In logic, there are no morals.* Everyone is at liberty to build up his own logic, i.e. his own form of language, as he wishes. All that is required of him is that ... he must state his methods clearly, and give syntactical rules instead of philosophical arguments. [*1937*, page 52]

Mathematics, then, consists of framework-truths: statements to whose acceptance any user of the linguistic framework is automatically committed. As such, mathematical truths do not describe or reflect any realm of fact; they are simply consequences of the decision to adopt one rather than another linguistic framework.

In *1953/9*, Gödel's concern is not with the details of Carnap's position, nor does he recognize differences between Carnap and Schlick or Hahn. "Rather my purpose is to discuss the relationship between syntax and mathematics from an angle which, I believe, has been neglected ..." (III, footnote 9). As we shall see, however, there are specific features of Carnap's view that do, arguably at least, affect the assessment of Gödel's criticism.

2

The heart of Gödel's criticism is an argument based on his Second Incompleteness Theorem. Gödel states, "a rule about the truth of sentences can be called *syntactical* only if it is clear from its formulation, or if it somehow can be known beforehand, that it does not imply the truth or falsehood of any 'factual' sentence." (III, §11; in a similar remark in V, page 3, he replaces " 'factual' sentence" by "proposition expressing an empirical fact".) This requirement will be met only if the rule of syntax is consistent, since otherwise the rule will imply all sentences, including factual ones. Hence, by the Second Theorem, mathematics not captured by the rule in question must be invoked in order to legitimize the rule, and so the claim that mathematics is solely a result of rules of syntax is contradicted.

This is a powerful argument, and Gödel invests some care in urging its force. He elaborates what would go wrong if the rule failed to have the conservativeness property, what kind of reasoning is needed to show consistency, and so on; he even considers a response that would rely on empirical evidence for consistency, rather than mathematical proof, despite the obvious uncongeniality of such a fallback to any logical positivist. However, an important presupposition of the argument is not made explicit. The argument depends on a realm of the "factual" or the "empirical" being available in advance, independently of and prior to the envisaged rules of syntax. For, as Gödel characterizes the "linguistic" view, first there are empirical sentences, which are true or false by virtue of facts in the world; mathematics is then added, by means of conventional syntactical rules. Gödel argues that the addition has to be known not to affect the empirical sentences given at the start; but to ascertain that requires more mathematics, and so there is a *petitio*.

The picture of the realm of the factual as fixed in advance of the linguistic rules can be seen in the ways Schlick and Hahn discuss logic and mathematics. Indeed, it underlies the positivist appropriation of Wittgenstein's word "tautology"; and so it can also be ascribed to Carnap in some early writings like *1930*. Consequently, Gödel's argument is very effective against the form of the linguistic conception of mathematics espoused in this period. Naturally, if *Logical Syntax of Language* is taken as just the technical working-out of this conception, then Gödel's argument works equally well against it. It appears that this is in fact how Gödel read *Logical Syntax*.

However, *Logical Syntax* can be read as presenting a different, more sophisticated position, signalled by Carnap's abandonment of the label "tautologous" and, most importantly, by the adoption of the Principle of Tolerance. In this view of Carnap, there is no notion of "fact" or "empirical world" that is given prior to linguistic frameworks. The rules of a language, and hence mathematics, have to be in place before sense can be made of such notions. That is, applied to *Logical Syntax*, Gödel's argument would presuppose a notion of empirical fact that transcends or cuts across different linguistic frameworks. However, as the Principle of Tolerance strongly suggests, it is central to the view of *Logical Syntax* that any such language-transcendent notion be rejected. Rather, the notion of empirical fact is given *by way of* the distinction between what follows from the rules of a particular language and what does not. Thus, on this reading Carnap's view undercuts the very formulation of Gödel's argument.[a]

As might then be expected, Carnap has little concern for consistency proofs in *Logical Syntax* (see *1937*, pages 129, 134). For Carnap, it is completely open for someone to propose an inconsistent linguistic framework. To be sure, such a framework will not be very useful, but this inexpediency is merely a pragmatic matter. Gödel cannot agree: he suggests that even Carnap would have to recognize an ontological distinction between consistent and inconsistent languages (V, page 10). Thus Gödel does not see the radical features Carnap's position has on the reading just suggested: he recognizes neither how pervasive is

---

[a]For a more detailed treatment of this point, see *Ricketts 1994*.

the relativity to linguistic frameworks that is claimed for notions like existence and factuality, nor how untrammeled the "liberty" in formulating frameworks that is expressed in the Principle of Tolerance. In this, Carnap took himself to be eliminating traditional philosophical oppositions like that of realism and idealism. But Gödel would surely dismiss such a position, taking it to deny obvious features of reality, and to amount to nothing more than extreme anti-realism.[b]

Now there is another way in which Carnap's treatment of mathematics can be charged with a *petitio*, without relying on the above argument based on the Second Incompleteness Theorem. Gödel mentions this, too, but does not make much of it. As has been noted, for Carnap the truths of mathematics are meant to be consequences of the adoption of a linguistic framework. "Consequence" here cannot be understood in a proof-theoretic sense, as Gödel's First Incompleteness Theorem shows that no deductive system will yield all mathematical truths. Rather, one needs the semantical notion of consequence in the metalanguage, and to define that one needs to make use of mathematics of some strength. In contrast, Gödel lays down, as a constraint on the syntactic approach, that "syntax" has to be finitary. Indeed, Gödel takes the point to be obvious: "The necessity of [this] requirement should be beyond dispute" (III, §18); for if nonfinitary reasoning is used, the program "is turned into its downright opposite: ... instead of justifying the mathematical axioms by reducing them to syntactical rules, these axioms (or at least some of them) are necessary in order to justify the syntactical rules" (III, §19).

This elementary point is completely convincing; what is interesting is that Carnap seems oblivious to it. He freely allows the introduction of nonfinitary syntactical notions (*1937*, page 165), and explicitly notes that his definition of mathematical truth for a formal language that includes classical mathematics requires means that outstrip what is formalizable in that language (*1937*, page 129). Carnap thus does not view the reduction of mathematics to syntax as providing a justification for mathematics; the identification of mathematics as framework-truths of the framework is not meant to legitimize them. Carnap could allow that, while mathematical truths come from syntactical rules, our recognition of particular truths or our trusting any particular formulation of what can be inferred from given syntactical rules requires more mathematics, or different mathematics, than that which those rules yield. (Of course, he would also assert, the additional or different mathematics we use is also the upshot of syntactical rules, albeit different ones.) In short, Carnap is not taking the clarification of the status of mathematics which

---

[b]See *Ricketts 1994* for a discussion of anti-realism and *Logical Syntax*.

Logical Syntax provides as addressing traditional foundational issues. Those issues are transformed, into questions of what can be done inside various linguistic frameworks or questions of what sort of frameworks are better for one or another purpose. As the Principle of Tolerance indicates, what remains of "foundations of mathematics" is done by describing, analyzing, and comparing different frameworks. Carnap's discussions of intuitionism (*1937*, §16), predicativity (§44), and logicism (§84) all exhibit this transformation.[c]

Gödel, on the other hand, always frames the issues in a foundational, epistemological fashion. In V, he even starts by identifying the syntactical view with the assertion that "mathematical intuition ... can be replaced by conventions about the use of symbols". Gödel is unquestionably correct in claiming that the positivists, including Carnap of *Logical Syntax*, do not provide a successful epistemological reduction of mathematics to syntax or conventions. On the interpretation just given, though, this task is simply not one that Carnap takes on.

There is a peculiar kind of standoff here. Carnap's position contains a circle, or, better, a regress: mathematics is obtained from rules of syntax in a sense that can be made out only if mathematics is taken for granted (in the metalanguage). Therefore, no full exhibition of the syntactical nature of mathematics is possible. This is not lethal, however, insofar as the structure of Carnap's views leaves no place for the traditional foundational questions that such an answer would certainly beg. But also, then, Carnap's position is not capable of convincing Gödel that a faculty of mathematical intuition is unnecessary, for Gödel's view is designed to address just the philosophical questions that Carnap discards. From Gödel's point of view, Carnap's position *is* viciously circular, or at best philosophically and mathematically empty.

3

The second principal object of Gödel's criticism is the positivist claim that the propositions of mathematics have no content. Gödel takes that claim to arise from, first, the recognition that mathematical propositions have no empirical content, in the sense that they hold irrespective of any empirical facts, and, second, the identification of content with empirical content. It is the latter step to which Gödel takes exception. "*Mathematical sentences have no content* only if the term 'content' is taken from the beginning in a sense acceptable only to empiricists and not well founded even from the empirical standpoint" (III, §5). Rather, Gödel

---

[c]This issue is discussed in greater detail in *Goldfarb and Ricketts 1992*.

holds, the notion of content has to include what he calls "conceptual content" as well, and mathematical propositions do have content of that sort (III, §34; V, page 9).

Gödel's first argument for the ill-foundedness of the positivists' notion rests on the impossibility of eliminating "mathematical intuition" that he takes himself to have shown, that is, the impossibility of giving an epistemological reduction of mathematics to syntax or conventions. As he puts it, "If the prima facie content of mathematics were only a wrong appearance, it would have to be possible to build up mathematics satisfactorily without making use of this 'pseudo' content" (III, §30; almost identical wording in V, page 5).

It is not easy to assess the force of this argument. The positivist denial of content to mathematics is based not simply on an identification of content with empirical content but on Wittgenstein's idea that content requires a contrast between what would be the case for a proposition to be true and what would be the case for it to be false. If a proposition is necessarily true, there is no such contrast, and hence it has no content. The positivists freely admit that it may not be a trivial matter to ascertain that a proposition is in fact contentless. (See *Carnap 1930* [page 143 in *Ayer 1959*] and *Hahn 1933* [page 159 in *Ayer 1959*].) In that sense a proposition may lack content but still require some epistemological grounding. Gödel's argument, then, is that if the grounding goes beyond a pure reduction, then it cannot be taken as showing that the proposition lacks content; rather, it must be viewed as a justification of the proposition itself, in which case, of course, the claim of lack of content must be abandoned.

The situation is a little clearer vis-à-vis *The Logical Syntax of Language*. There Carnap gives a technical definition, for any given language, of the content of a sentence. His definition has as a consequence that mathematical truths have null content and logically equivalent sentences have the same content (*1937*, page 175). In line with the Principle of Tolerance, Carnap could reply to Gödel's criticisms by inviting Gödel to give a technical definition of his own. For Carnap there would be no meaningful question of which definition is "really" correct. Rather, there are only questions of which precisely-defined notion reflects to a greater or a lesser extent some of the clearer uses of the informal notion of content that we think of ourselves as reconstructing. The answers will be pragmatic matters of degree rather than yes-or-no, and will be interest-relative. Given Carnap's transformation of epistemological questions in *Logical Syntax*, there is no reason to expect he would take the epistemological features Gödel adduces to be important in this regard, although of course he would also invite Gödel to argue otherwise. However, Gödel would reject this entire formulation of his point. In line with his conceptual realism, what is at stake for him is the actual constitution of a notion of content that does justice to the claims propositions make on an independent reality. Once again, there is a standoff: agreement is lacking even as to what the argument is about.

Gödel's other main argument against the positivists' identification of content with empirical content focuses on the relations between mathematical propositions and laws of nature. He points out that in the working of empirical theories, the contribution of mathematics cannot be separated from that of empirical laws. Thus he says, "Laws of nature without mathematics are exactly as 'void of content' . . . as mathematics without the laws of nature. The fact is that only laws of nature *together* with mathematics (or logic) have consequences verifiable by sense experience. It is, therefore, arbitrary to place all content in the laws of nature" (III, §34; see also V, page 8); and "For a certain kind of physical theory a new mathematical axiom . . . may lead to new empirically verifiable consequences exactly as a new law of nature" (V, page 9). He uses these and similar observations to support his claim that the positivist notion of content is deficient in that it "oblige[s] one to interpret objectively quite analogous situations differently" (V, page 2).

These sorts of arguments are close kin to those that were being formulated at about the same time by W.V. Quine as part of his attack on the analytic-synthetic distinction that is central to Carnap's system. (See, for example, *Quine 1936, 1951, 1960,* and *1960a*.) Now Quine takes the arguments to support the view that there is no philosophically interesting distinction to be made between mathematical truths and empirical ones: there are gradations of abstraction and of remove from the particularities of experience, but no sharp boundaries, and no qualitative differences or differences in type. Nor are there differences in principle between the objects of mathematics and the objects of physics, between numbers and molecules. Gödel, in contrast, takes the arguments to show only an *analogy* between empirical and mathematical laws. He insists on a sharp distinction between the two realms. "The syntactical point of view as to the nature of mathematics doubtless has the merit of having pointed out the fundamental difference between mathematical and empirical truth. This difference, I think rightly, is placed in the fact that mathematical propositions, as opposed to empirical ones, are true in virtue of the *concepts* occurring in them" (V, page 2). His disagreement with the positivists lies in his view of the realm of concepts as an independent reality, to which notions of object, fact and content apply just as much as they do to the empirical world. (See also V, page 9 and *1951*, page 30.)

Gödel does not lay out his view in any detail, but some features of it may be surmised. Although he is not explicit, there is some evidence that he uses "concept" not just for those like the concept of set and the concept of number, but also for what would more usually be called mathematical objects, like individual numbers and sets. (See, for example, III, §42 and footnote 45, and the end of III, §45.) If this is granted, then his statements of the role of concepts in mathematical truth become unproblematic expressions of his Platonism.

On the relationship between the conceptual realm and the physical world, Gödel says "with mathematical reason we perceive the most general (namely the 'formal') concepts and their relations, which are separated from space-time reality insofar as the latter is competely determined by the totality of particularities without any reference to the formal concepts" (III, §42). He also speaks of mathematical concepts as "referring to physical reality—to be more exact ... referring to combinations of things" (III, §34), and of "*concepts* in which we describe [physical] structures" (V, page 9). Thus, it appears, Gödel has a picture surprisingly like the one he ascribes to the positivists in his argument based on the Second Incompleteness Theorem (see section 2 above), namely, that the empirical world is fixed independently of mathematics. Mathematics applies to that independently determined physical structure only through the incorporation of mathematical concepts into the very terms in which the structure is described, e.g., in descriptions of combinations or series of things rather than of things individually. In that sense, mathematics enters by way of how the world is represented.

Although Gödel does not specify the "particulars" that, independently of mathematics, make up the physical world, it appears that he is not thinking of them as sense experiences. (See III, §42, where he speaks of "particular objects", and V, page 7). The following passage from *1964*, pages 271–272, would then mark something of a shift:

> Even our ideas referring to physical objects contain constituents qualitatively different from sensations or mere combinations of sensations, e.g., the idea of object itself .... Evidently the "given" underlying mathematics is closely related to the abstract elements contained in our empirical ideas.

The passage suggests that there is an interplay of empirical and purely conceptual elements in our notions of the physical world or physical structures, and so goes against the idea that space-time reality is "determined ... without any reference to formal concepts".[d]

Finally, on the question of how we gain knowledge of the mathematical realm, Gödel says little except that we do it by our faculty of mathematical intuition, which he also calls "mathematical reason" and

---

[d]It should be noted, too, that the view enunciated in III, §§41–42, is not expressed in any other of Gödel's writings of which I am aware.

sometimes simply "reason". He likens it to a physical sense (V, page 6), and suggests that it is not ordinarily considered to be an additional sense solely because its objects are different from those of the other senses (III, §42). He gives no further details about its structure; nor does he consider whether aspects of it are involved in other regions of our cognition, as would certainly be suggested by calling it "reason". Here perhaps it is most glaring that Gödel failed to arrive at the sort of "complete elaboration of the situation" that he sought. As his discussions of mathematical intuition in *1961/?* and in *1964* show, his thinking on the issue continued to evolve over the next few years.

Warren Goldfarb[e]

[e]I am indebted to Cheryl Dawson, John Dawson, Solomon Feferman, Charles Parsons and, especially, Thomas Ricketts for helpful comments and suggestions.

Is mathematics syntax of language?
(*1953/9–III*)

§1   Around 1930 R. Carnap, H. Hahn, and M. Schlick,[1] largely under the influence of L. Wittgenstein, developed a conception of the nature of mathematics[2] which can be characterized as being a combination of nominalism and conventionalism. It had been foreshadowed in Schlick's doctrine about implicit definitions.[3] Its main objective, according to Hahn and Schlick,[4]

---

[1]Cf. *Carnap 1935a*, p. 30; *Carnap 1934b*, *Hahn 1933*, *Carnap 1935b* and *Hahn 1935*; moreover, *Hahn 1930*, p. 96; *Hahn 1931*, p. 135; *Hahn 1933a*; *Schlick 1938*, p. 145, p. 222; *Wittgenstein 1922*.

[2]The terms "mathematical", "mathematics" throughout this paper are used as synonymous with "logico-mathematical", "logic and mathematics". Moreover, following Carnap, I am using the term "logic" as synonymous with "mathematics", but do not want to imply that no borderline between the two can be drawn. The term "axiom" is always used in the sense of "formal axiom or rule of inference or axiom rule", where an axiom rule is a procedure for producing infinitely many formal axioms.

[3]Cf. *Schlick 1918*, p. 30.

[4]Cf. *Hahn 1935*, pp. 13, 19 and *Schlick 1918*, p. 147.

was to conciliate strict empiricism[5] with the a priori certainty of mathematics. According to this conception (which, in the sequel, I shall call the syntactical viewpoint) mathematics can completely be reduced to (and in fact *is* nothing but) syntax of language.[6] I.e., the validity of mathematical theorems consists solely in their being consequences[7] of certain syntactical conventions about the use of symbols,[8] not in their describing states of affairs in some realm of things. Or, as Carnap puts it: *Mathematics is a system of auxiliary sentences without content or object.*[9]

[5]The tenet of empiricism in question evidently is that, in the last analysis, all knowledge is based on (external or internal) sense perceptions and that we do not possess an intuition into some realm of abstract mathematical objects. Since, moreover, because of the a priori certainty of mathematics, such a realm cannot be known empirically, it must not be assumed to exist at all. Therefore the objective of the syntactical program can also be stated thus: To build up mathematics as a system of sentences valid independently of experience, without using mathematical intuition or referring to any mathematical objects or facts.

[6]The formulation given in the papers cited reads: *Mathematics is syntax of language.* Since however mathematics is usually presented as a science of certain objects, about which certain propositions are asserted to be demonstrably true, the question is whether it can be reduced to or replaced by syntax, i.e., whether what is asserted in mathematics can be *interpreted* to be syntactical conventions and their consequences, and whether, on the basis of this interpretation, *the same conclusions as to ascertainable facts can be drawn*, if mathematical theorems are applied.

[7]Since the assertability of a mathematical theorem is a *logical* consequence of the syntactical rules and a few trivial general facts about physical symbols, and since, moreover, according to the syntactical viewpoint a logical consequence only repeats part of the content of the premises, one arrives at the conclusion that to assert the truth of a mathematical theorem in substance means nothing else but to repeat part of the conventions about the use of symbols. This shows most clearly the lack of objective content of mathematics on the basis of this view.

Viewing the situation from a different angle one may say that the semantical rules which determine under which circumstances a sentence of a certain form can be asserted, in certain limiting cases, have the consequences that a sentence can be asserted under *all* circumstances. This makes such sentences true, but void of content.

[8]E.g., according to *Hahn 1935*, p. 26, the laws of contradiction and of excluded middle express certain conventions about the use of the sign of negation.

[9]Cf. *Carnap 1935a*, p. 36; *Carnap 1935b*, p. 37. The whole passage reads as follows: "Wenn zu der Realwissenschaft die Formalwissenschaft hinzugefügt wird, so wird damit *kein neues Gegenstandsgebiet* eingeführt, wie manche Philosophen glauben, die den 'realen' Gegenständen der Realwissenschaft die 'formalen' oder 'geistigen' oder 'idealen' Gegenstände der Formalwissenschaft gegenüberstellen. *Die Formalwissenschaft hat überhaupt keine Gegenstände*; sie ist ein System gegenstandsfreier, gehaltleerer Hilfssätze." [In adjoining the formal sciences to the factual sciences *no new area of subject matter* is introduced, despite the contrary opinion of some philosophers who believe that the "real" objects of the factual sciences must be contrasted with the "formal", "*geistig*" or "ideal" objects of the formal sciences. *The formal sciences do not have any objects at all*; they are systems of auxiliary statements without objects and without content (translation from *Feigl and Brodbeck 1953*, p. 128).] I would like to say right here that Carnap today would hardly uphold the formulations I have quoted (cf. §45). Moreover some of them were given only by Hahn or Schlick, and probably would never have been subscribed to by Carnap. However, I am not concerned in this paper with a detailed

§2   The syntactical conventions concerned in this program are those by which
the use of some symbol a (or symbols a, b, etc.) is defined by stating rules
about the truth (or the assertibility) of sentences containing a (or a, b,
etc.), where these rules refer only to the outward structure of expressions,
not to their meaning, nor to anything else outside the expressions.

§3   Such rules, e.g., are: (1) For the symbol "2": If the sentence B contains
"2" in some place where A contains "$1 + 1$", and otherwise agrees with A,
then B is true if and only if A is true. (2) For the symbol "$=$": Every
sentence of the form $A = A$ is true. (3) For the symbol "$\exists$" (roughly
speaking): If some sentence of the form $\phi(a)$ is true then the sentence
$(\exists x)\phi(x)$ is true.

      The idea at the bottom of this interpretation of logical truth is the
following: The meaning of sentences is defined by (semantical) rules which
determine under which circumstances a given sentence can be asserted.
These rules, in certain limiting cases, may have the consequence that a
sentence can be asserted under *all* circumstances (e.g., it will rain or it will
not rain). This makes such sentences true, but void of content. Of course
rules especially referring to these limiting cases may then be formulated,
which is done in the axioms and rules of pure mathematics.

      Such rules may be called "syntactical" because they do not refer to
meaning and, therefore, assertions in conflict with them are excluded already because of their structure, exactly as assertions which do not conform
to the rules of grammar.

§4   In his *Logical Syntax of Language* ([[1937,]] pages 102–129) Carnap has
carried this program out. Another method of carrying it through can be
derived from a paper by F. P. Ramsey.[10] Finally much of the work of
the Hilbert school about the formalization and consistency of mathematics
can be interpreted to be a partial elaboration of this view,[11] although the
authors of these papers, for the most part, favor different philosophical
opinions.

§5   All these developments no doubt are interesting from a technical point
of view, moreover they have contributed much to the clarification of some

---

evaluation of what Carnap has said about the subject, but rather my purpose is to discuss the relationship between syntax and mathematics from an angle which, I believe,
has been neglected in the publications about the subject. For, while the syntactical
program itself and its elaboration, as far as it is possible, have been presented in detail
the negative results as to its feasibility in its most straightforward and philosophically
most interesting sense have never been discussed sufficiently.

[10]Cf. *Ramsey 1926*, p. 338.

[11]Because the axioms and rules of inference of formal systems can be interpreted
to be syntactical rules in the sense of §2 which stipulate that: (1) All formulae of a
certain structure are true; (2) Formulae obtained from true formulae by certain formal
operations are also true. Moreover consistency will turn out to be the key problem also
for the syntactical viewpoint.

fundamental concepts. However, they prove that: (1) *Mathematics can be interpreted to be syntax of language* only if the terms "language" or "syntax" or "interpreting" are taken in a very generalized or attenuated sense, or if only a small part of what is commonly regarded as "mathematics" is acknowledged as such. And they prove that (2) *Mathematical sentences have no content* only if the term "content" is taken from the beginning in a sense acceptable only to empiricists and not well founded even from the empirical standpoint. Thereby these results become unfit to serve the aforementioned purpose of the syntactical program or the support of the philosophical views in question (such as nominalism or conventionalism). On the other hand, if the terms occurring are taken in their ordinary sense (which also is the one required by the objective of the syntactical program and the philosophical questions involved), then assertion 1, except for a small section of mathematics, is disprovable. As to assertion 2, the examination of the syntactical viewpoint, perhaps more than anything else, leads to the conclusion that there *do* exist mathematical objects and facts which are exactly as objective (i.e., independent of our conventions or constructions) as physical or psychological objects and facts, although, of course, they are objects and facts of an entirely different nature. And this is true even for those sections of mathematics which *can* be reduced to syntax in the *strict* sense of the term. Therefore the view that physics and other empirical sciences describe some "realm of things", while mathematics does not (cf. the difference between "Realwissenschaft" and "Formalwissenschaft" in the passage quoted in footnote 9) seems hardly tenable.

§6   Starting with the first item, assertion 1 above, I shall now give a list of the meanings of the terms occurring in it, as they seem to me to be required by the objective of the syntactical program and the philosophical questions involved.

§7   1. Since the syntactical program aims at dispensing with mathematical intuition without impairing the usefulness of mathematics for the empirical sciences, it will have to be required of a satisfactory elaboration that mathematics is covered to the full extent to which it can be used in the empirical sciences (in particular for deriving verifiable consequences from laws of nature). This however is the case for all classical mathematics. Nor can certain parts of classical mathematics, such as the theory of the continuum, be discarded as belonging to physics. For they imply number-theoretical propositions without existential quantifiers, i.e., propositions whose purely mathematical character cannot be contested (cf. §25 and footnote 26). Therefore "mathematics" will have to mean classical mathematics. But nothing is changed in the results stated in the sequel if it means intuitionistic mathematics. In both cases mathematics (in assertion 1) must be considered to be a system of propositions knowable to be true. For the question (see §1 and footnotes 5 and 6) is exactly whether the intuitive content[12] of mathematics can be disregarded and nevertheless the theorems of mathematics be asserted and applied.[13]

§8   2. "Language" will have to mean some symbolism which can actually be exhibited and used in the empirical world. In particular it will have to be required that its sentences consist of a finite number of symbols. For sentences of infinite length, since they do not exist and cannot be produced in the empirical world, evidently are purely mathematical objects. Thus such objects, instead of being avoided, would be assumed right from the beginning.

§9   3. For the same reason it will have to be required of the *rules of syntax* that they be "finitary", i.e., that they must not contain phrases such as: "If there exists an infinite set of expressions with a certain property", nor even: "If all expressions of a certain infinite set have a certain property". For these phrases have no meaning unless use is made of some intuition for mathematical objects, such as the totality of all possible expressions, or the concept of a correct proof (in the sense of footnote 20), or the mathematical objects mentioned and their properties are introduced by special assumptions.[14]

---

[12]The existence, as a psychological fact, of an intuition covering the axioms of classical mathematics can hardly be doubted, not even by adherents of the Brouwerian school, except that the latter will explain this psychological fact by the circumstance that we are all subject to the same kind of errors if we are not sufficiently careful in our thinking.

[13]In the sense that not the theorems of classical mathematics, but only their derivability from certain axioms, is considered to be mathematical truth. K. Menger in his *1930a*, p. 324, basing this interpretation on a strictly formal concept of derivability, has extended it to *all* mathematics and has called this conception of mathematics "implicationism". However, thus interpreted, mathematics loses its applicability (cf. §§13–16) unless the consistency of the axioms is known, which however cannot be known on the basis of this standpoint. For a proposition stating the consistency of a system of axioms is not of the form: *B* follows from the axioms *A*. Therefore, if only implicationistic mathematics could be interpreted syntactically (which, of course, is possible since it *is* syntax), mathematics as to its applications could not.

[14]Carnap, to the objection that Platonism is implied by transfinite rules, replies (*1937*, p. 114) that one may know how to handle the transfinite concepts (in inferences, definitions, etc.) without making any metaphysical assumptions about the objective existence of the abstract entities concerned. This, of course, is true in the same sense as one also may know how to handle the concepts of physical objects without ascribing to these objects any existence in a metaphysical sense. But nevertheless, before one can rationally use them in science, he must assign to them existence at least in some immanent (Kantian) sense, in order to distinguish them from the objects assumed in some wrong (i.e., disprovable) physical theory. The same, therefore, applies to the transfinite mathematical entities, whose existence also can be disproved, namely, by an inconsistency derived from them. Hence, if mathematics is based on transfinite syntax, it is implied that there are specific mathematical entities which exist as objectively as physical entities. But this is exactly what the syntactical viewpoint denies (cf. fn. 9). To finitary rules evidently this argument does not apply, because the objects to which they refer may be looked upon as parts of the physical world (either existing already or producible).

§10　Of course the occurrence of the phrases mentioned also contradicts the ordinary concept of a syntactical rule, which requires that the rule should refer only to the structures of finite expressions.

§11　4. Moreover a rule about the truth of sentences can be called *syntactical* only if it is clear from its formulation, or if it somehow can be known beforehand, that it does not imply the truth or falsehood of any "factual" sentence (i.e., one whose truth, owing to the semantical rules of the language, depends on extralinguistic facts). This requirement not only follows from the concept of a convention about the use of symbols, but also from the fact that it is the lack of content of mathematics upon which its a priori admissibility in spite of strict empiricism is to be based. The requirement under discussion implies that the rules of syntax must be demonstrably consistent, since from an inconsistency *every* proposition follows,[15] all factual propositions included. Cf. also §19.

§12　5. According to what was said in the end of footnote 6, the phrase "Mathematics can be *interpreted* to be syntax of language" will have to mean: (1) that the formal axioms and the procedures of proof of mathematics can be deduced from suitably chosen rules of syntax, and (2) that the conclusions as to ascertainable facts which are obtained by applying mathematical theorems and which formerly were based on the intuitive truth of the mathematical axioms can be justified by syntactical considerations.[16] The second item, however, again requires a consistency proof for the syntactical rules, as can be seen from the following examples:

§13　If mathematics is interpreted to be a system of objectively true propositions, then, e.g., on the grounds of a proof for Goldbach's conjecture (which says that every even number is the sum of two primes), it can be predicted that a computing machine which is empirically known to work reliably will find two primes whose sum is some given large number N. In order to make the same prediction (for every N), if mathematics is interpreted syntactically, the consistency of the rules of syntax must be known. For a failure of the machine to furnish such a decomposition of N would entail the existence of a formal disproof of Goldbach's conjecture, provided that the formalism of finitary arithmetic follows from the rules of syntax.

---

[15]This holds both for classical and intuitionistic mathematics and, therefore, also for the syntactical rules replacing them (cf. §12). Note that under rather general assumptions about the logic chosen, consistency of the syntactical rules vice versa implies their compatibility with all possible sense experiences that can have occurred at any time.

[16]If mathematics is to be interpreted syntactically only as to its applications and only in the restricted sense that *each individual* observable fact which is a mathematical consequence of laws of nature, or of other observable facts and empirical induction, should also follow on the basis of the syntactical interpretation, then it is not necessary that the mathematical axioms themselves be derivable from the rules of syntax, since the same individual consequences as to observables may follow also with the help of different general concepts or axioms. As to this approach cf. fn. 42.

On the other hand, from the mere fact that Goldbach's conjecture follows from some arbitrarily assumed rules for handling certain symbols (even if they imply the formalism of finitary arithmetic) nothing whatsoever can be concluded about the result the machine will yield.

§14   The situation is quite similar for the prediction (on the basis of empirically known physical laws) that a bridge constructed in a certain manner will not break under a certain load, the parts of the bridge playing the same role as the elements of the computing machine.

§15   Similarly also in pure mathematics, if it is interpreted syntactically, one needs a consistency proof in order to draw the usual conclusions from mathematical theorems, e.g., in order to conclude from a proof of Goldbach's conjecture that a finitary procedure for decomposing an even number N into two primes will yield a positive result for every N. But this is exactly the question one is interested in if he wants to solve Goldbach's problem. (Note that the item discussed in this paragraph, viewed from the syntactical viewpoint, is not substantially different from that discussed in §13, since the activity of the mathematician is asserted in fact to be nothing but manipulation of objects of space-time reality [physical or psychical] by certain rules with a view to arrive at certain results.)

§16   Of course there is no question that by applying mechanically certain rules for handling symbols one arrives at the same sentences as by using mathematical intuition[17] (e.g., also at those sentences which express the two predictions and the mathematical conclusion dealt with in §§13–15). However, mathematical intuition in addition produces the conviction that, if these sentences express observable facts and were obtained by applying mathematics to verified physical laws (or if they express ascertainable mathematical facts), then these facts will be brought out by observation (or computation). Therefore syntax, if it is to be an acceptable substitute for mathematical intuition, must also yield sufficient reason for this expectation, and for this purpose a consistency proof is necessary. Mathematical intuition, it is true, is not acknowledged as a source of knowledge by proponents of the syntactical viewpoint. Therefore mathematics, as far as *by them* it is held to be well founded, can without difficulty be interpreted in terms of, or be replaced by, syntax.[18] However the original purpose and the chief interest of the syntactical interpretation refer to the question as

---

[17]Except that *all* mathematical intuition demonstrably cannot be expressed in *one* finite system of intuitively evident axioms. (Cf. the second paragraph of fn. 36 and fn. 43).

[18]In fact, according to the syntactical conception, intuitive mathematics itself, in truth, *is* only syntax of our thinking, where the mental images take the place of the symbols and the rules for their connection are afforded by (acquired or innate) associations between them. This is implied by the denial of the existence of objects corresponding to the mathematical thinking processes.

to whether (in particular in the applications of mathematics) it can *replace the belief in the correctness of mathematical intuition.*

§18   6. For the reasons stated in §§8–10 (i.e., in order not to substitute for intuitive mathematics a "syntax" as unacceptable for empiricists as intuitive mathematics itself) it will have to be required that: (1) not only in the rules of syntax, but also in the derivation of the mathematical axioms from them and in the proof of their consistency, only syntactical concepts in the sense of §§9–10 be used (i.e., only finitary concepts referring to finite combinations of symbols) and (2) only procedures of proof which cannot be rejected by anyone who knows how to apply these concepts at all.[19]

The necessity of the first requirement should be beyond dispute.

§19   For, if mathematical intuition and the assumption of mathematical objects or facts is to be dispensed with by means of syntax, it certainly will have to be required that the use of the "abstract" and "transfinite" concepts[20] of mathematics, which cannot be understood or used without mathematical intuition or assumptions about their properties, be based on considerations about finite combinations of symbols. If, instead, in the formulation of the syntactical rules some of the very same abstract or transfinite concepts are being used—or in the consistency proof, some of the axioms usually assumed about them—then the whole program completely

---

[19]I believe that what must be understood by "syntax", if the syntactical program is to serve its purpose, is exactly equivalent to Hilbert's "finitism", i.e., it consists of those concepts and reasoning, referring to finite combinations of symbols, which are contained within the limits of "that which is directly given in sensual intuition" ("das unmittelbar anschaulich Gegebene"). Cf. *Hilbert 1926*, pp. 171–173. The section of our knowledge thus defined is equivalent (by a one-to-one correspondence of its objects) with recursive number theory (cf. *Hilbert and Bernays 1934*, pp. 20–34 and pp. 307–346), except that it may rightly be argued (cf. *Bernays 1935*, p. 61) that combinatorial objects with an exorbitant number of elements or an exorbitant number of operations to be performed must not occur in finitary considerations. For these are purely theoretical constructions which are as far apart from the "immediately given" as the infinite and perhaps even go far beyond everything contained in space-time reality. Therefore the ideas of these objects or operations cannot be known to be meaningful or consistent unless we trust some mathematical intuition of things completely inaccessible to experience. If restrictions of this kind are introduced (which does not exclude that a theorem may refer to *all* integers or combinatorial objects of a certain kind that can be concretely given), then the negative results as to finitary consistency proofs mentioned in the sequel remain valid (cf. fn. 22) and stronger ones probably can be obtained (cf. §27).

[20]Abstract concepts, e.g., are: "proof" and "function", if these terms are understood in their original "contensive" meaning, i.e., if "proof" does not mean a sequence of expressions satisfying certain formal conditions, but a sequence of thoughts convincing a sound mind, and if "function" does not mean an expression of the formalism, but an understandable and precise rule associating mathematical objects with mathematical objects (in the simplest case integers with integers). An example for a transfinite (i.e., non-constructive) concept is "there exists", if this phrase means object existence irrespective of actual producibility. "Infinite set" is an abstract or transfinite concept according to whether potential or actual infinity is meant. The abstract and transfinite concepts together form the class of "non-finitary" concepts.

changes its meaning and is turned into its downright oppposite: instead
of clarifying the meanings of the non-finitary mathematical terms by explaining them in terms of syntactical rules, non-finitary terms are [used]
in order to formulate the syntactical rules; and, instead of justifying the
mathematical axioms by reducing them to syntactical rules, these axioms
(or at least some of them) are necessary in order to justify the syntactical
rules (as consistent).

As to the second half of requirement 6 it may be argued that, although
transfinite mathematical axioms clearly must not be used, it *is* permissible
to use empirical induction. E.g., consistency might be based on the fact
that no contradiction has arisen so far. Now it is true that, if consistency
is interpreted to refer to the handling of physical symbols, it is empirically
verifiable like a law of nature. However, if this empirical consistency is used,
mathematical axioms and sentences completely lose their "conventional"
character, their "voidness of content" and their "apriority" (in the sense of
footnote [7]) and rather become expressions of empirical facts. This can
be seen as follows:

If a syntactical convention is permissible only on the grounds of certain empirical facts, it frequently may in turn be used for deriving those
facts (or part of them). If, e.g., some operation is empirically known to
be associative, certain brackets may, by convention, be dropped. But from
this convention the associativity of the operation, i.e., an empirical fact,
can be derived. Similarly, if on the basis of its empirically known consistency C, some mathematical convention R is added to some system S of
mathematics and empirical science, then, though not C itself, still the empirical consistency C' of some slightly weaker convention will, in general, be
demonstrable in $S + R$. This refutes the argument that the mathematical
"conventions", although factual knowledge about symbols may be necessary for setting them up, do not *express or imply* any facts. In truth these
"conventions" are "void of content" only insofar as they *add* nothing to the
mathematics which, before they can be made, must be known already in
an empirical attire (i.e., mixed with synthetic facts).

§21   It might be said that substantially the same objection applies if a finitary consistency proof (in the sense of §18 and footnote 19) can be given,
since also in that case the proof has to make use of certain facts concerning combinations of symbols. Now it is true that to make any conventions
presupposes a knowledge of certain facts, because, e.g., it must be known
that the envisaged convention does not imply any new factual propositions.
Therefore, strictly speaking, one can speak of conventions only *relative to
a certain system of knowledge*, in the sense that they don't *add* anything
to this system, although they presuppose and possibly imply propositions
of this system and, therefore, considered in themselves *may* have content.
If one speaks of conventions and their voidness of content in an absolute
sense, this can only mean that they are conventions relative to that body

of knowledge which is indispensable for making any linguistic conventions at all. But this is exactly finitary combinatorics as defined in ⟦§18⟧ and footnote 19, provided the delimitation is not to separate things which evidently are of the same nature. But it must be kept in mind that, strictly speaking, not even such conventions are void of content.

In view of the reasons given in §§⟦7–19⟧ it can be said that *only if mathematics could be interpreted to be syntax in accordance with the requirements 1–6 could this fact be used in support of nominalism and conventionalism against realism.* For only in that case could a satisfactory foundation of mathematics be given independently of experience and without using mathematical intuition (see footnote ⟦5⟧), except, of course, that kind of experience or intuition which is necessary for making any linguistic conventions at all.

§23   Now, do the elaborations of the syntactical viewpoint which have actually been given satisfy the requirements 1–6? By no means. Ramsey's ideas necessitate admitting propositions of infinite (and even non-denumerable) length. Carnap uses non-finitary syntactical rules and arguments. Formalism under the requirements 2–6 has yielded a syntactical foundation only for a small part of mathematics. Only if one of the two requirements 5, 6 is substantially weakened (cf. the end of §24 and footnote 42) is there some hope that it may yield a foundation for all mathematics.

§24   Now the question arises whether this failure applies only to the particular attempts that have been made or has deeper reasons. The answer to this question depends on the precise extension of "finitary combinatorial reasoning". Now an explicit definition of this concept has not been given yet. But, in view of the work by G. Gentzen,[21] there can be no doubt that all such reasoning can be expressed in the formalism of classical number theory. Since, however, owing to a general theorem, a consistency proof for a system containing primitive recursive number theory can never be expressed in this system, it follows that not even classical number theory, still less any more comprehensive systems, can be proved consistent by finitary reasoning.[22] The fact that, in addition to the concepts directly referring to

---

[21]On the occasion of G. Gentzen's consistency proof for number theory (cf. *Gentzen 1938a*) it was ascertained up to which ordinal number definitions and proofs by transfinite induction can be expressed in the formalism of classical number theory. (Cf. *Hilbert and Bernays 1939*, pp. 360–374.) Thereby it became evident that those which cannot be so expressed are not finitary, while, on the other hand, all finitary proofs can be represented as inductions with respect to certain ordinal numbers.

[22]It may be argued that for all applications consistency up to some very great, but finite, number N of simple proof steps is sufficient and that, moreover, from a strictly empirical standpoint the question of consistency is meaningful only up to some number N of proof steps which is not absolutely beyond what can somehow be realized in our world. However, also in this weakened sense consistency cannot actually be proved by finitary reasoning, because the length of a finitary consistency proof in this sense would

combinations of symbols, certain abstract concepts[23] and evident propositions about them must be admitted, in order to carry through Hilbert's
ideas, was recognized also by leading proponents of formalism.[24]

§25   So, in carrying through the syntactical program, *the requirements 1–6
cannot be satisfied simultaneously,* nor even the requirements 1–3, 5 (part
2), 6; i.e., *the axioms about abstract transfinite concepts used in mathematics cannot be replaced by finitary considerations about combinations of
symbols and their properties and relations,* where "replacing" means that
the same consequences as to ascertainable facts (to be more exact, the
same universal propositions about them[25]) can be derived in either case,
and "ascertainable fact" here means any numerical equality for computable
functions (such as $5 + 7 = 12$) or equivalent combinatorial relations.[26]

§26   This holds no matter whether "mathematics" is understood to mean classical or intuitionistic or constructivistic mathematics, or even intuitionistic
number theory. Only if "mathematics" itself is confined to some level[27]
of finitary combinatorial reasoning, or contains non-finitary concepts and
their axioms only with artificial restrictions, or does not contain primitive
recursive number theory, can the syntactical program be carried through
under the requirements 2–6.

§27   Of these positive results the one covering first-order logic and its applications to finitary and empirical systems is the most interesting. For it shows
that the theory of quantification and of the propositional connectives (i.e.,
logic proper) actually *can* be interpreted syntactically[28] (although not al-

---

have to be of at least the same order of magnitude as N. Whether for each N there
actually exist such limited consistency proofs of this length is an interesting question
(cf. fn. 42). [Note that, in order that the length of a proof (i.e., the number of simple
proof-steps) be a true measure of the degree of complication, and for other reasons, it
must be stipulated that each definition and each assertion occurring in a proof contain
fewer than K symbols, where K is chosen once and for all.]

[23] Cf. fn. 20.

[24] Cf. *Bernays 1941a*, pp. 144, 147; *1935*, pp. 68, 69; *1935b*, p. 94; *1954*, p. 2;
also, *Gentzen 1937*, p. 203. In the more recent papers of the formalistic school the
term "finitary" has been replaced by "constructivistic", in order to indicate that it is
necessary to use certain parts of intuitionistic mathematics which are not contained
within the limits of that which is directly given in sensual intuition.

[25] Thereby I mean universal propositions all special instances of which are ascertainable facts, such as, e.g., Goldbach's conjecture.

[26] E.g., the statement that the formalized system of classical number theory is consistent is a universal proposition about ascertainable facts (concerning individual proof
figures) which can be proved from transfinite or abstract axioms, but not finitarily.

[27] Finitary mathematics can be split up into a hierarchy of levels in such manner that
the lower levels can be proved consistent in the higher ones.

[28] That nevertheless classical number theory cannot finitarily be proved consistent is
due to the fact that it is *more* than primitive recursive number theory with first-order
logic applied to it. For it also contains axioms which belong to neither one of these two
theories, but are true only in virtue of the meanings of both number-theoretical *and*
logical terms, namely, complete induction applied to quantified expressions.

ways in the strict sense of "syntax" explained in the end of footnote 19). The positive results as to finitary mathematics are of little interest, because even larger sections of finitary mathematics than those contained in the systems (or theories isomorphic to the former) are necessary for the consistency proofs. The circularity involved in founding finitary mathematics on syntax would disappear to a considerable extent if in this foundation only syntax in the strict sense (cf. footnote 19) were used. But this, very likely, is impossible, i.e., it probably is demonstrable that the consistency of exorbitantly great integers, even up to a moderate number of proof steps, cannot be proved in a feasible number of steps without using exorbitantly great integers in the (arithmetized) consistency proof. Therefore it would seem that, strictly speaking, not even finitary mathematics can be interpreted to be syntax of language.

§28   In general the concepts and axioms occurring in the section of mathematics proved consistent need not all occur among, or be derivable from, those sufficient for a formalized consistency proof. There exist certain possibilities of replacing some of them by others.[29] But the very fact that, if some of the essential concepts or axioms of the system are not contained in the system used for the consistency proof, others equally powerful (and equally problematic, if mathematical intuition is not trusted) must be present, reveals a principle which could perhaps be called the *non-eliminability of the mathematical content of an axiomatic system by the syntactical interpretation.* The phrase "equally powerful" can be made precise in several ways, e.g., thus: The axioms used in the consistency proof must have at least the same (in fact even a slightly greater) demonstrative power within finitary combinatorics (or finitary number theory) than those proved consistent; or thus: It must be possible, by means of the concepts and axioms used in the consistency proof, to construct a "model" for those proved consistent,[30] i.e., to define concepts demonstrably satisfying the given axioms and not satisfying any proposition disprovable from them.[31] So in this slightly weakened

---

[29] E.g., the non-constructive quantifiers and propositional connectives within number theory, roughly speaking, can be replaced by the concept of "intuitionistic number-theoretical proof", or by that of "computable function of finite type", or by that of "ordinal number $< \epsilon_0$". Also the general concept of "set" can be replaced by that of "ordinal number" in conjunction with that of "recursive function of ordinal numbers". The concept of "integer" (with the axiom of complete induction) can be replaced by the concept of "set" (and its axioms). Finally it is not impossible that the non-constructive concepts of mathematics can be replaced by constructive ones, provided "constructivity" is taken in a sufficiently wide sense.

[30] That this follows from a consistency proof was formulated as a conjecture by Carnap himself in the discussion at the *Deutsche Naturforschertagung* in Königsberg, 1930. (Cf. *Carnap 1931*, p. 143.)

[31] Without the second requirement a model always exists, even for inconsistent systems (on account of the fact that the sign of negation need not be interpreted by negation). But such models are of little interest, because they cannot replace the axioms under consideration as to their application in other fields or in empirical science.

sense the axioms used do imply those proved consistent. Finally also the above-mentioned theorem which says that the consistency of a system S of axioms containing arithmetic cannot be proved in any subsystem of S belongs to this order of ideas.

§29   On the grounds of these results it can be said that *the scheme of the syntactical program to replace mathematical intuition by rules for the use of symbols fails because this replacing destroys any reason for expecting consistency, which is vital for both pure and applied mathematics, and because for the consistency proof one either needs a mathematical intuition of the same power as for discerning the truth of the mathematical axioms or a knowledge of empirical facts involving an equivalent mathematical content.*

§30   This formulation of the non-feasibility of the syntactical program (which also applies to finitary mathematics) is particularly well suited for elucidating the question as to whether mathematics is void of content. For, if the prima facie content of mathematics were only a wrong appearance, it would have to be possible to build up mathematics satisfactorily without making use of this "pseudo" content.

§31   More precisely the situation can be described as follows: That mathematics *does* have content (in any acceptable sense of the term) appears from the fact that, in whatever way it, or any part of it, is built up, one always needs certain undefined terms and certain axioms (i.e., deductively unprovable assertions) about them.[32] *For these axioms there exists no other rational (and not merely practical) foundation except either that they*

---

[32] The question as to the existence of a content and as to the necessity of axioms (in the sense explained), of course, refers to mathematics as a system of sentences knowable or posited to be true, not as a hypothetico-deductive system. Some body of unconditional mathematical truth must be acknowledged, because, even if mathematics is interpreted to be a hypothetico-deductive system, still the propositions which state that the axioms imply the theorems must be unconditionally true.

The field of unconditional mathematical truth is delimited very differently by different mathematicians. At least eight standpoints can be distinguished. They may be characterized by the following catchwords: (1) classical mathematics in the broad sense (i.e., set theory included), (2) classical mathematics in the strict sense, (3) semi-intuitionism, (4) intuitionism, (5) constructivism, (6) finitism, (7) restricted finitism (cf. fn. 19), (8) implicationism (cf. fn. 13).

However, the conclusion that mathematics has content holds no matter which standpoint is taken. It even can be said that, if the most restricted standpoint (implicationism) is taken, it can be seen most easily that mathematical theorems express objective facts. For the singular combinatorial facts concerned are unequivocally ascertainable relations between the primitive terms of combinatorics, such as: "pair", "equality", "iteration", and they can least of all be eliminated by basing the use of those terms on conventions. For in order to apply (and know the consistency of) these conventions one needs an intuition or an empirical knowledge of facts involving the same or isomorphic concepts. On the other hand, to interpret the combinatorial facts to be nothing but physics of symbols amounts to an unjustifiable renunciation of an analysis of the situation and a return to Mill's conception of mathematics.

(or propositions implying them) can directly be perceived to be true (owing to the meaning of the terms or by an intuition of the objects falling under them), or that they are assumed (like physical hypotheses) on the grounds of inductive arguments, e.g., their success in the applications.[33] The former case would seem to apply at least to some mathematical axioms, e.g., the modus ponens and complete induction.[34] In the latter case (as far as it may be judged actually to occur) the mathematical character of the axioms, in spite of their inductive foundation, appears in the circumstance that they have consequences in that part of mathematics to which the former case applies, i.e., whose primitive terms have an immediately understandable clear meaning (e.g., the axioms of infinity mentioned in footnote 43 have number-theoretical consequences).

To eliminate mathematical intuition or empirical induction by positing the mathematical axioms to be true by convention is not possible. For, before any such convention can be made, mathematical axioms of the same power or empirical findings with a similar content are necessary already in order to prove the consistency of the envisaged convention. A consistency proof, however, is indispensable because it belongs to the concept of a convention that one knows it does not imply any propositions which can be falsified by observation (which, in the case of the mathematical "conventions", is equivalent with consistency[35] [cf. §11 and footnote 15]). Without a consistency proof the "convention" itself, since open to disproof, really

---

[33]By "success", within pure mathematics, of some non-evident mathematical axiom, I mean that many of its consequences can be verified on the basis of the evident axioms, the proofs however being more difficult, and that, moreover, it solves important problems not solvable without it. Note that also the consistency of such an axiom is indemonstrable on the basis of the evident axioms provided it solves any problem of Goldbach type.

[34]It seems arbitrary to me to consider the proposition "This is red" an immediate datum, but not so to consider the proposition stating modus ponens. For the difference only lies in the fact that in the first case a relation between an undefined concept and an individual object is perceived, while in the second case it is a relation between undefined concepts, and that, moreover, the second case relates to concepts of a different kind.

Note that such axioms, although they do not follow from definitions, nevertheless can be classed as analytic propositions. For "truth owing to the meaning of the undefined terms" has exactly the same significance for undefined terms that "truth owing to the definitions" has for defined terms.

Complete induction would seem to be an axiom (or a consequence of axioms) of the same kind. On the basis of the non-constructive standpoint this has been proved by Dedekind and Frege.

[35]Of course the consistency proof for the mathematical "conventions" cannot be said to consist simply in the fact that one can always maintain them, even after an inconsistency has arisen, by assuming that computational errors have been committed. For, if errors have been committed, the convention has not been applied, while what must be known before the convention can be made is that *if it is applied* no inconsistency will arise. Moreover it is no less false of the mathematical "conventions" than of laws of nature that we would retain them under all circumstances.

is an assumption (or else laws of nature could also be interpreted to be conventions).[36] Brought to its shortest form this proof runs as follows: If mathematical intuition is accepted at its face value, the existence of a content of mathematics evidently is admitted. If it is rejected, mathematical axioms become open to disproof and for this reason have content. (Cf. also §45.)

§34   It might be objected that this argumentation is anthropomorphic, because it is based on the possibility of a proof in a finite number of steps, and that, if the practical difficulty of running through infinite totalities is disregarded, mathematics *can* be founded on conventions and, therefore, *objectively* is void of content. The answer is this: If "practical" difficulties of this kind are disregarded, empirical science, too, can be so interpreted as to be void of content, namely, by defining empirical concepts by enumerating all objects falling under them.

It might secondly be objected that, in spite of the argument adduced, it can be maintained that mathematics *if consistent* has no content, i.e., that mathematics either is wrong or has no content, because if correct it is compatible with all possible sense experiences (cf. footnote 15). However, that this inference is not valid even from the empirical standpoint follows from the fact that laws of nature without mathematics are exactly as "void of content" (in this sense) as mathematics without the laws of nature. The fact is that only laws of nature *together* with mathematics (or logic) have

---

[36]This also shows directly that the objective of the syntactical program as formulated in the end of [§16] cannot be attained. A concept of objective truth which should be acceptable to proponents of the syntactical viewpoint is that of "empirical correctness", where a theory is called empirically correct if its consequences in the field of the directly observable agree with experience (provided, of course, that it has any such consequences). Now, in order to draw the conclusions discussed in §§13–16 one must know the empirical correctness, in this sense, of the theory consisting of mathematics (M) and the part of physics or psychology involved (W), or of W *and* the mathematical axioms used in the consistency proof for M. Similarly, in order to draw the conclusion discussed in §19, one must know the empirical correctness of the axioms used in the consistency proof for the syntactical system chosen in conjunction with certain simple laws of nature concerning the handling of symbols (L). *So, in spite of any syntactical interpretations that can be given, it is always necessary for the applications of mathematics to know the empirical correctness of some system which includes mathematics at least to a large extent.*

It is interesting to note that, even in cases where mathematical axioms *can* be introduced by conventions about the use of symbols, still the truth of mathematical theorems is not a bit more conventional than that of factual propositions. For even then what is conventional is solely which symbols are associated with which meanings; but once these conventions have been made, the truth of mathematical theorems is objectively determined. This follows from the fact that it is exactly by the rules of syntax that, according to the syntactical viewpoint, the meanings of the mathematical symbols are defined.

consequences verifiable by sense experience. It is, therefore, arbitrary to place all content in the laws of nature.[37] That mathematics does add something to them is particularly clear in such examples as that discussed in §[13]. Here physics describes the behaviour of the single electronic tube, while mathematics adds the laws as to how many tubes connected in a certain manner will react. E.g., Goldbach's conjecture evidently implies such a law (cf. §[13]). That the latter laws are not contained in the former is seen from the fact that (1) they involve concepts totally different from those occurring in the former and that (2) some of them might even be independent synthetic propositions in the sense of being knowable only by the help of a new inductive inference. What mathematics adds to the physical laws, it is true, are not any new properties of physical reality, but rather properties of the *concepts* referring to physical reality—to be more exact, of the concepts referring to combinations of things. But, as can be seen from the example given, such properties are something quite as objective as properties of physical reality and even verifiable by sense experience under the hypothesis that certain laws of nature which can be confirmed independently of mathematics proper hold good. In particular it is perfectly possible that properties of concepts (if they contain universal quantifiers) may *not* follow from the definitions or the meanings of the terms (as far as we are able to understand them) but still may be knowable in the same sense as laws of nature.[38] It is to be noted that the considerations given in this section about the content of mathematics are closely related with those of §[19]. For, if an axiom A is added to some system of axioms S it will, in general, produce new consequences of the kind discussed in this section if and only if it cannot be proved consistent, and, therefore, cannot be introduced as a mere symbolic convention, on the basis of the system S.

§35   All attempts of analyzing away the given or assumed facts expressed in the axioms of mathematics have failed and must necessarily fail: the axioms demonstrably cannot be replaced by definitions and the rule of substituting

---

[37]Laws of nature, it is true, in general have verifiable consequences due to the rules for the use of the symbols *occurring in them*, while mathematical axioms have not. However: (1) This is not true for *all* laws of nature (cf. the example given in [§13]); (2) In general *all* verifiable consequences of a law of nature cannot be obtained by the rules for the symbols occurring in it (cf. the examples referred to in [§13]).

[38]Whether such properties actually exist is not known, since this depends on whether there are enough evident axioms (cf. fn. 43). But at any rate there is no inconsistency involved in beings which can apply the mathematical concepts concerned in each special instance and also know how to use quantifiers, but cannot a priori prove all true propositions connecting these concepts.

the definiens for the definiendum.[39] If mathematics is reduced to logic (in the sense of the Frege-Russell system), then axioms about the primitive terms of logic must be assumed, some of which are so far from trivial that they are rejected as false or meaningless by many mathematicians. If the mathematical axioms are replaced by syntactical rules, one needs axioms of the same power about the primitive terms of syntax or about abstract or transfinite concepts to be used in the syntactical considerations.[40]

§36   All this also applies to finitary mathematics, which, therefore, likewise has content. Hence, if all mathematics could be reduced to finitary syntax (in accordance with the requirements 2–6), this would mean, not that it has no content, but at most that its content would not be larger than that of finitary combinatorics. (Cf. §21.) (Note that in this case a model for mathematics could be defined in terms of finitary combinatorics [cf. §[18]] and, given finitary combinatorics, the rest of mathematics could really be introduced by conventions.) In reality the content of mathematics turns out to be infinitely larger. It should be noted, however, that this is only the result of a mathematical investigation. A priori it is conceivable that the syntax necessary for building up and applying all mathematics would require only an insignificant part of finitary combinatorics, as would be the case, e.g., if all mathematics could be derived from explicit definitions and the law of identity. In that case there *would* be some reason for calling mathematics "void of content", although, strictly speaking, it would not be true even then.

[39]Neither classical nor intuitionistic mathematics, nor any system in which they can be proved consistent, nor even finitary mathematics, can be derived from definitions alone, where "definition" means an always applicable rule for eliminating the symbol defined from any grammatically well-formed expression, and "T" (truth by definition) and "F" (falsity by definition) are the only undefined terms. A reduction to definitions is impossible, because it would yield a decision procedure for all propositions occurring.

[40]It is to be noted that what can be shown is much more than that mathematics has content. From the considerations given above, it moreover follows that the content of mathematics is unlimited in the following sense: Outside every axiomatic system formalizing mathematical truth there exist propositions expressing new and independent mathematical facts in that they cannot be reduced to symbolic conventions on the basis of the axioms of the system. E.g., if the system contains arithmetic, the proposition stating the consistency of the system is of such nature.

The neglect of the conceptual content of sentences (i.e., of the "sense" according to Frege) as something objective (i.e., non-psychological) also is responsible for the wrong view that the conclusion in logical inference, objectively, contains no information beyond that contained in the premises. (Cf. fn. 7.) For the conclusion represents the empirical (or, more generally, the extra-logical) content of the premises, or part of it, in a conceptually different form and that the conclusion is implied by the premises is itself an objective fact concerning the primitive terms of logic occurring [in] and specific for these terms.

§37 It can be shown (see §§39, 41) that *the reasoning which leads to the conclusion that no mathematical facts exist is nothing but a petitio principii, i.e., "fact" from the beginning is identified with "empirical fact", i.e., "synthetic fact concerning sensations".*[41] In this sense the voidness of content of mathematics can be admitted, but it ceases to have anything to do with the philosophical questions mentioned in §1, since also Platonists should agree that mathematics has no content of this kind. For its content, according to Platonism, consists in relations between concepts or other abstract objects which subsist independently of our sensations, although they are perceived in a special kind of experience and although in conjunction with certain universally accepted laws of nature (L) they even have consequences verifiable by sense perception (cf. §⟦16⟧).

§38 All mathematics, as far as it has been developed or axiomatized, it is true, can be interpreted (in the sense of footnote 6 or §12) to be: (I) finitary inductive syntax (cf. §19); (II) finitary deductive syntax, if the applicability of mathematics (item 2 of §12) is renounced (where in both I and II "finitary" can be taken in an even narrower sense than that explained in the end of footnote 19); (III) almost all of mathematics that has been developed or axiomatized can be interpreted to be non-finitary syntax; (IV) some sections of it can even be interpreted to be finitary syntax in accordance with the requirements 2–6 (cf. §§26–27); (V) all of it can be interpreted in a certain weakened sense to be finitary syntax. This is accomplished by dropping the requirement (contained in §12) that the mathematical procedures of proof, too, should be derivable from the rules of syntax, and requiring only that this should be possible for each mathematical theorem, and by weakening item 2 of §12 in the manner indicated in footnote 16. Then there do exist syntactical systems which can finitarily be proved consistent and which (provided mathematics as defined in the beginning of this paragraph is consistent) allow ⟦us⟧ (at least theoretically) to derive all theorems demonstrable in it, and to make each individual application of the kind indicated in §16 (although the number of proof steps necessary may be far beyond anything actually feasible). These systems

---

[41]What induces to identifying "fact" and "synthetic fact" (i.e., to disregarding facts consisting in relations between concepts) is the circumstance that, due to definitions, relations between concepts apparently can be reduced to relations between logical concepts and that in empirical propositions the logical concepts in their regular use don't seem to belong to the subject matter of the proposition, but rather to be the means by which something is said about the subject matter. But, if a proposition is true due to nothing but properties of the means of expression, it cannot say anything about the subject matter; but neither about the means of expression, or they would not be means of expression, but subject matter. However, even if this is admitted, it only proves that mathematical propositions in certain formulations express no facts, because these are hidden in the means of expression. But this implies neither the non-existence nor the inexpressibility of mathematical facts, for concepts can also be made subjects of propositions.

are obtained by stipulating that a sentence A can be asserted if, for some n, A can be proved in n steps, and if, moreover, it can be proved finitarily that no inconsistency arises in n or fewer than n steps.[42]

§39   In all these schemes the syntactical rules replacing mathematics are really "void of content" in this sense that by themselves (cf. footnote 37) they do not imply the truth or falsehood of any sentence stating an empirical fact (i.e., they are compatible with all distributions of truth-values for atomic [i.e., logically indecomposable] empirical sentences, although in case I this is known only empirically). However, that this criterion for the existence of a content of propositions is inadequate even from the empirical standpoint has been pointed out in [[§19?]] and in [[§37?]]. That, moreover, if applied to the problem of the content of mathematical propositions, it involves a petitio principii can be brought to complete evidence as follows: Disregarding the additional requirements necessary in the cases IV, V the possibility of building up mathematics (or rather the parts of mathematics mentioned) in the ways described under I–V and the "lack of content" of mathematics in the sense explained are due solely to these two circumstances: (1) that pure mathematics implies nothing about the truth values of those sentences which contain no mathematical symbols (i.e., within the field of mathematics and natural science, the atomic empirical sentences), (2) that mathematics (or rather the parts of it mentioned) follow from a finite number of axioms and formal rules, known at the time when the lan-

---

[42]This scheme is substantially due to L. Kalmár. If the answer to the question raised in fn. 22 should turn out to be affirmative, the theorems of mathematics could actually be derived by this rule in a feasible number of steps. This scheme then would be the closest existing approximation to a satisfactory syntactical foundation of mathematics. But still syntax could not replace intuitive mathematics in the sense of fn. 6 or §12, because the inference illustrated by Goldbach's conjecture in §13, in general, could only be drawn for integers below some limit M, although M could be enlarged successively.

However, as indicated in the text, the scheme V combined with the treatment of empirical induction outlined in §[[16?]] opens up a possibility to set up a finitary syntactical system which can replace intuitive mathematics in the restricted sense explained in fn. 16. However, it very likely is demonstrable that, in any finitary syntax for which this is the case, the syntactical considerations for some possible applications necessarily go far beyond restricted finitism (cf. fn. 19 and §27). Therefore, even in case of an affirmative answer to the question of fn. 22, these syntactical schemes are open to the same objections as transfinite syntax (namely, that they have to refer to purely mathematical [or ideal] objects which cannot at all be realized in physical or psychical reality).

Incidentally, it must not be said that questions regarding the actual feasibility of proofs are of no philosophical interest. For the main function of mathematics (as of most of our conceptual thinking) is exactly to bring the vast manifold of possible situations and particularities of the existent under control. Therefore a mathematics that would be as unmanageable as the material to which it is applied would be without point. If, e.g., in some syntactical system, in order to apply some mathematical theorem to the number of electrons in our world, it were necessary to perform a number of operations about equal to the number of electrons, this system certainly could not replace intuitive mathematics even for empiricists.

guage is constructed,[43] or (in case III) that they have models consisting of (finite or infinite) combinations and iterated combinations, of finitely many primitive elements and combinatorial relations between them.

§41   All the conditions mentioned, however, might very well be satisfied also for some portion of empirical science with respect to the rest of it. We might, e.g., possess an additional sense that would show to us a second reality completely separated from space-time reality and moreover so regular that it could be described by a finite number of laws. We could then, by an arbitrary decision, recognize only the first reality as such and declare perceptions of the additional sense to be mere illusion, and sentences referring to the other reality to be without content and true only in consequence of syntactical conventions. These could be so chosen as to make exactly those sentences true which could be seen or inferred to be true with the help of the supposed additional sense.[44]

[43]Note that this condition is not satisfied for the whole of mathematics, but only for mathematics at certain stages of its historical development, and that, therefore, a definite finitary syntactical interpretation also can be given only to mathematics in this sense (cf. the second paragraph of fn. 36, and fn. 17). The "inexhaustibility" of mathematics makes the similarity between reason and the senses (cf. §42) still closer, because it shows that there exists a practically unlimited number of independent perceptions also of this "sense". Note that the inexhaustibility of mathematics appears, not only through foundational investigations, but also in the actual development of mathematics, e.g., in the unlimited series of axioms of infinity in set theory, which are analytic (and evident) in the sense that they only explicate the content of the general concept of set. That such series may involve a very great (perhaps even an infinite) number of actually realizable independent rational perceptions is seen from the fact that the axioms concerned are not evident from the beginning, but only become so in the course of the development of mathematics. E.g., in order only to understand the first transfinite axiom of infinity, one must first have developed set theory to a considerable extent. Moreover, if every number-theoretical question of Goldbach type (cf. fn. 25) is decidable by a mathematical proof, there *must* exist an infinite set of independent evident axioms, i.e., a set m of evident axioms which are not derivable from *any* finite set of axioms (no matter whether or not the latter axioms belong to m and whether or not they are evident). Even if solutions are desired only for all those problems of Goldbach type which are simple enough to be formulated in a few pages, there must exist a great number of evident axioms or evident axioms of great complication, in contradistinction to the few simple axioms upon which all of present day mathematics is built. (It can be proved that, in order to solve all problems of Goldbach type of a certain degree of complication k, one needs a system of axioms whose degree of complication, up to a minor correction, is $\geq k$.)

[44]It might be objected that the analogy between mathematical intuition and the supposed additional sense breaks down insofar as the general laws holding for the supposed second reality could be disproved by further observations. However, the same would happen for mathematics if an inconsistency arose. For a disproval of laws discovered by observation also is nothing else but an inconsistency between different methods of ascertaining the same thing, since empirical induction and the application of verified laws of nature also are such methods.

In reply to another possible objection it should be noted that the second reality, although completely separated from the first one, nevertheless might help us considerably in knowing the latter, e.g., if in some respects, or as to certain parts, it happened to be

§42   I even think that this comes pretty close to the true state of affairs, except that this additional sense (i.e., reason) is not counted as a sense, because its objects are quite different from those of all other senses. For, while through sense perception we know particular objects and their properties and relations, with mathematical reason we perceive the most general (namely the "formal") concepts and their relations, which are separated from space-time reality insofar as the latter is completely determined by the totality of particularities without any reference to the formal concepts.[45]

§43   But can it at least be said that, owing to the syntactical interpretations that are feasible, the intuitive content of mathematics can, without loss, be disregarded or be treated as a mere psychological adjunct without direct objective significance? It cannot. For either (1) the intuitive content of mathematics, to a very large extent, is necessary for the syntactical considerations replacing mathematics, or (2) mathematical theorems lose their applicability, either completely or in proportion as the intuitive content of mathematics is not used (cf. V), or (3) consistency, which is necessary for the application of mathematical theorems (cf. §§13–16), must be based on empirical induction. I believe that, at least for finitary and some parts of intuitionistic mathematics, practically everybody will agree that the consistency proof based on mathematical intuition is incomparably more convincing, and certainly not for the reason that we believe mathematical intuition to be derived from subconscious induction or Darwinian adaptation.

The arbitrariness involved in Carnap's definition of content consists in the fact that it is based on the relation of logical consequence, i.e., logically equivalent sentences by definition have the same content. The content thus defined, therefore, should be called "extralogical" or perhaps "empirical" content. By identifying this special kind of content with "content" one begs the question.

§44   This follows from the fact that, using other relations of consequence, one could similarly define various kinds of "mathematical content" (corresponding to the various sections of mathematics), e.g., in Carnap's language II, a "set-theoretical content" of analytic sentences by means of the relation

---

similar or isomorphic to it. In fact this would correspond closely to the manner in which mathematics is applied in theoretical physics.

[45]Mathematical concepts, it is true, can be introduced by rules for handling symbols, on the basis of empirical facts concerning physical symbols (cf. §[19]). However, in real fact the situation is just the opposite: the rules for the use of symbols are so chosen that they express properties of previously conceived mathematical concepts or objects.

However, it should be noted that the direct perceptibility of mathematical objects is by no means the only reason for asserting their existence. Even empiricists, who do not acknowledge this direct perception, nevertheless must assert the existence of mathematical objects, exactly as that of physical objects, if they consistently apply their own criteria of existence (cf. fn. 14 and §5).

of "consequence owing to number theory and the calculus of propositions and quantifiers", and similarly a "number-theoretical content" of analytic sentences which have no set-theoretical content. On the other hand, by using the relation of "consequence due to laws of nature", a concept of "accidental content" could be defined, according to which laws of nature would be "void of content".

§45   From later publications of Carnap it appears that today he would hardly uphold the formulation quoted in footnote 9. From what he says in *Carnap 1950*, p. 35, e.g., it follows that at present he does not object to associating, in scientific semantics, mathematical objects to formulae as their denotation. However he maintains that the philosophical question about the objective existence of mathematical objects does not refer to this "internal" existence, but means whether these objects formally introduced by axioms "really" exist. An answer to this question is asserted to have no "cognitive content", i.e., the question is considered to be meaningless, while formerly it was answered negatively; or else Carnap has changed his opinion about internal existence in mathematics.

At any rate Carnap's present standpoint (or, more generally, the positivistic standpoint consistently carried out) does not lend itself to founding any difference between *Real-* and *Formalwissenschaft* in the sense of the passage quoted in footnote 9. For the considerations of the preceding section literally apply also to the question of the existence of objects of physics or of everyday life. If nevertheless also from the positivistic standpoint, *Realwissenschaft* can be said to describe some realm of existent things, the sole reason is that the existential assumptions made in it (such as that of the existence of elementary particles or of some field) lead to correct consequences as to ascertainable facts, and that, moreover, a few of such existential assumptions lead to a great number of such consequences. But the same is true for the existential axioms of mathematics. E.g., the axioms about the existence of sets of integers underlying the theory of real numbers, due to analytic number theory, lead to consequences as to ascertainable mathematical facts in the sense of §[15]. In general, if by using concepts and their relations among each other and with the sensations, one arrives at verifiable consequences, it is exactly from the *existence of objects* having these relations that the verifiable consequences follow. In the case of mathematics, it is true, in order to obtain consequences verifiable by sense experience, one needs, in addition to the conceptual relations, certain laws of nature (L) about paper and pencil or about the subject matter to which it is applied. But this only means that mathematics is some kind of second-order theory, in the same sense as a physical theory which presupposes other parts of physics, e.g., a physical theory which does not imply physical geometry, but cannot be verified without a measuring of spatial magnitudes. That the existential assertions, also in mathematics, are not a mere "façon de parler" follows from the fact that they can be disproved (by inconsistencies derived from them) and that they have consequences as to ascertainable facts.

All this adds up to the conclusion that *also from the empirical standpoint, there is no reason to answer the question of the objective existence of mathematical and space-time objects differently.* As to the specific nature of these two kinds of objects and facts, it is true, there are profound differences. They become fully visible only if the *meaning* of the mathematical and empirical terms is considered. But they also appear from the syntactical point of view in the different role which these terms play in the formalism of science.

Is mathematics syntax of language?
(*1953/9–V)

It is well known that Carnap has carried through, in great detail, the conception that mathematics is syntax (or semantics) of language. However, not enough attention has been paid to the fact that the philosophical assertions which form the original content and the chief interest of this conception have by no means been proved thereby. Quite on the contrary, this, as well as any other possible execution of the syntactical scheme, rather tends to bring the falsehood of these assertions to light. I am speaking of the following assertions:

I. Mathematical intuition, for all scientifically relevant purposes, in particular for drawing the conclusions as to observable facts occurring in applied mathematics, can be replaced by conventions about the use of symbols and their application.

II. In contradistinction to the other sciences, which describe certain objects and facts, there do not exist any mathematical objects or facts. Mathematical propositions, because they are nothing but consequences of conventions about the use of symbols and, therefore, are compatible with all possible experiences, are void of content.

III. The conception of mathematics as a system of conventions makes the a priori validity of mathematics compatible with strict empiricism. For we know a priori, without having to appeal to any a priori intuition, that conventions about the use of symbols cannot be disproved by experience.

2 | The syntactical point of view as to the nature of mathematics doubtless has the merit of having pointed out the fundamental difference between

mathematical and empirical truth. This difference, I think rightly, is placed in the fact that mathematical propositions, as opposed to empirical ones, are true in virtue of the *concepts* occurring in them. However, by adopting the nominalistic point of view and identifying concepts with symbols, the syntactical conception transforms mathematical truth into conventions and, eventually, into nothingness. The expression thereof are the assertions I, II, III.

It seems to me that these assertions, for an adequate interpretation of the terms occurring in them (such as "content", "disprove", "replace", etc.), turn out to be wrong. Moreover, I believe, it can be shown directly that the arguments which may be adduced in favor of these assertions, including the existence of actual elaborations of the syntactical scheme, are all fallacious. The inadequacy of those meanings of the terms which make the assertions I – III true in particular appears in the circumstance that they oblige one to interpret objectively quite analogous situations differently.

| I. As far as assertion I is concerned, it is true that by applying certain   3
rules about the use of symbols (to be more precise, about the truth or false-hood of propositions containing them) one arrives at the same sentences, also in the applications, as by applying mathematical intuition. However, in order to have any reason for the expectation that, if these rules are applied to verified laws of nature (e.g., the primitive laws of elasticity theory), one will obtain empirically correct propositions (e.g., about the carrying power of a bridge), one evidently must know certain facts (at least with probability) concerning the rules of syntax. For to expect this for perfectly arbitrary rules about the truth or falsehood of propositions clearly would be folly. What must be known is that the rules, by themselves, do not imply the truth or falsehood of any proposition expressing an empirical fact. Such rules may be called "admissible". Admissibility, for *our* mathematics, entails consistency. For from an inconsistency *all* propositions, the empirical ones included, could be derived. But now it turns out that for proving the consistency of mathematics an intuition of the same power is needed as for discerning the truth of the mathematical axioms, at least in some interpretation. In particular the *abstract* mathematical concepts, such as "infinite set", "function", etc., cannot be proved consistent without again using abstract concepts, i.e., such as are not merely ascertainable properties or relations of finite combinations of symbols. So, while it was the primary purpose of the syntactical conception to justify the use of these problematic concepts by interpreting them syntactically, it turns out that quite on the con|trary, abstract concepts are necessary in order to justify the   4
syntactical rules (as admissible or consistent). If mathematics could be interpreted in terms of conventions which, for the proof of their admissibility, would require only the simplest kind of intuition referring to finite combinations of symbols, then the assertions I, II, III would be true, although not completely, still to a very large extent, i.e., the prima facie content of

mathematics would turn out for the most part to be an illusion. However, the fact is that, in whatever manner syntactical rules are formulated, the power and usefulness of the mathematics resulting is proportional to the power of the mathematical intuition necessary for their proof of admissibility. This phenomenon might be called "the non-eliminability of the content of mathematics by the syntactical interpretation". This vicious circle, it is true, can be escaped by founding consistency on empirical induction. But *at any rate it is clear that mathematical intuition cannot be replaced by conventions, but only by conventions plus mathematical intuition, or by conventions plus an empirical knowledge involving, in a certain sense, an equivalent mathematical content*. It is worth mentioning that the same result can also be obtained without any reference to the applications. For, strictly speaking, conventions cannot even be made without a foregoing proof of their admissibility, otherwise the term "convention" would not be appropriate. So the use of mathematical intuition is replaced by: (1) ascertaining the possibility of certain conventions, (2) making and using these conventions. Hence the syntactical interpretation of | mathematics does not relieve one of the necessity to acknowledge certain, by no means trivial, propositions of a mathematical character as true in a non-conventional sense.

II. The falsehood of assertion I points up the existence of a content of mathematics, i.e., of mathematical facts. For, if the prima facie content of mathematics were nothing but a false appearance, it would have to be possible to build up mathematics satisfactorily without making use of this "pseudo-content". Still more clearly the falsehood of assertion II appears from the following arguments:

1. Conventions about the use of symbols are void of content only in a *relative* sense, namely, insofar as they *add* nothing to a theory which implies their admissibility. If, e.g., on the ground of the empirically known associativity of some physical operation a convention about the dropping of brackets is introduced, then from this convention the associativity of the operation in question, i.e., an empirical proposition, follows. If some convention about the use of a mathematical symbol is introduced on the basis of its consistency, the situation is quite similar. For in general such a convention implies, although not its own consistency, still certain only slightly weaker propositions, i.e., substantially the same facts as those which justified its introduction. Generally speaking any law of nature can be interpreted to be a convention whose admissibility derives from this law of nature. It may be objected that, in order to derive the empirically verifiable fact of consistency, which | is concerned with the handling of symbols, the mathematical axioms by themselves are not sufficient, nor, therefore, the convention X in question. The answer is that nobody will call a law of nature (e.g., the primitive law of electrostatics) void of content, because it has verifiable consequences only in conjunction with other, independently

known, laws (such as those regarding physical space and mechanics). It may secondly be objected that the statement of consistency is for that reason void of content, because, unlike in a law of nature, all single instances concerning individual proof figures (which is all that is observable) can be derived from the axioms already before the convention X has been made. The answer is that the process of formal derivation in a theory is itself a kind of observation. So the objection is about the same as if it were said that a law of nature is void of content, because the single instances of it can be ascertained without its help, namely, by direct observation.

2. The fact that the syntactical program can be carried through (in the sense of assertion I above) is due solely to these two circumstances: (i) the true mathematical propositions follow from a comparatively small number of primitive ones; (ii) they are, in a sense, separable from other propositions, because no synthetic (empirical) propositions follow from them. Therefore, if we had a physical sense whose objects were of a similar regularity and similarly separated from those of the other senses, we could interpret also the propositions based on impressions of *this* sense to be syntactical conventions without content and associate no facts or objects with them or their constituents. The similarity between mathematical intuition and a physical sense is very striking. | It is arbitrary to consider "This is red" an   7 immediate datum, but not so to consider the proposition expressing modus ponens or complete induction (or perhaps some simpler propositions from which the latter follows). For the difference, as far as it is relevant here, consists solely in the fact that in the first case a relationship between a concept and a particular object is perceived, while in the second case it is a relationship between concepts. Moreover, there do exist a great number of independent impressions also of mathematical intuition, although, for practical purposes in the present state of science, a few are sufficient. One might say that, in contradistinction to other sciences, the experiences of mathematical intuition are not the *object* of mathematics and that for *this reason* the assertion quoted in footnote [?] is true. However, in truth, experiences are not the object of most other sciences either. E.g., animals seen in hallucinations are not objects of zoology. On the other hand, a general mathematical theorem, in a sense, has the mathematical experiences relating to the special cases as its object. Hence, again, there is no substantial difference between mathematics and other sciences.

3. Even if mathematics is interpreted syntactically, this makes it not a bit more "conventional" (in the sense of "arbitrary") than other sciences. For the rules for the use of a symbol, according to the syntactical conception, are the definition of its meaning, so that different rules simply introduce different concepts. But the choice of the concepts is arbitrary also in other sciences. Everything else, however, namely, what can be asserted on the basis of the definitions, is exactly as objectively determined in mathematics as in other sciences. Viewed from this angle the content

8   of mathematics | appears in the fact that definitions implicitly assert the
existence of the object defined. In particular, if a symbol is introduced by
stating rules as to which sentences containing it are true, then from these
rules much the same conclusions can be drawn as could be from the assumption of the existence of an object satisfying those rules. Only in special
cases, such as explicit definitions, is the consistency of such an assumption
trivial.

4. If it is argued that mathematical propositions have no content because, by themselves, they imply nothing about experiences, the answer is
that the same is true of laws of nature. For laws of nature without mathematics or logic imply as little about experiences as mathematics without
laws of nature. That mathematics, at least in most applications, does add
something to the content of the laws of nature is best seen from examples
where one has very simple laws about certain elements, e.g., those about
the reactions of electronic tubes. Here mathematics clearly adds the general laws as to how systems of tubes connected in a certain manner will
react. That the latter laws are not contained in the former is seen from the
following facts: (a) The latter laws may contain concepts not definable in
terms of those occurring in the former (e.g., the concept of a combination
of any finite number of elements). (b) In order to understand the laws of
nature it is sufficient, as far as the mathematical concepts occurring are
concerned, to know rules which decide on their applying or not applying in
each particular case. But such rules by no means imply the general laws
governing them. (c) These general laws may even require new empirical
inductions, namely, in case the mathematical problem in question should
9   be unsolvable. E.g., this may occur in a case | like Goldbach's Conjecture,
which evidently implies a certain law about the reactions of a computing
machine. Note that the *general* mathematical laws may even be required
for predicting the result of a *single* observation, e.g., in case the latter depends on an infinity (e.g., a continuum) of physical elements. Therefore, for
a certain kind of physical theory a new mathematical axiom (which would
solve problems of mathematical physics formerly undecidable) may lead
to new empirically verifiable consequences exactly as a new law of nature.
Mathematical propositions, it is true, do not express physical properties
of the structures concerned, but rather properties of the *concepts* in which
we describe those structures. But this only shows that the properties of
those concepts are something quite as objective and independent of our
choice as physical properties of matter. This is not surprising, since concepts are composed of primitive ones, which, as well as their properties, we
can create as little as the primitive constituents of matter and their properties. However, in spite of the objective character of conceptual truth,
it is quite necessary to distinguish sharply these two kinds of content and
facts as "factual" and "conceptual". What Carnap calls "content" really
is "factual content".

5. Even disregarding the possibility that a mathematical axiom may be disproved by wrong empirical consequences derived from it in conjunction with well-verified laws of nature, it is in any event disprovable by an inconsistency derived from it. If an inconsistency is not recognized as a disproval, but only as a proof for the "inexpediency" of the "convention" in question, the same can be done for laws of nature, which also can be interpreted to be conventions which become "inexpedient" in case a counterexample is met with. | Note that an inconsistent mathematical axiom would imply   10
also wrong empirical propositions, and therefore, before the inconsistency is discovered, would work out in the applications exactly as a wrong law of nature. If the possibility of a disproval of mathematical axioms frequently is disregarded, this is due solely to the convincing power of mathematical intuition. But it is the very starting point of the syntactical conception to reject mathematical intuition. From the disprovability of mathematical axioms it follows that, if mathematics is dealt with from the positivistic point of view without prejudice, some immanent existence should be attributed to the objects of a correct mathematics, in contradistinction to those of ⟦an⟧ inconsistent one, exactly as is done for the objects of a correct physics. Those objects, e.g., infinite sets or properties of properties, are of a specific nature, i.e., different from those of other sciences. Note that these mathematical objects and facts cannot be eliminated (as, e.g., infinite points in geometry can), since there always remain primitive mathematical terms and axioms about them, either in the scientific language or the metalanguage, where "axiom" here means a proposition assumed on the ground of its intuitive evidence or because of its success in the applications. Positivists, least of all, have any reason to attribute to the mathematical objects, in contradistinction to others, only some "pseudo-existence", for the difference between the two primarily lies in their different *intuitive character*, while the role they play in the *formalism* of science is very similar (at least in comparison to the more abstract | physical objects, such as fields   11
of force, atoms, etc.). Therefore the attitude positivists should take on the basis of their own point of view would be to consider the assumption of mathematical objects and axioms to be irreducible hypotheses of science, exactly as the assumption of a field and of the laws governing it.

III. As to assertion III it suffices to say that non-admissible syntactical rules lead to empirical consequences and, therefore, can be disproved by experience. On the other hand, an a priori certainty that the syntactical rules replacing mathematics are admissible (hence *a fortiori* consistent) can only be obtained by extensive use of mathematical intuition.

IV. What psychologically plays a large part in whatever plausibility the thesis of the voidness of content of mathematical propositions may have, and what moreover constitutes the grain of truth contained in the syntactical conception of mathematics, are these circumstances:

1. There is no possible state of affairs which is excluded by a logically

true proposition, while the content of a proposition seems to consist in the very fact that it excludes certain possibilities.

2. The logical concepts in their use in empirical propositions don't seem to belong to the subject matter of the proposition, but rather seem to be means of expression. However, if a proposition is true already due to properties of the means of expression, it cannot say anything about the subject matter, but neither about the means of expression, or they would not be means of expression, but subject matter.

12 | 3. The meaning of sentences is defined by semantical rules which determine under which circumstances a given sentence can be asserted. These rules, in certain limiting cases, may have the consequence that a sentence can be asserted under *all* circumstances (e.g., "It will rain or it will not rain tomorrow"). Such sentences are true but void of content. The negations of such sentences are excluded as incorrect already on the basis of their structure, irrespective of any meaning, exactly as sentences which do not conform to the rules of grammar, so that logic can be regarded as part of syntax.

However, as to item 1 it can be answered that there are different levels of possibility, as appears already in the distinction between physical and logical possibility. Item 2, even if the antecedent is admitted, does not exclude that the logical concepts may be made the subject matter of non-empirical propositions. Moreover, what is regarded as the content of a proposition largely is a question of what one is interested in. E.g., one may very well say that the proposition mentioned above, although it says nothing about rain, does express a property of "not" and "or".

Kurt Gödel at his office desk, 9 May 1958

Veli Valpola

Introductory note to *1961/?

This text is based on a shorthand draft from late 1961 or the subsequent years for a lecture that was never given. The manuscript was found in an envelope from the American Philosophical Society postmarked 13 December 1961, and labelled by Gödel "Vortrag, Konzept". Gödel was elected to membership in that society in April 1961 and signed the roll book the following November. After the November meeting he wrote to G.W. Corner, the executive secretary of the Society, "I would first like to say that I enjoyed the meeting on November 9 very much and that some of the lectures were very interesting to me." On December 13 Corner then wrote Gödel to advise him of the custom that new members were invited to give twenty-minute talks on topics of their own choosing, and he suggested that Gödel could do so either in April or November 1963. Since this letter of Corner's was found in the envelope with Gödel's manuscript, it is reasonable to assume that the latter is a draft for such a lecture. However, there is no indication that Gödel ever replied to the invitation.

The aim of the lecture is to describe in philosophical terms the development of the study of the foundations of mathematics in our century and fit it into a general scheme of possible philosophical *Weltanschauungen*. Gödel places the *Weltanschauungen* and also specific philosophical views on a scale according to their affinity with or distance from metaphysics or religion. On the left, furthest from metaphysics, he places skepticism, materialism, positivism, empiricism and pessimism, on the right, spiritualism, idealism, theology, apriorism and optimism. Gödel notes that the development of philosophy since the Renaissance largely has been from the right to the left on this scale.

Gödel makes clear that his sympathies lie with the right. This is in keeping with his statements to various interlocutors; compare the picture of Gödel's world view that one obtains from *Wang 1987*. This sympathy is muted and qualified in Gödel's publications and lectures; in the present text, Gödel gives more open expression to this attitude than in anything else intended for public presentation. Still, Gödel says that "the truth lies in the middle or consists of a combination of the two conceptions" (page 7), and much of the lecture is a development of that thought.

Gödel observes that mathematics is a discipline that long has withstood the *Zeitgeist*'s move to the left; in particular, Mill's attempt to develop an empirical theory of mathematics never found much support. Also, mathematics became more abstract and reached more clarity concerning its foundations, e.g., through the exact founding of the infinitesimal calculus and of the complex numbers.

Around the turn of the century mathematics got into trouble. In particular, the antinomies of set theory were seized upon by the skeptics and used as a pretext for a more "leftist" view on mathematics. However, Gödel regards their reaction as unwarranted, for two reasons: (1) the contradictions do not occur within mathematics, but on its extreme border towards philosophy; and (2) they have been resolved in a manner that is completely satisfactory and which seems almost obvious to everyone who understands the theory.

However, Gödel's incompleteness theorem shows that the rightward conception of mathematics as a complete system of truths cannot be rescued by appeal to a set of axioms and formal rules. On the other hand, he also rejects the leftward conception of mathematics according to which

> the truth of the axioms from which mathematics starts out cannot be justified or recognized in any way, and therefore the drawing of consequences from them has meaning only in a hypothetical sense, whereby this drawing of consequences itself (in order to satisfy even further the spirit of the time) is construed as a mere game with symbols according to certain rules, likewise not [supported by] insight [page 5].

Gödel inveighs against this leftist conception of mathematics because it

> gives up aspects whose fulfillment would in any case be very desirable and which have much to recommend themselves: namely, on the one hand, to safeguard for mathematics the certainty of its knowledge, and on the other, to uphold the belief that for clear questions posed by reason, reason can also find clear answers [page 6].

Expressing his view that "the truth lies in the middle", Gödel retains the rightward idea of mathematics as a system of truth, which is complete in the sense that "every precisely formulated yes-or-no question in mathematics must have a clear-cut answer." However, he rejects the idea, shared by rightists and leftists alike, that the basis of these truths is their derivability from axioms. He points out that Hilbert's attempt to combine these two ideas proves to be impossible in view of his own incompleteness theorem and the corollary concerning consistency proofs. (Characteristically, Gödel does not mention that these results are his own.) Instead of trying to secure the certainty of mathematics by manipulation of physical symbols, Gödel proposes "cultivating (deepening) knowledge of the abstract concepts themselves which lead to the setting up of these mechanical systems" (page 7).

This alternative approach to the foundations of mathematics is connected with the second reason Gödel gave for not being disturbed by the antinomies of set theory, *viz.*, that they have been resolved in a manner that is completely satisfactory and which seems almost obvious to everyone who understands the theory. According to Gödel, there is a certain kind of obviousness that accrues to mathematical statements once their meaning has been properly clarified. This clarification cannot merely consist in giving explicit definitions of some concepts in terms of others, but requires, at least to a large extent, an elucidation of meaning that does not appeal to other, ultimately undefinable concepts.

The method for this clarification of meaning Gödel finds in Husserl's phenomenology. Gödel describes Husserl's method as "focusing more sharply on the concepts concerned by directing our attention ... onto our own acts in the use of these concepts, onto our powers in carrying out our acts, etc." It is "a procedure or technique that should produce in us a new state of consciousness in which we describe in detail the basic concepts we use in our thought, or grasp other basic concepts hitherto unknown to us" (page 8).

Gödel sees no objective reason for rejecting phenomenology and presents as a consideration in its favor that the development of children proceeds in two directions. They do not only experiment with the objects of the external world and develop their sensory and motor organs. They also come to a better understanding of language and of the basic concepts on which it rests, such as that of a logical inference. Gödel observes that if one views the development of empirical science as a systematic extension of what the child does when it develops in the first direction, then given the astonishing success of this procedure, it seems quite possible that a systematic and conscious advance in the second direction will also far exceed the expectations one may have a priori (pages 8–9).

The central piece of support that Gödel gives in favor of his view is the following example of a development in the second direction:

> ... in the systematic establishment of the axioms of mathematics, new axioms, which do not follow by formal logic from those previously established, again and again become evident. It is not at all excluded by the negative results mentioned earlier that nevertheless every clearly posed mathematical yes-or-no question is solvable in this way. For it is just this becoming evident of more and more new axioms on the basis of the meaning of the primitive notions that a machine cannot imitate [page 9].

Except for the reference to the incompleteness results, which were attained after Husserl had developed his phenomenology, this is all very similar to what Husserl wrote about mathematical evidence. Gödel states that it also agrees in principle with Kant's conception of mathematics (page 9).[a]

This is all Gödel says in this manuscript concerning what the phenomenological method consists in and the reasons why we should accept it. It may be the paucity of particulars and argument in this part of the manuscript that made him decide not to give the lecture. However, Gödel had a very thorough knowledge of phenomenology and there is much more to be found on phenomenology elsewhere in his writings.

Gödel did not start to study Husserl until 1959 (*Wang 1981*, page 658 and *Wang 1987*, page 12 ff.), but he soon became quite absorbed by the reading of Husserl's writings. He owned all of Husserl's main works,[b] and his underlining and comments in the margins indicate that

---

[a]In remarking that "ever newer axioms that are logically independent from the earlier ones" must be "intuitively grasped", Gödel says that this is in general agreement with Kant's view of mathematics. He says that Kant's claims taken literally are false

> since Kant asserts that in the derivation of geometrical theorems we always need new geometrical intuitions, and that therefore a purely logical derivation from a finite number of axioms is impossible [[p. 9]].

Gödel no doubt has in mind the well-known passage in the *Critique of pure reason* where Kant discusses the proof that the sum of the angles of a triangle is two right angles and remarks that the theorem is arrived at "through a chain of inferences guided throughout by intuition" (A 717/B 745). The interpretation of this as implying that the theorem cannot be derived logically from the axioms is not obvious but is central to some influential modern interpretations of Kant's conception of the role of intuition in mathematics, in particular in *Beth 1956/7*; *Hintikka 1965, 1967*, and other writings; and *Friedman 1985, 1992*. These interpretations emphasize, as Gödel does not in his remark, that the formal logic available to Kant is much weaker than modern logic. Gödel's partial agreement with this line of interpretation is noteworthy; at the time of working on the present text, he could have known Beth's paper but in all probability not the writings of Hintikka.

Note that the word translated "intuitions" in the above quotation from Gödel is "Intuitionen" and not the Kantian "Anschauungen", no doubt because the intuitions in question are propositional, and Kant uses "Anschauung" in relation to *objects*. The question of the relation of Kant's and Gödel's conceptions of intuition is a complicated one. Gödel certainly recognized that they were not identical.

[b]Gödel owned the *Logische Untersuchungen* (in the edition *1968*), *Ideen*, Book I (*1950–*, volume III, first edition), *Cartesianische Meditationen und Pariser Vorträge* (*1950–*, volume I, second edition), *Die Krisis der europäischen Wissenschaften und die transzendentale Phänomenologie* (*1950–*, volume VI, second edition). He also owned *Husserl 1965* and both volumes of *Spiegelberg 1965*. All of these books are heavily annotated by Gödel, with the exception of *Husserl 1968*. In the latter were found, however, several pages of Gödel's shorthand notes, referring to page numbers in the text, indicating that Gödel probably had based his study of that text on a borrowed copy. The work was out of print for many years before 1968. The notes indicate that the text Gödel had used was the second edition (*1913a, 1921*). References to Husserl in the following will be, as usual, to page and line in *Husserl 1950–*.

he studied them carefully. Most of his comments are positive and expand upon Husserl's points. However, sometimes he is critical, notably in his notes to the *Logische Untersuchungen* (*Husserl 1900, 1901*) and to some sections of Husserl's last work, the *Krisis* (*Husserl 1950–* , Volume VI, one part published in 1936). Generally, Gödel is most appreciative of the *Ideen* (*1913*) and the other works written after Husserl's "idealist" conversion around 1907.

Gödel had expressed views on the philosophy of mathematics similar to those of Husserl long before he started to study him. What he found in Husserl was not radically different from his own view; what impressed him seems to have been Husserl's general philosophy, which would provide a systematic framework for a number of his own earlier ideas on the foundations of mathematics.

Two main points in Husserl's philosophy of mathematics that were central to Gödel even before he read Husserl are the following:

Realism. Gödel had held realist views on mathematical entities since his student days (*Wang 1974*, pages 8–11), or more exactly, since 1921–22.[c] It has often been noted that his essay "Russell's mathematical logic" (*1944*) contains a number of strongly realist statements, such as the well-known passage in which he says that the assumption of classes and concepts is in the same sense necessary to obtain a satisfactory system of mathematics as physical bodies are necessary for a satisfactory theory of our sense perceptions (page 137).

Gödel expresses similar views in his Gibbs lecture (**1951*) and also in his essay "Is mathematics syntax of language?" (**1953/9*, both in this volume). In the latter he writes the following about *concepts* and their properties:

> Mathematical propositions, it is true, do not express physical properties of the structures concerned [in physics], but rather properties of the *concepts* in which we describe those structures. But this only shows that the properties of those concepts are something quite as objective and independent of our choice as physical properties of matter. This is not surprising, since concepts are composed of primitive ones, which, as well as their properties, we can create as little as the primitive constituents of matter and their properties [**1953/9-V*, page 9].

[c] *Feferman 1984a*, pp. 549–552; see, however, also Feferman's introductory note to Gödel **1933o* in this volume, p. 39, where Feferman comments on Gödel's apparently skeptical statement about Platonism on p. 19 of that manuscript.

Note the comparison between concepts and physical objects in this passage and in that from *1944* referred to above. Gödel does not say straightforwardly that the properties of concepts are objective, but that they are *as objective as* the physical properties of matter. This brings him close to Husserl, who in the *Ideen* and his other "idealist" works maintained that physical objects, as well as concepts and mathematical objects, are objective, but not in the straightforward, realist sense. Husserl called himself an "idealist". However, his view is much more realist and Platonist than traditional idealist points of view, and as Husserl later acknowledged, "idealism" is a very misleading label for his position.[d]

In a letter of 3 February 1959, responding to an inquiry from Paul Arthur Schilpp about the state of his contribution *1953/9* to his volume on Carnap (*Schilpp 1963*), Gödel says:

> ... a complete elucidation of the situation turned out to be more difficult than I had anticipated, doubtless in consequence of the fact that the subject matter is closely related to, and in part identical with, one of the basic problems of philosophy, namely the question of the *objective reality of concepts* and their relations [emphasis mine].

Also in his Supplement to the second edition of "What is Cantor's continuum problem?" (*1964*, page 272) Gödel compares mathematical and physical objects and notes that the question of the objective existence of the objects of mathematical intuition is an exact replica of the question of the objective existence of the outer world. He then goes into the issue of mathematical intuition. This issue is intimately connected with the question of the nature and existence of mathematical entities. On this point, too, Gödel's views are similar to those of Husserl, and the similarity may help us to see what they both are getting at.

Intuition. Already in *1944*, Gödel talks about elementary mathematical evidence or mathematical "data" and compares it to sense perception (page 128; cf. pages 137, 142). Gödel's own notion of mathematical intuition is discussed in all the other three papers mentioned, and in each of them it is compared to perception. First, in his Gibbs lecture, Gödel states:

---

[d]In 1934 he wrote in a letter to Abbé Baudin: "No ordinary 'realist' has ever been as realistic and concrete as I, the phenomenological 'idealist' (a word which by the way I no longer use)." Quoted in *Kern 1964*, p. 276 n., my translation.

What is wrong, however, is that the *meaning* of the terms (that is, the concepts they denote) is asserted to be something man-made and consisting merely in semantical conventions. The truth, I believe, is that these concepts form an objective reality of their own, which we cannot create or change, but only *perceive* and describe 〚**1951*, page 30, emphasis mine〛.

In his essay for the Carnap volume (**1953/9-V*, pages 6–7), Gödel writes:

The similarity between mathematical *intuition* and a *physical sense* is very striking. It is arbitrary to consider "This is red" an immediate datum, but not so to consider the proposition expressing modus ponens or complete induction (or perhaps some simpler propositions from which the latter follows). For the difference, as far as it is relevant here, consists solely in the fact that in the first case a relationship between a concept and a particular object is perceived, while in the second case it is a relationship between concepts 〚emphasis mine〛.

Finally, there is the famous passage from *1964* (page 271) where Gödel compares mathematical intuition to sense perception and says that we have something like a *perception* of the objects of set theory, as is seen from the fact that the axioms force themselves upon us as being true.

Gödel leaves it unclear here whether he thinks that the objects of mathematical intuition are propositions, as in the quotation from **1953/9*, concepts (first quotation, from **1951*), sets and concepts (as would naturally be inferred from the passage in *1964*, page 271), or all three. An answer emerges from Husserl. According to him, intuition, as well as perception, is of objects. There are two kinds of intuition, according to Husserl: perception, where the object intuited is a physical object, and categorial, or eidetic intuition, where the object is an abstract entity. The object, whether it be concrete or abstract, is always intuited as having various properties and bearing relations to other objects. These properties and relations can be singled out for our attention in acts of judgment, in which the object is judged to have such and such features. Intuition of objects is hence more basic for Husserl than judgments: intuition provides the evidence for judgments. It is not clear whether Gödel shared this view that intuition of objects is more basic.

Neither perception nor categorial intuition is an infallible source of evidence. Both always involve anticipations concerning aspects of their objects that have not yet been explored, and which may turn out to be wrong. Since there is no other source of evidence, Husserl acknowledges that errors are always possible,[e] even in mathematics and logic.[f]

Skeptics may be doubtful that we can experience any other objects than physical objects. However, Husserl points out that already in the perception of a physical object many features are involved that go beyond mere sensation, thus, e.g., the reification that is involved in the notion of object itself. These features, and especially those involved in reification, are those that are studied in mathematics. Gödel seems to have something very similar in mind in when he writes in the supplement to *1964*:

> That something besides the sensations actually is immediately given follows (independently of mathematics) from the fact that even our ideas referring to physical objects contain constituents qualitatively different from sensations or mere combinations of sensations, e.g., the idea of object itself .... Evidently the "given" underlying mathematics is closely related to the abstract elements contained in our empirical ideas. It by no means follows, however, that the data of this second kind, because they cannot be associated with actions of certain things upon our sense organs, are something purely subjective, as Kant asserted. Rather they, too, may represent an aspect of objective reality, but, as opposed to the sensations, their presence in us may be due to another kind of relationship between ourselves and reality [pages 271–2].

This goes far beyond what Gödel says about Husserl in the present manuscript. However, Gödel's very high praise of Husserl in this manuscript indicates that we may expect a high degree of similarity in their views.[g]

Gödel's characterization of the relationship between Kant and Husserl is interesting. Here, as in the case of the rightist and the leftist views on mathematics, where Gödel found the truth in the middle, he recognizes "deep truths" in the old view, and seeks to extricate these insights from their unclear and incorrect formulations. This extrication and further development of Kant's fundamental insights Gödel finds in Husserl:

---

[e]See *Husserl 1913*, §87 (*1950–* , III/1, 201.34–35).

[f]E.g., in *Husserl 1929*, §58 (*1950–* , XVII, 164.32–34), also *1950–* , VI, 270.28–29 (§73).

[g]This is confirmed through Gödel's very extensive remarks in his copies of Husserl's books.

In particular, the whole phenomenological method, as I sketched it above, goes back in its [central] idea to Kant, and what Husserl did was merely that he first formulated it more precisely, made it fully conscious and actually carried it out for particular domains. Indeed, just from the terminology used by Husserl, one sees how positively he himself values his relation to Kant.

... just because of the lack of clarity and the literal incorrectness of many of Kant's formulations, quite divergent directions have developed out of Kant's thought—none of which, however, really did justice to the core of Kant's thought. This requirement seems to me to be met for the first time by phenomenology, which, entirely as intended by Kant, avoids both the death-defying leaps of idealism into a new metaphysics as well as the positivistic rejection of all metaphysics. But now, if the misunderstood Kant has already led to so much that is interesting in philosophy, and also indirectly in science, how much more can we expect it from Kant understood correctly? [pages 9–10]

Gödel's comments in the margins of Husserl's works indicate that he regarded many of Husserl's observations as important insights. A central point is the one he speaks of in the passage above: Husserl's very special version of idealism, which incorporates both the constructivist idea that we "constitute" reality and the Platonistic idea that this reality, the mathematical as well as the physical, is there independently of us and is not "something purely subjective, as Kant asserted" (*Gödel 1964*, page 272, quoted above).

In spite of this disagreement, Gödel closes the present text with remarks about Kant that are a striking expression of the idea that the truth lies in the middle between "rightward" and "leftward" conceptions. One might have expected him to choose rather Leibniz as a point of reference in the past, as he in effect did in *1944*. Yet Gödel says that the phenomenological method goes back in its central idea to Kant and that phenomenology is the first philosophy really to do justice to "the core of Kant's thought". It may be that the study of Husserl made Gödel more sympathetic toward Kantian philosophy; however, the question of Gödel's understanding of and attitude toward Kant is a large question which has not been much explored.[h]

Dagfinn Føllesdal[i]

Transcription of *1961/?* is by Cheryl Dawson in consultation with Eckehart Köhler and William Craig. The translation was drafted by Eckehart Köhler and Hao Wang and revised by John Dawson, Charles Parsons and William Craig.

---

[h]On some particular aspects of this matter, however, see §§6–8 of the introductory note to *1946/9* in this volume.

[i]I am grateful to Charles Parsons for numerous suggestions for improvements and for contributing footnote a. I am also grateful to Solomon Feferman, Cheryl Dawson, John Dawson, Richard Tieszen, Dag Prawitz, Per Martin-Löf, Dick Haglund, and Herman Ruge Jervell for their valuable comments on earlier versions of this introductory note and to the Axel o Margaret Ax:son Johnson Foundation for support of this and other work.

The modern development of the foundations
of mathematics in the light of philosophy
(*1961/?)

Ich möchte hier versuchen, die Entwicklung der mathematischen Grundlagenforschung seit etwa der Jahrhundertwende in philosophischen Begriffen zu beschreiben und in ein allgemeines Schema von möglichen philosophischen Weltanschauungen einzuordnen. Dazu ist es zunächst nötig sich
über dieses Schema selbst klar zu werden. Ich glaube, das fruchtbarste
Prinzip, um eine Übersicht über die möglichen Weltanschauungen zu gewinnen, werden wird, daß man sie nach dem Grad und der Art ihrer Affinität
zu, bzw. Abkehr von, der Metaphysik (oder Religion) einteilt. So bekommt
man sofort eine Einteilung in zwei Gruppen: auf deren einer Seite Materialismus, Skeptizismus, Positivismus, auf deren anderer, Spiritualismus,
Idealismus, Theologie stehen, und man sieht auch sofort Gradunterschiede
in dieser Reihe, indem der Skeptizimus noch weiter von der Theologie steht
als der Materialismus, währenddes anderseits der Idealismus, z. B. in seiner
pantheistischen Form, eine Abschwächung der Theologie im eigentlichen
Sinn ist.

Aber auch für die Analyse speziell zulässiger philosophischer Lehren, erweist sich dieses Schema als fruchtbar, indem man sie entweder in diesem
Sinn einordnet oder in Mischfällen ihre materialistischen und spiritualisti-
2 schen Elemente aufsucht. So wird man z. B. sagen, daß der| Apriorismus im
Prinzip auf die rechte und der Empirismus auf die linke Seite gehören. Aber
anderseits gibt es auch solche Mischformen wie eine empiristisch begründete
Theologie. Man sieht ferner auch, daß der Optimismus im Prinzip nach
rechts und der Pessimismus nach links gehört. Denn der Skeptizismus ist
ja ein Pessimismus hinsichtlich der Erkenntnis. Der Materialismus neigt
dazu, die Welt als einen ungeordneten und daher sinnlosen Haufen von
Atomen zu betrachten. Außerdem erscheint ihm der Tod als endgültige und
vollständige Vernichtung, während anderseits Theologie und Idealismus in
Allem Sinn, Zweck und Vernunft sehen—anderseits ist der Schopenhauer-
sche Pessimismus eine Mischform, nämlich ein pessimistischer Idealismus.
Ein anderes Beispiel einer offentsichtlich rechtsseitigen Theorie ist die des
objektiven Rechts und objektiver aesthetischer Werte, während die Deutung der Ethik und Aesthetik auf Grund von Gewohnheit, Erziehung, etc.
nach links gehört.

Nun ist es eine bekannte Tatsache, es ist auch eine Platitüde, daß die
Entwicklung der Philosophie seit der Renaissance im großen und ganzen
von rechts nach links gegangen ist, nicht in einer geraden Linie, sondern
mit Rückschlag; aber doch im ganzen. Insbesondere hat diese Entwick-

374

The modern development of the foundations of mathematics in the light of philosophy (*1961/?)

I would like to attempt here to describe, in terms of philosophical concepts, the development of foundational research in mathematics since around the turn of the century, and to fit it into a general schema of possible philosophical world-views [[Weltanschauungen]]. For this, it is necessary first of all to become clear about the schema itself. I believe that the most fruitful principle for gaining an overall view of the possible world-views will be to divide them up according to the degree and the manner of their affinity to or, respectively, turning away from metaphysics (or religion). In this way we immediately obtain a division into two groups: skepticism, materialism and positivism stand on one side, spiritualism, idealism and theology on the other. We also at once see degrees of difference in this sequence, in that skepticism stands even farther away from theology than does materialism, while on the other hand idealism, e.g., in its pantheistic form, is a weakened form of theology in the proper sense.

The schema also proves fruitful, however, for the analysis of philosophical doctrines admissible in special contexts, in that one either arranges them in this manner or, in mixed cases, seeks out their materialistic and spiritualistic elements. Thus one would, for example, say that apriorism belongs in principle on the right and empiricism on the left side. On the other hand, however, there are also such mixed forms as an empiristically grounded theology. Furthermore one sees also that optimism belongs in principle toward the right and pessimism toward the left. For skepticism is certainly a pessimism with regard to knowledge. [Moreover,] materialism is inclined to regard the world as an unordered and therefore meaningless heap of atoms. In addition, death appears to it to be final and complete annihilation, while, on the other hand, theology and idealism see sense, purpose and reason in everything. On the other hand, Schopenhauer's pessimism is a mixed form, namely a pessimistic idealism. Another example of a theory evidently on the right is that of an objective right and objective aesthetic values, whereas the interpretation of ethics and aesthetics on the basis of custom, upbringing, etc., belongs toward the left.

Now it is a familiar fact, even a platitude, that the development of philosophy since the Renaissance has by and large gone from right to left—not in a straight line, but with reverses, yet still, on the whole. Particularly

lung in der Physik gerade in unserer Zeit einen Höhepunkt erreicht, indem
weitgehend die Möglichkeit einer Erkenntnis der objektivierbaren Sachver-
halte bestritten [[wird,]] und [[es]] behauptet wird, daß man sich | begnügen
muß, Beobachtungsresultate vorauszusagen; was eigentlich das Ende jeder
theoretischen Wissenschaft im üblichen Sinne ist (obwohl dieses Voraus-
sagen für praktische Zwecke wie Fernsehapparaten oder Atombomben vollkommen hinreichend sein kann).

Es müßte ein Wunder sein, wenn diese, ich möchte sagen rabiate, Entwicklung sich nicht auch in der Auffassung der Mathematik geltend gemacht hätte. Tatsächlich hat die Mathematik ihrer Natur nach als aprio-
rische Wissenschaft selbst überall eine Neigung nach rechts und hat daher
dem Zeitgeist, der seit der Renaissance herrschte, lang widerstanden, d. h.,
die empiristische Theorie der Mathematik, wie sie z. B. von Mill aufgestellt
wurde, fand keinen rechten Anklang. Ja, die Mathematik selbst entwickelt
sich zu immer höheren Abstraktionen weg von der Materie und zu immer
größerer Klarheit in den Grundlagen (etwa durch exakte Begründung der
Infinitesimalrechnung, der komplexen Zahlen) also weg vom Skeptizismus.

Aber schließlich um die Jahrhundertwende hatte auch ihre Stunde ge-
schlagen, insbesondere waren es die Antinomien der Mengenlehre, Widersprüche, die angeblich innerhalb der Mathematik aufgetreten waren, welche
in ihrer Bedeutung von Skeptikern und Empiristen übertrieben und zum
Vorwand für den Umsturz | nach links verwendet wurden. Ich sagte "an-
geblich" und "übertrieben", weil (1) diese Widersprüche nicht in der Mathematik sondern an ihrer äußersten Grenze zur Philosophie zu auftraten und
(2) sie in einer vollständig befriedigenden sowie für jeden, der diese Theorie versteht, beinahe selbstverständlichen Weise aufgelöst wurden. Aber
solche Argumente helfen nichts gegen den Zeitgeist und so war das Resultat, daß von vielen oder den meisten Mathematikern geleugnet wurde,
daß die Mathematik, wie sie sich vorher entwickelt hatte, ein System von
Wahrheiten darstellt; vielmehr wurde das nur für einen (je nach Temperament größeren oder kleineren) Teil der Mathematik anerkannt und alles
übrige wurde bestenfalls in einem hypothetischen Sinn beibehalten, indem
nämlich die Theorie mit Recht nur besagt, daß man aus gewissen (nicht
zu rechtfertig[[end]]en) Annahmen mit Recht gewisse Folgerungen ziehen
kann. Dabei schmeichelte man sich, daß man eigentlich alles wesentliche
beibehalten hätte. [Indem ja was den Mathematiker interessiert neben dem
Folgern aus jenen Annahmen, ist, was man vollziehen kann.] In Wahrheit
aber wird dadurch die Mathematik zu einer empirischen Wissenschaft, denn
wo ich etwa aus den willkürlich angenommenen Axiomen beweise, daß jede
natürliche Zahl Summe von vier Quadraten ist, so folgt daraus gar nicht
mit Sicherheit, daß ich nie ein Gegenbeispiel gegen dieses Theorem finden
werde, denn meine Axiome könnten | ja widerspruchsvoll sein und ich kann
höchstens sagen, es folge mit einer gewissen Wahrscheinlichkeit, da ja trotz
vieler Folgerungen bisher kein Widerspruch entdeckt wurde. Außerdem

in physics, this development has reached a peak in our own time, in that, to a large extent, the possibility of knowledge of the objectivizable states of affairs is denied, and it is asserted that we must be content to predict results of observations. This is really the end of all theoretical science in the usual sense (although this predicting can be completely sufficient for practical purposes such as [making] television sets or atom bombs).

It would truly be a miracle if this (I would like to say rabid) development had not also begun to make itself felt in the conception of mathematics. Actually, mathematics, by its nature as an a priori science, always has, in and of itself, an inclination toward the right, and, for this reason, has long withstood the spirit of the time [*Zeitgeist*] that has ruled since the Renaissance; i.e., the empiricist theory of mathematics, such as the one set forth by Mill, did not find much support. Indeed, mathematics has evolved into ever higher abstractions, away from matter and to ever greater clarity in its foundations (e.g., by [giving] an exact foundation of the infinitesimal calculus [and] the complex numbers)—thus, away from skepticism.

Finally, however, around the turn of the century, its hour struck: in particular, it was the antinomies of set theory, contradictions that allegedly appeared within mathematics, whose significance was exaggerated by skeptics and empiricists and which were employed as a pretext for the leftward upheaval. I say "*allegedly*" and "*exaggerated*" because, in the first place, these contradictions did not appear within mathematics but near its outermost boundary toward philosophy, and secondly, they have been resolved in a manner that is completely satisfactory and, for everyone who understands the theory, nearly obvious. Such arguments are, however, of no use against the spirit of the time, and so the result was that many or most mathematicians denied that mathematics, as it had developed previously, represents a system of truths; rather, they acknowledged this only for a part of mathematics (larger or smaller, according to [their] temperament) and retained the rest at best in a hypothetical sense—namely, [one] in which the theory properly asserts only that from certain assumptions (not themselves to be justified), we can justifiably draw certain conclusions. They thereby flattered themselves that everything essential had really been retained. [Since, after all, what interests the mathematician, in addition to drawing consequences from these assumptions, is what can be carried out.] In truth, however, mathematics becomes in this way an empirical science. For if I somehow prove from the arbitrarily postulated axioms that every natural number is the sum of four squares, it does not at all follow with certainty that I will never find a counterexample to this theorem, for my axioms could after all be inconsistent, and I can at most say that it follows with a certain probability, because in spite of many deductions no contra-

verlieren durch diese hypothetische Auffassung der Mathematik viele Fragen die Form: Gilt der Satz A oder gilt er nicht? Vollkommen im Sinn dann von willkürlichen Annahmen kann ich ja nicht erwarten, daß sie die sonderbare Eigenschaft haben, immer gerade entweder A oder $\sim A$ zu implizieren.

Obwohl diese nihilistischen Folgerungen dem Zeitgeist sehr gut entsprechen, so setzte hier eine Reaktion, selbstverständlich nicht seitens der Philosophie, wohl aber seitens der Mathematik ein, die, wie ich ja schon sagte, ihrer Natur nach dem Zeitgeist gegenüber sehr widerspenstig ist. Und so entstand jenes merkwürdige Zwitterding, welches sowohl dem Zeitgeist als der Natur der Mathematik gerecht zu werden suchte, welches der Hilbertsche Formalismus darstellt. Dieser besteht in folgendem: es wird einerseits in Übereinstimmung mit den in der heutigen Philosophie herrschenden Ideen anerkannt, daß die Axiome, von denen die Mathematik ausgeht, sich in ihrer Wahrheit in keiner Weise begründen oder erkennen lassen und daher das Ziehen von Folgerung[en] aus ihnen nur in einem hypothetischen Sinn Bedeutung hat, wobei man dieses Folgern selbst (um dem Zeitgeist noch weiter Genüge zu tun) als ein bloßes Spiel mit Symbolen nach gewissen, ebenfalls nicht einsichtigen, Regeln auffaßt.

Anderseits aber hielt man an dem der früheren rechtsseitigen Philosophie der Mathematik und dem Instinkt des Mathematikers entsprechenden Glauben fest, daß[a] ein Beweis für die Richtigkeit eines solchen Satzes, wie der Darstellbarkeit von allen Zahlen als Summen [von] Quadraten, eine sichere Begründung für diesen Satz geben muß; und ferner auch daran, daß jede präzis formulierte Ja- oder Nein-Frage in der Mathematik eine eindeutige Antwort haben muß, d. h., also man geht darauf aus, für an sich unbegründete Regeln des Zeichenspiels als eine ihnen sozusagen zufällig zukommende Eigenschaft zu beweisen, daß von zwei Sätzen A, $\sim A$ immer genau einer abgeleitet werden kann. Daß nicht beide abgeleitet werden können, beinhaltet die Widerspruchsfreiheit, und daß immer einer wirklich abgeleitet werden kann, bedeutet eindeutige Beantwortbarkeit der durch A ausgedrückten mathematischen Frage. Natürlich muß man, wenn man diese

6   beiden Aussagen mit mathematischer Sicherheit begründen will, | einen gewissen Teil der Mathematik im Sinne der alten rechtsseitigen Philosophie als wahr anerkennen. Aber das ist ein Teil, der dem Zeitgeist viel weniger zuwider ist als die hohen Abstraktionen der Mengenlehre. Er bezieht sich nämlich nur auf konkrete und endliche Objekte im Raum, nämlich die Zeichenkombinationen.

---

[a]The word *man* has been omitted here between *daß* and *ein*. See the last section of textual notes for this article.

diction has so far been discovered. In addition, through this hypothetical conception of mathematics, many questions lose the form "Does the proposition *A* hold or not?"   [For,] from assumptions construed as completely arbitrary, I can of course not expect that they have the peculiar property of implying, in every case, exactly either *A* or ∼*A*.

Although these nihilistic consequences are very well in accord with the spirit of the time, here a reaction set in—obviously not on the part of philosophy, but rather on that of mathematics, which, by its nature, as I have already said, is very recalcitrant in the face of the *Zeitgeist*. And thus came into being that curious hermaphroditic thing that Hilbert's formalism represents, which sought to do justice both to the spirit of the time and to the nature of mathematics. It consists in the following: on the one hand, in conformity with the ideas prevailing in today's philosophy, it is acknowledged that the truth of the axioms from which mathematics starts out cannot be justified or recognized in any way, and therefore the drawing of consequences from them has meaning only in a hypothetical sense, whereby this drawing of consequences itself (in order to satisfy even further the spirit of the time) is construed as a mere game with symbols according to certain rules, likewise not [supported by] insight.

But, on the other hand, one clung to the belief, corresponding to the earlier "rightward" philosophy of mathematics and to the mathematician's instinct, that a proof for the correctness of such a proposition as the representability of every number as a sum of four squares must provide a secure grounding for that proposition; and furthermore, also that every precisely formulated yes-or-no question in mathematics must have a clear-cut answer. I.e., one thus aims to prove, for inherently unfounded rules of the game with symbols, as a property that attaches to them so to speak by accident, that of two sentences *A* [and] ∼*A*, exactly one can always be derived. That not both can be derived constitutes consistency, and that one can always actually be derived means that the mathematical question expressed by *A* can be unambiguously answered. Of course, if one wishes to justify these two assertions with mathematical certainty, a certain part of mathematics must be acknowledged as true in the sense of the old rightward philosophy. But that is a part that is much less opposed to the spirit of the time than the high abstractions of set theory. For it refers only to concrete and finite objects in space, namely the combinations of symbols.

Was ich bisher sagte sind ja eigentlich nur Selbstverständlichkeiten, an die ich bloß erinnern wollte, weil sie für das Folgende wichtig sind. Aber der nächste Schritt in der Entwicklung ist nun der, daß sich zeigt, [[daß]] es nicht möglich ist, die alten rechtsseitigen Aspekte der Mathematik auf eine solche mit dem Zeitgeist mehr oder weniger in Übereinstimmung stehende Weise zu retten. Sogar wenn man sich auf die Theorie der natürlichen Zahlen beschränkt, ist es unmöglich, ein System von Axiomen und formalen Regeln zu finden, aus dem für jeden zahlentheoretischen Satz immer entweder A oder $\sim A$ ableitbar wäre, und es ist ferner unmöglich, für einigermaßen umfassende Axiome der Mathematik einen Widerspruchsfreiheitsbeweis durch bloße Betrachtung der konkreten Zeichenkombinationen ohne Einführung abstrakterer Elemente zu führen. Die Hilbertsche Kombination von Materialismus und den Aspekten der klassischen Mathematik erweist sich also als unmöglich.

Daher bleiben nur zwei Möglichkeiten offen; man muß entweder die alten rechtsseitigen Aspekte der Mathematik aufgeben oder man muß versuchen, diese im Widerspruch mit dem Zeitgeist aufrechtzuerhalten. Offenbar ist der erste Weg der einzige, der in unsere Zeit paßt und der daher auch meistenteils eingeschlagen wird. Aber man sollte sich bewußt bleiben, daß das eine rein negative Haltung ist. Man gibt einfach Aspekte auf, deren Erfüllung doch jedenfalls sehr wünschenswert wäre und die viel für sich haben, nämlich einerseits der Mathematik die Sicherheit ihrer Erkenntnisse zu wahren und anderseits den Glauben aufrechtzuerhalten, daß für von der Vernunft gestellte klare Fragen die Vernunft auch klare Antworten finden kann. Und man beachte, man gibt diese Aspekte auf, nicht weil die erzielten mathematischen Resultate dazu zwingen, sondern weil das die einzige Möglichkeit ist, trotz dieser Resultate mit der herrschenden Philosophie in Übereinstimmung zu bleiben.

Nun kann man sich ja den großen Fortschritten, welche unsere Zeit in vieler Hinsicht aufzuweisen hat, keineswegs verschließen und man kann mit einem gewissen Recht geltend machen, daß diese Fortschritte ebendiesem 7 Linksgeist in der Philosophie und Weltanschauung zu danken seien. | Aber anderseits, wenn man die Sache in richtiger historischer Perspektive betrachtet, muß man sagen, daß die Fruchtbarkeit des Materialismus zum Teil nur auf den Exzessen und der falschen Richtung der vorhergehenden rechtsseitigen Philosophie beruht. Was Recht und Unrecht bzw. Wahrheit und Falschheit dieser beiden Richtungen betrifft, so scheint mir die richtige Einstellung die zu sein, daß Wahrheit in der Mitte liegt oder in einer Kombination dieser beiden Auffassungen besteht.

Nun hatte ja im Fall der Mathematik gerade Hilbert eine solche Kombination versucht, aber diese offenbar zu primitiv und zu stark nach einer Richtung hinneigend. Jedenfalls ist kein Grund, dem Zeitgeist blindlings zu vertrauen und es ist daher zweifellos der Mühe wert, einmal die andere der oben genannten Alternativen, welche die genannten Resultate übriglassen,

What I have said so far are really only obvious things, which I wanted to recall merely because they are important for what follows. But the next step in the development is now this: it turns out that it is impossible to rescue the old rightward aspects of mathematics in such a manner as to be more or less in accord with the spirit of the time. Even if we restrict ourselves to the theory of natural numbers, it is impossible to find a system of axioms and formal rules from which, for every number-theoretic proposition A, either A or $\sim A$ would always be derivable. And furthermore, for reasonably comprehensive axioms of mathematics, it is impossible to carry out a proof of consistency merely by reflecting on the concrete combinations of symbols, without introducing more abstract elements. The Hilbertian combination of materialism and aspects of classical mathematics thus proves to be impossible.

Hence, only two possibilities remain open. One must either give up the old rightward aspects of mathematics or attempt to uphold them in contradiction to the spirit of the time. Obviously the first course is the only one that suits our time and [is] therefore also the one usually adopted. One should, however, keep in mind that this is a purely negative attitude. One simply gives up aspects whose fulfillment would in any case be very desirable and which have much to recommend themselves: namely, on the one hand, to safeguard for mathematics the certainty of its knowledge, and on the other, to uphold the belief that for clear questions posed by reason, reason can also find clear answers. And as should be noted, one gives up these aspects not because the mathematical results achieved compel one to do so but because that is the only possible way, despite these results, to remain in agreement with the prevailing philosophy.

Now one can of course by no means close one's eyes to the great advances which our time exhibits in many respects, and one can with a certain justice assert that these advances are due just to this leftward spirit in philosophy and world-view. But, on the other hand, if one considers the matter in proper historical perspective, one must say that the fruitfulness of materialism is based in part only on the excesses and the wrong direction of the preceding rightward philosophy. As far as the rightness and wrongness, or, respectively, truth and falsity, of these two directions is concerned, the correct attitude appears to me to be that the truth lies in the middle or consists of a combination of the two conceptions.

Now, in the case of mathematics, Hilbert had of course attempted just such a combination, but one obviously too primitive and tending too strongly in one direction. In any case there is no reason to trust blindly in the spirit of the time, and it is therefore undoubtedly worth the effort [at least] once to try the other of the alternatives mentioned above—which the results cited leave open—in the hope of obtaining in this way a

zu versuchen in der Erwartung, auf diese Weise eine brauchbare Kombination zu erhalten. Das heißt offenbar die Sicherheit der Mathematik nicht dadurch sicherzustellen, daß man gewisse Eigenschaften in Projektion auf materielle Systeme, nämlich das Umgehen mit physischen Symbolen, beweist, sondern dadurch, daß man die Erkenntnis der abstrakten Begriffe selbst, welche zur Aufstellung jener mechanischen Systeme führt, kultiviert (vertieft) und daß man ferner nach dem gleichen Verfahren Einsichten über die Lösbarkeit und über tatsächliche Methoden zur Lösung aller sinnvollen mathematischen Probleme zu gewinnen sucht.

Auf welche Weise aber ist es möglich, die Kenntnis jener abstrakten Begriffe zu erweitern, d. h., also diese Begriffe selbst zu präzisieren und umfassende und sichere Einsicht über die für sie bestehenden Grundrelationen, d. h., die für sie geltenden Axiome, zu gewinnen? Offenbar nicht dadurch oder jedenfalls nicht ausschließlich dadurch, daß man versucht, explizite Definitionen für Begriffe und Beweise für Axiome zu geben. Denn dann braucht man ja dafür offenbar andere undefinierbare abstrakte Begriffe und für sie geltende Axiome. Sonst hätte man ja nichts, woraus man definieren oder beweisen könnte. Das Verfahren muß also wenigstens zum 8 großen Teil | in einer Sinnklärung bestehen, die nicht in Definieren besteht.

Nun gibt es ja heute den Beginn einer Wissenschaft, welche behauptet, eine systematische Methode für eine solche Sinnklärung zu haben, und das ist die von Husserl begründete Phänomenologie. Die Sinnklärung besteht hier darin, daß man die betreffenden Begriffe schärfer ins Auge faßt, indem man die Aufmerksamkeit in einer bestimmten Weise dirigiert, nämlich auf unsere eigenen Akte bei der Verwendung dieser Begriffe, auf unsere Mächte bei der Vollführung unserer Akte, etc. Man muß sich dabei klar darüber sein, daß diese Phänomenologie nicht eine Wissenschaft im selben Sinn ist wie die anderen Wissenschaften. Sie ist vielmehr [oder sollte jedenfalls sein] ein Verfahren oder Technik, welches in uns einen neuen Bewußtseinszustand hervorbringen soll, in dem wir die von uns verwendeten Grundbegriffe unseres Denkens detaillieren oder andere bisher uns unbekannte Grundbegriffe erfassen. Ich glaube, es besteht gar kein Grund, ein solches Verfahren von vornherein als aussichtslos abzulehnen. Am wenigsten Grund dafür haben natürlich Empiristen, denn das würde heißen, daß ihr Empirismus in Wahrheit ein Apriorismus mit dem verkehrten Vorzeichen ist.

Aber nicht nur besteht kein objektiver Grund der Ablehnung [der Phänomenologie], sondern im Gegenteil, man kann Gründe zugunsten angeben. Wenn man die Entwicklung eines Kindes betrachtet, so sieht man, daß diese in zwei Richtungen vor sich geht; einerseits besteht sie in einem Experimentieren mit den Gegenständen der Außenwelt und mit seinen Sinnen- und Bewegungsorganen, anderseits in einem besseren und besseren Verstehenlernen der Sprache und das heißt, sobald das Kind über die primitivste Form des Bezeichnens hinaus ist, der Grundbegriffe auf denen sie beruht. Hinsichtlich der Entwicklung in dieser zweiten Richtung kann man

workable combination. Obviously, this means that the certainty of mathematics is to be secured not by proving certain properties by a projection onto material systems—namely, the manipulation of physical symbols—but rather by cultivating (deepening) knowledge of the abstract concepts themselves which lead to the setting up of these mechanical systems, and further by seeking, according to the same procedures, to gain insights into the solvability, and the actual methods for the solution, of all meaningful mathematical problems.

In what manner, however, is it possible to extend our knowledge of these abstract concepts, i.e., to make these concepts themselves precise and to gain comprehensive and secure insight into the fundamental relations that subsist among them, i.e., [into] the axioms that hold for them? Obviously not, or in any case not exclusively, by trying to give explicit definitions for concepts and proofs for axioms, since for that one obviously needs other undefinable abstract concepts and axioms holding for them. Otherwise one would have nothing from which one could define or prove. The procedure must thus consist, at least to a large extent, in a clarification of meaning that does not consist in giving definitions.

Now in fact, there exists today the beginning of a science which claims to possess a systematic method for such a clarification of meaning, and that is the phenomenology founded by Husserl. Here clarification of meaning consists in focusing more sharply on the concepts concerned by directing our attention in a certain way, namely, onto our own acts in the use of these concepts, onto our powers in carrying out our acts, etc. But one must keep clearly in mind that this phenomenology is not a science in the same sense as the other sciences. Rather it is [or in any case should be] a procedure or technique that should produce in us a new state of consciousness in which we describe in detail the basic concepts we use in our thought, or grasp other basic concepts hitherto unknown to us. I believe there is no reason at all to reject such a procedure at the outset as hopeless. Empiricists, of course, have the least reason of all to do so, for that would mean that their empiricism is, in truth, an apriorism with its sign reversed.

But not only is there no objective reason for the rejection [of phenomenology], but on the contrary one can present reasons in its favor. If one considers the development of a child, one notices that it proceeds in two directions: it consists on the one hand in experimenting with the objects of the external world and with its [own] sensory and motor organs, on the other hand in coming to a better and better understanding of language, and that means—as soon as the child is beyond the most primitive designating [of objects]—of the basic concepts on which it rests. With respect to the development in this second direction, one can justifiably say that the child

mit Recht sagen, daß das Kind Bewußtseinszustände verschiedener Höhen durchläuft; z. B. kann man sagen, daß ein höherer Bewußtseinszustand erreicht wird, wo das Kind zuerst den Gebrauch des Wortes lernt und ebenso in dem Augenblick, wo es zum ersten Mal eine logische Schlußfolgerung versteht. Nun kann man ja die ganze Entwicklung der empirischen Wissenschaft als eine systematische und bewußte Erweiterung dessen, was das Kind tut, wenn es sich in der ersten Richtung entwickelt, auffassen. | Der Erfolg dieses Verfahrens ist aber ein erstaunlicher und weitaus größerer als man a priori erwarten würde. Er führt ja zur ganzen technologischen Entwicklung der neueren Zeit. Das läßt es also als durchaus möglich erscheinen, daß auch ein systematisches und bewußtes Weitergehen in der zweiten Richtung die Erwartungen, die man a priori haben kann, weit übertreffen wird.

Tatsächlich hat man Beispiele wo, sogar ohne ein systematisches und bewußtes Verfahren anzuwenden, sondern ganz von selbst, eine beträchtliche weitere Entwicklung in der zweiten Richtung, über den "gesunden Menschenverstand" hinaus, stattfindet. Es zeigt sich nämlich, daß bei einem systematischen Aufstellen der Axiome der Mathematik immer wieder neue und neue⟦re⟧ Axiome evident werden, die nicht formallogisch aus den bisher aufgestellten folgen. Es ist durch ⟦die⟧ früher erwähnten negativen Resultate gar nicht ausgeschlossen, daß trotzdem auf diese Weise jede klar gestellte mathematische Ja- oder Nein-Frage lösbar ist, denn eben dieses Evidentwerden immer neuerer Axiome auf Grund des Sinnes der Grundbegriffe ist etwas, was eine Maschine nicht nachahmen kann.

Ich möchte darauf aufmerksam machen, daß dieses intuitive Erfassen immer neuerer und von den früheren logisch unabhängiger Axiome, welches zur Lösbarkeit aller Probleme selbst eines sehr eingeschränkten Gebiets nötig ist, prinzipiell mit der Kantschen Auffassung der Mathematik übereinstimmt. Allerdings sind die diesbezüglichen Äußerungen von Kant wörtlich verstanden unrichtig, denn Kant behauptet, daß wir zur Ableitung der geometrischen Theoreme immer neue geometrische Intuitionen benötigen und also eine rein logische Ableitung aus einer endlichen Zahl von Axiomen unmöglich ist. Das ist nachweislich falsch. Wenn wir aber in dieser Aussage den Term "geometrisch" durch "mathematisch" oder durch "mengentheoretisch" ersetzen, dann wird es eine nachweislich richtige Aussage. Ich glaube, es ist eine allgemeine Eigenschaft vieler Kantschen Behauptungen, daß sie wörtlich verstanden falsch sind, aber in einem allgemeineren Sinn tiefe Wahrheiten enthalten. Insbesondere geht die ganze phänomenologische Methode,| wie ich sie vorhin skizzierte, der Idee nach auf Kant zurück und was Husserl getan hat, war bloß, daß er sie zuerst präziser formuliert und völlig bewußt gemacht und für einzelne Gebiete wirklich durchgeführt hat. Man sieht ja schon aus der von Husserl angewendeten Terminologie, wie positiv er selbst seine Beziehung zu Kant wertet.

Ich glaube, daß gerade darauf, daß die Kantsche Philosophie letzten Endes, allerdings in einer nicht völlig klaren Weise, auf der Idee der Phäno-

passes through states of consciousness of various heights, e.g., one can say that a higher state of consciousness is attained when the child first learns the use of words, and similarly at the moment when for the first time it understands a logical inference.

Now one may view the whole development of empirical science as a systematic and conscious extension of what the child does when it develops in the first direction. The success of this procedure is indeed astonishing and far greater than one would expect a priori: after all, it leads to the entire technological development of recent times. That makes it thus seem quite possible that a systematic and conscious advance in the second direction will also far exceed the expectations one may have a priori.

In fact, one has examples where, even without the application of a systematic and conscious procedure, but entirely by itself, a considerable further development takes place in the second direction, one that transcends "common sense". Namely, it turns out that in the systematic establishment of the axioms of mathematics, new axioms, which do not follow by formal logic from those previously established, again and again become evident. It is not at all excluded by the negative results mentioned earlier that nevertheless every clearly posed mathematical yes-or-no question is solvable in this way. For it is just this becoming evident of more and more new axioms on the basis of the meaning of the primitive notions that a machine cannot imitate.

I would like to point out that this intuitive grasping of ever newer axioms that are logically independent from the earlier ones, which is necessary for the solvability of all problems even within a very limited domain, agrees in principle with the Kantian conception of mathematics. The relevant utterances by Kant are, it is true, incorrect if taken literally, since Kant asserts that in the derivation of geometrical theorems we always need new geometrical intuitions, and that therefore a purely logical derivation from a finite number of axioms is impossible. That is demonstrably false. However, if in this proposition we replace the term "geometrical" by "mathematical" or "set-theoretical", then it becomes a demonstrably true proposition. I believe it to be a general feature of many of Kant's assertions that literally understood they are false but in a broader sense contain deep truths. In particular, the whole phenomenological method, as I sketched it above, goes back in its [central] idea to Kant, and what Husserl did was merely that he first formulated it more precisely, made it fully conscious and actually carried it out for particular domains. Indeed, just from the terminology used by Husserl, one sees how positively he himself values his relation to Kant.

I believe that precisely because in the last analysis the Kantian philosophy rests on the idea of phenomenology, albeit in a not entirely clear

menologie beruht und eben dadurch etwas völlig neues und eben für jede
echte Philosophie charakteristisches in das Denken eingeführt hat, daß gerade darauf der ungeheure Einfluß beruht, den Kant auf die ganze folgende
Entwicklung der Philosophie ausübte. Es gibt ja kaum irgendeine spätere
Richtung, die nicht in irgendeiner Weise sich auf Kantsche Ideen bezieht.
Anderseits haben aber eben wegen der Unklarheit und im wörtlichen Sinn
Unrichtigkeit vieler Kantscher Formulierungen sich ganz entgegengesetzte
philosophische Richtungen aus [[dem]] Kantschen Denken entwickelt, von
denen aber keine dem Kantschen Denken in seinem Kern wirklich gerecht
wurde. Dieser Forderung scheint mir erst die Phänomenologie zu genügen,
welche ganz im Sinne Kants sowohl dieselben Salto mortale des Idealismus
in eine neue Metaphysik als auch die positivistische Ablehnung jeder Metaphysik vermeidet. Wenn nun aber schon der falsch verstandene Kant zu so
vielem Interresanten in Philosophie und indirekt auch in der Wissenschaft
geführt hat, wieviel mehr kann man es von dem richtig verstandenen Kant
erwarten?

way, and has just thereby introduced into our thought something completely new, and indeed characteristic of every genuine philosophy—it is
precisely on that, I believe, that the enormous influence which Kant has
exercised over the entire subsequent development of philosophy rests. Indeed, there is hardly any later direction that is not somehow related to
Kant's ideas. On the other hand, however, just because of the lack of
clarity and the literal incorrectness of many of Kant's formulations, quite
divergent directions have developed out of Kant's thought—none of which,
however, really did justice to the core of Kant's thought. This requirement
seems to me to be met for the first time by phenomenology, which, entirely
as intended by Kant, avoids both the death-defying leaps of idealism into
a new metaphysics as well as the positivistic rejection of all metaphysics.
But now, if the misunderstood Kant has already led to so much that is
interesting in philosophy, and also indirectly in science, how much more
can we expect it from Kant understood correctly?

Introductory note to *1970

1. Background

Gödel showed his *1970 to Dana Scott, and discussed it with him, in February 1970. Gödel was very concerned about his health at that time, feared that his death was near, and evidently wished to insure that this proof would not perish with him. Later in 1970, however, he apparently told Oskar Morgenstern that though he was "satisfied" with the proof, he hesitated to publish it, for fear it would be thought "that he actually believes in God, whereas he is only engaged in a logical investigation (that is, in showing that such a proof with classical assumptions [completeness, etc.], correspondingly axiomatized, is possible)."[a]

Scott made notes on the proof and presented a version of the argument to his seminar on logical entailment at Princeton University in the fall of 1970. Through this presentation and the recollections and notes of those who attended the seminar, Gödel's ontological proof has become fairly widely known. Discussion of the proof, thus far, has been based largely on Scott's version of it (*Scott 1987*), which differs somewhat in form from Gödel's own memorandum. The latter is published here—though not for the first time; like Scott's version, it was published as an appendix to *Sobel 1987*, pages 256–7.

Gödel had devised his ontological proof some time before 1970. Other, presumably earlier, versions of it have been found among his papers. A sheet of paper headed "Ontological Proof" (in German), and dated, in Gödel's own hand, "ca. 1941", contains some but not all of the ideas of the proof. Extensive preparatory material is contained in the philosophical notebook "Phil XIV". The first page of this notebook bears a notation indicating that it was written during the period "Ca. July 1946–May 1955". The last page of the notebook contains the note "Asbury Park 1954 p. 100 ff.", which presumably applies to the pages (103–109) pertaining to the ontological proof. Other documents, including letters, indicate that Gödel intended to leave Princeton for the shore 9 August 1954, was vacationing in Asbury Park on 25 August 1954, and was probably back in Princeton by 3 October 1954. We may reasonably assume, then, that the notebook pages on the ontological proof were written in the late summer and early fall of 1954 and were completed at any rate

---

[a]Morgenstern's diary for 29 August 1970, Box 15 of the Oskar Morgenstern Papers, quoted by courtesy of the Special Collections Department, Duke University Library, Durham, North Carolina. I am indebted to John Dawson for noticing and communicating this item.

by May 1955.[b] Relevant excerpts from the notebook, and two of the (presumably earlier) loose sheets headed "Ontological Proof", including the one dated "ca. 1941", are published in Appendix B to this volume.

Among the historic sponsors of the ontological argument, it is not to Anselm or Descartes but to Leibniz that the parentage of Gödel's proof belongs, as scholars interested in the proof have long recognized (see, e.g., *Sobel 1987*, page 241). The study of Leibniz is known to have been a major intellectual preoccupation for Gödel during the 1930's (*Menger 1981*, §§8, 12), and especially during 1943–46 (*Wang 1987*, pages 19, 21, 27). Little discussion of Leibniz's treatment of the ontological argument as such has been found in Gödel's papers, but he must have known two things about it:

1. Leibniz held that Descartes's ontological proof is incomplete. It does succeed in proving the conditional proposition that *if* God's existence is so much as possible, then God actually (and indeed necessarily) exists. But it assumes without proof that God's existence is possible; and that, Leibniz argues, must be proved in order to complete the demonstration. Leibniz says this in many places in his writings, some of them so familiar to students of Leibniz that Gödel can safely be assumed to have known them (e.g., *Leibniz 1969*, pages 292–3). In January 1678 Leibniz wrote down an elaborate and interesting proof of the conditional proposition (*Leibniz 1923–*, II, i, 390–1), but I have seen no specific evidence that Gödel was familiar with that text.

2. Leibniz also held that the ontological proof can be completed by proving the possibility of God's existence. His main attempt to accomplish this is based on a conception of God as *Ens perfectissimum*, a being whose attributes are all the perfections, where perfections are identified with simple, purely positive qualities, and where a purely positive quality cannot be limited and therefore cannot be an inferior degree of any quality. Leibniz argues that purely positive qualities must all be consistent with each other, so that no inconsistency can arise from the conception of an *Ens perfectissimum*, which must therefore be a possible being. This argument is most fully developed in texts Leibniz wrote in 1676 (*Leibniz 1923–*, VI, iii, 395–6, 571–79). It recurs with almost cryptic brevity at the end of his life in §45 of the famous "Monadology" of 1714, but it has become known mainly through one of the 1676 texts, "That an *Ens Perfectissimum* Exists", which has been generally accessible, both in Latin and in English translation, since the end of the nineteenth century (*Leibniz 1923–*, VI, iii, 578–9 = *1969*, pages 167–8).

---

[b]I am indebted to John and Cheryl Dawson for the information on dating cited here.

This text at least, and the "Monadology", were surely known to Gödel, whose ontological proof is built around an idea of positive properties.

Gödel's treatment of the ontological proof resembles Leibniz's on both of these points. The first point will be the subject of §2 of this introduction. The second will occupy us in §§3–4.

2. If possible, then actual

Gödel resembles Leibniz in making the ontological proof proceed by way of the conditional thesis that if the divine existence is so much as possible, then it is actual, and indeed necessary. In his **1970*, this thesis occurs as the line

$$M(\exists x)G(x) \supset N(\exists y)G(y),$$

which I shall be calling (iii). (I follow Gödel in using M and N as possibility and necessity operators, respectively.) As noted above, however, Gödel shows no clear influence of Leibniz's fullest argument for the thesis, which turns on a rather different conception of "essence" from Gödel's.

The grounds Gödel gives for the conditional thesis show more affinity with a type of "ontological argument" based on modern modal logic which has gained currency in the last thirty years. Charles Hartshorne published a proof of this type in his *1962* (pages 50–53), and subsequent discussion has established its logical properties quite clearly (see *Lewis 1970*; *Adams 1971*; *Plantinga 1974*, pages 196–221). In a presentation approximating Hartshorne's, the first part of the proof has the following steps, which are found also in Gödel's proof:

(i) $N[(\exists x)G(x) \supset N(\exists y)G(y)]$
(ii) $M(\exists x)G(x) \supset MN(\exists y)G(y)$
(iii) $M(\exists x)G(x) \supset N(\exists y)G(y).$

Step (i) is the necessitation of the line immediately following, and inferred from, the theorem following Axiom 4 in Gödel's **1970*. Since Gödel takes this line to be entailed by a theorem, he may be presumed to accept its necessitation.[c] Step (i) is the thesis that it is necessary that if God exists at all, God exists necessarily—or, more briefly, that it is impossible for God to exist contingently. Some philosophers have thought that the concept of God is a concept of a Necessary Being, and that (i) follows straightforwardly from the concept of God (cf. *Hartshorne 1962*,

---

[c]The presumption is made explicit in Sobel's reconstruction of the proof (*1987*, pp. 247–8).

page 41; *Findlay 1965*). Gödel gives a more complicated derivation of (i), which hinges on the claim (Axiom 4) that necessary existence is a positive property. Since he has made it true by definition, and hence necessarily true, that God (if God exists) has all positive properties, and since (by Axiom 3) any property that is positive is necessarily positive, it follows that God (if God exists) has necessary existence. That is, (i) follows from these assumptions. This strategy for proving (i) is obviously akin to the attempts that have been made, in the history of the ontological argument, to derive something equivalent to (i) from the claim that necessary existence is a "perfection", it being assumed that God, by definition, possesses all perfections.[d]

Step (ii) is inferred from the line that corresponds to (i) in Gödel's proof, and does indeed follow from (i) by the principle

(iv) $N(p \supset q) \supset (Mp \supset Mq)$,

which would be an axiom or theorem in any system of modal logic that would be likely to be used in this context. The inference from (ii) to (iii), on which Gödel also relies, depends on a more controversial principle,

(v) $MNp \supset Np$,

which is a form of the characteristic axiom of S5, the most powerful of the standard systems of modal propositional logic. One of the firm results of recent studies of modal versions of the ontological argument is that (iii) does follow from (i) in S5.[e] Whether it is appropriate in this context to rely on S5, and particularly on (v), is certainly open to question, but several philosophers have believed that it is appropriate.[f] Gödel must apparently be counted among them, though he may have had some reservation on this point.[g]

There is no evidence that Gödel was influenced by the recent work of others on modal ontological proofs. The derivation of (iii) from (i) in S5 had been published by Hartshorne in 1962 and was attracting the attention of other students when Gödel showed his ontological proof to Scott in 1970. But, as already noted, Gödel had developed his proof

---

[d]See *Anselm 1974*, pp. 94–5 (*Proslogion*, chapter 3); *Malcolm 1960*, p. 46.

[e]See *Hartshorne 1962*, pp. 39–40, 51–53; *Plantinga 1974*, pp. 213–17. In one sense, S5 is more than is needed. A similar modal ontological proof can be constructed in the somewhat weaker modal system sometimes called "Brouwerian", in which (v) is replaced by the axiom $p \supset NMp$ (*Adams 1971*, pp. 40–48). But there is no strong reason for thinking the Brouwerian system more acceptable than S5 in this context.

[f]*Hartshorne 1962*, pp. 39–40, 51–53; *Adams 1971*, pp. 42, 45–6; *Plantinga 1974*, p. 215; *Sobel 1987*, p. 246.

[g]Morton White (in personal correspondence) reports that Gödel expressed "reservations about his ontological proof because of his doubt about using some principle in modal logic", but that Gödel did not specifically mention S5 or its characteristic axiom. So far as I am aware, this is the only point in the proof about which Gödel is known to have expressed a reservation.

some years earlier. His notebook entries on the proof, from 1954 or 1955, do not articulate the modal logic used in the proof, but there is no reason to doubt that he was already consciously relying on (v) or on something equivalent to it. One of Gödel's early sketches for the ontological proof, dating perhaps from the 1940's, ends with an inference precisely from (ii) to (iii), in which he must be relying implicitly on (v) as a principle.[h] Gödel may in fact have been the first student of modern logic to see that this principle could be used to prove that "if the concept of necessary existence is consistent, then there are things to which it applies", as he put it in that early sketch.

One problem about the logical apparatus of Gödel's *1970* should be noted. Definition 2 fails to imply that every essence of x must be true of x. It implies, indeed, that if there is any property that is necessarily false of everything, it is an essence of x. Then from the definition of "$E(x)$", with the assumption that there is a property that is necessarily false of everything (an assumption that Gödel seems to make in *1970*, since he treats "$x \neq x$" as expressing a [negative] property), we could further infer that "$E(x)$" is not true of anything. The latter conclusion is obviously contrary to Gödel's intent in the proof. Moreover, the claim in footnote 3, that "any two essences of x are *necessarily equivalent*", also seems to presuppose that every essence of x must be true of x. Scott (*1987*, page 258) doubtless represents Gödel's intention correctly when he adds "$\varphi(x)$" as a conjunct to the right side of the definition of "φ Ess x".[i] It is interesting that the page on which Gödel wrote the early sketch of his ontological proof mentioned in the previous paragraph ends with a note in which Gödel proposes a definition of essence whose right side is like that of Definition 2 of *1970* except that "$\varphi(x)$" is added as a conjunct, so that the definition does imply that every essence of x is true of x.[j]

3. Leibniz's possibility proof

Accepting the conditional thesis that if God's existence is possible, then God exists, one needs only the further premise that God's existence is possible in order to detach the consequent and infer by modus

---

[h]This sketch is printed in Appendix B to this volume. Of the two such documents reproduced there, it is the one not dated by Gödel.

[i]Scott is followed in this by Sobel (*1987*, page 244) and Anderson (*1990*, page 292). I assume that a quantifier, elsewhere in the right side of Scott's definition of "φ Ess x", printed in Sobel's appendix as "$\forall x$", should be "$\forall \psi$".

[j]Most of the observations in this paragraph are due to Charles Parsons.

ponens that God actually exists. But how to justify the possibility premise? Possibility is often assumed rather easily, but should not be in this case, for at least two reasons. One reason, emphasized by Leibniz, is that the concept of God is the concept of a sort of maximum (a maximum of perfection), and the concept of a maximum can seem innocent at first glance, while representing something really impossible (e.g., "the largest number"; see *Leibniz 1969*, page 211). Another reason, not noted by Leibniz but prominent in recent discussion of modal ontological arguments, is that at the point in such an argument at which a possibility premise is required, it is typically supposed to have been proved that the existence of God is either impossible or necessary ("$M(\exists x)G(x) \supset N(\exists y)G(y)$" in Gödel's proof). In this context, assuming the possibility of God's existence commits one quite directly to the impossibility of God's nonexistence. But why shouldn't the possibility of God's nonexistence be assumed as easily as the possibility of God's existence? (Cf. *Adams 1988*.) So it would be important, in completing a modal ontological proof, to give a *proof* that God's existence is possible.

Leibniz's attempt to accomplish this begins with a conception of God as a being that possesses all perfections. "A perfection", he says, is what he calls "every simple quality that is positive and absolute, or [*seu = that is*] that expresses without any limits whatever it expresses."[k] Three points about this definition claim our attention. (1) Perfections are *qualities*. What is meant here may not be precisely the Aristotelian category of "quality", but it is surely something narrower than we might mean by "property". For instance, it presumably does not include relations. The divine nature is constituted by *internal* properties. (2) The *simplicity* of the perfections plays a part in the best-known formulation of Leibniz's possibility proof, excluding any analysis of them (*Leibniz 1969*, page 167). But this is superfluous, as Leibniz recognized (*1923–*, VI, iii, 572). Pure positiveness is the only characteristic of the perfections that is really needed for the proof, and the only one that appears in the brief version of the proof in the "Monadology" (§45). (3) The final clause of the definition indicates that "absolute" is being used to mean *unlimited*, not qualified by any limitation. And limitation is understood here as a partial negation. "Absolute" is therefore an intensification of "positive": a perfection is a *purely* positive quality, a quality that involves no negation at all. What sort of involvement of negation is excluded will become clear as we examine the strategy of Leibniz's argument.

Leibniz argues that all *simple* positive qualities are mutually compatible, on the ground that if they were not, "one would express the

---

[k]I give my own translation from *Leibniz 1923–*, VI, iii, 578–9. An English translation of the whole text is found in *Leibniz 1969*, pp. 167–8.

exclusion of the other, and so one of them would be the negative of the other, which is contrary to the hypothesis, for we assumed that they are all affirmative." He argues further that it follows that any conjunction of purely positive qualities is possible, "for if individual [attributes] are thus compatible, pluralities will be too, and therefore also composites" (*1923–*, VI, iii, 572). His argument for possibility depends on the exclusion of negation from the construction of any purely positive quality. It seems to presuppose a conception according to which a purely positive quality must either be a simple positive quality or, if complex, must be constructible from simple positive qualities without the aid of negation.[1] That is the sense in which, for Leibniz, a purely positive quality cannot involve negation.

Leibniz assumes that the only way in which a conjunction of qualities could be impossible is by having, when fully analyzed, two conjuncts, of which one is formally the negation of the other. But a conjunction of purely positive qualities cannot be impossible in this way. For it cannot, when fully analyzed, have any conjunct that is formally the negation of anything.

Since perfections are purely positive qualities, Leibniz infers that the conjunction of all perfections cannot be impossible, and therefore is possible. Treating the possibility of a conjunction of qualities as equivalent to the possibility of the existence of a being possessing all those qualities, he infers that the existence of a being possessing all perfections is possible. And since such a being would satisfy his definition of God, he infers that the existence of God is possible.

Two possible difficulties for this argument may be noted here.

(1) One might question the assumption that the only way in which a conjunction of qualities can be impossible is by a contradiction involving formal negation, occurring between the qualities or arising in their analysis into a conjunction of simpler qualities. There is ample basis in Leibniz's writings for ascribing to him such a formalistic conception of impossibility; but it is not obvious that even his own statements and arguments are always in keeping with it. And some philosophers, including Descartes (*1985*, pages 45–6), for example, have maintained that simple properties can necessarily exclude each other without either of them being analyzable at all, and without either being the formal negation of the other.

(2) A proof that God's existence is possible will not satisfy the needs of a modal ontological argument unless the God whose existence is proved possible is one that must exist necessarily if at all. But why couldn't the

---

[1] It may even be that he thought the construction must involve no other logical operation besides conjunction.

conjunction of all perfections be exemplified contingently? The obvious move for Leibniz to attempt in response to this question is to hold that necessary existence is one of the perfections. And in the best-known version of his possibility proof he does say at least that *existence* is one of the perfections (*Leibniz 1969*, page 167). But this response is attended with problems. One which soon occurred to Leibniz himself is that it may be doubted whether existence is a *quality*, as perfections must be (*Leibniz 1923–*, II, i, 313); and presumably this doubt would apply to necessary existence as well.

4. Gödel's possibility proof

Gödel's *1970* contains a strategy for proving possibility that differs from Leibniz's in ways that may help Gödel to deal with both of these difficulties, but that may also bring compensating disadvantages in their train. This is because Gödel's *1970* uses a conception of a positive property that is quite different from Leibniz's conception of a perfection. Two differences may be noted here, having to do with the notions of properties and of positiveness, respectively.

1. Gödel's *1970* speaks of the entities in the domain of the predicate variable φ simply as "properties". This category seems not to be restricted to what Leibniz would count as qualities. Gödel's definitions of G and E, and his syntactical treatment of them and of $x = x$ and $x \neq x$, suggest that he was pretty generally willing here to postulate properties corresponding to propositional functions of a single individual variable. Perhaps Gödel would restrict the applicability of his notion of properties more narrowly than this suggests, but no such restriction is found in the text; in particular, nothing excludes relational properties corresponding to propositional functions of several variables.

This certainly makes it easier for Gödel to defend the thesis (his Axiom 4) that necessary existence is a positive property, which he uses, as noted in §2 above, in arguing that God's existence is necessary if possible. For necessary existence, as Gödel understands it, clearly does correspond to a propositional function of one individual variable. (It is necessary exemplification of the individual's essence(s).[m]) And Gödel's

---

[m]By relating the necessity thus indirectly to the individual, Gödel avoids quantifying, with an individual variable, into a modal context. Sobel (*1987*, p. 246) cites one exception to the proof's avoidance of this controversial type of quantitification, but the exception is in Scott's version, not in Gödel's *1970*, which uniformly avoids such quantification.

notion of properties is not restricted in its application to any category from which there is an obvious reason for excluding necessary existence.

Of course it does not immediately follow that necessary existence is indeed positive, but there is nothing in Gödel's apparatus to exclude its positiveness. In his *1970* it is asserted as an axiom, but Gödel's notebooks contain at least two arguments for it ("Phil XIV", pages 103–4, 106–7). They are similar to each other; the simpler asserts as axioms that "the necessity of a perfective is a perfective, and being is a perfective" ("Phil XIV", page 106), where "perfective" plays the part played by "positive" in *1970*. From these axioms (fairly plausible on Gödel's assumption that every property, in a broad sense, is either positive or negative), it immediately follows that necessary being is a perfective (positive).

2. Gödel offers several interpretations of the meaning of "positive" (or "perfective"). Only the one that is farthest from his *1970* agrees fully with Leibniz's central idea of the purely positive as involving no negation at all in its construction from simple positive properties. According to the interpretation that seems intended to go with the 1970 proof, "positive means positive in the moral aesthetic sense (independently of the accidental structure of the world)." This classifies "positive" as a value predicate, and indicates that what is positive is necessarily positive,[n] as claimed in Axiom 3 of *1970*. But it does not identify logical properties of positiveness that are likely to be of much help in proving the mutual consistency of all positive properties.

This interpretation also is disturbingly similar to one that is rejected in one of Gödel's notebooks: "The interpretation of 'positive property' as 'good' (that is, as one with positive value) is impossible, because the greatest advantage + the smallest disadvantage is negative" ("Phil XIV", page 105). The reason given for the rejection, however, is not directed at the assumption that "positive" is a value predicate. The objection is rather that "good" does not express a sufficiently demanding standard of value. That is made clear by the amendment that Gödel goes on to propose: "It is possible to interpret the positive as perfective; that is, 'purely good', that is, such as implies no negation of 'purely good'" ("Phil XIV", page 105). This amendment makes clear that "positive" is to mean *purely* positive or *purely* good, and not just positive or good to some degree.

---

[n] I take the parenthetical phrase, "independently of the accidental structure of the world", to apply to the positiveness of the positive properties. Perzanowski (*1991*, page 628) seems to take it to apply to any thing's possession of a positive property, for he writes, "According to Gödel, positive means: independent of the accidental structure of the world."

It also specifies an important logical property of (pure) positiveness. Unlike Leibniz, who defined perfections, and purely positive qualities more generally, in terms of the role that negation does not play in their *internal* logical structure, Gödel here characterizes purely positive properties, or "perfectives", in terms of what they *imply*. The importance of this for his ontological proof is underlined as he goes on in his notebook to say, "The chief axiom runs then (essentially): A property is a perfective if and only if it implies no negation of a perfective" ("Phil XIV", page 106). This axiom (or the "only if" half of it) reappears as Axiom 5 in Gödel's **1970*. (The "if" half follows from Axiom 5 together with Axiom 2.) We may reasonably infer that "positive" means *purely* positive in **1970*, and that the "moral aesthetic" explanation of the sense of "positive" given there does not share the feature to which Gödel objected in the rejected explanation in the notebook.

This way of specifying the concept of a (purely) positive property generates the proof of the possibility of God's existence in Gödel's **1970*. Gödel assumes that the sum of all positive properties is itself a positive property (Axiom 1), and that positive properties imply only positive properties (Axiom 5). From these assumptions it follows that "the system of all positive properties is compatible", and hence that the existence of God, as the possessor of all positive properties, is possible.

This possibility proof does not depend on the controversial Leibnizian assumption that the only way in which properties can be incompatible is by formal contradiction arising from negation involved in their construction. That advantage may be outweighed by a major disadvantage, however. If Leibniz's assumptions are accepted, they give a *reason* for believing that all purely positive qualities are mutually consistent, and a sort of explanation of *why* they are consistent, showing that there is no way in which they could be mutually inconsistent. But Gödel's **1970* provides no such explanation, and the axioms from which the mutual compatibility of all purely positive properties is inferred in **1970* are too close to the conclusion to have much probative force to establish it. Of the axiom that "a property is a perfective if and only if it implies no negation of a perfective", Gödel himself, in his notebook, states that it "says essentially that the positive properties form a maximal compatible system" ("Phil XIV", page 106). It seems as fair to say that about Axiom 5 in the 1970 proof. But then is it not question-begging to rely on Axiom 5 to prove that "the system of all positive properties is compatible"?

At the end of **1970* Gödel tersely suggests an alternative, more Leibnizian interpretation of positiveness and a corresponding strategy of proof. Positive, he says, "may also mean pure 'attribution' as opposed to 'privation' (or *containing* privation)". By itself this may be a cryptic formulation, but a footnote explains that what is meant is that "the disjunctive normal form [of a purely positive property] in terms of elementary properties contains a member without negation". Gödel adds that "this interpretation" supports a "simpler proof", but he does not give the proof.

The central idea of the suggested proof is presumably that there is no way in which properties can be mutually inconsistent if the disjunctive normal form of each, in terms of elementary properties, contains at least one member without negation. It must be assumed here that the elementary properties are positive. They correspond to the simple, positive properties of Leibniz's scheme. Gödel sees all other properties as constructed out of them by operations of disjunction (*inclusive* disjunction must be meant here) and negation. Leibniz (if I understand him aright) had allowed no negation at all in the construction of purely positive qualities from simple, positive qualities. Gödel is more liberal on this point, seeing that as long as each purely positive property has in its disjunctive normal form at least one disjunct that involves no negation in its construction, no formal inconsistency can arise among purely positive properties, even if negation is involved in the construction of other disjuncts. In this way he has accomplished an improvement in Leibniz's proof, for the suggested proof seems to have all the advantages of Leibniz's argument, with a less restrictive conception of the purely positive. On the other hand, it depends no less than Leibniz's proof on the controversial assumption that the only way in which properties can be incompatible is by formal contradiction arising from negation involved in their construction.

An even more Leibnizian conception of the purely positive is suggested in Gödel's notebook, when he proposes the theorem: "The positive properties are precisely those that can be formed out of the elementary ones through application of the operations &, $\vee$, $\supset$" ("Phil XIV", page 108). On this construal the purely positive properties will be those that involve no negation at all in their construction from elementary properties (provided the disjunction operation here too is inclusive).

5. Discussion of Gödel's proof, 1970–1991

There is a small but growing secondary literature on Gödel's ontological proof. It has been pointed out that "Gödel's theory is certainly [formally] consistent, having a monistic model comprising one object, one atomic property, hence one [possible] world and, of course, one God."[o] In unpublished work Petr Hájek has proposed proofs of mutual

[o] *Perzanowski 1991*, p. 629.

independence of some of the axioms in Dana Scott's version of the proof.

The first full publication of Gödel's ontological proof was in *Sobel 1987*. Sobel reproduces both Dana Scott's version and Gödel's own **1970*, but discusses chiefly Scott's version. Sobel criticizes the proof as a piece of philosophical theology. One of his main criticisms is that "a being that was *God-like* in the sense of the system would, in connection with many religiously important properties, have not them but their negations." His reason for this claim is that he thinks that some of the traditional attributes of God are incompatible with necessary existence. He deems it "obvious" that no necessarily existing being "would be *sentient* or *cognizant* .... It is at least a firm modal intuition of *mine*", he says, "that there are possible worlds in which there are ... no *sentient* or *cognizant*" things (*Sobel 1987*, pages 249–50).

Sobel's intuitions on this point are shared by many philosophers, but consciously rejected by virtually all partisans of the ontological argument. It would be naive to expect the latter to accept Sobel's objection and conclude that God is not a cognizant being. The form of Sobel's objection is therefore somewhat misleading. Friends of the ontological argument are bound to see it as merely a repackaging of a familiar empiricist objection, based on the claim (consciously rejected by them) that a being possessing the sort of reality generally ascribed to God could not exist necessarily.

It remains a serious question, however, whether the being whose existence is purportedly proved by Gödel's ontological proof is the God of traditional theism. Despite its role in the philosophical theologies of Leibniz, Wolff, and Kant, and its resonance with many medieval philosophical theologies, it is not immediately obvious that the concept of a being possessing the sum of all purely positive properties (or qualities) is a concept of God. Any employment of Gödel's ontological proof in philosophical theology would require further argument on this point, with particular attention to Gödel's conception of positive properties.

Sobel's other main objection is that the assumptions of Gödel's ontological proof generate a proof that all truths are necessary truths. For on a liberal construal of the notion of a property, "if something is true, then ... a God-like being [if one actually exists] has *the property of being in the presence of this truth*. But every property of a God-like being [i.e., every actual property of God] is necessarily instantiated, from which it follows that this truth [i.e., any actual truth] is a necessary truth."[P] (That every actual property of God is necessarily instantiated follows from $N(\exists y)G(y)$, the conclusion of Gödel's ontological proof, since G,

---

[P] *Sobel 1987*, p. 253. Sobel also gives a formal proof based on the ideas contained in this informal exposition.

as the "essence" of God, in Gödel's sense, entails all of God's actual properties.)

It is characteristic of Leibnizian philosophical theology to be in some danger of leaving no truths contingent (see *Adams 1977*). And it is not altogether clear that Gödel was determined to avoid such a necessitarian conclusion. He wrote a notebook entry about the ontological proof in which he seems quite favorable to the thesis that "for every compatible system of properties there is a thing" ("Phil XIV", page 107). That thesis looks strongly necessitarian, but the interpretation of the entry containing it is not obvious; one may wonder, for instance, whether merely possible objects count as "things" here.[q]

Another relevant entry comes at the very end of the notebook section devoted to the ontological proof. Gödel had written that propositions "of the form $\varphi(a)$" are "the only synthetic propositions" because "they depend not on God, but on the thing a" ("Phil XIV", page 108). In this context a must be an individual other than God. Such individuals, and truths about them, do depend causally on God, according to traditional theism. The independence Gödel has in mind here is presumably logical rather than causal. Later, at the bottom of the following page, with a line indicating insertion at this point, or reference to it, Gödel wrote ("Phil XIV", page 109):

> This doesn't work, because then God would have an imperfective, which consists in the fact that imperfectives are possible. Everything that follows from a perfective, such as something good, that is a perfective, is.

This correction, I think, must have arisen from something like the following train of thought: If my having gray hair is synthetic and logically independent of God, then it is contingent, and both it and its falsity are possible, and likewise for your having whatever color of hair you have. Gödel seems to take these possibilities as implying "that imperfectives are possible"—presumably on the ground that of the two properties, (1) having gray hair and (2) not having gray hair, one must be a perfective and the other an imperfective.

But why would "the fact that imperfectives are possible" constitute an imperfective that God would have? Here I suppose we must invoke something like Sobel's assumption that, for every truth, God has the property of coexisting with that truth, or perhaps the traditional theistic assumption that, for every truth, God has the property of knowing that truth. (These assumptions imply that my having gray hair is not,

[q]As Charles Parsons has suggested to me might be the case.

after all, logically independent of all of God's properties; perhaps the independence Gödel had in mind is only a logical independence from God's internal, nonrelational properties; or perhaps it is an independence from God's necessary properties, assuming for the sake of the present argument that God may have some contingent properties.) Then since I in fact have gray hair, God has in fact the property of coexisting with my having gray hair. And that property must be a perfective if all God's (actual) properties are perfectives. Its negation, the property of not coexisting with my having gray hair, must then be an imperfective, given Gödel's assumption that every negation of a perfective is an imperfective. But if my having gray hair is possibly false, then God has the property of possibly not coexisting with my having gray hair. And this possibility will be an imperfective; for, as Gödel maintains in his notebook ("Phil XIV", page 103n; cf. page 107), the possibility of a negative is negative, and presumably the possibility of an imperfective must also be imperfective. So if God has no imperfectives, as Gödel's definition of deity requires, my having gray hair (when I do) must not be possibly false, and in general, "everything that follows from a perfective, such as something good", must be—a conclusion of Leibnizian optimism, and perhaps more than Leibnizian necessitarianism.

I grant that the suggestion that Gödel would have accepted the sweeping necessitarian implication with which Sobel charges him is somewhat speculative. In any event, there are possible modifications of Gödel's assumptions that avoid the sweeping necessitarianism without undermining his ontological proof. Axiom 2 of his *1970* is equivalent to the conjunction of two conditionals:

(A) If a property is positive, then its negation is not positive.

(B) If a property is not positive, then its negation is positive.

Anderson (*1990*) has pointed out that of these conditionals, only (A) is required for Gödel's ontological proof, but (B) is required for the proof that all truths are necessary. He argues that (B) is less plausible than (A), as (B) "seems to overlook a possibility: that both a property and its negation should be *indifferent*". He sets out a revised version of Gödel's ontological proof, which has (A) but not (B) as an axiom, and which still has the conclusion that "the property of being God-like* is necessarily exemplified". Anderson's version of the proof also differs from Gödel's in not requiring an essence of a thing to entail all the actual properties of the thing, but only a subset classified as "essential" to the thing, and in defining a God-like* being as one that has all and only the positive properties as *essential* properties, and not merely as properties. Anderson proposes a "possible worlds" model to prove that the assumptions of this proof are consistent with there being contingent truths (*Anderson 1990*, pages 295–97).

Another way of avoiding the sweeping necessitarian conclusion is to use a more restrictive notion of a property than Sobel does. In deriving the necessitarian conclusion, he relies on the very strong assumption that "properties" include all those abstracted in accordance with the principle

$$\hat{\beta}[\varphi](\alpha) \equiv \varphi',$$

"where β is an individual variable, α is a term, φ is a formula, and φ' is a formula that comes from φ by proper substitution of α for β" (*Sobel 1987*, page 251). This assumption is not part of Gödel's argument, and Hájek, in the unpublished work cited above, has argued that if it is replaced with certain weaker assumptions about properties, the axioms of (Scott's version of) Gödel's ontological proof can be shown by a "possible worlds" model to be consistent with the existence of contingent truths.

One way of filling out Hájek's suggestion would be to restrict the category of "properties" to *nonrelational* properties for purposes of the ontological proof. This could be accomplished by restricting it to properties that Leibniz would have counted as *qualities*, but it might not be necessary to go that far. The important point is that if relational properties are not counted as properties for purposes of the argument, then such "properties" as that of "being in the presence of this truth", which are relational, will not be among the actual properties of God that must be necessarily instantiated according to the argument. If the properties of God that are necessarily instantiated are exclusively nonrelational, then their necessity will not imply the necessity of truths about other beings.

Robert Merrihew Adams[r]

[r]I am indebted to Charles Parsons for helpful comments on an earlier version of this note, and to Jay Atlas, Dana Scott, and Morton White for sharing their recollections bearing on the history of Gödel's ontological proof.

Ontological proof
(*1970)

Feb. 10, 1970

$P(\varphi)$ φ is positive   (or $\varphi \in P$).

Axiom 1. $P(\varphi).P(\psi) \supset P(\varphi.\psi)$.[1]

Axiom 2. $P(\varphi) \vee P(\sim\varphi)$.[2]

Definition 1. $G(x) \equiv (\varphi)[P(\varphi) \supset \varphi(x)]$ (God)

Definition 2. $\varphi \, \text{Ess.} \, x \equiv (\psi)[\psi(x) \supset N(y)[\varphi(y) \supset \psi(y)]]$. (Essence of x)[3]

$$p \supset_N q \;=\; N(p \supset q). \quad \text{Necessity}$$

Axiom 3. $P(\varphi) \supset NP(\varphi)$
$\sim P(\varphi) \supset N\sim P(\varphi)$

because it follows from the nature of the property.[a]

Theorem. $G(x) \supset G \, \text{Ess}.x$.

Definition. $E(x) \equiv (\varphi)[\varphi \, \text{Ess} \, x \supset N(\exists x) \, \varphi(x)]$. (necessary Existence)

Axiom 4. $P(E)$.

Theorem. $G(x) \supset N(\exists y)G(y)$,
  hence $(\exists x)G(x) \supset N(\exists y)G(y)$;
  hence $M(\exists x)G(x) \supset MN(\exists y)G(y)$.   ($M$ = possibility)
    $M(\exists x)G(x) \supset N(\exists y)G(y)$.

| $M(\exists x)G(x)$ means the system of all positive properties is compatible.   2
This is true because of:
Axiom 5. $P(\varphi).\varphi \supset_N \psi :\supset P(\psi)$, which implies

$$\begin{cases} x = x & \text{is positive} \\ x \neq x & \text{is negative.} \end{cases}$$

[1] And for any number of summands.

[2] Exclusive or.

[3] Any two essences of x are *necessarily equivalent*.

---

[a] Gödel numbered two different axioms with the numeral "2". This double numbering was maintained in the printed version found in *Sobel 1987*. We have renumbered here in order to simplify reference to the axioms.

But if a system S of positive properties were incompatible, it would mean that the sum property s (which is positive) would be $x \neq x$.

Positive means positive in the moral aesthetic sense (independently of the accidental structure of the world). Only then [are] the axioms true. It may also mean pure "attribution"[4] as opposed to "privation" (or *containing* privation). This interpretation [supports a] simpler proof.

If φ [is] positive then *not*: $(x)N{\sim}\varphi(x)$. Otherwise: $\varphi(x) \supset_N x \neq x$; hence $x \neq x$ [is] positive, so $x = x$ [is] negative, contrary [to] Axiom 5 or the existence of positive properties.

---

[4]I.e., the disjunctive normal form in terms of elementary properties[b] contains a member without negation.

---

[b]Here Gödel uses the abbreviation "prop.", which could be read, in isolation, either as "properties" or "propositions". In the context, however, it is clear that it is properties whose positiveness is under discussion. The related discussion in the excerpts from "Phil XIV" in the appendix, below, explicitly concerns "positive properties". With regard to fn. 4, where the reference to "disjunctive normal form" might lead us to think first of propositions, note that in "Phil XIV", p. 108, Gödel speaks explicitly of properties ("Eigenschaften") that are "members of the conjunctive normal form" of complex properties. An interpretation of fn. 4 is offered in the introductory note, pp. 397–398 above.

Introductory note to *Gödel *1970a,*
**1970b,* and **1970c*

1. Introduction

Gödel *1970a is a handwritten document that was sent to Alfred Tarski for submission to the *Proceedings of the National Academy of Sciences.* It lists four axioms and claims to deduce from them that $2^{\aleph_0} = \aleph_2$. **1970b* and **1970c* are two other handwritten documents that bear on **1970a.* **1970b* claims, on the contrary, to deduce the continuum hypothesis from some of the axioms mentioned in **1970a.* **1970c* is a letter to Tarski (apparently never sent; cf. Gregory Moore's introductory note to *Gödel 1947* and *1964*, these *Works*, Volume II, page 175) which acknowledges serious errors in **1970a.*[a]

Upon receiving **1970a*, Tarski asked the author to referee the manuscript. After a careful study of the manuscript, I was unable to follow the argument. I reported back to Tarski that if the author were anyone but Gödel, I would certainly recommend that the manuscript be rejected. (The manuscript has never been published hitherto.) Subsequently, D. A. Martin showed that a key argument of the paper was demonstrably wrong. (See section 6 below.)[b]

In the remainder of this introductory note, I shall describe the proof in **1970a* to the extent that I now understand it. I shall also discuss a model of set theory whose study sheds considerable light on **1970a.* What comments I make on the other papers will be in the course of discussing **1970a.*

The rest of this note is organized as follows: Section 2 gives my reconstruction of the precise formulation of the four axioms from which **1970a* purports to deduce that $2^{\aleph_0} = \aleph_2$. Section 3 contains an outline of Gödel's alleged proof. Section 4 contains the proof that Axioms 3 and 4 entail $2^{\aleph_0} = 2^{\aleph_1}$. Section 5 presents the main result of **1970b*: The rectangular axiom $A(\aleph_1, \aleph_0)$ implies $2^{\aleph_0} = \aleph_1$. Section 6 is devoted

---

[a]The three papers bear (in Gödel's handwriting) the notations "I Fassung", "II Fassung", and "III Fassung". ("Fassung" is German for "version".) Thus it seems likely that Gödel viewed them as different installments or versions of a single paper. In addition, **1970b* has written at its top "Nur für mich geschrieben" ("Written only for me").

[b]For further discussion of the history of **1970a*, we recommend Moore's introductory note to *Gödel 1947* and *1964*, especially pp. 173–175.

The paper *Ellentuck 1975* discusses many of the issues discussed in the present note, and the interested reader might wish to read it. In addition, *Takeuti 1978* explores some interesting generalizations of the arguments of section 5.

to the discussion of various "random real" models. They show among
other things that the first three of Gödel's axioms *do not* entail a bound
on the continuum. Section 7 gives a brief list of open questions. Finally,
I close in section 8 with some speculations on the provenance of Gödel's
errors.

2. The axioms

The axioms used by Gödel are reviewed in this section. For the first
two axioms, it is merely a question of restating them and introducing
some useful terminology. Axiom 3 is slightly misstated in *1970a* and
Axiom 4 is not spelled out in detail. In each case, I give what I believe
is the intended formulation.[c]

The standard conventions concerning ordinals and cardinals are followed: each ordinal α is identified with the set of ordinals smaller than
α and each cardinal $\aleph$ is identified with the least ordinal α of that cardinality. κ^+ is the least cardinal greater than κ. I write ω_n for the cardinal
$\aleph_n$ when stressing its "ordinal aspects".

The first two axioms concern the set $^{\kappa}\lambda$ of all functions from κ to λ
(for various infinite cardinals κ and λ).

Two partial orderings on $^{\kappa}\lambda$ are relevant here: If $f, g \in {}^{\kappa}\lambda$, then f
dominates g, (notation: $f \gg g$) iff $(\forall \alpha < \kappa) \, f(\alpha) > g(\alpha)$; f *eventually*
dominates g (notation: $f > g$) iff $(\exists \eta < \kappa)(\forall \alpha > \eta) f(\alpha) > g(\alpha)$.[d]

For $S \subseteq {}^{\kappa}\lambda$, we let $S^\star = \{ f \restriction \alpha \mid f \in S \text{ and } \alpha < \kappa \}$. The principle
$A(\kappa, \lambda)$ asserts the following: There is a set $C \subseteq {}^{\kappa}\lambda$ such that:

1. C has cardinality κ^+.
2. C is cofinal in $^{\kappa}\lambda$ with respect to the partial order $\ll$. That is, for
   every $f \in {}^{\kappa}\lambda$, there is a $g \in C$ such that $f \ll g$.
3. $C^\star$ has cardinality κ.

Gödel's Axioms 1 and 2 are easily seen to be equivalent to the assertion that $A(\aleph_n, \aleph_n)$ holds for all n such that $0 \leq n < \omega$.

I shall use the terminology of *Ellentuck 1975* and refer to the princi-
ples $A(\aleph_n, \aleph_n)$ as the *square axioms*, and the principles $A(\aleph_n, \aleph_m)$ (for
$m < n$) as the *rectangular conjectures*.

---

[c]Conversations with Stevo Todorcevic were extremely helpful in puzzling out the
precise meaning of Axiom 4. I had also asked Gaisi Takeuti what he believed Axiom
4 to be, and he independently conjectured the interpretation presented below.

[d]Gödel uses the phrase "majorizes by end pieces" for this notion.

Two weakenings of the principle $A(\kappa, \lambda)$ will also be considered: $A_1(\kappa, \lambda)$ asserts that there is some family of functions C which is cofinal in ${}^\kappa\lambda$ (with respect to $\gg$), and which is of cardinality κ^+. $A_2(\kappa, \lambda)$ asserts that there is a family of functions C which is cofinal in ${}^\kappa\lambda$ (with respect to $\gg$) such that C^* has cardinality κ.[e]

Gödel's Axiom 3 reads as follows:

There exists a *complete*, i.e., not extendable, scale of functions (N → Reals) in which every ascending or descending sequence has cofinality ω_1.

Here: (1) What Gödel almost certainly means by a "complete scale" is a maximal linearly ordered subset of ${}^\omega\mathbb{R}$ with respect to the partial order of "eventual domination".

(2) The last symbol of the axiom is rendered as ω_1 in the handwritten manuscript and as ω in a typed version of the manuscript. It seems likely that the rendering in the typed version is a typographical error. For, assuming (1), any complete scale is easily seen to be an η_1 set,[f] and any η_1 set contains increasing sequences of length ω_1.

(3) The axiom is still slightly defective, even if the last symbol was intended to be ω_1, since clearly an increasing sequence of length ω_1 will contain an initial subsequence of length ω. What I think Gödel intended to say for Axiom 3 was: There is a maximal linearly ordered subset, S, of ${}^\omega\mathbb{R}$ in which every well-ordered ascending or descending sequence (with respect to the $<$ order) has length at most ω_1.

Gödel's Axiom 4 asserts that the S of Axiom 3 satisfies "the Hausdorff continuity axiom". I have not been able to locate this phrase in the literature. However, in *Hausdorff 1914*, pages 90–91 one finds the notion of an ordered set being *stetig* (continuous) if it has more than two points, is densely ordered, and is order-complete (i.e., satisfies the l. u. b. axiom).

There is some evidence that this is what Gödel means. He makes an appeal to the order-completeness of the scale in an insertion to the second page of the manuscript.[g] However, it is easy to see that no η_1 set can be order-complete. (An ascending ω sequence in such a set cannot have a least upper bound.) Hence this formulation of the axiom must be rejected.

---

[e]The proposition $A(\kappa, \lambda)$ is in fact equivalent to the conjunction of $A_1(\kappa, \lambda)$ and $A_2(\kappa, \lambda)$.

[f]That is, no countable subset is either cofinal or coinitial in the set, and if A and B are countable subsets such that every member of A is less than every member of B (in symbols, $A < B$), then there is an element c such that $A < c < B$, i.e., lies between the elements of A and the elements of B.

[g]The passage in question is here the third paragraph after Axiom 4 and begins with the words: "Next construct a measure of zero sets ...."

The following weaker version of Axiom 4 (cf. footnote c above) suffices to derive that $2^{\aleph_0} \geq \aleph_2$ (which Gödel asserted to follow from Axiom 4):

> The ordered set S of Axiom 3 may be chosen so as to contain no (ω_1, ω_1^*) gaps. That is, if A and B are subsets of S such that $A < B$ and A has order type ω_1 and B has order type ω_1^* then there is an element c such that $A < c < B$.[h]

3. Outline of the purported proof

As I have indicated in the introduction, there are parts of the purported proof that I do not understand, and parts that are demonstrably wrong. For brevity, I shall use words like "proof" and "shows" rather than "purported proof" and "attempts to show".

The argument in question has three phases:

1. One shows that $2^{\aleph_0} \geq \aleph_2$.

This, indeed, follows from Axioms 3 and 4 and was presumably "well known". I will briefly sketch this part of the argument below. It establishes, in fact, that the stated axioms entail Lusin's hypothesis: $2^{\aleph_0} = 2^{\aleph_1}$.

2. One shows that $2^{\aleph_0} \leq \aleph_2$.

I am unable to follow this part of the proof. Gödel asserts that this inequality actually follows from Axioms 1 through 3, but this assertion is definitely wrong, as we shall see below in section 6. However, his proof in fact makes an appeal to Axiom 4 (though one that seems unjustified to me).

This portion of the proof also makes major use of the rectangular conjecture $A(\aleph_2, \aleph_1)$, which is claimed to follow from the axioms (see phase 3 of the proof below). In the models studied in section 6, this principle as well as Axioms 1–3 hold, and yet the continuum can be made "arbitrarily large".

Nevertheless, a variant of the argument (given in *1970b*) does establish that the principle $A(\aleph_n, \aleph_0)$ (for $0 < n < \omega$) entails $2^{\aleph_0} \leq \aleph_n$. I present that argument in section 5.

3. Finally, one establishes that Axioms 1 and 2 entail the principles $A(\aleph_n, \aleph_m)$ for $m < n < \omega$.

---

[h] Here ω_1^* is the order-type which is the reverse of ω_1.

Again, I am unable to follow the argument, and it is this argument that Gödel repudiates in his letter to Tarski, *1970c*. In fact, the assertion cannot be right since it follows from what has already been said that the principle $A(\aleph_1, \aleph_0)$ is false in the models studied in section 6. This was first observed by Martin.

4. Axioms 3 and 4 entail Lusin's hypothesis

The idea behind Axioms 3 and 4 is that the scale S should be viewed as a higher-order analogue of the reals. Axiom 3 corresponds to the fact that ω_1 cannot be embedded order-isomorphically into $\mathbb{R}$. Axiom 4 corresponds to the order completeness of the reals. In the proof that follows, the set W_0 plays a role analogous to the rational numbers, while W_1 plays a role analogous to the irrationals.

We have already remarked that it follows easily from the fact that S is a maximal linearly ordered subset of $^\omega\mathbb{R}$ that S is an η_1 set.

Let W be the set of all functions from ω_1 to 2, ordered lexicographically. Let W_0 consist of those functions f in W such that there is a largest ordinal $\gamma < \omega_1$ such that $f(\gamma) = 1$. (Hence each $f \in W_0$ is eventually 0.)

Since S is an η_1 set, it is easy to construct an order-preserving map Φ from W_0 into S: one defines Φ in ω_1 stages. At stage α, one handles those $f \in W_0$ such that α is the largest point at which f is non-zero. Of course, one preserves the inductive hypothesis that Φ is order-preserving on its current domain.

Now consider the subset W_1 of W consisting of those f for which the set of α such that $f(\alpha) = 0$ and the set of α such that $f(\alpha) = 1$ are both unbounded. Clearly W_1 has cardinality $2^{\aleph_1}$, and between any two distinct elements of W_1 there is an element of W_0.

Each element of W_1 determines an (ω_1, ω_1^*) gap in W_0. Using the fact that, by Axiom 4, S has no (ω_1, ω_1^*) gaps, we can easily define a map Ψ of W_1 into S, by sending $f \in W_1$ into some element of S that fills the Φ-image of the gap determined by f in W_0. Since W_0 is order-dense in W_1, it is clear that Ψ is order-preserving (and hence one-to-one).

We have mapped a set of cardinality $2^{\aleph_1}$ injectively into a set of size $2^{\aleph_0}$. Hence $2^{\aleph_0} = 2^{\aleph_1}$.

5. Rectangular axioms and bounds on the continuum

In this section, we outline the (correct) proof from *1970b* that $A_2(\aleph_1, \aleph_0)$ entails $2^{\aleph_0} = \aleph_1$. The proof will be based on a construction of a matrix $\langle F(\alpha, n) \rangle$ of closed subsets of $\mathbb{R}$, defined for $\alpha < \omega_1$ and

$n < \omega$. For each fixed α, the sequence of closed sets $\langle F(\alpha, n) \mid n \in \omega \rangle$ will be increasing with union $\mathbb{R}$. It follows that

$$\mathbb{R} = \bigcup_{f \in {}^{\omega_1}\omega_0} \bigcap_{\alpha < \omega_1} F(\alpha, f(\alpha)).$$

Now let C be a family of functions from ω_1 to ω_0 provided by $A_2(\aleph_1, \aleph_0)$. Since for fixed α the sequence $\langle F(\alpha, n) \mid n \in \omega \rangle$ is increasing, and the family C is cofinal in ${}^{\omega_1}\omega_0$, we obtain:

$$\mathbb{R} = \bigcup_{f \in C} \bigcap_{\alpha < \omega_1} F(\alpha, f(\alpha)).$$

Next, to each $h \in C^*$, we associate a closed set $G(h)$ as follows:

$$G(h) = \bigcap_{\alpha \in \operatorname{dom}(h)} F(\alpha, h(\alpha)).$$

Now if $h \in C$, the sequence of closed sets $\langle G(h \restriction \alpha) \mid \alpha < \omega_1 \rangle$ is decreasing and hence is eventually constant. Thus there is a restriction h_0 of h such that

$$\bigcap_{\alpha < \omega_1} F(\alpha, h(\alpha)) = G(h_0). \tag{1}$$

Since there are only $\aleph_1$ possibilities for h_0, we have achieved a representation of $\mathbb{R}$ as a union of $\aleph_1$ closed sets.

Our plan now is to use our freedom in defining the matrix F to achieve the goal that each of the closed sets $G(h_0)$ is countable. (In fact, the argument succeeds in making them finite.) It will follow immediately that the cardinality of $\mathbb{R}$ is $\aleph_1$, completing the present proof.

We need the following elementary lemma.

Lemma 1. *Let G be an infinite closed set. Then there is an increasing sequence of closed sets $\langle F_n \mid n \in \omega \rangle$ with union $\mathbb{R}$ such that $F_n \cap G \neq G$ for all $n \in \omega$.*

Proof: If G is unbounded, then we simply take $F_n = [-n, n]$. Otherwise, G has an accumulation point p. Take $F_n = (-\infty, p - 1/n] \cup \{p\} \cup [p + 1/n, \infty)$. This suffices.

Now fix an enumeration $k : \omega_1 \to C^*$ such that each element of C^* is listed ω_1 times. This is used to define the matrix F as follows. Suppose that we are at stage α and wish to define $F(\alpha, n)$ for $n \in \omega$. Let $g = k(\alpha)$. There are three cases:

Case 1: Domain $g > \alpha$.
Then "do nothing", i.e., we set $F(\alpha, n) = \mathbb{R}$ for all n.

Case 2: Case 1 does not apply and $G(g)$ is finite.

Again we "do nothing".

Case 3: Case 1 does not apply and $G(g)$ is infinite.

In this case, apply Lemma 1 to $G(g)$ to obtain a sequence of closed sets F_n with union $\mathbb{R}$; then set $F(\alpha, n) = F_n$.

This completes our description of the construction. It is now routine to see that for any h in C, the set $G(h_0)$ is finite.

In fact, if it were infinite, there would be a stage $\gamma < \omega_1$ such that for every $n \in \omega$, $G(h_0) \cap F(\gamma, n)$ is a proper subset of $G(h_0)$. But this contradicts (1).

By comparison with the proof just given, I can now explain my difficulties with the corresponding proof in *1970a* that the four axioms imply that $2^{\aleph_0} \leq \aleph_2$.

1. In the proof just given, the fact that a decreasing sequence of closed subsets of $\mathbb{R}$ of length ω_1 must eventually become constant was used in a crucial way. Gödel seems to use a similar principle in his argument, but it is not clear what that principle is. Is it true that if $2^{\aleph_0} = 2^{\aleph_1} = \aleph_2$, then a decreasing sequence of length ω_2 of $F_{\aleph_1}$ sets must eventually become constant (modulo the ideal of sets of cardinality at most $\aleph_1$)?[i]

2. Some substitute for Lemma 1 is needed. The substitute would say something of the form: If G is a set which is not of "strong measure zero" then there is an increasing sequence of sets having a certain property X (where precisely what the property X should be is not clear to me) of length $\aleph_1$ such that G is not contained in any member of the sequence, and the union of the sequence is all of $\mathbb{R}$.

3. In his construction of a "measure of zero sets" (cf. footnote g above), Gödel uses a trivial variant of the scale S (of axioms 3 and 4) which is a maximal linearly ordered subset of ${}^\omega \mathbb{R}^+$. He describes a certain Dedekind cut in this scale, and wishes to assert that it is the cut determined by an element. That would follow if the set S were order-complete, but as I have already remarked this leads easily to a contradiction. I do not see how to repair this step.

As Gödel remarks in *1970b* and *1970c*, it is possible to generalize the arguments of this section to show that the generalized continuum hypothesis is equivalent to certain of its rectangular axiom consequences.

---

[i] An $F_{\aleph_1}$ set is a set which is the union of at most $\aleph_1$ closed sets.

Using the results of *Takeuti 1978*, it is not hard to show that the *GCH* is equivalent to the assertion that $A(\kappa^+, \kappa)$ holds for all infinite cardinals κ. (Gödel makes in **1970c* an analogous claim that the assertion that $A(\kappa^+, \aleph_0)$ holds for all infinite κ is equivalent to *GCH*. I am unable to verify this latter claim.)

6. An illuminating model

6.1. There is a standard forcing extension whose study sheds considerable light on Gödel's purported proof. In the following discussion I take for granted the rudiments of forcing as presented, for example, in *Jech 1978*.

The ground model M is to be a countable transitive model of $ZFC + GCH$.[j] In M, let κ be an uncountable cardinal and let X be I^κ, the product of κ many copies of the unit interval $[0, 1]$. X is of course a compact Hausdorff space. Moreover, X carries a natural Borel measure μ which is the product of the various copies of Lebesgue measure in the different factors.

Let $\mathcal{B}$ be the measure algebra of X. One concrete description of $\mathcal{B}$ is that it is the quotient of the σ-algebra of Baire sets modulo the σ-ideal of sets of measure zero.[k]

Let N be obtained from M by adding an M-generic homomorphism $h : \mathcal{B} \to \mathbf{2}$ to the model M. The Boolean algebra $\mathcal{B}$ satisfies c. c. c., and hence M and N have the same cardinals and cofinality function.

6.2. The construction just described is a familiar one (due to the present writer) that is usually described as "adjoining κ random reals to the model M". The sequence of random reals determined by h can be described as follows:[l]

The model N has its own version of X, which we sometimes refer to as X^N. (Similarly, we sometimes refer to "M's version of X" as X^M.) There is a natural coding (say by elements of $^\omega\kappa$) of the Baire subsets of X which works in both models and has the following absoluteness properties.

---

[j]There is a little fudging here since ZFC can't prove the existence of a model of ZFC. Part of knowing the rudiments of forcing is knowing how to get around this problem!

[k]Here a Baire set can be described as a member of the σ-algebra generated by sets of the form $\{f \mid f(\alpha) \in U\}$, where $\alpha < \kappa$ and U is an open subset of $[0, 1]$.

[l]We refer the reader to *Solovay 1970* where the analogous result when X is the unit interval is worked out in detail.

1. An element of M codes a Baire subset of X in M iff it does so in N.

2. If two codes represent the same set in M they code the same set in N. Thus there is a natural way of extending a Baire subset of X^M to a Baire subset of X^N.

3. Let B be a Baire subset of X^M and B' be its canonical extension to a Baire subset of X^N. Then $B' \cap X^M = B$.

4. Let B, B' be as above. Then the measure of B as computed in M is the same as the measure of B' as computed in N.

We can now characterize the κ-sequence z associated to h as follows; it is the unique element of X^N such that for any Baire set B of M,

$$z \in B' \quad \text{iff} \quad h([B]) = 1.$$

Here B' is the canonical extension of B to a Baire subset of X^N, and $[B]$ is the image of B in the Boolean algebra $\mathcal{B}$.

Conversely, we can recover h from z using the displayed equivalence.

The element z is random over M in the sense that it lies in no Baire set of X of measure zero with code in M. Conversely, if $z \in X$ lies in no Baire set of measure zero with code in M, then it arises from an h in the way just indicated.[m]

6.3. There is a version of the product lemma which holds for sequences of random reals, though its proof owes more to Fubini's theorem than to the usual product lemma. To state it, let α and β be ordinals of the model M, and let z be an $\alpha + \beta$-sequence of random reals over the model M (in the sense just described). In an obvious way, z can be viewed as the concatenation of sequences z_1 and z_2 where z_1 is an α-sequence of reals and z_2 is a β-sequence of reals. The result in question asserts that z_1 is a random sequence of reals of length α over M, and that z_2 is a random sequence of reals of length β over $M[z_1]$.

There is also a converse. If z_1 is a random sequence of reals of length α over M and z_2 is a random sequence of reals of length β over $M[z_1]$, then the concatenation of z_1 and z_2 is a random sequence of reals of length $\alpha + \beta$ over the model M.

Let N be as above. The computation of cardinal exponentiation in the model N is straightforward. In particular, if $\aleph$ is an infinite cardinal

---

[m] A little care must be used here in explicating what the meaning of X is in V. The ordinal κ is, of course, an ordinal in V. If one takes the product of κ copies of $[0, 1]$ (as computed in V) as V's version of X, it may be that $\kappa = \aleph_2$ in M. Since M is a countable model, κ is certainly not equal to $\aleph_2$ in V, though it will equal $\aleph_2$ in N, since the forcing extension N/M satisfies c. c. c.

of M which is less than the cofinality of κ (as computed in M), then in N we have $2^{\aleph} = \kappa$.

Thus, if we choose κ to have cofinality (in M) greater than $\aleph_1$, then Lusin's hypothesis holds in N, and in fact we have:

$$2^{\aleph_0} = 2^{\aleph_1} = \kappa.$$

6.4. We turn now to a discussion of the principles $A(\aleph_n, \aleph_m)$. Consider first the case when $1 \leq m \leq n < \omega$. We show that in this case, the principle is valid in N. The cofinal collection of functions that witnesses the truth of the axioms is the set W of those functions from $\aleph_n$ to $\aleph_m$ which lie in M.

It is clear (since M is a model of GCH and cardinals are absolute between M and N) that, in N, W has cardinality $\aleph_{n+1}$ and the set $W^\star$ of initial segments of functions of W has cardinality $\aleph_n$. It remains to show that if $f : \aleph_n \to \aleph_m$ lies in N, then there is a $g \in M$ such that $g \gg f$. But this is easily seen by a "possible values" argument, using the fact that the forcing extension satisfies c. c. c. and that the cofinality of $\aleph_n$ is greater than ω.

6.5. Consider next $A(\aleph_0, \aleph_0)$. Again this is true in N, but the proof is slightly more difficult.

We can obtain N from M by forcing with the following set P of conditions: A typical element $p \in P$ is a Baire subset of X of positive measure. The condition p' extends (i.e., gives more information than) p iff $p' \subseteq p$.

Again, take the set W that witnesses the truth of the proposition to be the set of all functions from ω to ω which lie in M. We have only to show that if $f : \omega \to \omega$ is a function lying in N then there is a function g lying in M such that $g \gg f$.

Fix a term $\mathbf{f}$ that denotes f in N. We may assume that it is forced (by every condition) that $\mathbf{f}$ is a function from ω to ω, and (towards obtaining a contradiction) that some condition B forces that $\mathbf{f}$ is not dominated by any function in the ground model.

Working in the ground model M, it is not hard to define (by induction on n) a sequence $\langle B_n \mid n \in \omega \rangle$ of conditions and a sequence $\langle g(n) \mid n \in \omega \rangle$ of integers so that:

1. $B_0 = B$.
2. $B_{n+1} \subseteq B_n$.
3. $\mu(B_n - B_{n+1}) < 10^{-(n+1)} \cdot \mu(B)$.
4. B_{n+1} forces "The value of $\mathbf{f}(n)$ is at most $g(n)$".

Let $B' = \bigcap_{n=0}^{\infty} B_n$. Then our construction insures that:

5. $\mu(B') > 0$. (So B' is a condition.)

6. $B' \subseteq B$.

7. B' forces that the function $\mathbf{f}$ is dominated by the function g of M.

This is the desired contradiction with the assumed properties of the condition B, and so completes the proof that $A(\aleph_0, \aleph_0)$ holds in N.

6.6. Having now checked that Axioms 1 and 2 hold in the model N, we wish to show next that if $n \geq 1$ and $\kappa > \aleph_{n+1}$, then the principle $A(\aleph_n, \aleph_0)$ fails in N. So, towards a contradiction, suppose that the family $\langle h_\alpha \mid \alpha < \aleph_{n+1} \rangle$ is a dominating family witnessing the truth of $A(\aleph_n, \aleph_0)$ in N.

Our first goal is to establish the following. There is a countable transitive model M_1 of ZFC and a κ-sequence of reals z_1 such that:

1. M_1 and M have the same ordinals, cardinals, and cofinality function. (In fact, M_1 is a c. c. c. forcing extension of M.)

2. $N = M_1[z_1]$. The κ-sequence of reals, z_1, is random over M_1.

Remark: The basic facts about randomness do not require that the ground model satisfy GCH. The assumption that M satisfied GCH was used only in computing the powers of cardinals in N.

3. The family $\langle h_\alpha \mid \alpha < \aleph_{n+1} \rangle$ lies in M_1.

4. The fact that this family dominates is forced by the empty condition (viewing N as a forcing extension of M_1 obtained by adjoining a κ-sequence of random reals.)

For the moment, we work in M. If E is a subset of κ, we let $\mathcal{B}(E)$ be the complete Boolean subalgebra of $\mathcal{B}$ generated by elements of the form $\{f \mid f(\alpha) \in U\}$, where $\alpha \in E$ and U is a Borel subset of $[0,1]$.

Since $\mathcal{B}$ satisfies c. c. c., it follows that every element $b \in \mathcal{B}$ lies in $\mathcal{B}(E)$ for some at most countable subset E of κ.

We fix in M a sequence of terms $\langle \mathbf{h}_\alpha \mid \alpha < \aleph_{n+1} \rangle$ such that, for every $\alpha < \aleph_{n+1}$:

1. $\mathbf{h}_\alpha$ is a name for h_α.

2. The empty condition forces that $\mathbf{h}_\alpha$ is a map from $\aleph_n$ to $\aleph_0$.

Let B be a condition, true in N, that forces the statement that the sequence $\langle \mathbf{h}_\alpha \mid \alpha < \aleph_{n+1} \rangle$ dominates.

Still working in M, we can find a set E of cardinality $\aleph_{n+1}$ such that:

3. The condition B lies in $\mathcal{B}(E)$.

4. If $\alpha < \aleph_{n+1}$, $\beta < \aleph_n$, and $n \in \omega$, then the Boolean value of the statement "$\mathbf{h}_\alpha(\beta) = n$" lies in $\mathcal{B}(E)$.

Next let π be a permutation of κ, lying in M, such that $\pi[E] \subseteq \aleph_{n+1}$. Let z be the canonical random κ-sequence of reals such that $N = M[z]$. We define a new κ-sequence of reals, z^*, by the prescription:

$$z^*_{\pi(\alpha)} = z_\alpha.$$

(The prescription determines z^* completely since π is a permutation.)

It is clear that $N = M[z^*]$ and that $z \upharpoonright E \in M[z^* \upharpoonright \aleph_{n+1}]$. Moreover, it is clear that z^* is random over M. (Use the characterization of random sequences over M as those that lie in no set of measure zero with code in M.)

We write z^* as the concatenation of the sequences z_0 and z_1, where $z_0 = z^* \upharpoonright \aleph_{n+1}$ and z_1 is the tail of z^* remaining after z_0 is deleted. We set $M_1 = M[z_0]$. Then the first three claims (at the start of this subsection) about M_1 and z_1 are evident. For the final claim, note that the assertion that the family $\langle h_\alpha \mid \alpha < \aleph_{n+1} \rangle$ dominates has all its parameters in the ground model M_1. Hence its Boolean value must be fixed by all the automorphisms of the relevant Boolean algebra and (since that algebra is homogeneous) must be either $\mathbf{0}$ or $\mathbf{1}$. But it can't be $\mathbf{0}$, since the assertion whose Boolean value is being determined holds in N.

6.7. Now reinitialize the notation so that the symbol M now refers to the M_1 constructed in section 6.6, and z denotes the sequence there denoted by z_1. The result is that we have a model N which is obtained from the model M by adjoining a random κ-sequence of reals, z, such that every map of $\aleph_n$ into ω lying in N is everywhere dominated by a function in the ground model M.

We now show that this situation is absurd. Define a function $g : \aleph_n \to \omega$, lying in N, as follows:

$$g(\alpha) = n \quad \text{iff} \quad \frac{1}{n+1} \geq z_\alpha > \frac{1}{n+2}.$$

Suppose that h is a function in M mapping $\aleph_n$ to ω. We claim that the measure of the Boolean value of "h dominates g everywhere" is zero, and hence that h does not dominate g. Since the argument works for any such h, we will have contradicted the assumption that g is dominated everywhere by some $h \in M$.

Since the domain of h is uncountable, there clearly is some $j \in \omega$ such that $h(\alpha) = j$ for infinitely many α. Let $\lambda = (j+1)/(j+2)$. Then clearly, if A is a finite subset of $\aleph_n$ of size r, the measure of the Boolean value of the assertion that "$g(\alpha) \leq j$ for all $j \in A$" is λ^r. It follows that the measure of the Boolean value of the assertion "g is dominated by h" is 0, so the proof that $A(\aleph_n, \aleph_0)$ fails in N if $n > 0$ and $\kappa > \aleph_{n+1}$ is complete.[n]

6.8. That Axiom 3 holds in the model N is a consequence of the following well-known result: Let $\langle h_\alpha \mid \alpha < \omega_2 \rangle$ be a sequence of distinct functions in N from ω to $\mathbb{R}$; then there are $\alpha < \beta < \omega_2$ such that h_β does not eventually dominate h_α. That the fact just recalled entails Axiom 3 is clear. It immediately rules out increasing ω_2-sequences in a scale S, and by exploiting the fact that R is order antiisomorphic to itself, it rules out decreasing ω_2-sequences as well.

Here is a brief sketch of the proof of the stated result. Fix a sequence of names, lying in M, say $\langle \mathbf{h}_\alpha \rangle$, for the sequence of functions $\langle h_\alpha \rangle$. Towards a contradiction, let B be a condition that forces that $\langle \mathbf{h}_\alpha \mid \alpha < \omega_2 \rangle$ is an increasing sequence with respect to "eventual domination".

We associate to each h_α a countable subset S_α of κ such that that $\mathbf{h}_\alpha \in M[z \upharpoonright S_\alpha]$ is forced.

By a "Δ-system argument",[o] we can pass to a subset of $\aleph_2$ of size $\aleph_2$ and achieve the following: There is a fixed "kernel" K such that $S_\alpha = K \cup D_\alpha$. Moreover, the D_α's are pairwise disjoint.

By an argument similar to one just given,[p] we can absorb $\langle z_\alpha \mid \alpha \in K \rangle$ into M. After this reorganization, the S_α's are pairwise disjoint. Since K is countable, our new M is still a model of GCH. We may suppose as well that the support of the condition B is absorbed into M.

By passing to a subsequence and relabeling, we can assume that S_α has the same order type for all α. That order type can't be 0, since there are only $\aleph_1$ maps of ω into $\mathbb{R}$ in the model M. By an automorphism of B, we can easily reorganize $\aleph_0$ random reals into one. (This uses the usual bijection of ω with ω^2.) The upshot is that (by dropping to a submodel if necessary) we can assume that $h_\alpha = \tau_\alpha(z_\alpha)$.[q] Here τ_α is a canonical term for a real in the forcing extension that adds a single random real. Moreover, the fact that "$h_\alpha < h_\beta$ for $\alpha < \beta$" is forced by the empty condition. (This is because we absorbed the support of the condition that forces "the family $\langle h_\alpha \mid \alpha < \aleph_2 \rangle$ is increasing (with respect to $<$)" into the ground model during our reorganizations.)

---

[n]I suspect that the proof just given is, essentially, Martin's unpublished proof that Axioms 1 and 2 do not entail the rectangular axioms $A(\aleph_n, \aleph_m)$ for every n and m with $m < n < \omega$.

[o]Cf. *Jech 1978*, p. 225, Lemma 22.6

[p]Cf. the "reorganization" in our proof that $A(\aleph_{n+1}, \aleph_0)$ fails in N.

[q]After the drop to a submodel, we get a new version of N which is obtained from M by adding a sequence of reals of length precisely $\aleph_2$.

Let us call two terms τ and τ' equivalent if (with respect to the forcing that adds a single random real x) the empty condition forces that "$\tau(x) = \tau'(x)$". Then, since M satisfies CH, there are precisely $\aleph_1$ equivalence classes of terms. Thus by passing to a subsequence again and relabeling, we may assume that $h_\alpha = \tau(z_\alpha)$ for all $\alpha < \aleph_2$ and some fixed term τ.

Now we can easily reach a contradiction as follows: Recall that after our various reorganizations, the sequence z has length $\aleph_2$. Define a new sequence of reals z^* as follows:

$$z_\alpha^\star = \begin{cases} z_1 & \text{if } \alpha = 0, \\ z_0 & \text{if } \alpha = 1, \\ z_\alpha & \text{otherwise.} \end{cases}$$

It is easy to see that the sequence z^* is random over M. Since it is forced (by the empty condition) that $\tau(z_0) < \tau(z_1)$, it is also forced that $\tau(z_0^\star) < \tau(z_1^\star)$. That is, $\tau(z_1) < \tau(z_0)$. But clearly it can't happen in the same model that h_0 eventually dominates h_1 and that h_1 eventually dominates h_0. This completes the proof that Axiom 3 holds in N (provided the ground model M satisfies CH).

6.9. Conclusions

1. Axioms 1 through 3 imply no bound on the size of the continuum. This is relevant, since Gödel says in *1970c* that

> Axiom 4 is rather doubtful, while Axioms 1–3 seem *extremely likely* to me. But probably they can be proved *not* to imply $2^{\aleph_0} \leq \aleph_2$.

What we have sketched above is precisely such a proof.

2. As Martin observed (and as we have discussed in the current section), the square axioms (Axioms 1 and 2 of Gödel) do not entail (in ZFC) the rectangular conjectures.

7. Open questions

The following questions (so far as I know) remain open:

1. Are Axioms 1 through 4 consistent with ZFC? If so, do they imply a bound on the size of the continuum?
2. Are Axioms 3 and 4 consistent with the axioms of ZFC?

That proving the consistency of Axioms 3 and 4 may be somewhat delicate is indicated by the existence of "Hausdorff gaps". In *Hausdorff 1936* it is shown how one can construct in any model M of set theory sequences of functions $\langle f_\alpha \mid \alpha < \aleph_1 \rangle$ and $\langle g_\alpha \mid \alpha < \aleph_1 \rangle$ that form an (ω_1, ω_1^*) gap which is indestructible in any extension of the model that preserves $\aleph_1$. More precisely:

1. $\langle f_\alpha \rangle$ is an ω_1-sequence of functions from ω to ω which is strictly increasing with respect to the "eventual domination" order $<$.
2. $\langle g_\alpha \rangle$ is an ω_1-sequence of functions from ω to ω which is strictly decreasing with respect to the "eventual domination" order $<$.
3. If $\alpha, \beta < \omega_1$, then $f_\alpha < g_\beta$.
4. There is no function $h : \omega \to \omega$ such that $f_\alpha < h < g_\alpha$ for all $\alpha < \omega_1$.
5. In fact, if N is a model of ZFC which is an extension of M and $\aleph_1^N = \aleph_1^M$, then there is no h which "splits the gap" in N.[r]

Thus in building up the set S (to witness Axioms 3 and 4) in some model of set theory, we would need to have some mechanism for avoiding Hausdorff gaps.

8. Some final remarks

It seems clear to this writer, based both on the detailed study of *1970a* described above and on the evidence of *1970c*, that Gödel *did not* have a proof of $2^{\aleph_0} = \aleph_2$ from his Axioms 1 through 4. The question of whether there is such a proof (and indeed of whether these axioms are consistent) remains open.

Even if there is such a proof, there remains the question of the plausibility of Axioms 1 through 4, clearly a much more subjective issue. I do not even find Axiom 1 convincing. The truth of Axioms 3 and 4 seems quite problematical.

The question remains as to how Gödel, indisputably one of the greatest logicians of all time, could write a manuscript such as *1970a*. My speculation, to some extent supported by *1970c*, goes as follows:

Gödel had suspected for some time that axioms of "growth" such as Axioms 1 and 2 above were plausible and should lead to some bound

---

[r] Of course, Hausdorff did not talk of models of set theory or prove absoluteness results. He gave a direct construction in ZFC of what is now called a Hausdorff gap and proved properties 1 though 4 for his construction. The notion of a Hausdorff gap and the proof that Hausdorff gaps are indestructible under cardinal preserving extensions (property 5 above) are due to Kunen (unpublished). An exposition of these results can be found in *Dales and Woodin 1987*.

on the continuum. The idea of decomposing the real line into a small number of small sets (indexed by the initial segments of some scale) led him to the proof-attempts of *1970a* and *1970b*. Ill health both made it seem urgent that his ideas be communicated to the world and made it impossible for him to carry out his usual scrupulously careful presentation and checking of the details.

Robert M. Solovay[s]

[s]Conversations and correspondence with R. Laver, D. A. Martin, G. Takeuti and S. Todorcevic were very helpful in preparing this note. Comments on an earlier draft of this note by J. Dawson, S. Feferman, D. A. Martin and G. Moore were also extremely helpful.

Some considerations leading to the probable conclusion that the true power of the continuum is $\aleph_2$ (*1970a*)

Consider the following four axioms:

1. There exists a scale of functions $\omega_n \to \omega_n$ of type ω_{n+1} majorizing by end pieces (b.e.p.) every function $\omega_n \to \omega_n$. (It follows that there exists a set M of power $\aleph_{n+1}$ majorizing *everywhere* any function $\omega_n \to \omega_n$.)[1]

2. The total number of initial segments [of] all the functions of this scale and of M is $\aleph_n$.

(This is trivial in the lowest case and probable in higher cases because the initial segments are of small importance in a scale by end pieces and can be changed arbitrarily without undesirable effects, moreover because the distinguished functions are somehow to be "smooth functions" [such as generalizations of totally positive functions].)

2   3. There exists a *complete*, i.e., not extendible, scale of | functions ($N \to$ Reals) in which every ascending or descending sequence has cofinality ω_1.

4. The Hausdorff continuity axiom for this complete scale.

[1]The functions of this set are called "distinguished".

From the first two axioms follow similar statements for the functions $\omega_n \to \omega_m$ $(m < n)$. In order to show that Axioms 1 and 2 also hold for functions $\omega_1 \to \omega_0$ (and generally $\omega_m \to \omega_n$, $m > n$) associate with each $\alpha \in \omega_1$ an $\alpha_n \to \alpha$.

From these axioms it follows that $2^{\aleph_0} = \aleph_2$ and from Axioms 1–3 that $2^{\aleph_0} \leq \aleph_2$. For the proof we can assume that $2^{\aleph_0} > \aleph_1$ by Axiom 4 and carry on most considerations neglecting $\aleph_1$ points.

Next construct a measure of zero sets whose values are elements of the complete scale, by considering the Dedekind section created by the property, of a function $N \to$ Reals, of covering[2] the given zero set (*and any end segment of it*) by a sequence of intervals of these lengths. An absolute zero set is one coverable in this manner by any function $N \to$ Reals.

We associate with each initial segment $\alpha \to \omega_1$ of a "distinguished" function $g: \omega_2 \to \omega_1$ an $f(\alpha) < \omega_2$ [and] $> \alpha$, different [ordinals] to different [segments] (Axiom 2). Next we associate a decomposition of the interval into an increasing ω_1 sequence $s(\beta, x)$, $x < \omega_1$, of sets to each $\beta < \omega_2$ by stipulating that each set of the decomposition at $\beta = f(\alpha)$ should (by intersecting) decrease the measure (real measure or zero set measure accordingly) of the set $\Delta_{\chi<\alpha} s(\chi, g(\chi))$. | Then by the second half   **3** of Axiom 3 any sequence $s(\chi, g(\chi))$ for *any* g must end after $\aleph_1$ steps [and must] end with an absolute zero set. Then the sum of all these zero sets (whose number by Axiom 2 evidently is $\leq \aleph_2$) is the interval. Moreover in this whole process no other sets have to be used but $G_{\aleph_1}$ sets (i.e., intersections [of] $\aleph_1$ open sets) which turn out (neglecting, as always in equalities,[3] sets of power $\leq \aleph_1$) to be the same as the $F_{\aleph_1}$ sets (i.e., sums of $\aleph_1$ closed sets). Hence any such set either contains a perfect subset (and then, as can easily be seen, is not an absolute zero set) or has power $\leq \aleph_1$. Hence the interval has power $\leq \aleph_2 \cdot \aleph_1 = \aleph_2$. That each $G_{\aleph_1}$ set $\doteq$ to an $F_{\aleph_1}$ set follows thus: Replace the open sets G_α yielding the $G_{\aleph_1}$ (call the latter H) by the corresponding sets G'_α of closed intervals, where $g(\alpha, n)$ is to denote the sum of the first n of these intervals $(G'_a = \sum_{n=1}^{\infty} g(\alpha, n))$. | Then for each $f: \omega_1 \to N$, the sequence[a] $\Delta_\beta = \Delta_{\alpha<\beta} g(\alpha, f(\alpha))$ must   **4** become constant at a certain $\beta_f < \omega_1$. Hence

$$\Delta_{\alpha<\omega_1} g(\alpha, f(\alpha)) = \Delta_{\alpha<\beta_f} g(\alpha, f(\alpha))$$

and H is the sum over all f of the right-hand sides of these equations. Since confining oneself to distinguished sequences f does not change the result, it follows by Axiom 2 that $H \doteq$ the sum of $\aleph_1$ closed sets.

---

[2]Or covering up to $\aleph_1$ points.

[3]We use the symbol $\doteq$ for this equality.

---

[a]Here Gödel uses the symbol Δ for intersection; a modern writer would use $\bigcap$.

5 | Next consider the function $h(\alpha) = \omega^\alpha (\alpha < \omega_1)$. If g is the given function $\omega_1 \to \omega$, form the function $h_1(\alpha) = h(\alpha)_{g(\alpha)}$. Then for each such g the associated $h(\alpha) - h_1(\alpha)$ is a function $\omega_1 \to \omega_1$ which converges toward ω_1, since each end segment of $h(\alpha)$ is similar to $h(\alpha)$. Now take any distinguished function k smaller [than] $h(\alpha) - h_1(\alpha)$ and form $r(\alpha) = \underline{h(\alpha) - k(\alpha)}$. This is a distinguished function for which $\underline{h(\alpha) > h(\alpha) - k(\alpha) > h(\alpha) - (h(\alpha) -}$ $\underline{h_1(\alpha)) = h_1(\alpha)}$, i.e., [$r(\alpha)$ lies] between $h(\alpha)$ and $h_1(\alpha)$. Now form $p(\alpha) =$ [the] smallest integer n such that $h(\alpha)_n \geq r(\alpha)$. Then $p(\alpha) \geq g(\alpha)$, since $h(\alpha)_{p(\alpha)} \geq r(\alpha) \geq h_1(\alpha) = h(\alpha)_{g(\alpha)}$. That the set of functions $p(\alpha)$ satisfies Axioms 1 and 2 follows from the fact that the functions $k(\alpha)$ do, from which the $p(\alpha)$ are derived.

A proof of Cantor's continuum hypothesis from a highly plausible axiom about orders of growth
(*1970b)

Emile Borel (see [Note II, page 117,] of *Borel 1898*) considers as self-evident the fact that the scale of the orders of growth of the functions $\omega \to \omega$ has order type ω_1, i.e., that there exists a set S_0 of order type ω_1 of such functions, ordered by the relation "greater from a certain point on", which, for *any* function $f: \omega \to \omega$, contains a function g majorizing f from a certain point on. Analogous axioms for $\aleph_n$ (or even any regular ordinal) are highly plausible, much more so than the continuum hypothesis. From these axioms follows easily the existence of sets R_n of power $\aleph_{n+1}$ such that *any* function $f: \omega_n \to \omega_n$ is majorized for *every* value of the argument by some $g \in R_n$.

2 | In addition, however, the following axiom seems very likely: *The set of all initial segments of functions of S_n (or R_n) has power $\aleph_n$.* Its likelihood derives from the following reasons:

    1. It is trivially true for $n = 1$.

    2. The initial segments of the sets of S_n can, to a large extent, be changed arbitrarily without in any way disturbing the character of S_n as a scale, so that it should be possible to use only functions of a very restricted class, perhaps the totally monotonic, in some generalized sense.

From the above axioms follow similar axioms for the sets of functions $\omega_n \to \omega_m$, $m < n$, namely that there are sets R_m^n of power $\aleph_{n+1}$ of majorizing functions, whose total number of initial segments is $\aleph_n$.

Proof for $n = 1$, $m = 0$:

| The proof of $2^{\aleph_0} = \aleph_1$ follows from this theorem (or rather the second [3] half of it) and therefore also from the axioms about $\omega_1 \to \omega_1$ functions corresponding to it.

We have at first this trivial *Lemma*: For any infinite closed set M in the interval $I \ (= [p, q])$ there exists a sequence M_i of closed sets such that $I = \sum M_i$, $M_{i+1} \supset M_i$, $MM_i \subset M$.

Proof: Put $M_i = [p, a_i] + [a, q]$, where $a_i \in M$, $a_i \to a$, $a_i < a$ (similarly if $a_i > a$).

Associate a unique sequence $A_i(M)$ to each infinite M_i and set $A_i(M) = I$ for finite M.

Now define by induction on β a sequence g_β $(\beta < \omega_1)$ of such decompositions of I (i.e., $\sum_i g_\beta^i = I$, $g_\beta^{i+1} \supset g_\beta^i$): First associate with each initial segment $f \restriction \alpha$ of a function $f \in R_0^1$ an ordinal $\beta = H(f \restriction \alpha) > \alpha$ [with] different β [associated] to different $f \restriction \alpha$, and in such manner that all ordinals $< \omega_1$ [and] > 0 occur as values of H. Next, for $\beta > 0$, put $g_\beta^i = A_i(\Delta_\alpha^f)$, where Δ_α^f is an abbreviation for $\Delta_{\xi \leq \alpha} g_\xi^{f(\xi)}$ and α is determined by $\beta = H(f \restriction \alpha)$, and set $g_0^i = I$. | Because of $MM_i \subset M$ in [4] the lemma, we have $g_\beta^i \cdot \Delta_\alpha^f \subset \Delta_\alpha^f$ provided Δ_α^f is infinite. Hence also $\Delta_{\beta+1}^f \subset \Delta_\alpha^f$ provided Δ_α^f is infinite. Therefore Δ_α^f for every $f \in R_0^1$ must become finite for some α (denoted by $\alpha(f)$ in the sequel) and remain constant from there on. On the other hand $I = \sum_{f \in R_0^1} \Delta_{\xi < \omega_1} g_\xi^{f(\xi)}$, because *every* function $h : \omega_1 \to \omega$ is majorized by some $f \in R_0^1$. But for each $f \in R_0^1$, $\Delta_{\xi < \omega_1} g_\xi^{f(\xi)} = \Delta_{\xi \leq \alpha(f)} g_\xi^{f(\xi)}$, where the right-hand side is a finite set. However, as f runs through the elements of R_0^1, there can be only $\aleph_1$ different $\Delta_{\xi \leq \alpha(f)} g_\xi^{f(\xi)}$ because this term for two f's has the same value if the two initial segments $f \restriction \alpha(f)$ are equal. So it follows that I has power $\aleph_1$.

| It seems to me this argument gives *much* more likelihood to the truth of [5] Cantor's continuum hypothesis than any counterargument set up to now gave to its falsehood, and it has at any rate the virtue of deriving the power of the set of *all* functions $\omega \to \omega$ from that of certain *very* special sets of these functions. Of course the argument can be applied to higher cases of the generalized continuum hypothesis (in particular to all $\aleph_n$). It is, however, questionable whether the whole generalized continuum hypothesis follows.

[Unsent letter to Alfred Tarski]
(*1970c)

Dear Alfred,

Many thanks for returning my paper and for your letter of May 19. I am sorry my answer comes so late. But I wanted first to think matters over thoroughly. Unfortunately my paper, as it stands, is no good. I wrote it in a hurry shortly after I had been ill, had been sleeping very poorly and had been taking drugs impairing the mental functions.

So it is not surprising it contains a serious mistake. The derivation of Axioms 1 and 2 for $\omega_1 \to \omega_0$ functions given on page 5 is wrong, because $h(\alpha) - h_1(\alpha) = h(\alpha)$. The existence of distinguished functions between $h(\alpha)$ and $h_1(\alpha)$ cannot be proved, although, strangely enough, this would be quite easy if h *were* $< h_1$ (instead of $>$), where h is distinguished. I now believe that Axioms 1 and 2 for $\omega_\alpha \to \omega_\beta$ functions ($\alpha > \beta$) don't follow from those for $\omega_\alpha \to \omega_\alpha$ functions and that, in fact, they are wrong.

2 On the other hand if they did follow, my | paper would be no good either, because then $2^{\aleph_0} = \aleph_1$ would follow easily from the proof of my paper by considering closed sets instead of $F_{\aleph_1}$ sets, $\omega_1 \to \omega_0$ functions instead of $\omega_2 \to \omega_1$ functions, and the relation "proper subset" instead of "of smaller measure". In fact the generalized continuum hypothesis could be proved in the same manner by considering higher continua. But I still believe it is false (as well as Cantor's), because the $\omega_\alpha \to \omega_\beta$ axioms are not evident in the least for $\alpha > \beta$.

What can be obtained by the method of my paper is a nice equivalence result for the generalized continuum hypothesis, namely, its equivalence with this statement: There exists for every isolated ordinal α a set M

3 of power $\aleph_{\alpha+1}$* of $\omega_\alpha \to \omega_0$ functions such that (1) the set of all | initial segments of the elements of M has power $\aleph_\alpha$ and (2) *any* $\omega_\alpha \to \omega_0$ function is majorized (except perhaps for fewer than $\aleph_\alpha$ argument values) by *some* function of M.

This shows that the generalized continuum hypothesis follows from certain very special and weak cases of it.

My conviction that $2^{\aleph_0} = \aleph_2$ of course has been somewhat shaken. But it still seems plausible to me. One of my reasons is that I don't believe in any kind of irrationality such as, e.g., random sequences in an absolute sense. Perhaps $2^{\aleph_0} = \aleph_2$ does follow from my Axioms 1–4, but unfortunately

---

* "Of power $\aleph_{\alpha+1}$" may be omitted without destroying the equivalence.

424

Axiom 4 is rather doubtful, while Axioms 1–3 seem *extremely likely* to me. But probably they can be proved *not* to imply $2^{\aleph_0} \leq \aleph_2$.

I hope this letter will convince you that my state of mind has improved since I sent you the paper.

Sincerely yours,

| P.S. A measure theory of zero sets would be very interesting, but I am   4 doubtful the definition given in my paper is the one to be chosen.

Appendix A
Excerpt from *1946/9-A*[a]

As to its positive part, there also is agreement insofar as in a relative sense a time isomorphic with Newtonian time exists in relativity theory. But (as noted above) there is a certain difference as to the question relative to what such a Newtonian time exists. This difference, it seems to me, does not so much lie in the fact that time is defined in relativity theory also relative to other things than an observer (for, in whatever way time may depend on the perceiving subject it will always be possible to characterize it in some abstract way without mentioning the subject explicitly); but rather the difference consists in this, that in relativity theory time, insofar as it depends on the observer (i.e., is experienced and measured by him), depends only on certain very general and purely physical characteristics of his, namely, the world line he describes. According to Kant, on the other hand, time (and space) are based upon "the relation of the objects to our sensibility", which seems to be something entirely different.

Closer examination however shows that these two things are not so far apart from each other as it looks at first sight. For the relation of the things to sensibility (if, for the time being, we confine attention to the outer sense) is given by the manner in which our sensations and images of outer objects depend on these objects. The most conspicuous characteristic of this dependence however is that those processes outside our consciousness which determine our sensations immediately (and which | mediate all other connection between our consciousness and the outer world) very nearly lie in one world line and in such and such a particular world line.

This characteristic, it is true, gives only a very general and incomplete description of the relation of the things to sensibility; moreover it concerns only the first part of sensibility (the faculty of receiving sensations (cf [[?]])[b] and only the outer, not the inner, sense. So there is a very considerable difference between Kant and relativity theory in this respect, but it is not a difference in principle but rather in degree, insofar as for Kant time (and

---

[a]On the different versions of *Gödel *1946/9*, see the textual notes to *1946/9* as well as fn. a and §2 of the introductory note. The passage printed here begins after a break after the first line of p. 8 of *A* and runs to the end of p. 11. It corresponds to p. 8 of *B2* and thus develops its thought at considerably greater length. (*A* continues on p. 12 with a discussion that begins with a sentence agreeing verbatim with the first sentence of p. 9 of *B2*.) We do not know Gödel's reasons for so considerably abbreviating this discussion in the later versions.

[b]It is not clear what Gödel might have intended to refer to here. He does not give earlier in *A* an explanation of the "first part of sensibility" (and what is given in *B2* [p. 4b] is hardly more extensive than what is said here). Possibly he intended to refer to a passage in Kant such as the following:

space) exist only relative to the special structure of our organs of sense and imagination and the special manner of their being influenced by outer and inner objects, in relativity theory however they exist relative to our body insofar as it is (irrespective of its structure) a kind of sense organ owing to the general laws of physical influence. So the "only-relative" existence of time is asserted in a stronger sense by Kant than in relativity theory. A similar relationship subsists between the two theories also in other respects (see the paragraph about space below), so that it may be said in general that (as far as the nature of space and time is concerned) Kant and relativity theory go in exactly the same direction, but relativity theory goes only one step in the direction indicated by Kant.

It might however be objected that even the equality in direction is only fictitious, and that in relativity theory time has not any more to do with sensibility than with any other of our or our bodies' properties, which are all localized in a sense along one world line, and therefore allow time | to be   10 defined relative to them; and this view seems to be supported by the fact that time also exists relative to lifeless objects, which have no sensibility at all.

I think however that this objection is not justified and that as to the question under consideration there really is a deep-rooted affinity in principle, although not in detail, between Kant and relativity theory; and that for the following reasons: (1) There exists a very natural generalization of "sensibility", which belongs to lifeless objects as well, namely the faculty of being physically influenced in the inner state by outer objects. (2) The relation between sensibility and its location in space-time is not a merely accidental one (as in the case of most other properties of ours), but quite on the contrary the very meaning of being at a certain place is to have certain relations of mutual physical influence to other objects (namely, roughly speaking, to influence and be influenced by near objects in a higher degree than by far ones), so that our describing a certain world line directly expresses a relation of the things to our sensibility.

That the difference subsisting in this respect between Kant and relativity theory is essentially only a matter of degree is seen very clearly from the fact that, if the situation subsisting in relativity theory that time depends on the observer is somewhat extrapolated, the agreement with certain formulations of Kant becomes very striking indeed. For such an extrapolation

---

The capacity (receptivity) for receiving representations through the mode in which we are affected by objects, is entitled *sensibility* .... The effect of an object upon the faculty of representation, so far as we are affected by it, is entitled *sensation* (A 19–20/B 33–4, translation from *Kant 1933*).

On Gödel's problematic conception of the first and second part of sensibility, see §8 of the introductory note to *1946/9*.

leads to the following views: (1) Not only the temporal relations between individual events, but also the general character of time (which Kant evidently had in mind in the first place, when he spoke of a dependence of time on the relation of things to our sensibility) depends | on the nature of the set of points of immediate sensorial contact we have with reality, e.g., the one-dimensionality of time on the one-dimensionality of the set. (2) For a being which had no sensibility at all (i.e., no contact through sensations with reality) but only "pure understanding", no time at all would exist. But this is exactly what Kant means, who frequently explains the things in themselves as the things such as pure understanding would perceive them. (This, of course, does not apply to human understanding, which is not pure insofar as it apprehends concepts only with the help of sensual images and symbols and therefore depends on sensibility and on time).

These considerations also answer the objection that Kant and relativity theory have for that reason nothing to do with each other because one speaks of a relativity to the body, the other to the mind. In reality Kant speaks of a relativity to sensibility, and having a sensibility is something closely related to having a body. The analogy, it is true, breaks down insofar as Kant bases the idea of time not on the relation of physical realities [to outer sense, but of psychic realities]^c to inner sense, i.e., to self consciousness (only the special manner in which individual outer events are projected on this time scheme depends of course on their relation to outer sense). But this difference does not impair the equality of principle, since according to Kant the psychic realities in us are for our self-consciousness also a kind of outer world which we know only as an appearance, i.e., incompletely and indirectly through the medium of sensations.

---

^{c}These words are supplied from *MS*. Their omission from the typescript *A* was very probably a typist's error.

Appendix B
Texts relating to the ontological proof

[Included in this appendix are the contents of two small pieces of paper found accompanying the ontological proof and three excerpts from the "Max-Phil" notebooks.

The two small pieces of paper, only one of which was dated, are part of document number 060565. With the exception of the original title, *Ontolog. Beweis*, which we have translated here, the dated sheet needs no translation. The translation of other material here is by Robert M. Adams.

The "Max-Phil" notebooks consist in a series of small notebooks labelled variously "Max" and "Phil". The series begins with "Max 0," but the following notebooks are numbered with roman numerals. A note inside the front cover of "Max II" says "Das war ursprünglich das erste Max Heft neben Ph. Heft; später Max und Ph. zusammengezogen." ("That was originally the first Max volume apart from [the] Ph. volume; later Max and Ph. joined.") They continue to be labelled "Max" until "Phil XIV", where a note on the first page says "Heft [volume] XIII (= Max XIII) (VI 45 – IV 46) wurde im April 1946 verloren [was lost in April 1946]". The title page of the last volume says "Max XV 1955–end of ??" and, on two separate lines, "Phil." and "V 1955". The notebooks contain a mix of notes (mostly in shorthand) on other people's lectures, programs of study and action he apparently intended to follow, and, increasingly as the numbers of the volumes increase, sections titled "Bem" (remark), "Phil" or "Math". The later volumes appear to be his own writing and it is from these that we have printed the excerpts below.

In order to preserve the character of these notes, instead of numbering footnotes we have reproduced Gödel's footnote flags as he wrote them. Our editorial and textual notes remain lettered. The various sections are separated by a double horizontal line, and Gödel's footnotes pertinent thereto are printed just above that line and below a single dashed line extending the full width of the page. Other lines are Gödel's own.]

Ontological Proof   ca. 1941

$$\Phi(\varphi) \equiv N(\exists x)\varphi(x)$$
$$A(x) \equiv (\exists x)[(x)[\varphi(x) \equiv_N A(x)].\Phi(\varphi)]$$
$$G(x).\varphi\epsilon+ \rightarrow \varphi(x)$$
$$(\exists x)G(x) \rightarrow N(\exists x)G(x)$$
$$G = \hat{x}[(\varphi)\varphi\epsilon+ \rightarrow \varphi(x)]$$
$$\psi\epsilon+ \rightarrow \psi(G)$$

$$\text{Ontologischer Beweis} \quad G(x) =_{Df} x$$

$NE(x) \equiv_{Df} N(\exists y)\text{Ess}_x(y)$ notwendige Existenz

$G(x) \supset NE(x)$ da NE eine pos. Eigenschaft ist

$G(x) \supset .\text{Ess}_x \supset G^\otimes$ (gilt für jede Eigenschaft statt G)

$G(x) \supset N(\exists y)G(y)$ folgt aus den 3 vorhergehenden)

$(\exists x)G(x) \supset N(\exists y)G(y)$

$M(\exists x)G(x) \supset MN(\exists y)G(y)$ (beiderseitige Hinzufügung von M)

daher $\supset N(\exists y)G(y)$

Ebenso folgt: Wenn der Begriff notwendiger Existenz widerspruchsfrei ist, so gibt es Dinge, für die er gilt.

- -

$^\otimes$Dazu braucht man, das alle Eigenschaften Gottes durch eine Eigenschaft 2-ten Typs definiert sind. [Das musste überhaupt die Df. der Essenz sein.] Oder Ess_x definiert^a durch

$$\varphi\epsilon\text{Ess}_x \equiv (\psi)\{\psi(x) \supset N(x)[\varphi(x) \supset \psi(x)]\}.\varphi(x)$$

Excerpts from "Max XI"

[From page 97:]

Bemerkung (Philosophie): Wenn der ontologische Beweis richtig ist, so kann man a priori die Existenz (Wirklichkeit) eines nichtbegrifflichen Gegenstandes einsehen. Ist vielleicht ein anderer Beweis der: Wenn es nichts wirklich gebe, so wäre eben dies tatsächlich etwas wirkliches (was über [einen] blossen Begriff hinausgeht)?

[From page 149:]

Bemerkung (Theologie): Die Überlegung: nach dem Satz vom zureichenden Grund muß die Welt eine Ursache haben. Dies muß an sich notwendig sein (sonst würde sie wieder eine Ursache verlangen). Beweis der Existenz eines apriorischen Gottbeweises (der darin enthaltene ist es nicht).

---

^{a}In the formula below, we have removed a dot just before the N.

$$\text{Ontological Proof} \quad G(x) =_{Df} x$$

$NE(x) \equiv_{Df} N(\exists y)\text{Ess}_x(y)$ necessary existence

$G(x) \supset NE(x)$ since NE is a positive property

$G(x) \supset .\text{Ess}_x \supset G^{\otimes}$ (holds for any property in place of G)

$G(x) \supset N(\exists y)G(y)$ (follows from the 3 preceding)

$(\exists x)G(x) \supset N(\exists y)G(y)$

$M(\exists x)G(x) \supset MN(\exists y)G(y)$ (addition of M on both sides)

hence $\quad \supset N(\exists y)G(y)$

Likewise it follows that: If the concept of necessary existence is consistent, then there are things for which it holds.

- -

$^{\otimes}$For this it is required that all the properties of God are defined by a second-order property. [That must in general be the Df. of the essence.] Or Ess_x is defined[a] by

$$\varphi\epsilon\text{Ess}_x \equiv (\psi)\{\psi(x) \supset N(x)[\varphi(x) \supset \psi(x)]\}.\varphi(x)$$

Excerpts from "Max XI"

[From page 97:]

Remark (Philosophy): If the ontological proof is correct, then one can obtain insight a priori into the existence (actuality) of a non-conceptual object. Is this perhaps another proof: If there were nothing actual, then precisely this would in fact be something actual (which goes beyond a mere concept)?

[From page 149:]

Remark (Theology): The reflection: according to the Principle of Sufficient Reason the world must have a cause. This must be necessary in itself (otherwise it would require a further cause). Proof of the existence of an a priori proof of the existence of God (the proof it contains fails to be one).

Excerpt from "Phil XIV"

103 |*Philosophie*: Beschäftigung mit Philosophie, selbst wenn keine positiven Ergebnisse herauskommen (sondern ich ratlos bleibe), ist auf jeden Fall wohltätig. Es hat die Wirkung (daß "die Farbe heller"), d. h., *daß die Realität deutlicher als* solche erscheint.

Philosophie: Der ontologische Beweis muß auf den Begriff des *Wertes* (p besser als $\sim p$) gegründet werden, und auf die Axiome:[x]

1. Logisch Äquivalente haben den gleichen Wert (daher immer entweder φ oder $\sim\varphi$ positiv).

2. Wenn p, q negative Werte haben, so auch $p \vee q$.[b]

3. Np, Mp negativ, wenn p negativ.

4. Das Sein ist positiv.

Dann folgt, I. daß logische Folgen aus positiven positiv sind;

II. daß notwendige Existenz positiv ist, weil nämlich:

104 | $$x = x.(\exists y)\varphi(y) \,.\, \equiv \,.\, \varphi(x) \vee [(\exists y)\varphi(y).x = x]$$

Die positiven und die wahren Sätze sind aus verschiedenen Gründen dasselbe—vergleiche übernächste Anmerkung?

- -

[x]Er kann nur auf *Axiome* und *nicht* auf eine Definition (= Konstruktion) von "positiv" gegründet werden, denn die Konstruktion ist mit beliebigem Verhältnis verträglich.

Philosophie: Der philosophische Grundbegriff ist die Ursache. Diese involviert: Wille[x], Kraft, Genuß[x], Gott, Zeit, Raum*. Das Bejahen des Seins ist die Ursache der Welt. Das erste Geschöpf: es kommt zum Sein das Bejahen des Seins dazu. Daraus folgt weiter, daß möglichst viele Sein entstehen werden—und dies ist der letzte Grund der Verschiedenheit (variatio delectat).

Die Harmonie bedeutet mehr Sein als Disharmonie, denn der Gegensatz der Teile hebt ihr Sein auf. Regelmäßigkeit besteht in Übereinstimmung, z.
105 B.: bei gleichem Winkel, ist die gleiche Farbe. |Vielleicht kann man die anderen Kantschen Kategorien (d. h., die logischen, einschließlich Notwendigkeit) aus der Ursache definieren und die logischen (mengen-theoretischen) Axiome aus den Axiomen für Ursache ableiten. [Eigenschaft = Ursache

---

[b]The following two axioms were originally written at the bottom of the page as an insertion.

Excerpt from "Phil XIV"

Philosophy: Engaging in philosophy is salutary in any case, even when no positive results emerge from it (and I remain perplexed). It has the effect that "the color [appears] brighter," that is, that *reality* appears *more clearly as* such.

Philosophy: The ontological proof must be grounded on the concept of *value* (p better than $\sim p$) and on the axioms:[x]

1. Logical equivalents have the same value (so that always either φ or $\sim\varphi$ [is] positive).
2. If p, q have negative values, so too does $p \vee q$.[b]
3. Np, Mp are negative, if p is negative.
4. Being is positive.

Then follows, I. that logical consequences of positives are positive;

  II. that necessary existence is positive, because:

$$x = x.(\exists y)\varphi(y) . \equiv . \varphi(x) \vee [(\exists y)\varphi(y).x = x]$$

The positive and the true sentences are the same, for different reasons—compare the note after next?

- -

[x]It can be grounded only on *axioms* and *not* on a definition (= construction) of "positive," for a construction is compatible with an arbitrary relationship.

Philosophy: The fundamental philosophical concept is cause. It involves: will,[x] force, enjoyment,[x] God, time, space.[*] The affirmation of being is the cause of the world. The first creature: to being is added the affirmation of being. From this it follows further that as many beings as possible will be produced—and this is the ultimate ground of diversity (variety delights).

Harmony implies more being than disharmony, for the opposition of the parts cancels their being. Regularity consists in agreement, for example: at the same angle, there is the same color. Perhaps the other Kantian categories (that is, the logical [categories], including necessity) can be defined in terms of causality, and the logical (set-theoretical) axioms can be derived from the axioms for causality. [Property = cause of the difference of

der Verschiedenheit von Dingen.] Außerdem wäre zu erwarten, daß die
analytische Mechanik aus einem solchen Axiom folgen würde.

----

*Nahe sein = Möglichkeit der Einwirkung.
ˣDaher das Leben und die Bejahung und Verneinung.

====

Philosophie: Ontologischer Beweis

1. Die Interpretation von "positiver Eigenschaft" als "guter" (d. h., einer
mit positivem Wert) ist unmöglich, weil der größte Vorteil + dem kleinsten
Nachteil negativ ist.

2. Es ist möglich, die positive als perfectiv zu interpretieren, d. h., "rein
106 gut", d. h., solche, welche keine Negation von "rein gut"ˣ impliziert. | Das
Hauptaxiom läuft dann (im wesentlichen): Eine Eigenschaft ist eine Perfective, dann und nur dann wenn sie keine Negation einer Perfectiven impliziert: neben diesem braucht man nur noch die Axiome:⁺ Die Notwendigkeit einer Perfectiven ist eine Perfective$^\phi$ und das Sein ist eine Perfective.*

3. Wenn man die positiven als Assertionen (+ Tautologien), die negativen als Privationen^X (+ Kontradiktionen) interpretiert,^{c} laufen^d die Axiome genau wie in 2. Das erste dieser Axiome sagt im wesentlichen, daß
die positiven Eigenschaften ein maximal kompatibles System bilden.^{e} Das
zweite Axiom ist plausibeler für die Interpretation 3 (für 2 nur, wenn man
107 die Widerspruchsfreiheit von $N\varphi(x)$ | voraussetzt). Das dritte Axiom ist
überflüßig.

4. Daß die Notwendigkeit einer positiven Eigenschaft positiv ist, ist
die wesentliche Voraussetzung für den ontologischen Beweis. Wenn man
annimmt $\varphi(x) \supset N\varphi(x)$ [weil aus dem Wesen von x folgend], dann ist es
leicht beweisbar, daß es für jedes kompatibele System von Eigenschaften
ein Ding gibt, aber das ist *der schlechte Weg*. Vielmehr will $\varphi(x) \supset N\varphi(x)$
erst aus der Existenz Gottes folgen.

ˣKann nicht durch "gut" ersetzt werden.
⁺Es braucht *nicht* angenommen zu werden, daß immer entweder φ oder
$\sim \varphi$ positiv ist.
$^\phi$Oder wenn $M\varphi$ eine Perfective, dann auch φ.
*Oder : es gibt eine Perfective. Daraus folgt, daß das Sein eine Perfective
ist, da es die Möglichkeit dieser Perfectiven impliziert.
^{X}D. h.: teilweise Privationen (eventuell kombiniert mit Assertionen).

----

^{c}Here Gödel surely intended "interpretiert" ("interprets"), although the shorthand
clearly says "impliziert" ("implies").

things]. Moreover it should be expected that analytical mechanics would follow from such an axiom.

----

*Being near = possibility of influence.
×Hence life and affirmation and negation.

===

Philosophy: Ontological Proof

1. The interpretation of "positive property" as "good" (that is, as one with positive value) is impossible, because the greatest advantage + the smallest disadvantage is negative.

2. It is possible to interpret the positive as perfective; that is, "purely good," that is, such as implies no negation of "purely good."× The chief axiom runs then (essentially): A property is a perfective if and only if it implies no negation of a perfective. All that is needed besides this is the axioms:+ The necessity of a perfective is a perfective,ϕ and being is a perfective.*

3. If the positives are interpreted[c] as assertions (+ tautologies) and the negatives as privations[X] (+ contradictions), the axioms run[d] exactly as in 2. The first of these axioms says essentially that the positive properties form a maximal compatible system.[e] The second axiom is more plausible for interpretation 3 (for 2, only if the consistency of $N\varphi(x)$ is presupposed.) The third axiom is superfluous.

4. That the necessity of a positive property is positive is the essential presupposition for the ontological proof. If $\varphi(x) \supset N\varphi(x)$ is assumed [as following from the essence of x], then it is easily provable that for every compatible system of properties there is a thing, but that is *the inferior way*. Rather $\varphi(x) \supset N\varphi(x)$ should follow first from the existence of God.

×Cannot be replaced by "good."

+It need *not* be assumed that always either φ or $\sim\varphi$ is positive.

ϕOr if $M\varphi$ is a perfective, then φ is too.

*Or: there is a perfective. From this it follows that being is a perfective, since it implies the possibility of this perfective.

XThat is, partial privations (possibly combined with assertions).

---

dBeside the lines "3. Wenn man ... laufen" ("3. If the ... run") were two heavy vertical bars.

eBeside the lines "dieser Axiome ... bilden" ("of these axioms ... form") was a single vertical bar with the words "*Das ist das Beste*" ("*That is the best*") written along the bar.

Philosophie: Für alle Eigenschaften, die aus endlich vielen $\varphi_1, \ldots, \varphi_n$ durch
108 Konjunktion und Disjunktion gebildet sind, sind die elementaren | Eigen-
schaften (d. h. die schwächsten) die Glieder der konjunktiven Normalform,
zunächst nur eine negativ, alle übrigen positiv. Die negative ist die schwä-
chste negative Eigenschaft. *Durch Hinzufügung von positiven Gliedern wird
sie stärker* (weil nämlich jede positive Eigenschaft andere impliziert* und
daher durch eine negative viele positive ausgeschlossen werden). Theorem:
Die positiven Eigenschaften sind genau die, welche aus den elementaren
durch Anwendung der Operationen $\&$, $\vee$, $\supset$ gebildet werden können.

----

*D. h., es tritt nicht einfache Addition der Elementen ein.

Philosophie:[?] Die einzigen synthetischen Sätze sind die von der Form $\varphi(a)$
(z. B.: *Ich* habe diese Eigenschaft), denn diese haben keine objektive Bedeutung, oder: Sie hängen nicht von Gott, sondern von dem Dinge a ab.

[From the bottom of page 109:][f]

Geht nicht, weil dann Gott Imperfektiv hätte, die darin besteht, daß Im-
perfektive möglich sind. Alles, was aus einer Perfectiven folgt, wie etwas
Gutes, das ist eine Perfective, ist.

---

[f]This text was connected to the bottom of page 108 by an arrow, and was apparently
meant to be inserted here.

Philosophy: For all properties that are formed out of finitely many $\varphi_1, \ldots, \varphi_n$ through conjunction and disjunction, the elementary properties (that is, the weakest) are the members of the conjunctive normal form, in the first instance only one negative, all the rest positive. The negative is the weakest negative property. *Through addition of positive members it becomes stronger* (because every positive property implies others[*] and therefore through one negative many positive properties are excluded). Theorem: The positive properties are precisely those that can be formed out of the elementary ones through application of the operations &, $\vee$, $\supset$.

----

[*]That is, it is not a matter of simple addition of elements.

Philosophy:[?] The only synthetic propositions are those of the form $\varphi(a)$ (for example: *I* have this property), for these have no objective meaning, or: They depend not on God, but on the thing a.

[[From the bottom of p. 109]][f]

This doesn't work, because then God would have an imperfective, which consists in the fact that imperfectives are possible. Everything that follows from a perfective, such as something good, that is a perfective, is.

Textual notes

The individual copy-texts and their concomitant textual issues are discussed under the individual works.

In these notes, the pairs of numbers on the left indicate page and line in this volume. (Line numbers do not count titles at the top of a page.) We have sometimes deemed it important to indicate material that was crossed out by Gödel. Such material is printed in these notes with an overstrike. In the notes for items not themselves originally in shorthand, we have indicated material transcribed from shorthand (e.g., Gödel's own annotations) by slant roman type.

Gödel *1930c

The copy-text for *1930c is a ten-page typescript, *Nachlass* item 040009, bearing handwritten insertions and annotations. Prior to cataloguing of the *Nachlass*, it was one of several items in an envelope that Gödel labelled "Manuskripte Korrekt *der* 3 Arbeiten in Mo[nats]H[efte] + *Wiener Vorträge über die ersten zwei*" (manuscripts, proofs for the three papers in *Monatshefte* [1930, 1931, and 1933i], plus Vienna lectures on the first two.) On the basis of that label, *1930c ought to be the text of Gödel's presentation to Menger's colloquium on 14 May 1930—the only occasion, aside from the meeting in Königsberg, on which Gödel is known to have lectured on his dissertation results (see the chronology in Volume I of these *Collected Works*). Internal evidence, however, especially the reference on the last page to the incompleteness discovery, suggests that the text must be that of the later talk. Since no other lecture text on this topic has been found, it may well be that Gödel used the same basic text on both occasions, with a few later additions.

To avoid confusion among formulas, Gödel's numbering scheme has here been slightly altered; in particular, roman rather than arabic numerals have been used to label the axioms from *Principia mathematica*.

	Original	Replaced by
18, 31	*und*	oder
18, 34	Aussagen und für	Aussagen, für
22, 8	"[Es] wird zu zeigen sein"	[moved from after 3.]
22, 28	Satz	Sätze
22, 35–6	Aussage- und Funktions-variablen	Funktionsvariablen
24, 13	die vorgelegte Formel 3	der] vorgelegten Formel (3)
24, 27	etc.	usw.

In addition, the following annotations and emendations, by Gödel but not printed in the text, should be noted.

18, 7   Inserted at end of line: "*und d. h.* [[?]] *tautologisch*" ("and that is to say [[?]] is tautological").

18, 14   Inserted after "versteht." ("understand") (with reference to the term "engerem Funktionenkalkül" ["restricted functional calculus"]): "*Der Ausdruck stammt aus den Grundzüge der theoretischen Logik von H.A.*" ("The expression [['restricted functional calculus'] stems from *Grundzüge der theoretischen Logik* by H[[ilbert and]] A[[ckermann, their *1928*].").

20, 19   Above the period are the words "d. h. wenn" ("that is, if").

20, 21   The words "Vorsetzen eines Negationszeichens" ("prefixing a negative sign") were crossed out on the typescript. Above them was written "*mit d. Neg.* [[?]] *entsteht*" ("results in the negative [[?]]".

20, 24-5   The words "hier mit II bezeichneten" ("here denoted by II") were crossed out and replaced by "*folgenden*" ("the following"), but subsequent references to II were not altered.

20, 27-8   Above the words "oder widerlegbar. Nehmen" ("or refutable. Let") is written "*Überzeugen wir uns, daß dieser Satz dasselbe bezeichnet wie die früher ausgesprochenen.*" ("We may convince ourselves that this proposition denotes the same as that articulated earlier.").

20, 28   After the period Gödel inserted "Also die frühere Fassung gilt." ("Thus the earlier version holds."), a passage repeated five lines below.

22, 14   "*Wenn diese drei Sätze* [[be]]*wiesen sind*" ("If these three propositions are proved") is an insertion to the text (and thus the "Es" was changed here to "es").

22, 29   After "Zunächst Punkt 2" ("First, point 2") before the period, Gödel inserted the words "das ist der Satz" ("that is, the proposition"), apparently to distinguish the second proposition from a similarly numbered formula. In the present text this ambiguity has been resolved by the omission of parentheses around proposition numbers.

24, 27   Above "Nun betrachten" ("Now consider") there is a problematical shorthand insertion, reading "*die* [[?]] *Variabeln* [[*bezeichne?*]] *ich mit* $\varphi \vee \psi$, [[?]]" ("the [[?]] variables I denote by $\varphi \vee \psi$, [[?]]").

26, 36-7   Above the words "der Axiome" ("of the axioms") is "*obige 6 Axiome und Grundregeln*" ("above 6 axioms and rules"). Above the following "des" ("of the"), is a "u.s." (probably abbreviating "und so" ["and so"]).

In addition to the annotations noted above, the manuscript includes a variant, apparently earlier, version of (Gödel's) page 6. In this version, the text from (22, 33), beginning with "Die vorgelegte Formel ...", through (24, 15) ("... was ich hier nicht näher ausführe.") is replaced by the following (in which all abbreviations have been expanded). A translation (by John Dawson) follows the German text.

Die hier angeschriebene Formel (3) hat den zweiten Grad. $A(x, y, z, u)$ ist dabei aus Aussage- und Funktionsvariablen allein mittels der Operationen des Aussagenkalküls aufgebaut zu denken, also ohne "alle" und "es gibt". Ich zeige zunächst, daß diese Formel mit der darunterstehenden gleichzeitig erfüllt ist. Dabei ist in der zweiten Formel unter F eine zweistellige Funktionsvariable zu verstehen: ~~die in $A(x,y,z,u)$ nicht vorkommt~~. Nehmen wir an, die Formel (3) sei erfüllt. Ich will zeigen, daß dann auch die 2. erfüllt ist. Denken wir uns in Formel (3) die Aussagen und Funktionen eingesetzt, welche sie zu einem wahren Satz machen. Läßt man in diesem Satz die beiden ersten Zeichen des Präfixes weg (allgemein gesprochen: das Präfix bis zum zweiten Wechsel zwischen All- und E-Zeichen), so entsteht ein Ausdruck mit zwei freien Individuenvariablen, das heißt, eine ganz bestimmte Relation, die wir mit R bezeichnen.

Setzt man nun in der zweiten Formel für F die eben definierte Relation R und in A dieselben Aussagen und Funktionen ein wie vorhin in die obere Formel, so ist der entstehende Satz wahr, denn der zweite Teil nach dem Konjunktionszeichen ist einfach die hier stehende Definition für R, wenn man die Definiens durch Aequivalent ersetzt, und der erste Teil wird offenbar nach der Definition von R mit der oberen Formel gleichbedeutend, das heißt, genauer mit dem wahren Satz, der aus ihr durch Einsetzung entstanden ist. Wenn also Formel (3) erfüllt ist, dann auch Formel (4). Man überzeugt sich eben so leicht von der Umkehrung und ferner auch davon, daß, wenn die zweite Formel widerlegbar ist, daß heißt, ihre Negation beweisbar, daraus ein Beweis für die Negation der ersten folgt, was ich hier nicht näher ausführen kann. Bringt man nun die Formel (4) auf die Normalform, was hier in Formel geschehen ist, so zeigt sich, daß sie einen um eins geringeren Grad hat, nämlich den ersten.

That is,

The formula (3), written here, has second degree. In it, $A(x, y, z, u)$ is to be thought of as built up from propositional and functional variables alone, by means of the operations of the propositional calculus, hence without "for all" and "there exists". I show first of all that this formula is satisfied simultaneously with the one written below it. Therein, by F in the second formula is to be understood a two-place function variable ~~which does not occur in $A(x,y,z,u)$~~. We assume formula (3) to be satisfied. I will show that then the second formula is also satisfied. Into formula (3) we imagine the propositions and functions to have been substituted that make it into a true statement. If in this statement the first two quantifiers of the prefix be left out (stated in general: the prefix up to the second alternation between universal and existential quantifiers), there results an expression with two free individual variables, that is, a quite definite relation that we denote by R.

Now if the relation just defined be substituted for F in the second formula and the same propositions and functions substituted into A as in the formula above, the resulting statement is true, because the second part, after the conjunction sign, is simply the definition of R written here, if one replaces the definiens by its equivalent; and according to the definition of R, the first part is obviously synonymous with the formula above, that is, precisely with the true statement that resulted from it through substitution. Thus if formula (3) is satisfied, so is formula (4). One can just as easily convince oneself of the converse, and also, furthermore, that if the second formula is refutable, that is, its negation is provable, a proof of the negation of the first formula follows from it—a matter I can't go into in greater detail here. If one now brings formula (4) into the normal form that appears in the formula here [pointing to it], it can be seen that it has a degree one smaller, namely, the first.

*Gödel *1931?*

As found in the *Nachlass* (item 040405.00), the typescript was quite complete and required only minor copy-editing as described in the introduction to these notes.

*Gödel *1933o*

The manuscript as found in the *Nachlass* (item 040114) was titled "Vortr. Cambridge". The title used in this volume was taken from the report of the meeting at which the lecture was given.

	Original	Replaced by
46, 5	axiomatischen der Mengenlehre	axiomatischen Mengenlehre
46, 15	2.	second
46, 26; 29	type	types
46, 33	3.	third
47, 20	and	an
47, 27	of given of given	of given
48, 3	, and and	, and,
49, 10	led in to contradiction	led to contradictions
49, 12	inferences	inference
49, 26	anyone	anyone's
49, 34	1.	first
50, 2	like in the case	as in the case
51, 34	concerning, it	concerning it, it
52, 31	system	systems
53, 3	anyone	anyone's
53, 11	to give	in giving
53, 16	, which is obvious	. This is obvious
53, 39–40	of a doubtful	of doubtful
53, 43	them and this	them. This

Gödel *1933?

The original typescript of *1933?* (*Nachlass* item 040410) had almost no equations written in. The equations and symbols used here were filled in by I. Halperin, with the exception of the first half of the first paragraph on Gödel's page 3, where a portion of the typescript had been completed by Gödel himself, thus providing a clue to the intended notation. Abbreviations for "Hilfssatz" and "Beweis" have been filled in.

	Original	Replaced by
56, 26	bilden	bildet
56, 30	vorkommen	vorkommt
58, 12	$\lim_{i\to\infty} w_k^i$	$\lim_{i\to\infty} w_i^k$
58, 19	sei $\mathfrak{w}_1$	sei w_1
58, 19	$= \mathfrak{w}$	$= w_1$
58, 35	konvergieren	konvergiert
60, 20	Andererseits ist	Andererseits sind

*Gödel *1938a*

In addition to the manuscript (*Nachlass* item 040148) transcribed from
shorthand and published here, another shorthand manuscript (*Nachlass*
item 040147), labelled by Gödel "Konz."—apparently a draft (*Konzept*)
or outline—was found in the same envelope. We have made use of it in
some of our editorial additions or emendations; in particular, the bracketed
passage on page 98, beginning on line 16 and ending on line 18 was taken
from *Konzept*, page 6, and replaces the following:

Beweisskizze:
1. $p \supset q . \supset . \neg q \supset \neg p \supset : q \supset a . \supset . p \supset a$ Transitivität
Folgt: Beweis

Both because of the shorthand and because of the note-like character of
the manuscript, some uncertainties (and a few outright errors) have had
to be confronted. We have made several kinds of corrections, some grammatical (relating both to syntax and word-order) and some mathematical.
Because the shorthand need not indicate pluralization unless necessary for
clarity, we have added or changed to plurals where deemed necessary. In
addition, we have changed declensions of articles and adjectives, conjuga-
tion of verbs, and word order to fit the grammar, even when the shorthand
was clear. All such changes are listed here, in case a better reading that
has escaped us may occur to the reader.

	Shorthand original	Replaced by
88, 1–2	aufstellen	aussprechen [Cf. *Hilbert 1928*]
90, 2	schon nicht	nicht schon
90, 2	Bedingung	Bedingungen
90, 19	geschlossen werden aus $A(x, f(x))$	aus $A(x, f(x))$ geschlossen werden
92, 6	*dies*	*dieser*
92, 12	durchwegs	durchweg
92, 12–13	Sicherheit herzustellen	Sicherheit des Schließens herzustellen
92, 13	der niederen	der gewöhnlichen niederen
94, 14	wesentlich besteht	besteht wesentlich
102, 27	dem System	des Systems
102, 30	ist	sind
112, 7	hinausgegangen ist,	hinausgegangen wird,
112, 12	von ihr reduziert	von sich reduziert
112, 25	den Gentzenschen	dem Gentzenschen

Some errors in the writing of shorthand text are the result of the writer's having mistakenly written one symbol for another, or of carelessness in the formation or the positioning of symbols.

	Shorthand original	Probable intention	Type of error
90, 3	"an" or "ein"	sein	position
90, 19	das	darf	position
92, 12–13	herauszustellen	herzustellen	careless formation of an extra loop
94, 25	wir	werden	position
96, 16	*nicht*	*nichts*	missing loop
98, 31	*Daß wir so*	*Daß das so*	position, missing loop
104, 30	der	das	lack of curve on downstroke
108, 29	sich	sieht	lengthened downstroke

On page 90, line 1, we chose "Rahmendefinition" over "Raumdefinition" on the basis of meaning, as the shorthand is ambiguous here. On page 94, line 26, a mysterious character, initially thought to resemble a ψ, was finally deemed to be a "für" with a line through it. We have restored that "für" to make sense of the "die" that follows it. On page 106, line 1, a word which looks like "erhalten" could be read as "innerhalb"; we have chosen the latter as preferable on the basis of structure and meaning.

In some cases—usually in the neighborhood of a deletion, erasure or insertion made by Gödel himself—words intended for deletion were apparently left in unintentionally. The following is a list of such occurrences, with our deletion printed with an overstrike.

88, 13–15	Teilfragen. ~~Daraus~~ in der
90, 24	*Axiome* ~~jedenfalls~~: jedenfalls
94, 28	nur ~~solche~~ Typen
96, 28	, ~~man~~ denn man
102, 28	Ordinalzahlen ~~definieren~~ der zweiten
108, 20	Eindruck ~~mit~~, daß
112, 24	— auch | ~~auch~~ noch

On page 100, after line 28, which corresponds to the upper right-hand corner of Gödel's page 9, there occurred a symbol that looked more like the numeral "10" than a shorthand symbol. Since it appeared to be extraneous, it was removed. On page 110, between lines 6 and 7, an extraneous "$\sim\varphi(x).\varphi(f(x))$" near the left margin was removed as being unneeded and potentially confusing.

At the end of section IV there was a shorthand note, probably intended as a footnote, for which no corresponding footnote citation appeared in the text. It read "bei der Zahlentheorie unentsch." (The last word is conjectural, especially since Gödel rarely abbreviated in shorthand except for a few standard personal abbreviations, such as "fu" for "Funktion".) We have deleted this note, as we could not find an appropriate referent for it.

We have attempted to keep all other changes to a minimum, but some were necessary for the sense of the text or for correction of mathematical grammar. The following is a list of these minor changes.

	Original	Replaced by
88, 15	dieser Teilantworten	der Teilantworten
92, 25–27	$W(T)$	$Wid(T)$
	$W(A)$	$Wid(A)$
94, 11–12	eingeführt). 4	eingeführt), ⟦Systeme⟧ 2
94, 20	$\Phi(n, f)$	$\Phi(f, n, k)$
98, 1	$p \supset q \supset : q \supset r . \supset . p \supset r$	$p \supset q . \supset : q \supset r . \supset . p \supset r$
98, 2	$p :\supset . q \supset r . \supset : q . \supset p \supset r$	$p \supset . q \supset r :\supset : q \supset . p \supset r$
98, 4	$p :\supset . p \supset q . \supset q$	$p \supset : p \supset q . \supset q$
100, 13	$BBa \supset Ba$	$Bp \rightarrow BBp$
	$Ba \supset a$	$Bp \rightarrow p$
100, 25	$f(z, u)Bp \supset r$	$f(z, u)Bp, r$
100, 27	qBa	aBq
100, 28	$aB[(u)u \sim B(0 = 1)]$	$aB[(u)\sim uB(0 = 1)]$
104, 4	beweisbar im System S_2;	beweisbar im System S_1;
106, 10–11	nötig ... Reduzibilitätsaxiom	nötig ... Reduzibilitätsaxiom)
106, 27	$x < n$ R_o	$x < a$ R_o
106, 27	abzuleiten $E(n)$	abzuleiten $E(a)$
108, 33	$\neg\Phi \quad \neg\neg f$	$\neg\neg\Phi$
112, 13	2. Es wäre	(B) Es wäre

In two places—page 110, sections 17 and 19—a numbered section seemed to begin in the middle of a line although the number lay in the margin. The lines, however, are not aligned precisely, so a good argument can be made that the beginning of the line belongs with the preceding section; we have printed it that way.

*Gödel *1939b*

This lecture, *Nachlass* item 040237, was transcribed from shorthand. We have replaced Gödel's dashes at the ends of sentences with periods, and some new paragraph indentations were made for clarity. There were many abbreviations in longhand among the shorthand words. These have been filled in without specific mention, with two exceptions mentioned below. Gödel was not consistent about his spelling of the word "Kontinuumhypothese", sometimes abbreviating it, sometimes putting part of it in shorthand and sometimes hyphenating it. We have spelled it uniformly as shown. There were four sheets of equations accompanying this lecture and some gaps in the lecture with indications of what should be inserted from these sheets. We have filled in from the sheets as indicated. Two readings from the shorthand were slightly questionable. One occurs on page 128, line 13, where the words "streng genommen" are the closest we could come, and the other occurs on page 136, line 15, where the word "Raum" was our best guess. We have tried to uniformize his definitional practice by using double quotation marks, but these have been superseded by italics where Gödel had underlined a word or phrase.

Individual changes:

	Original	Replaced by
128, 16	anwendbar ist,	anwendbar,
130, 28	welches des sich	welches es sich
134, 8	implic.	implizit
136, 2	ad. transf.	ad transfinitum
138, 9	$(\exists y)y \in x \ . \ y \in a$	$(\exists y)[y \in x \ . \ y \in a]$
138, 33	$\sum M_\alpha$	$\sum_{\alpha < \beta} M_\alpha$
142, 42	daß das das fundamentale	daß das fundamentale
146, 30	für jedes	jedem
150, 20	1 ist von K	1 von K

*Gödel *193?*

The original manuscript (*Nachlass* item 040411) of *193?* was a handwritten notebook with left-hand pages numbered 1 through 30, and with what we assume were notes for the blackboard on the facing right-hand pages. These have been incorporated into the present text between horizontal brackets within the numbered page that they faced in the original notebook. Most of the German in these notes was transcribed from shorthand, as was the word "verschachtelte" on page 173, line 23 (see below). Gödel also used the right-hand side as extra room for writing insertions and making corrections, which here have been incorporated directly into the text. On page 29 of the manuscript (page 174 of our text, line 21), Gödel had corrected the numerals '0' and '1' to '1' and '2', respectively. We have restored them to the original '0' and '1'. We have also uniformly changed "according as to" to "according to".

Individual changes:

	Original	Replaced by
164, 29	lost and it	lost. It
164, 29–30	case. Since	case, since
165, 35	[illegible letter] problems	a problem
165, 35	A are	A is
166, 7–8	in the order of magnitude	on the order of magnitude
166, 42	1 and 2	0 and 1
167, 1	recursion if	recursion. If
167, 5	&	and
167, 17	a first sight	at first sight
167, 38	equal [both occurrences]	equals
168, 7	n_2 [twice]	n_r
168, 16	mechanic computability	mechanical computability
168, 20	procedure of	procedure for
168, 29	n_k	n_r
168, 43	1 and 2	0 and 1
169, 28	contradiction but	contradiction. But
170, 3–4	The words "can actually be written down" were restored from a passage crossed out on a facing page.	
170, 16	u_n	u_1
170, 19	for any arbitrary	for arbitrary
171, 10	$\varphi(x,y)$ outside	$\varphi(x,y)$. Outside
172, 12	formulas.	formulas).
172, 25	any arbitrary function	an arbitrary function
172, 31	relative-prime	relatively prime
172, 39	$1 + l!x$	$1 + N!x$
	$0 \le x < l$	$0 \le x < N$

The following were translated from German; as mentioned above, slant roman indicates that the original was in shorthand.

| 169, 16 | Konstruktion des unentsch. Satzes | Construction of the undecidable sentence |
| 169, 17–18 | *Angenommen, K ist die Klasse der Nummern n für welche* $\sim(\exists x, z)R(x,z,n,n)$ *beweisbar* | Assume K is the class of numbers n for which $\sim(\exists x, z)R(x,z,n,n)$ is provable. |

169, 19	*und N die Nummer von*	and N is the number of
	B durch	B by
173, 23	*verschachtelte*	nested
173, 29	*non-verschachtelte*	unnested

*Gödel *1940a*

The title was given originally as "Vortrag Widfr. Continuum Harvard and Brown" ("Lecture consistency continuum Harvard and Brown"); however, as far as is known, Gödel never spoke at Harvard. It is possible that the reference to Harvard is to the 1940 International Congress of Mathematicians (see page 163 above) and that Gödel intended to use the present text (and not *193?) for his address to the congress. The original manuscript (*Nachlass* item 040262) was written in a notebook with text on the left-hand pages; the right-hand pages were generally reserved for corrections and insertions, but there were a few places where material was written with no specific insertion indicated. As in some other items in this volume, it seems a reasonable surmise that the material on the right-hand side of the notebook not specifically intended as insertions or corrections may have been written on the board. Since it was relatively obvious in this lecture where that material should be inserted, we have done so. On page 179, lines 22 and 23, the original manuscript reads "I set $S(x_1 x_2 \ldots x_n) \equiv$ a given propositional function φ in $R_1 \ldots R_n$ and S conf. $x_1 \ldots$", with the equivalence

$$S(x_1 \ldots x_n) \equiv \varphi(S \restriction x_1 R_1 \ldots R_n)$$

on the right-hand side. In our text as presented here, the right-hand side of the equivalence was inserted and the formula displayed, to read "I set

$$S(x_1, \ldots, x_n) \equiv \varphi(S \restriction x_1, R_1, \ldots, R_n),$$

φ a given propositional function in $R_1, \ldots, R_n$ and S confined to $x_1 \ldots$".

On page 182, lines 2–8, the original read "... and let this be the corresponding chain of recursive definitions leading up to S. So each $R_i \ldots$", with the list of equivalences on the right as follows:

order μ

$$\sum \begin{cases} (x_i)[R_1(x_1 \ldots x_{r_1}) \equiv \varphi_1(<, R_1 \restriction x_1)] \\ (x_i)[R_2(x_1 \ldots x_{r_2}) \equiv \varphi_2(<, R_1, R_2 \restriction x)] \\ (x_i)[S(x_1 \ldots x_r) \equiv \varphi_{n+1}(<, R_1 \ldots R_n, S \restriction x_1)] \\ \text{constants } \alpha_1 \ldots \alpha_m (< \mu) \end{cases}$$

In the text here, the list of equivalences was inserted and displayed between "leading up to S." and "So each R_i ...".

On page 182, lines 34–37, the original read "... below them. Therefore ...", with the two lines of corresponding objects written on the right. These lines were inserted between the two sentences. Also in the original notebook between Gödel's page 36 and his page 37, on the right-hand side, the following equivalences occurred:

$$n \,\epsilon\, K \equiv (\exists K) \left[\frac{2k+1}{2^n} < K < \frac{2K+2}{2^n} \right]$$
$$P > 0 \equiv (\exists c) c^2 P = 1)$$
$$P \neq 0 \equiv P > 0 \vee -P > 0$$

It was judged that these were not actually part of this lecture and so they were omitted.

Other editorial emendations are as follows:

	Original	Replaced by
175, 23	to give	in giving
176, 25	to well-order	in well-ordering
177, 1	equivalences	equivalents
177, 7	mathematics and	mathematics. And
177, 11	model. And	model; and
177, 31	M_n actually new	M_n new
177, 32	will occur	will actually occur
177, 32	this reason that	the reason that,
178, 7	using	use
178, 16	to use	in using
178, 19	we can prove	we could prove
178, 23	to produce	producing
178, 43	have once decided	have decided
179, 7	relation	relations
179, 13–14	Here are some examples.	⟦Here some examples were apparently given.⟧
179, 17	Examples	⟦Examples were again given.⟧
179, 28	on several places	in several places
179, 35–36	Examples.	⟦An example was presumably again given.⟧
180, 6	3. such that each	3. each
180, 16	arbitrary	arbitrarily
180, 32	logistic	logical
180, 35	k-tupels	k-tuples
180, 42	the I lemma	the first lemma

	Original	Replaced by
181, 4	2. The next lemma No. 2 is	The next lemma is
181, 5–6	.... It says: a recursive	.... It says: 2. A
181, 12	3. Finally the third	Finally, the third
181, 15	1.)	(i)
181, 16	2.)	(ii)
181, 18–19	lemma 3 says: there	lemma 3 says: 3. There
182, 23	the smaller relation	the "smaller [than]" relation
182, 30–31	the smaller relation	the "smaller [than]" relation
183, 18	subsist (under quotation marks)	"subsist"
183, 22	"subsists" under quotation marks	"subsists"
183, 31	requires this that	requires that
183, 36	of	from
183, 36	logistic	logical
183, 38	lemma I	lemma 1
184, 25	this sense	the sense
184, 32; 33	<	less than
185, 4	is this that	is that
185, 15	a_i	n_i
185, 15–16	to prove	in proving
185, 18	this: that either	this: Either
185, 22	is this that	is that
185, 32	unenumerably	non-enumerably

Gödel *1941

The original manuscript (*Nachlass* item 040263) for *1941 was a note-book handwritten on left-hand pages with the right-hand pages reserved for insertions and formulas. We have included some of these formulas where they were needed. Except for a number of abbreviations and insertions, the text was relatively clean. A few formulas have been corrected; these have been noted below. Large parentheses surrounding two separate passages were removed, as we felt that the sole reason for their presence was to indicate a passage that might be left out if time pressed. The first of these two passages begins on page 198, line 5 with "In" and ends on page 198, line 26 with "B". The second begins on page 200, line 13 with "Also" and ends on page 200, line 22 with "found".

	Original	Replaced by
189, 25	The first	First
189, 26	the second	second
190, 8	contradiction which	contradiction, which
190, 13	this equivalence	the equivalence
190, 32	this: that the	this: the
190, 35	P and $\neg P$	P, and $\neg P$
191, 10	requirement	requirements
191, 12	functions, which	functions which
191, 12	introduces, must	introduces must
192, 15	be in	be at
192, 25	assumptions	assumption
192, 40	depend of	depend on
193, 17	satisfied for	satisfied by
193, 22	say that	say,
194, 6	succeeded to prove	succeeded in proving
194, 9	the set-theoretical	set-theoretical
194, 21	So written in one equality that means this. So this	So this
194, 35	First, 1.	1.
194, 37	so on, that this is	so on—is
194, 40	2. and second	2.
195, 1	$T_2(x, g(x))$	$T_2(x, G(x))$
195, 31	$M(f_i, g_i)$	$M(f_i, g_j)$
196, 10	First 1.	1.
196, 11	Second 2. that	2.
196, 29	$(\exists u)(v)N(xy)$	$(\exists u)(v)N(u, v)$
198, 3	$A, A \rightarrow B$ is	$A, A \rightarrow B$ are
198, 22	rule of	rule for
200, 14	is this that	is that
200, 24	transfinite type).	transfinite) type.

In the passages quoted at the end of the introductory note, the following abbreviations were filled in as indicated and the following corrections made:

	Original	Replaced by
188, 37	quantif.	quantifiers
189, 2	log.	logical
189, 3	expression	expressions
189, 3	ordinal nu.	ordinal numbers
189, 4	ord. nu.	ordinal numbers
189, 9	induct	induction
189, 9	ord	ordinals
189, 10	ϵ-nu.	ϵ-number

Gödel *1946/9

Texts for *1946/9

The texts available to the committee for this article were the following:

MS — A completely handwritten manuscript in a spiral-bound notebook with text on left-hand pages and emendations and insertions on the right-hand pages, this text comprises text pages numbered 1–71 and footnote pages I-XLI. The footnotes are numbered in the order in which he added them to the manuscript and thus are not in numerical order in the paper. *MS* was found in three separate packages in the *Nachlass*, the first of which is now numbered 040416 and contains pages 1–51 along with some crossed-out pages. The second in number is 040417, which contains pages XIII-XLI (i.e., the last half of the footnotes), while the third is 040418, which contains a large number of deleted pages followed by the second half of the text and the first half of the footnotes. There was a note on the first page of *MS* saying "4 Carbons— Omit all footnote numbers—leaving space". A second note gave a symbol for leaving extra space for subtitles, which apparently never got added to any of the versions, unless all he intended were the roman numerals that appear in the margins of *B2*.

A — Labelled by Gödel himself "Man A", this typescript is unmarked except that page 19, though found between pages 18 and 20, has been renumbered as page 25. But neither *A* nor either of the *B* versions has page numbers above 23. Moreover, this renumbered page contains two footnote numbers (his 55 and 60) that are higher than any that occur in versions *B1* and *B2*. (Version *C1* *does* have even higher page numbers, and, in fact, the underlying typescript of page 25 in *C1* is the same as that of this page.) The text of *A* is a verbatim transcript of *MS* with the exception that there are no footnote citations in the paper and the footnotes are renumbered in a list at the end. Comparison with the text of the footnotes in *MS* and their position there shows that in the list at the end of *A* they have been reordered to agree with the order of their occurrence in *MS*. However, the list of footnotes in *A* does not include all that were in pp. I-XLI of *MS*. It is likely that *A* should have had at least one more page of footnotes, for the citations in *MS* include two last footnotes that do not appear in the list at the end of *A*. These last two do appear in the lists of footnotes for the two *B* versions, which lists are identical with the list for *A* so far as that for *A* goes.

The "B" versions: The underlying typescripts of the two versions labelled
"Man B" by Gödel are the same, consisting partially of pages from the same
typing that produced *A* and partially of pages typed anew (in carbon copy).
The only differences between the two versions lie in the corrections thereto.
In particular, except for the end of his footnote 32 in *B1*, in every place
where there is a correction in *B1*, there is also a correction—not necessarily
the same correction—in *B2*, but not conversely. Thus we presume that *B2*
is the latest of the three versions discussed so far. Some details of each
version are worth noting.

B1 — Labelled by Gödel "Man B (= Man A korrigiert)" [Man A cor-
rected], the text of this version is preceded by several pages with
annotations, the first of which has a note (partially in shorthand)
that reads "Kant u. Rel. Th. (*längere Form*) *ältere Fassung*[[en]]
A,B" ["Kant a[[nd]] Rel. Th. (longer form) older version[[s]] A,B"].
That is followed by a page which says (in shorthand) "*Korrek-
tur: Der Raum existiert nicht neben und unabhängig von der
Materie in* Newt. *und* Einst." ["Correction: Space does not exist
apart from and independent of matter in Newt. and Einst."]. A
third page contains a few equations, after which the text finally
begins. At the top of the first page of text, in addition to the
above-mentioned label, Gödel has written "*genau in der Form
1949 (nichts später geändert)*" ["exactly in the form 1949 (nothing changed later)"]. The typescript of footnotes 1–43 agrees
with that of A, with some handwritten additions and deletions;
the text of footnotes 44 and 45 agrees with that of the corresponding footnotes in *MS*; the text for 46–51 is new. There is also a
loose sheet with the same *Nachlass* number that says "Kant u.
Einstein (*längere Fassungen:* Man A,B *mehrere Kopien*) Man B
scheint die beste Fassung zu sein" ["Kant a[[nd]] Einstein (longer
versions: Man A,B several copies) Man B appears to be the best
version"].

B2 — Labelled by Gödel "Man B (= *Korr. von A*)" [(= Corr. of A)],
this appears to be the original typescript of the pages used for
B1. *B2* has roman numerals pencilled in the margins, which we
have set as section numbers in the version presented here.

The "C" versions: By this time, Gödel himself had become confused
enough by his various versions to make up concordances among them. One
of those apparently describes the structure of *C2*, using roman numerals
that appear in the margins thereof (corresponding to those in *B2*). An
annotation at the side, partially in shorthand, says "*Sonst nichts* Man.
C (*gedruckte* version)" ["Otherwise nothing Man. C (typed version)"]; another at the bottom, completely in shorthand, says "[*Diese*[[s]] *Kapitel nicht*

folgend]". The latter can be translated either as "These chapters not following" or "Not following these chapters". The context is not quite sufficient to determine the ending on "Diese", so that it could be the singular "Dieses" (see "Gödel's Gabelsberger shorthand", this volume, page 7), although the singular seems less likely. There is also a possibility that Gödel wrote "nicht" where he intended "nichts", as the difference would be only a small loop at the bottom of the shorthand symbol. If so, the intended meaning would have been "Diese Kapitel nichts folgend" ["Nothing following these chapters"].

C1 — Gödel had written at the top left of page 1, in shorthand, "*letzte längere Formulierung (der Anfang rein geschrieben)*" ["last longer formulation (the beginning cleanly written)"] and at the top right, partially also in shorthand, "Man C *handgeschriebene* [handwritten] version" "p. 1–30, I–XIV". But, in fact, the first page is typewritten. The manuscript as a whole consists of a mélange of handwritten pages, pages taken from the *B* typescripts—some of which were the underlying typescript pages from *A*—and two other typewritten pages that were typed on the same typewriter that produced *C2*. All three kinds of pages have emendations or insertions and there are new footnotes for numbers 51 through 68.

C2 — Gödel wrote on this one "Man C (*später als* B) *gedruckte* version" ["Man C (later than B) typed version"]. The typescript as found contained only the first fourteen pages and appears to be part of a completely new typescript done on a pica typewriter. (The earlier versions were in elite type except for the two pages mentioned above under *C1*.) There are not only handwritten emendations and deletions but also some typewritten ones. Where the text of *C2* overlaps with that of *C1*, it agrees with the content of *C1* for the most part, but the order in which various parts occur differs greatly from that in *C1*. A note on the right-hand side of page 1 says "p. 1-14, *keine Fußnoten*" ("p. 1–14, no footnotes)") and below that "(~p. 22.1 *der handgeschriebenen* version)" ["(~p. 22.1 of the handwritten version)"]. Roman numerals in the margins appear intended to indicate correspondences with the *B* versions.

The texts as presented in this volume

Citations were uniformized with standard practice in these volumes, except in the case of Kant's *Critique of pure reason* and *Prolegomena*, where they were altered to agree with standard practice of Kant scholarship;

see the editors' note following the introductory note as well as the preface
to the volume as a whole.

The footnotes were renumbered to put them in numerical order, as
Gödel had added and deleted some numbers in the course of revision; some
changes in wording of the text as well as the footnotes were made to reflect
our reordering or citation practice. In addition, some footnotes in version
B2 were simply eliminated by the decision to use the standard method of
citing *The Critique of Pure Reason* in text. Because the footnotes were numerous and many were quite long, we give below concordances of Gödel's
original footnote numbers with those used here, in order to aid the reader
in comparing the text as given here with the manuscripts.

It should also be noted that Gödel himself deleted pages 18 and 19 of
version *C1*. Thus the numbers of the manuscript pages as shown here in
the margins skip from 17 to 20.

Specific changes to version *B2*: (See also the explanations following this
list.)

	Original	Replaced by
230, 7	neither is	is neither
230, fn 1, 6	arising by	arising in
230, fn 4, 1	it is	is it
231, 11	means by it that	means that
231, fn 5, 2	entity the ... is used	entity is the ... used
232, 14	but that on	but on
233, 6	*itself*	*themselves*
233, 20	itself	themselves
233, 24	such an adequacy could	could such an adequacy
233, 31	is insofar not	is not
233, 31	latter as,	latter insofar as,
234, 9	By relativity	In relativity
234, 10	what of time remains	what remains of time
234, 14	part of the pairs	some of the pairs
234, fn 13, 3–4	all that in relativity theory remains	all that remains in relativity theory
234, fn 13, 5	exact, that every	exact, every
235, 1	absolute before	absolute "before"
235, 6	part of the pairs	some of the pairs
235, fn 14, 1	independent from	independently of
236, 1–2	what of time remains	what remains of time
236, 7–8	outspoken	[pronounced]
236, 14;19	by Kant	for Kant
236, 19	to a large extent is	is to a large extent
238, 25	for part of	for some of

	Original	Replaced by
238, 27	remaining part	remaining ones
239, fn 22, 3	and that in general	and in general
240, fn 24, 8	, (see below p. 20	; see below p. 20)
240, fn 24, 12	present)	present,
241, 13	theory subsists only	theory only
241, 34	in this point	at this point
241, fn 27, 1	Riemannian world)	Riemannian) world
242, fn 27, 2–3	4) consequences	four) would consequences
242, fn 27, 3	kind would become	kind become
243, 1	space, in that since	space, since
243, 2	not assert sense that	not assert that
243, 3	a priori can only be known by experience, provided a reasonable meaning is given to this phrase.	a priori.
243, 9	in that sense that	in this sense, that
243, fn 33, 3	Labatchevskyan	Lobachevskian
244, 26	Democrit	Democritus
245, 17	One may confer	One may compare
245, 21	quoted in footnote 44	cited above

The change on page 236, lines 7–8 was made on the assumption that Gödel had in mind the German "ausgesprochen", which may be translated either as "outspoken" or "pronounced". We have chosen the latter over the former as fitting the context here.

The deletion noted above on page 241, line 13 was necessitated by a revision in which we assume that Gödel simply forgot to delete the word "subsists". The original read "... that this incompatibility of the Kantian a priori with relativity theory subsists only ..." and Gödel had struck out the words "this incompatibility of" in order to insert "is incompatible" after "a priori", thus leaving no function for "subsists" in the sentence.

The changes on page 243, lines 1–3 listed above deserve some explanation, lest they look like a complete editorial revision. For such an explanation, see the introductory note to this item, in particular, pages 224–226.

Specific changes to version *C1*:

	Original	Replaced by
247, 5	neither is	is neither
247, fn 4	by relativity	in relativity
248, 17/249, 1	by Kant	for Kant
249, 15	subject and etc. ...	subject ...

249, fn 10, 1	near the end.	[p. 286].
249, fn 10, 1–2	referred to in [47]	from B 69n. referred to above
250, 27	by relativity	in relativity
250, 30	nor allows time	nor [does it] allow time
250, 32	does not exclude yet	does not yet exclude
251, 15	as it is	as is
251, 26	of time remains	remains of time
252, 1	doubtlessly	doubtless
252, 3	outspoken	[pronounced]

(The change above was made for the same reason as that in *B2*, 236, 7–8.)

252, 4	does (*in the R-worlds*)	does
252, 13	by Kant	for Kant
252, 20	it is by	it is in
252, 30	by Kant	for Kant
253, 25	it was spoken of time only	only time was spoken of
253, 37	orthogonal on	orthogonal to
253, 40	as this is	as is
255, 19	in this point	at this point
255, fn 19, 1	Riemannian world)	Riemannian) world
256, fn 24, 2–3	space. For	space, for
258, 25	confer	compare
258, 28	quoted in	cited above
258, fn 28, 8	by relativity	in relativity

In addition, a parenthetical remark at the end of footnote 24 on page 256 was omitted. The remark said "(vgl. [cf.] p. 25a)". It turns out that page 25a was nothing more than the short insertion to page 25 in which this very footnote occurred!

Footnote concordances

For each of the versions published here, we first give a list to enable the reader to recover the number of a footnote in Gödel's original from the number in this volume. It is followed by a table in which Gödel's footnote numbers are correlated with those used here or their disposition explained.

Version *B2*:

In this list, numbers are given in pairs, with the first being the number in this volume, the second the number in Gödel's original typescript, with remarks following.

(1:1); (2:47); (3:2); (4:46); (5:3); (6:5); (7:7); (8:8); (9:10); (10:49); (11:48); (12:15); (13:13); (14:14); (15:18); (16:19); (17:20); (18:22); (19:51); (20:50); (21:25); (22:26); (23:28); (24:42); (25:29); (26:30); (27:31); (28:32); (29:33); (30:34); (31:37); (32:38); (33:39); (34:40); (35:41); (36: reference to 42); (37:45)

Originally, footnote 11 (Gödel's 48) followed a phrase that was crossed out in version *B2*; here it was placed after "upon us," since the footnote itself was not crossed out.

Footnote 12 (Gödel's 15) originally referred to a previous footnote which was here obviated by the system of reference to the *Kritik der reinen Vernunft*. We have reworded footnote 12 accordingly. Footnote 24 was originally an insertion that simply said "cnf. ^{42}". Since this preceded the actual occurrence of footnote 42, we used that text for footnote 24. The flag to Gödel's footnote 42 was renumbered as 36 and we simply cited footnote 24 at that point.

Correlation of Gödel's numbers with the disposition of footnotes in the text as given here:

Gödel's footnote number	Our footnote number or location of deleted flag	Replaced by
1	1	
2	3	
3	5	
4	231, 17, words	231, 18 and 19, (B 69 n.)
4a	231, 21, says	231, 23, (B 69 [[emphasis Gödel's]])
5	6	
6	231, 24, that	231, 26, (B 67)
7	7	
8	8	
9	Deleted by Gödel	
10	9	
11	233, 12, says	233, 16, (B 53)
12	Deleted by Gödel	
13	13	
14	14	
15	12	
16	Deleted by Gödel	
17	Deleted by Gödel	
18	15	
19	16	
20	17	

21	Deleted by Gödel	
22	18	
23	238, 21, relations	238, 21, (B 42 a)
24	Deleted by Gödel	
25	21	
26	22	
27	Deleted by Gödel	
28	23	
29	25	
30	26	
31	27	
32	28	
33	29	
34	30	
35	Deleted by Gödel	
36	Deleted by Gödel	
37	31	
38	32	
39	33	
40	34	
41	35	
42	24	
43	245, 19, *reason*	245, 17, B xxi, Bxxii n. and Bxxvi n.
44	245, 21, knowledge	245, 19, Bxxvi n.
45	37	
46	4	
47	2	
48	11	
49	10	
50	20	
51	19	
52	Deleted by Gödel	

Version *C1*:

As for version *B2*, numbers are given in pairs, with the first being the number in this volume, the second the number in Gödel's original.

(1:1); (2:4); (3:66); (4:56); (5:65); (6:64); (7:51); (8:2); (9:7); (10:8); (11:10); (12:61); (13:19); (14:59); (15:53); (16:57); ([[17]]:63); (18:58); (19:31); (20:68); (21:33); (22:55); (23:62); (24:60); (25:39); (26:40); (27:42); (28:67); (29:45)

Correlation of Gödel's numbers with the disposition of footnotes in the text as given here:

Gödel's footnote number	Our footnote number or location of deleted flag	Replaced by
1	1	
2	8	
3	Deleted by Gödel	
4	2	
5	Number crossed out by Gödel	
6	249, 16, says	249, 19, (B 69 ⟦emphasis Gödel's⟧)
7	9	
8	10	
9	Number crossed out by Gödel	
10	11	
11	Number crossed out by Gödel	
12	Number crossed out by Gödel	
13	Number crossed out by Gödel	
14	Number crossed out by Gödel	
15	250, 11, passage referred to in	250, 11, passage ⟦from B54⟧ referred to on page 2′ above
16	Number crossed out by Gödel	
17	Number crossed out by Gödel	
18	Number crossed out by Gödel	
19	13	
20	Number crossed out by Gödel	
21	Number crossed out by Gödel	
22	Number crossed out by Gödel	
23	Number crossed out by Gödel	
24	Number crossed out by Gödel	
25	Number crossed out by Gödel	
26	Number crossed out by Gödel	
27	Number crossed out by Gödel	
28	Number crossed out by Gödel	
29	Number crossed out by Gödel	
30	Number crossed out by Gödel	
31	19	
32	Number crossed out by Gödel	
33	21	
34	Number crossed out by Gödel	
35	Number crossed out by Gödel	
36	Number crossed out by Gödel	

37	Number crossed out by Gödel	
38	Number crossed out by Gödel	
39	25	
40	26	
41	Number crossed out by Gödel	
42	27	
43	258, 26, in the preface to the the 2nd ed. of the Critique of pure reason	258, 26, at B xxi, B xxi n. and B xxvi n.
44	258, 27, knowledge	258, 28, (B xxvi n.)
45	29	
46	Number crossed out by Gödel	
47	249, 13, words	249, 16, (B 69 n.)
48	Number crossed out by Gödel	
49	Number crossed out by Gödel	
50	Missing from manuscript	
51	7	
52	Missing from manuscript	
53	15	
54	Missing from manuscript	
55	22	
56	4	
57	16	
58	18	
59	14	
60	24	
61	12	
62	23	
63	[[17]]	
64	6	
65	5	
66	3	
67	28	
68	20	

*Gödel *1949b*

Gödel's manuscript (*Nachlass* item 040279) for this lecture had many abbreviations, and he had written the equations on a separate set of sheets, apparently for the purpose of writing them on the blackboard. In many passages, these equations were referred to in a manner which would be meaningless in a written transcript. Therefore, in the text presented here, the abbreviations have been expanded, and the equations have been taken from the accompanying sheets and placed in their proper place in the text.

More substantive, but necessary, changes include correction of a small sign mistake in one of Gödel's formulae (see note for page 272, line 28) and an inconsistency in one of his notation conventions. Because of that inconsistency, all equations have been recast in terms of coordinates x_0 to x_3. Gödel sometimes inserted a comma in writing some of the Christoffel symbols. These commas have been omitted here for consistency with subsequent usage, for example, in his expression for R_{ijkl}. On page 276, line 3, the phrase "(If you wish, I can carry out the computation in detail later on.)" has been replaced by some calculations from the sheets. On page 277, line 6 a clause crossed out by Gödel has been restored from the original manuscript in order to make sense of the passage following. In particular, in the original manuscript, "Before going on with the discussion of the properties of this" had been crossed out, with "entering in" written above "going on with". We have restored this clause with "entering into a" replacing "going on with the".

The following list is a compilation of all other changes to the manuscript.

Uniform changes

Original	Replaced by
in the point	at the point
in the (or whose) origin	at the (or whose) origin
orthogonal on	orthogonal to
i.e. to say	that is to say

Individual changes

	Original	Replaced by
269, 22	like the square	as the square
270, 7	places of space	places in space
270, 16	principle like this	principle [followed by displayed equation]
270, 23–4	assumption V= [blank space]	assumption [followed by displayed equation]
270, 34	place of the world	place in the world
271, 27	have these values	have values [followed by displayed equation]
272, 9–12	velocity vector has these values here	velocity vector [followed by displayed equation] has values [followed by displayed equation]

272, 16;27	these formulae here	the formulae [followed by displayed equation]
272, 28	$\omega_1 = (1/2)(-\Gamma^2_{30} - \Gamma^3_{20})$	$\omega_1 = (1/2)(-\Gamma^2_{30} + \Gamma^3_{20})$
273, 9	these formulae	the formulae displayed
273, 13	expression is written up here	expression is [followed by displayed equation]
273, 16	a_{ikl} as you can verify immediately	As you can verify immediately, a_{ikl}
274, 14	simultaneousness	simultaneity
274, 19–20	working about ... and, ..., about	working on ... and, ..., on
274, 35	this expression here	the expression [followed by displayed equation]
274, 37	reads like this.	reads like this: [followed by displayed equation]
274, 40/275, 1	the 3-dim. linear element which I have underlined in red,	the underlined three-dimensional linear element
275, 3	the 3-space underlined in red	the three-space with the underlined metric
275, 6	this expression	the expression [followed by displayed equation]
275, 10	this term $(g_{t\varphi})$	the term $g_{t\varphi}$
275, 15	a kind of helices, as it	helices of a sort, as they
275, 20	the four transformations written up here	the four transformations [followed by displayed equations]
275, 36	the linear tr. written here	the linear transformation [followed by displayed equation]
276, 2	this formula	the formula [followed by displayed equation]
276, 7	(i.e., these three here)	[other than 3]
276, 12	R_{00} is different $\neq 0$ namely $=1$	R_{00} is not 0, and it is equal to 1
276, 13;16	this equality	the equality [followed by displayed equation]
276, 22	equality with	equality by
276, 22	subtract it from the first one.	subtract the result from the equality above:
277, 4	one page in print	one printed page
277, 7–8	by which this is easiest arrived at	by which this [solution] is most easily obtained

278, 2	accomplish that ... theory should	ensure that ... theory will
278, 18	this formula	the formula
278, 26	these values	the values
278, 34	this form here	the form
278, 41	three last ones	last three
279, 5	will remain	remained
279, 6	are as good	would be as good
279, 6	we may	we might
279, 16;26	this expression here	the expression [followed by displayed equation]
279, 20	this equality here	the equalities here
279, 30	multiplied with	multiplied by
280, 4	symmetry this is only possible in that way that it is	symmetry, it must be
280, 9	digit of the diagonal	entries on the diagonal
280, 24	these two forms	the two forms
280, 33	this equation here	the equation [followed by displayed equation]
280, 38	owing to this equation	owing to the preceding equation
280, 39	conjugate complex	complex conjugate
280, 40	this sum	the sum
281, 3	itself and this	itself. This
281, 5	this equation	the equation
281, 11	to prove it	to prove this
281, 13–15	$p^{-1} = \bar{p}$ & between $p, \bar{p}, 1$ subsists a linear relation, ..., where α is a real number. But the geodesics	(a) $p^{-1} = p$, (b) there subsists a linear relation among $p, \bar{p}$ and $1, \ldots,$ (here α is a real number), and (c) the geodesics
281, 22	this transformation here	the transformation [followed by displayed equation]
281, 29	distance of two points	distance between two points
281, 32–34	defined by this expression here. Only	defined by [followed by displayed equation]; only
282, 1	given by this expression here.	given by [followed by displayed equation]
282, 8	algebra are these four here	algebra are [followed by displayed equation]

282, 10–11	denote them by ... these rules of multiplication	denote ⟦the latter three⟧ by ... the rules of multiplication ⟦followed by displayed equations⟧
282, 25	for quat $\neq 0$	for non-zero quaternions
282, 30	these three quaternions	the three quaternions ⟦followed by displayed expressions⟧
282, 32	rules written up here	rules ⟦followed by displayed equation⟧
283, 1	element which is written up here	element ⟦followed by displayed equation⟧
283, 5	three space underlined in red	underlined three space
283, 10	this form here	the form ⟦followed by displayed equation⟧
283, 26	line	lines
283, 26	expressible in this way	expressible as ⟦followed by displayed equation⟧
283, 34–35	to another one	to another
284, 1	that not only	not only that
284, 29	element here	element, ⟦followed by displayed equation⟧
284, 33	inequality here	inequality, and
286, 1–3	line it will always look this and never like this	line will always look like that in figure 1, and never like that in figure 2,
287, 6	e.g.	for example

Gödel *1951

Gödel's handwritten text for this lecture (*Nachlass* item 040293–5) consists of pages numbered 0–39, plus additional pages 2a, ad 19, and 29′ (the latter containing insertions to the text). In addition, there are 27 pages of footnotes (numbered 1–59′) and 18 pages (numbered I–XVIII) of numbered interpolations to the text. Original page numbers given here refer only to the basic 39 pages of text.

Gödel's footnote numbering is chaotic; he apparently both added and deleted footnotes in the course of later revisions to the manuscript. Moreover, two of his footnotes later were converted to insertions and one insertion was marked to be used as a footnote. Consequently, we have consecutively renumbered the footnotes relevant to the text as presented here. At the end of these notes we have provided a concordance with Gödel's original number, for the convenience of those dedicated scholars determined to check the original texts. (In a few cases, Gödel indicated a footnote without supplying a number.)

Gödel uses the abbreviation "prop." both for "proposition" and (infrequently) for "property". We believe the contexts make clear which he intended, but a few cases may perhaps be debatable. The reader should be alert for these words. Because Gödel's usage of some Latin abbreviations was often not truly acceptable English style, we have here replaced "i.e." by "that is", "e.g." by "for example", and "cnf." by "see". Because of the elaborate revisions undergone by this text, nesting of parenthetical expressions became almost inevitable. We have here observed the usual practice of using square brackets inside parentheses.

	Original	Replaced by
305, 37	satisfies	satisfy
306, fn 6, 3	always the set	the set
306, fn 6, 3	is to be	is always to be
307, fn 7, 4-5	theorem, without ... consistency holds;	theorem holds, without ... consistency;
313, 16–17	by no means is	is by no means
313, 32	empiristic	empiricist
315, fn 23, 19	view considering	view that considers
316, 28	(and that the	and the
316, 29	so derived)	so derived
317, fn 25, 2	$\neq$	$=$
318, fn 27, 1	Examples for	Examples of
319, 25	by which we believe to express	which we believe express
319, 32	The trouble only is	The only trouble is
320, 6–7	arbitrary	arbitrarily

One insertion, numbered insertion 50, was flagged to be inserted in two different places, one on Gödel's page 18 and the other on his page 23. We have chosen the location on page 18 as a more fitting context. The passage in question begins on our page 312, line 16 with "The same would be true ..." and runs to the end of that paragraph. The other place marked by Gödel as a possible location for this passage is on our page 316, line 3 just after "'All stallions are horses'".

On page 315, line 7, just before the material noted (||) as to be omitted, the words "asserts mathematical propositions to be true solely due to certain arbitrary rules about the use of symbols" were inserted, presumably as a bridge over that omitted text. That has been deleted here.

On page 320, line 5, the word "constructed" is written above "built up", but the latter has not been crossed out. On the assumption that Gödel simply forgot to complete the deletion, "built up" has been replaced here by "constructed".

In footnotes 15 and 22, Gödel's intended cross-reference is unclear. Rather than impose a reference that may not have been what he wanted, we have left these cross-references uncompleted. (In fact, a shorthand annotation at the top of the first page of footnotes says, "*Welche Fußnote in 18 zitieren?*" ["Which footnote in 18 [our 22] to cite?"].)

In footnote 23, line 20, the "tionship" in "relationship" has been crossed out and something illegible has been written above it as a substitute. We have restored the "relationship".

On page 316, in line 29, just after "so derived", Gödel wrote (at the end of an arrow) "Cnf. footnote 56". But footnote 56 had also been marked to be incorporated into his footnote 39 (our 23), along with footnote 59. We surmise that Gödel intended to replace that reference with footnote 57, which became our footnote 24. (See the footnote concordance below.)

Transcription [translation] of shorthand annotations
referred to in editorial footnote g

"*Ist das nicht eine* red. ad abs."? ["Isn't that a *reductio ad absurdum?*"]

"*Außerdem: was aus Math folgt ist wahr (gleichgültig was Math ist)*". ["Moreover: What follows from mathematics is true (it doesn't matter what mathematics is)."]

"*Ferner*: contin. creation" ["Furthermore: contin[uous] creation"]

"*Ferner: math Faktoren* intuit. nicht *überflußig (weil nötig um das zu beweisen) Kalmár argument*" ["Furthermore: mathematical factors [are] intuit[ively] *not superfluous* (because [they're] necessary in order to prove that)—Kalmár's argument"]

"*Ferner: unzuverläßig, weil auf int. Faktoren basiert. Wenn man jeden Begriff durch Aufzählung definiert, wird alles analyt. (wahr auf Grund von Regeln über des Gebrauch von Symbolen)*". ["Furthermore: unreliable, because based on intuitive factors. If one defines every notion by means of enumeration, everything becomes analytic (true on the basis of the rules about the use of symbols)".]

Between Gödel's page 29 and his page 30, a trail of revisions caused him to bypass the following passage:

> This view can be refuted by the demonstrable fact that any alleged proof of it (not by any chance for all languages which of course is imp[ossible] but for one specially contr[ucted] lan-
> 30 guage) must presuppose & | use the very same math[ematical] facts, whose non-existence it is supposed[58] to show. Moreover it is easily seen that for any division of the empirical facts in two classes A, B such that the facts of B imply nothing about those of [A] then, using the facts of B one could constr[uct] a language in which the propositions expressing the facts of B would be "void of content" and true solely due to semantical rules.

Footnote 58 reads as follows:

> One might ask: isn't it sufficient, at least a refut[ation] of realism, that the tautological character of math[ematics] can be concluded from math[ematics] itself? For this inference although not binding for nominalists (who have to leave the validity of math[ematics] in abeyance until they succeed to derive it on the basis of their phil[osophical] presup[positions]) will have to be acknowledged at least by Realists & hence will implicate them into self-contradiction. This conclusion would be correct if "tautolog[ous]" in this connection really meant "void of content". However what it means (by def[inition]) is only that *there exists a language* in which math[ematics] is void of cont[ent] *insofar* as [it] follows from the rules of syntax.

A heavy line drawn across the page at this point seems to indicate that Gödel intended footnote 58 to end here, although more text follows below the line.

Footnote concordance

Numbers are given in pairs, with the first being the number in this volume, the second the number in Gödel's original manuscript. The second position is blank if the note had no number and contains multiple numbers if several footnotes were incorporated into one by his revision.

(1:1); (2:36); (3:2); (4:54); (5:5); (6:4); (7:6); (8:52); (9:51); (10:15); (11:16); (12:7); (13:17); (14:28); (15:43); (16:29); (17:25); (18:8,40); (19:55); (20:59′); (21:38); (22:18); (23:39,56,59,20); (24:57); (25:31); (26:); (27: insertion 8); (28:57′); (29:49); (30:34)

The text for footnotes 26 and 27 came from following the trail of insertions that began at insertion 13. The material for footnote 26 was marked to become footnotes, but had not been numbered as such. That for footnote 27 was marked "insertion 8, footnote".

As for the rest of Gödel's numbered footnotes, 13 and 32 were marked to be converted to insertions; 12 and 27 were actually crossed out by Gödel, while 3, 14 and 27 were on a list he made of footnotes to be deleted. As mentioned above, 58 was flagged in bypassed material, while 10, 19, 33, 42, 45, 50 and 53 were flagged in deleted material. The remainder were apparently never flagged in the text that we have.

Gödel *1953/9-III

We have not renumbered the sections of this paper (*Nachlass* item 040436), in spite of the fact that some section numbers were missing; in fact, in a set of notes on version III also found in the *Nachlass*, there was an instruction "p. 7 §17 weglassen" ("omit p. 7 §17"). Other sections were crossed out on the typescript or missing altogether.

Because Gödel had done many revisions of this paper, some of the pages of footnotes had been cannibalized for other versions. We have restored footnote 9 by taking the corresponding footnote from version II, namely, footnote 8, which occurs after the same text in that version as the text containing the flag for footnote 9 in version III. We restored the flag for footnote 18 (which became our footnote 19) as well as the section number for section 18 by comparison with a passage that had originally been numbered §18 and then crossed out. Since the footnote had not been similarly deleted and was cross-referenced in several places, we felt it should be restored. There were two flags for his footnote 19 (which became our footnote 20); we have replaced the second one by a footnote 23 referring the reader to footnote 20.

On the front of his typescript, Gödel had written "p. 24 p. 37 und p. 40 sind bei der 4. Fassung" ("p. 24 p. 37 and p. 40 are with the 4th version"). Page 24 would have contained footnote 10 and part of 11. No page 24 or 36 was found with version IV, but a typed page containing the text of some footnotes numbered 9, 10, 11, 11a and 12 was found with version III. Basing our action upon content, we have used footnote 9 on that page to provide our footnote 10 and the first line of footnote 10 of that page to complete footnote 11, since the remaining text was identical to the second part of footnote 11 as found in version III. Page 37 was found at the end of version IV and provided us with the missing footnote 37 (here renumbered as footnote 41) of version III. Page 40 was also found with version IV and provided us with the end of section 43 and the entire section 44 as well as footnote 41, which has been renumbered here as footnote 44. For a concordance of footnotes, see the end of this note.

On page 348, footnote 36, line 10, a reference to §17 has been changed to §19, as we have no §17 and §19 appears to contain the relevant material. Also on page 348, line 6, a reference to §46 has been altered to §45 on the basis of comparison of text with other versions. For the same reason, on page 351, line 1, since we were missing §40, we have altered "§40–41" to read "§39, 41", and in line 15, a missing §20 has caused us to revise "§20" to "§19", while in line 31, "§17" has been changed to "§16", again on the basis of content. On page 340, footnote 17, line 2, as well as page 353, footnote 43, line 4, Gödel refers to "the 3rd paragraph of footn. 34" (our 36). As that footnote had no third paragraph and the second paragraph seems appropriate, we have altered this to read "the second paragraph of footnote 36". On page 354, footnote 45, line 9, a reference to the missing §§46–50 has been altered to refer to §5, once more on the basis of content.

The following corrections have also been made, either as a correction of fact, as a correction for English grammar or for editorial consistency:

	Original	Replaced by
335, fn 5, 1	evidently is this that,	evidently is that,
335, fn 6, 1	papers quoted	papers cited
336, 24	(p. 101–129)	([[*1937*,]] pages 102–129)
337, 6	"content" from the beginning is taken	"content" is taken from the beginning
337, 13	assertion No. 1	assertion 1
337, 14	assertion No. 2	assertion 2
337, 25	assertion No. 1	assertion 1
337, 35	Nor can can certain	Nor can certain
337, 41–42	assertion No. 1	assertion 1
337, 43	6) exactly is whether	6) is exactly whether
338, fn 14, 13	this exactly is	this is exactly
339, 25	on the ground	on the grounds
340, 13	this exactly is	this is exactly
341, 5	required: 1. that	required that: (1)
341, 10	who at all knows ... these concepts.	who knows ... these concepts at all.
341, fn 20, 9	according as to	according to
342, 12	to refer to to	to refer to
342, 18	on the ground	on the grounds
342, 26	consistency of C'	consistency C'
343, 2	this exactly is	this is exactly
343, 9	case a	case could a
343, 10	mathematics could be	mathematics be

343, 20	38) there is some	42) is there some
344, 5–6	requirements 1-3, 5 , 6	requirements 1-3, 5 (part 2), 6
344, 18	the syntactical program can	can the syntactical program
346, 4	belongs into this	belongs to this
346, 5	On the ground	On the grounds
347, 3	*on the ground*	*on the grounds*
347, fn 33, 5	of the Goldbach	of Goldbach
348, fn 36, 10	§17 one must	§19, one must
352, fn 42, 19	thinking) exactly is	thinking) is exactly
353, fn 43, 10	in this sense that that they	in the sense that they
354, 21	for this reason	for the reason

On page 346, line 8, the word "vital" was written above the word "fundamental" without deletion of the latter. We have chosen to go with the replacement word on the theory that he simply neglected to cross out the original.

For lack of reference, some phrases were struck from the text by the editors, as we could not fill the omissions they contained. These were located as follows:

	Following	Omitted phrase
338, fn 14, 13	entities	(cnf. §)
356, 8	considered	(cnf.)

Footnote concordance

Numbers are given in pairs, with the first being the number in this volume, the second the number in Gödel's original manuscript. The second position contains multiple numbers if several footnotes were incorporated into one by his revision.

(1:1), (2:2), (3:3), (4:4), (5:5), (6:6), (7:7), (8:8), (9:Version II-8), (10:9), (11: 10, 11), (12:12), (13:13), (14:14), (15:15), (16:16), (17:17), (18:16a), (19:18), (20:19), (21:20), (22:21), (23:19), (24:22), (25:23), (26:24), (27:25), (28:26), (29:27), (30:28), (31:29), (32:30), (33:31,43), (34:32), (35:33), (36:34), (37:44), (38:45), (39:35), (40:36), (41:Version IV-37), (42:38), (43:39), (44:41), (45:42)

Gödel *1953/9-V*

The typescript of version V (*Nachlass* item 040446) is relatively straight-forward, with no footnotes, and required only the following in addition to routine copy-editing.

For reasons of English grammar or content, the following changes were made:

	Original	Replaced by
359, 11	below	above
359, 33	not an object	not objects
360, 6	the ... assumption is	is the ... assumption
360, 31	theories	theory
360, 38–39	concept	concepts

At page 359, line 19, after "constituents", an instruction to "insert page 30" was disregarded, as we could find no page 30, either in version V or in version IV. In addition, for lack of information, several phrases were struck from the text by the editors, as we could not fill the omissions they contained. These were located as follows:

	Following	Omitted phrase	Reason for omission
359, 28	sufficient	(see footn.)	No footnote was found.
361, 8	with	(see)	Lack of reference
361, 9	propositions	(see)	Lack of reference
361, 20	sciences	", in contrast to the assertions quoted in footn. "	No footnote was found.
361, 36	rules	(see)	Lack of reference
361, 39	intuition	(see)	Lack of reference

Gödel *1961/?*

The original manuscript of *1961/?* is a shorthand draft (*Nachlass* item 040411.5) and thus had run-on paragraphs, many abbreviations in long-hand and one occurrence of the insertion symbol most frequently used by Gödel. While not marked with the same symbol, the sentence beginning on page 380, line 1 was found on a separate sheet labelled "Einschaltungen" ("insertions"), and we have inserted it in the text where the symbol occurred. For better readability, we have introduced paragraph breaks. We have filled in all abbreviations; where these were absolutely clear as to their meaning, we have left the completion unmentioned. Where there was any doubt, we have noted it below, except for the many cases of "Ph.". Depending on the context, this abbreviation could stand either for "Philosophie" or for "Phänomenologie", and we have filled in whichever seems to fit. (In general, Gödel tended to use "Phä" or "Phän." when he intended "Phänomenologie", but the *Nachlass* does contain a letter to Bernays in which the actual letter says "Phänomenologie" and the shorthand draft has, in longhand, "Ph.")

Since this was transcribed from Gödel's shorthand, it is subject to the usual possible errors both in writing and in transcription. We have made several kinds of corrections. Not listed below are necessary changes of case of individual longhand letters (from lower-case to upper and vice-versa), and pluralization when needed; as the shorthand need not indicate pluralization unless needed for clarity, we have pluralized without mention where we deemed it necessary.

We have also changed declensions of articles and adjectives, conjugations of verbs, and word order to fit the grammar whenever it was clear what was intended, even when the shorthand was unambiguous. A few changes in word choice were made for reasons of sense. One word we did *not* change occurs on page 386, line 11, where the shorthand clearly has "dieselben Salto mortale". Even though the plural may not have been intended, we have retained it and translated it accordingly. We list all our changes here in case a better reading should occur to the reader. Changes on the basis of possible error in the shorthand or due to an apparently faulty revision by Gödel are listed separately.

	Shorthand original	Replaced by
374, 7	werden will	werden wird
374, 9	auf deren einerseits	auf deren einer Seite
374, 10	auf der anderen	auf deren anderer
376, 17	Inf. Rechnung [or "Richtung"]	Infinitesimalrechnung
376, 21	Skept. und Emp.	Skeptikern und Empiristen
376, 31	Teil der Mathematik)	) Teil der Mathematik
378, 2	und	oder
378, 2	in Sinn	im Sinn
378, 21	dem Inst. des Math.	dem Instinkt des Mathematikers
380, 16	aufgeben und man	aufgeben oder man
380, 35	Exz.	Exzessen
382, 29	welches uns in	welches in uns
382, 39-40	exp.	Experimentieren
384, 21	Ja- und Nein-Frage	Ja- oder Nein-Frage
384, 43	darin	darauf

Some errors creep into the writing of shorthand either by the mistaking of one symbol for another or by the positioning of the symbols, either of which turns one word into another. Thus:

	Shorthand original	Probable intention	Type of error
376, 4	dessen or muß	muß	position, omission of ending
376, 10	Tatsäche	Tatsächlich	omission of ending loop
376, 25	sie ist einer	sie in einer	position
376, 30	Wahrheit den	Wahrheiten	position of plural ending
376, 36–7	neben das Folgern	neben dem Folgern	wrong form of article
378, 6	den Zeitgeist	dem Zeitgeist	position (symbol rotated)
378, 8	Mathematik sein	Mathematik ein	position
380, 35	die falschen	der falschen	wrong symbol (similar form)
382, 8	wurde	über	position
382, 16–17	Begriffe aus	Begriffe und	careless formation of an extra loop
382, 26	klarer	klar	extension of stroke for 'r'

Since this was a draft, there was an occasional word or passage that was apparently left in unintentionally after Gödel made a revision, e.g., a deletion, a replacement of a phrase, an insertion, etc. The following is a list of such occurrences, with the word we have deleted printed with an overstrike.

	Original
374, 34-5	Tatsache, ~~daß die Entwicklung der Philosophie seit der~~, es ist auch eine Platitüde, daß die Entwicklung der Philosophie seit der Renaissance
376, 11	Wissenschaft ~~sich~~ selbst
376, 24–25	~~und andererseits~~ und (2)
378, 22	, daß ~~man~~ ein
378, 24	und ~~daß~~ ferner
384, 36	daß ~~das~~ sie

Gödel *1970

The original manuscript of *1970* occurs in Gödel's *Nachlass* as item 060566.

As usual for these texts, we have filled in abbreviations where the completion was obvious. The reader should take special note that in footnote 4, the abbreviation "prop" was completed to "properties", although the word "proposition" might very well have been meant. The title has also been translated to "Ontological proof" from "*Ontologischer Beweis*". On page 403, line 1, a shorthand "*oder*" was replaced by "or", because the article was written in English. On page 403, lines 15 and 16, Gödel's double quotation marks used to indicate repetition from the line above have been replaced by the word or formula that was to be repeated. In each case, the word or formula was the first thing on that line.

Gödel *1970a

This paper (*Nachlass* item 040511) was found in two forms in the *Nachlass*, one manuscript (which had written at the top left "p 1–6 an Tarski: I Fassung" ["p 1–6 to Tarski: I version"] and at the top right, "Proc. Nat. Ac.", both underlined) and one typescript. As might be expected, the manuscript had a rather elaborate trail of insertions for such a short paper and the typist had not produced all the sections in the order requested. The manuscript was used as copy text for the version here, but the following list will include the typescript for the benefit of those who might have a copy of the typescript.

On page 420, line 20, the original manuscript read "segments of this scale" with the words "all the functions" inserted above the line between 'segments' and 'of'. The typist had typed "segments [of] all functions by this scale". We have replaced both by "segments ⟦of⟧ all the functions of this scale."

Following "$(m < n)$." on page 421, line 2, the manuscript had an arrow followed by the words "p. 4" On page 4 of the manuscript there was another arrow near the bottom of the page pointing to the sentence beginning with "In order to show ...". This was the last material not crossed out on the page (with the exception of an insertion to that sentence at the very bottom). The typist had typed that sentence in place, beginning a new paragraph after "H = the sum of $\aleph_1$ closed sets." We have moved that sentence to the indicated place on (his) page 2. Moreover, on the last page of the manuscript (which was unnumbered and contained only insertions), the first text was labelled "ad p. 2" and consisted of the paragraph beginning "Next construct ...". On the typescript, it had been inserted before the paragraph beginning "From these axioms ...". We have moved this material to precede the last paragraph on (his) page 2, as that seemed to be the only place it fit. The words "(or covering up to $\aleph_1$ points)" were given at the bottom of the page as an insertion where we have placed the flag for footnote 2. We have made this into a footnote in order to facilitate a reordering of phrases at the end of the sentence. Originally, the end of that sentence read "covering, $^\forall$ by a sequence ... lengths (*and any end segment of it*) the given zero set", where "$\forall$" indicated the insertion mentioned above. In the interests of coherence, we have rearranged the phrases to read "covering^2 the given zero set (*and any end segment of it*) by a sequence of intervals of these lengths".

In addition to the usual routine corrections of spelling and grammar, the following changes have been made:

	Manuscript	Typescript	Here replaced by
420, 27–28	$(N \to \text{Real})$	$(N \to \text{Real})$	$(N \to \text{Reals})$
420, 28	confinality	cofinality	cofinality
420, 28	ω_1	ω	ω_1
421, 10; 12	$N \to \text{Real}$	$N \to \text{Real}$	$N \to \text{Reals}$
421, 14	$< \omega_2, > \alpha$ and different	$> \omega_2, \alpha$ and different	$> \omega_2$ [[and]] $> \alpha$, different
421, 15	ones	ones	[[segments]]
421, 20–21	end	and	[[and must]] end
422, 3	$\omega_1 \to \omega_1$ which converges	$\omega_1 \to \omega_1$ converges	$\omega_1 \to \omega_1$ which converges
422, 4	disting.	distinguished	distinguished
422, 5	smaller $h(\alpha)$	smaller [than] $h(\alpha)$	smaller [[than]]$h(\alpha)$
422, 6–7	$(h(\alpha) - h_1(\alpha) =$	$(h(\alpha) - h_1(\alpha) =$	$(h(\alpha) - h_1(\alpha)) =$

Gödel *1970b

This manuscript occurs in the *Nachlass* as item 040512. A line at the top of the page read (in longhand German) "p 1–5 II Fassung Nur für mich geschrieben" ("p 1–5 II version Written only for me").

	Original	Replaced by
422, 20	functions g	function g
423, 19	$< \omega_1, > 0$	$< \omega_1$ [[and]] > 0
423, 24	remains	remain

Gödel *1970c

The letter as found in the *Nachlass* was unsigned. Written at the top in longhand was "p. 1–5 III Fassung" ("p. 1–5 III version") followed in shorthand by "*nicht abgeschickt*" ("unsent").

	Original	Replaced by
424, 5	purely	poorly

Appendix A

In addition to the words on page 428, line 20 (see editorial footnote c), we have inserted one left parenthesis on the basis of the original manuscript. This occurs on page 428, line 2, before "which". (For a description of the textual sources for *1946/9* see the textual notes for that article.)

	Original	Replaced by
426, 29	by Kant	for Kant
427, 28	to the other	to other
427, 29	by the near	by near
427, 30	by the far	by far

Appendix B

We have filled in many abbreviations from the small amount of longhand in the manuscript. On page 434, lines 10 and 12, "Negation" was chosen over "Negative" to fill in the abbreviation "Neg". We have attempted to refrain from making too many changes, but those that were necessary for grammatical reasons or to make sense of a passage are listed below.

	Original	Replaced by
434, 2	aus einer solchen	aus einem solchen
434, 6	positiver Eigenschaft	"positiver Eigenschaft"
434, fn x	ersetzt werden durch "gut".	durch "gut" ersetzt werden.

References

Ackermann, Wilhelm
   1928   Zum Hilbertschen Aufbau der reellen Zahlen, *Mathematische Annalen 99*, 118–133; English translation by Stefan Bauer-Mengelberg in *van Heijenoort 1967*, 493–507.
   See also Hilbert, David, and Wilhelm Ackermann.

Adams, Robert Merrihew
   1971   The logical structure of Anselm's arguments, *The philosophical review 80*, 28–54; reprinted in *Adams 1987*, 221–242.
   1977   Leibniz's theories of contingency, *Rice University studies 63* (*4*), 1–41.
   1987   *The virtue of faith and other essays in philosophical theology* (New York and Oxford: Oxford University Press).
   1988   Presumption and the necessary existence of God, *Noûs 22*, 19–32.

Anderson, C. Anthony
   1990   Some emendations of Gödel's ontological proof, *Faith and philosophy 7*, 291–303.

Ando, Tsuyoshi
   See Halperin, Israel, and Tsuyoshi Ando.

Anselm of Canterbury
   1974   Proslogion (English translation by Jasper Hopkins and Herbert Richardson), in *Hopkins and Richardson 1974*, 91–112.

Ayer, Alfred Jules
   1959   (ed.) *Logical positivism* (Glencoe, Ill.: Free Press).

Banaszczyk, Wojciech
   1987   The Steinitz constant of the plane, *Journal für die reine und angewandte Mathematik 373*, 218–220.
   1990   The Steinitz theorem on rearrangement of series for nuclear spaces, *ibid. 403*, 187–200.

Bar-Hillel, Yehoshua
   1970   (ed.) *Mathematical logic and foundations of set theory* (Amsterdam: North-Holland).

Barwise, Jon
1977   (ed.) *Handbook of mathematical logic* (Amsterdam: North-Holland).

Becker, Oskar
1930   Zur Logik der Modalitäten, *Jahrbuch für Philosophie und phänomenologische Forschung 11*, 497–548.

Beeson, Michael J.
1985   *Foundations of constructive mathematics: metamathematical studies* (Berlin: Springer).

Behrend, Felix A.
1954   The Steinitz-Gross theorem on sums of vectors, *Canadian journal of mathematics 6*, 108–124.

Behrmann, Jörn
1976   Biobibliographische Notiz, in *Zilsel 1976*, 44–46.

Bell, David, and Wilhelm Vossenkuhl
1992   (eds.) *Wissenschaft und Subjektivität/Science and subjectivity* (Berlin: Akademie Verlag).

Benacerraf, Paul, and Hilary Putnam
1964   (eds.) *Philosophy of mathematics: selected readings* (Englewood Cliffs, N. J.: Prentice-Hall; Oxford: Blackwell).
1983   Second edition of *Benacerraf and Putnam 1964* (Cambridge, U.K.: Cambridge University Press).

Bergström, Viktor
1931   Zwei Sätze über ebene Vektorpolygone, *Abhandlungen aus dem mathematischen Seminar der Hamburgischen Universität 8*, 206–214.

Bernays, Paul
1922   Über Hilberts Gedanken zur Grundlegung der Arithmetik, *Jahresbericht der Deutschen Mathematiker-Vereinigung 31*, 10–19.
1935   Sur le platonisme dans les mathématiques, *L'enseignement mathématique 34*, 52–69; English translation by Charles D. Parsons in *Benacerraf and Putnam 1964*, 274–286.
1935b   Quelques points essentiels de la métamathématique, *ibid. 34*, 70–95.
1941a   Sur les questions méthodologiques actuelles de la théorie hilbertienne de la démonstration, in *Gonseth 1941*, 144–152.

1954   Zur Beurteilung der Situation in der beweistheoretischen For-
        schung, *Revue internationale de philosophie 8*, 9–13.
1967   Hilbert, David, in *Edwards 1967*, vol. 3, 496–504.
See also Hilbert, David, and Paul Bernays.

Beth, Evert Willem
1956/7   Über Lockes "allgemeines Dreieck", *Kant-Studien 48*, 361–380.

Betsch, Christian
1926   *Fiktionen in der Mathematik* (Stuttgart: Fr. Frommanns Ver-
        lag [H. Kurtz]).

Bollert, Karl
1921   *Einsteins Relativitätstheorie und ihre Stellung im System der
        Gesamterfahrung* (Leipzig: Theodor Steinkopff).

Borel, Emile
1898   *Leçons sur la théorie des fonctions* (Paris: Gauthier-Villars).

Brodbeck, May
See Feigl, Herbert, and May Brodbeck.

Browder, Felix E.
1976   (ed.) *Mathematical developments arising from Hilbert prob-
        lems*, Proceedings of symposia in pure mathematics, vol. 28
        (Providence, R. I.: American Mathematical Society).

Buchholz, Wilfried
1986   A new system of proof-theoretical ordinal functions, *Annals of
        pure and applied logic 32*, 195–207.

Buchholz, Wilfried, Solomon Feferman, Wolfram Pohlers and Wilfried Sieg
1981   *Iterated inductive definitions and subsystems of analysis: re-
        cent proof-theoretical studies*, Lecture notes in mathematics,
        vol. 897 (Berlin: Springer).

Buldt, Bernd
See Schimanovich et alii.

Burkhardt, Hans, and Barry Smith
1991   (eds.) *Handbook of metaphysics and ontology* (Munich: Philo-
        sophia Verlag).

Carnap, Rudolf

1930   Die alte und die neue Logik, *Erkenntnis 1*, 12–26; English translation by Isaac Levi in *Ayer 1959*, 133–145.

1931   Die logizistische Grundlegung der Mathematik, *Erkenntnis 2*, 91–105; English translation by Erna Putnam and Gerald J. Massey in *Benacerraf and Putnam 1964*, 31–41.

1931a   Diskussion zur Grundlegung der Mathematik (Carnap's remarks in *Hahn et alii 1931*), *Erkenntnis 2*, 141–146; English translation by John W. Dawson, Jr. in *Dawson 1984*, 120–124.

1934   Die Antinomien und die Unvollständigkeit der Mathematik, *Monatshefte für Mathematik und Physik 41*, 263–284.

1934a   *Logische Syntax der Sprache* (Vienna: Springer); translated into English by Amethe Smeaton as *Carnap 1937*.

1934b   Die Aufgabe der Wissenschaftslogik, *Einheitswissenschaft 3* (Vienna: Gerold); French translation by Ernest Vouillemin in *Carnap 1935b*, 1–27; English translation by Hans Kaal in *McGuinness 1987*, 46–66.

1935a   Formalwissenschaft und Realwissenschaft, *Erkenntnis 5*, 30–37; French translation by Ernest Vouillemin in *Carnap 1935b*, 29–37; English translation by Herbert Feigl and May Brodbeck in *Feigl and Brodbeck 1953*, 123–128.

1935b   *La problème de la logique de la science. Science formelle et science du rèel*, Actualités scientifiques et industrielles 291; French translation of *Carnap 1934b* and *Carnap 1935a* (Paris: Hermann et Cie.).

1937   *The logical syntax of language* (London: Kegan Paul, Trench, Trubner; New York: Harcourt, Brace); English translation of *Carnap 1934a*, with revisions.

1947   *Meaning and necessity, a study in semantics and modal logic* (Chicago: University of Chicago Press).

1950   Empiricism, semantics, and ontology, *Revue internationale de philosophie 4*, 20–40; reprinted in *Carnap 1956*, 205–221.

1956   Second, enlarged edition of *Carnap 1947*.

See also Hahn et alii.

Chakrabarti, Sandip K., Robert Geroch and Can-bin Liang

1983   Timelike curves of limited acceleration in general relativity, *Journal of mathematical physics 24*, 597–598.

Chandrasekhar, Subrahmanyan, and James P. Wright

1961   The geodesics in Gödel's universe, *Proceedings of the National Academy of Sciences, U.S.A. 47*, 341–347.

Chen, Kien-Kwong
    1933   Axioms for real numbers, *Tôhoku mathematical journal 37*, 94–99.

Church, Alonzo
    1932   A set of postulates for the foundation of logic, *Annals of mathematics (2) 33*, 346–366.
    1933   A set of postulates for the foundation of logic (second paper), *ibid. 34*, 839–864.
    1934   The Richard paradox, *American mathematical monthly 41*, 356–361.
    1935   A proof of freedom from contradiction, *Proceedings of the National Academy of Sciences, U.S.A. 21*, 275–281.
    1960   The consistency of primitive recursive arithmetic (unpublished notes).

Clark, Peter, and Bob Hale
    1994   (eds.) *Reading Putnam* (Oxford: Basil Blackwell).

Dales, H. Garth, and W. Hugh Woodin
    1987   *An introduction to independence for analysts*, London Mathematical Society lecture note series, no. 115 (Cambridge, U. K.: Cambridge University Press).

Darboux, Gaston
    1912   *Éloges académiques et discours* (Paris: Librairie scientifique A. Hermann et fils).

Davis, Martin
    1950   *On the theory of recursive unsolvability* (doctoral dissertation, Princeton University).
    1953   Arithmetical problems and recursively enumerable predicates, *The journal of symbolic logic 18*, 33–41.
    1965   (ed.) *The undecidable: basic papers on undecidable propositions, unsolvable problems and computable functions* (Hewlett, N. Y.: Raven Press).
    1973   Hilbert's tenth problem is unsolvable, *The American mathematical monthly 80*, 233–269.
    1982   Why Gödel didn't have Church's thesis, *Information and control 54*, 3–24.

Davis, Martin, Yuri Matiyasevich and Julia Robinson
    1976   Hilbert's tenth problem. Diophantine equations: positive aspects of a negative solution, in *Browder 1976*, 323–378.

Davis, Martin, and Hilary Putnam
    1958   Reductions of Hilbert's tenth problem, *The journal of symbolic logic 23*, 183–187.

Davis, Martin, Hilary Putnam and Julia Robinson
    1961   The decision problem for exponential diophantine equations, *Annals of mathematics (2) 74*, 425–436.

Dawson, John W., Jr.
    1984   Discussion on the foundation of mathematics, *History and philosophy of logic 5*, 111–129.
    1984a   Kurt Gödel in sharper focus, *The mathematical intelligencer 6 (4)*, 9–17.

Dekker, Jacob C. E.
    1962   (ed.) *Recursive function theory*, Proceedings of symposia in pure mathematics, vol. 5 (Providence, R. I.: American Mathematical Society).

Denef, Jan
    1980   Diophantine sets over algebraic integer rings II, *Transactions of the American Mathematical Society 257*, 227–236.

Denef, Jan, and Leonard Lipschitz
    1978   Diophantine sets over some rings of algebraic integers, *The journal of the London Mathematical Society, series 2, 18*, 385–391.

Descartes, René
    1985   *The philosophical writings of Descartes*, vol. I, English translation by John Cottingham, Robert Stoothoff and Dugald Murdoch (Cambridge, U. K.: Cambridge University Press).

Diller, Justus, and Kurt Schütte
    1971   Simultane Rekursionen in der Theorie der Funktionale endlicher Typen, *Archiv für mathematische Logik und Grundlagenforschung 14*, 69–74.

Dingler, Hugo
    1931   *Philosophie der Logik und Arithmetik* (Munich: Reinhardt).

Dvořak, Johann
    1981   *Edgar Zilsel und die Einheit der Erkenntnis* (Vienna: Löcker).

Dvoretzky, Aryeh
    1961   Some results on convex bodies and Banach spaces, in *Proceedings of the international symposium on linear spaces. Hebrew University of Jerusalem, 1960* (Jerusalem: Jerusalem Academic Press; Oxford: Pergamon Press), 123–160.

Edwards, Paul
    1967   (ed.) *The encyclopedia of philosophy* (New York: Macmillan and the Free Press).

Einstein, Albert
    1921   Geometrie und Erfahrung, *Sitzungsberichte der Preussischen Akademie der Wissenschaften*, 123–130; English translation by G. B. Jeffery and W. Perrett in *Einstein 1922*, 25–56.
    1922   *Sidelights on relativity* (London: Methuen).
    1983   Reprint of *Einstein 1922* (New York: Dover).

Einstein, Albert, Boris Podolsky and Nathan Rosen
    1935   Can quantum-mechanical description of physical reality be considered complete?, *The physical review, series 2, 47*, 777–780.

Ellentuck, Erik
    1975   Gödel's square axioms for the continuum, *Mathematische Annalen 216*, 29–33.

Feferman, Solomon
    1962   Transfinite recursive progressions of axiomatic theories, *The journal of symbolic logic 27*, 259–316.
    1964   Systems of predicative analysis, *ibid. 29*, 1–30.
    1977   Theories of finite type related to mathematical practice, in *Barwise 1977*, 913–971.
    1981   How we got from there to here, in *Buchholz et alii 1981*, 1–15.
    1984a   Kurt Gödel: conviction and caution, *Philosophia naturalis 21*, 546–562.
    1985a   A theory of variable types, *Revista colombiana de matemàticas 19*, 95–106.
    1988   Weyl vindicated: "Das Kontinuum" 70 years later, in *Temi e prospettive della logica e della filosofia della scienza contemporanee, vol. I–Logica* (Bologna: CLUEB), 59–93.
    1988a   Hilbert's program relativized: proof-theoretical and foundational reductions, *The journal of symbolic logic 53*, 364-384.
    See also Buchholz et alii.

Feferman, Solomon, and Wilfried Sieg
    1981   Iterated inductive definitions and subsystems of analysis, in *Buchholz et alii 1981*, 16–77.

Feigl, Herbert, and May Brodbeck
   1953   (eds.) *Readings in the philosophy of science* (New York: Apple-
             ton-Century-Crofts).

Findlay, John N.
   1948   Can God's existence be disproved?, *Mind 57*, 176–183.
   1965   Revised version of *Findlay 1948*, in *Plantinga 1965*, 111–122.

Flagg, Robert C.
   1986   Integrating classical and intuitionistic type theory, *Annals of
             pure and applied logic 32*, 27–51.

Flagg, Robert C., and Harvey M. Friedman
   1986   Epistemic and intuitionistic formal systems, *Annals of pure
             and applied logic 32*, 53–60.

Fraenkel, Abraham A.
   1922   Der Begriff 'definit' und die Unabhängigkeit des Auswahlaxi-
             oms, *Sitzungsberichte der Preussischen Akademie der Wissen-
             schaften, Physikalisch-mathematische Klasse*, 253–257; English
             translation by Beverly Woodward in *van Heijenoort 1967*, 284–
             289.

Friedman, Harvey M.
   See Flagg, Robert C., and Harvey M. Friedman.

Friedman, Michael
   1985   Kant's theory of geometry, *The philosophical review 94*, 455–
             506; reprinted in revised form in *Friedman 1992*, 55–95.
   1992   *Kant and the exact sciences* (Cambridge, Mass.: Harvard Uni-
             versity Press).

Gabelsberger, Franz Xaver
   1834   *Anleitung zur deutschen Redezeichenkunst oder Stenographie*
             (Munich: Georg Franz); republished in 1908 (Wölfenbüttel:
             Heckner).

Gamow, George
   1946   Rotating universes? (letter to the editor), *Nature 158*, 549.

Gandy, Robin O.
   1988   The confluence of ideas in 1936, in *Herken 1988*, 55–111.

Gentzen, Gerhard

1936   Die Widerspruchsfreiheit der reinen Zahlentheorie, *Mathematische Annalen 112*, 493–565; English translation by M. E. Szabo in *Gentzen 1969*, 132–213.

1937   Unendlichkeitsbegriff und Widerspruchsfreiheit der Mathematik, in *Travaux du IXe Congrès International de Philosophie*, vol. VI, Actualités scientifiques et industrielles 535 (Paris: Hermann et Cie), 201–205.

1938   Die gegenwärtige Lage in der mathematischen Grundlagenforschung, in *Forschungen zur Logik und zur Grundlegung der exakten Wissenschaften, Neue Folge 4* (Leipzig: S. Hirzel), 5–18; reprinted in 1970 (Hildesheim: Verlag Dr. H. A. Gerstenberg); English translation by M. E. Szabo in *Gentzen 1969*, 234–251.

1938a   Neue Fassung des Widerspruchsfreiheitsbeweises für die reine Zahlentheorie, in *Forschungen zur Logik und zur Grundlegung der exakten Wissenschaften, Neue Folge 4* (Leipzig: S. Hirzel), 19–44; reprinted in 1970 (Hildesheim: Verlag Dr. H. A. Gerstenberg); English translation by M. E. Szabo in *Gentzen 1969*, 252–286.

1943   Beweisbarkeit und Unbeweisbarkeit von Anfangsfällen der transfiniten Induktion in der reinen Zahlentheorie, *Mathematische Annalen 119*, 140–161; English translation by M. E. Szabo in *Gentzen 1969*, 287–308.

1969   *The collected papers of Gerhard Gentzen*, edited and translated into English by M. E. Szabo (Amsterdam: North-Holland).

Geroch, Robert
See Chakrabarti, Sandip K., Robert Geroch and Can-bin Liang.

Gödel, Kurt

1929   *Über die Vollständigkeit des Logikkalküls* (doctoral dissertation, University of Vienna).

1930   Die Vollständigkeit der Axiome des logischen Funktionenkalküls, *Monatshefte für Mathematik und Physik 37*, 349–360.

1930a   Über die Vollständigkeit des Logikkalküls, *Die Naturwissenschaften 18*, 1068.

1930b   Einige metamathematische Resultate über Entscheidungsdefinitheit und Widerspruchsfreiheit, *Anzeiger der Akademie der Wissenschaften in Wien 67*, 214–215.

1931   Über formal unentscheidbare Sätze der *Principia mathematica* und verwandter Systeme I, *Monatshefte für Mathematik und Physik 38*, 173–198; translated into English by Jean van Heijenoort as *Gödel 1967*.

1931a Diskussion zur Grundlegung der Mathematik (Gödel's remarks in *Hahn et alii 1931*), *Erkenntnis 2*, 147–151; English translation by John W. Dawson, Jr. in *Dawson 1984*, 125–128.

1931b Review of *Neder 1931*, *Zentralblatt für Mathematik und ihre Grenzgebiete 1*, 5–6.

1931c Review of *Hilbert 1931*, *ibid. 1*, 260.

1931d Review of *Betsch 1926*, *Monatshefte für Mathematik und Physik (Literaturberichte) 38*, 5.

1931e Review of *Becker 1930*, *ibid. 38*, 5–6.

1931f Review of *Hasse and Scholz 1928*, *ibid. 38*, 37.

1931g Review of *von Juhos 1930*, *ibid. 38*, 39.

1932 Zum intuitionistischen Aussagenkalkül, *Anzeiger der Akademie der Wissenschaften in Wien 69*, 65–66; reprinted, with additional comment, as *1933n*.

1932a Ein Spezialfall des Entscheidungsproblems der theoretischen Logik, *Ergebnisse eines mathematischen Kolloquiums 2*, 27–28.

1932b Über Vollständigkeit und Widerspruchsfreiheit, *ibid. 3*, 12–13.

1932c Eine Eigenschaft der Realisierungen des Aussagenkalküls, *ibid. 3*, 20–21.

1932d Review of *Skolem 1931*, *Zentralblatt für Mathematik und ihre Grenzgebiete 2*, 3.

1932e Review of *Carnap 1931*, *ibid. 2*, 321.

1932f Review of *Heyting 1931*, *ibid. 2*, 321–322.

1932g Review of *von Neumann 1931*, *ibid. 2*, 322.

1932h Review of *Klein 1931*, *ibid. 2*, 323.

1932i Review of *Hoensbroech 1931*, *ibid. 3*, 289.

1932j Review of *Klein 1932*, *ibid. 3*, 291.

1932k Review of *Church 1932*, *ibid. 4*, 145–146.

1932l Review of *Kalmár 1932*, *ibid. 4*, 146.

1932m Review of *Huntington 1932*, *ibid. 4*, 146.

1932n Review of *Skolem 1932*, *ibid. 4*, 385.

1932o Review of *Dingler 1931*, *Monatshefte für Mathematik und Physik (Literaturberichte) 39*, 3.

1933 Untitled remark following *Parry 1933*, *Ergebnisse eines mathematischen Kolloquiums 4*, 6.

1933a Über Unabhängigkeitsbeweise im Aussagenkalkül, *ibid. 4*, 9–10.

1933b Über die metrische Einbettbarkeit der Quadrupel des R_3 in Kugelflächen, *ibid. 4*, 16–17.

1933c Über die Waldsche Axiomatik des Zwischenbegriffes, *ibid. 4*, 17–18.

1933d Zur Axiomatik der elementargeometrischen Verknüpfungsrelationen, *ibid. 4*, 34.

1933e   Zur intuitionistischen Arithmetik und Zahlentheorie, *ibid. 4*, 34–38.

1933f   Eine Interpretation des intuitionistischen Aussagenkalküls, *ibid. 4*, 39–40.

1933g   Bemerkung über projektive Abbildungen, *ibid. 5*, 1.

1933h   (with K. Menger and A. Wald) Diskussion über koordinatenlose Differentialgeometrie, *ibid. 5*, 25–26.

1933i   Zum Entscheidungsproblem des logischen Funktionenkalküls, *Monatshefte für Mathematik und Physik 40*, 433–443.

1933j   Review of *Kaczmarz 1932, Zentralblatt für Mathematik und ihre Grenzgebiete 5*, 146.

1933k   Review of *Lewis 1932, ibid. 5*, 337–338.

1933l   Review of *Kalmár 1933, ibid. 6*, 385–386.

1933m   Review of *Hahn 1932, Monatshefte für Mathematik und Physik* (*Literaturberichte*) *40*, 20–22.

1933n   Reprint of *Gödel 1932*, with additional comment, *Ergebnisse eines mathematischen Kolloquiums 4*, 40.

1934   *On undecidable propositions of formal mathematical systems* (mimeographed lecture notes, taken by Stephen C. Kleene and J. Barkley Rosser); reprinted with revisions in *Davis 1965*, 39–74.

1934a   Review of *Skolem 1933, Zentralblatt für Mathematik und ihre Grenzgebiete 7*, 97–98.

1934b   Review of *Quine 1933, ibid. 7*, 98.

1934c   Review of *Skolem 1933a, ibid. 7*, 193–194.

1934d   Review of *Chen 1933, ibid. 7*, 385.

1934e   Review of *Church 1933, ibid. 8*, 289.

1934f   Review of *Notcutt 1934, ibid. 9*, 3.

1935   Review of *Skolem 1934, ibid. 10*, 49.

1935a   Review of *Huntington 1934, ibid. 10*, 49.

1935b   Review of *Carnap 1934, ibid. 11*, 1.

1935c   Review of *Kalmár 1934, ibid. 11*, 3–4.

1936   Untitled remark following *Wald 1936, Ergebnisse eines mathematischen Kolloquiums 7*, 6.

1936a   Über die Länge von Beweisen, *ibid. 7*, 23–24.

1936b   Review of *Church 1935, Zentralblatt für Mathematik und ihre Grenzgebiete 12*, 241–242.

1938   The consistency of the axiom of choice and of the generalized continuum hypothesis, *Proceedings of the National Academy of Sciences, U.S.A. 24*, 556–557.

1939   The consistency of the generalized continuum hypothesis, *Bulletin of the American Mathematical Society 45*, 93.

1939a   Consistency proof for the generalized continuum hypothesis, *Proceedings of the National Academy of Sciences, U.S.A. 25,* 220–224; errata in *1947,* footnote 23.

1940   *The consistency of the axiom of choice and of the generalized continuum hypothesis with the axioms of set theory* (lecture notes taken by George W. Brown), Annals of mathematics studies, vol. 3 (Princeton: Princeton University Press); reprinted with additional notes in 1951 and with further notes in 1966.

1944   Russell's mathematical logic, in *Schilpp 1944,* 123–153; reprinted, with some alterations, as *Gödel 1964a* and as *Gödel 1972b.*

1946   Remarks before the Princeton bicentennial conference on problems in mathematics; first published in *Davis 1965,* 84–88; reprinted, with some alterations, as *Gödel 1968.*

1947   What is Cantor's continuum problem?, *American mathematical monthly 54,* 515–525; errata, *55,* 151; revised and expanded as *Gödel 1964.*

1949   An example of a new type of cosmological solutions of Einstein's field equations of gravitation, *Reviews of modern physics 21,* 447–450.

1949a   A remark about the relationship between relativity theory and idealistic philosophy, in *Schilpp 1949,* 555–562.

1952   Rotating universes in general relativity theory, *Proceedings of the International Congress of Mathematicians; Cambridge, Massachusetts, U.S.A. August 30-September 6, 1950,* I (Providence, R.I.: American Mathematical Society, 1952), 175–181.

1955   Eine Bemerkung über die Beziehungen zwischen der Relativitätstheorie und der idealistischen Philosophie (German translation of *Gödel 1949a* by Hans Hartmann), in *Schilpp 1955,* 406–412.

1958   Über eine bisher noch nicht benützte Erweiterung des finiten Standpunktes, *Dialectica 12,* 280–287; revised and expanded in English as *Gödel 1972;* translated into English by Wilfrid Hodges and Bruce Watson as *Gödel 1980.*

1962   Postscript to *Spector 1962,* 27.

1964   Revised and expanded version of *Gödel 1947,* in *Benacerraf and Putnam 1964,* 258–273.

1964a   Reprint, with some alterations, of *Gödel 1944,* in *Benacerraf and Putnam 1964,* 211–232.

1965   Expanded version of *Gödel 1934,* in *Davis 1965,* 39–74.

1967   English translation of *Gödel 1931,* in *van Heijenoort 1967,* 596–616.

1968   Reprint, with some alterations, of *Gödel 1946,* in *Klibansky 1968,* 250–253.

1972   On an extension of finitary mathematics which has not yet been used (to have appeared in *Dialectica*; first published in *Gödel 1990*, 271–280), revised and expanded English translation of *Gödel 1958*.

1972a   Some remarks on the undecidability results (to have appeared in *Dialectica*; first published in *Gödel 1990*, 305–306).

1972b   Reprint, with some alterations, of *Gödel 1944*, in *Pears 1972*, 192–226.

1974   Untitled remarks, in *Robinson 1974*, x.

1980   On a hitherto unexploited extension of the finitary standpoint, English translation of *Gödel 1958*, *Journal of philosophical logic 9*, 133–142.

1986   *Collected works*, volume I: *Publications 1929–1936*, edited by Solomon Feferman, John W. Dawson, Jr., Stephen C. Kleene, Gregory H. Moore, Robert M. Solovay, and Jean van Heijenoort (New York and Oxford: Oxford University Press).

1990   *Collected works*, volume II: *Publications 1938–1974*, edited by Solomon Feferman, John W. Dawson, Jr., Stephen C. Kleene, Gregory H. Moore, Robert M. Solovay, and Jean van Heijenoort (New York and Oxford: Oxford University Press).

Goldfarb, Warren and Thomas Ricketts
1992   Carnap and the philosophy of mathematics, in *Bell and Vossenkuhl 1992*, 61–78.

Gonseth, Ferdinand
1941   (ed.) *Les entretiens de Zurich, 6-9 decémbre 1938* (Zurich: Leemann).

Goodman, Nicolas D.
1970   A theory of constructions equivalent to arithmetic, in *Myhill, Kino and Vesley 1970*, 101–120.

1973   The arithmetic theory of constructions, in *Mathias and Rogers 1973*, 274–298.

Grinberg, Victor S., and Sergei V. Sevast'yanov (Гринберг, Виктор С., и Сергей В. Севастьянов)
1980   Value of the Steinitz constant (Russian), *Akademiia Nauk S.S.S.R. Funktsional'nyi analiz i ego prilozheniya 14* (2), 56–57; English translation in *Functional analysis and its applications 14*, 125–126.

Groß, Wilhelm
1917   Bedingt konvergente Reihen, *Monatshefte für Mathematik und Physik 28*, 221–237.

Guard, James R.
1961   *The independence of transfinite induction up to ω^ω in recursive
           arithmetic* (doctoral dissertation, Princeton University).

Hahn, Hans
1930   Die Bedeutung der Wissenschaftlichen Weltauffassung, insbe-
           sondere für Mathematik und Physik, *Erkenntnis 1*, 96–105;
           English translation by Hans Kaal in *Hahn 1980*, 20–30.
1931   Diskussion zur Grundlegung der Mathematik (Hahn's remarks
           in *Hahn et alii 1931*), *Erkenntnis 2*, 135–141, 145; English
           translation by John W. Dawson, Jr. in *Dawson 1984*, 116–120,
           123–124.
1932   *Reelle Funktionen* (Leipzig: Akademische Verlagsgesellschaft).
1933   *Logik, Mathematik und Naturerkennen, Einheitswissenschaft 2*
           (Vienna: Gerold); French translation by Ernest Vouillemin in
           Hahn 1935; English translation of sections 1–4 by Arthur Pap
           in *Ayer 1959*, 147–161; English translation by Hans Kaal in
           McGuinness 1987, 24–45.
1933a   Die Krise der Anschauung, in *Krise und Neuaufbau in den
           exakten Wissenschaften, fünf Wiener Vorträge* (Leipzig and
           Vienna: F. Deuticke); English translation in *Newman 1956*,
           vol. 3, 1956–1976; English translation in *Hahn 1980*, 73–102.
1935   *Logique, mathématiques et connaissance de la réalité*. Ac-
           tualités scientifiques et industrielles 226 (Paris: Hermann et
           Cie.); French translation of *Hahn 1933*.
1980   *Empiricism, logic, and mathematics: philosophical papers*,
           edited by Brian McGuinness (Dordrecht: Reidel).

Hahn, Hans, Rudolf Carnap, Kurt Gödel, Arend Heyting, Kurt
Reidemeister, Arnold Scholz, and John von Neumann
1931   Diskussion zur Grundlegung der Mathematik, *Erkenntnis 2*,
           135–151; English translation by John W. Dawson, Jr. in *Daw-
           son 1984*, 116–128.

Hájek, Petr
199?   Der Mathematiker und die Frage der Existenz Gottes (be-
           treffend Gödels ontologischen Beweis), in *Schimanovich et alii
           199?*, to appear.

Hale, Bob
See Clark, Peter, and Bob Hale.

Halperin, Israel
1986   Sums of a series, permitting rearrangements, *Comptes rendus mathématiques de l'Académie des Sciences/Mathematical reports of the Academy of Science (Canada) 8*, 87–102.

Halperin, Israel, and Tsuyoshi Ando
1989   *Bibliography: series of vectors and Riemann sums* (Hokkaido University, Sapporo 060, Japan).

Hartshorne, Charles
1962   *The logic of perfection and other essays in neoclassical metaphysics* (Lasalle, Ill.: Open Court).

Hasse, Helmut, and Heinrich Scholz
1928   *Die Grundlagenkrisis der griechischen Mathematik* (Charlottenburg: Metzner).

Hausdorff, Felix
1914   *Grundzüge der Mengenlehre* (Leipzig: Veit); reprinted in 1949 (New York: Chelsea).
1935   *Mengenlehre*, third, revised edition of *Hausdorff 1914* (Berlin and Leipzig: Walter de Gruyter).
1936   Summen von $\aleph_1$ Mengen, *Fundamenta mathematicae 26*, 241–255.

Heckmann, Otto, and Engelbert L. Schücking
1955   Bemerkungen zur Newtonschen Kosmologie. I., *Zeitschrift für Astrophysik 38*, 95–109.

Herbrand, Jacques
1930   *Recherches sur la théorie de la démonstration* (doctoral dissertation, University of Paris); also *Prace Towarzystwa Naukowego Warszawskiego, wydział III*, no. 33; reprinted in *Herbrand 1968*, 35–153; English translation by Warren Goldfarb in *Herbrand 1971*, 44–202.
1931   Sur la non-contradiction de l'arithmétique, *Journal für die reine und angewandte Mathematik 166*, 1–8; reprinted in *Herbrand 1968*, 221–232; English translation by Jean van Heijenoort in *van Heijenoort 1967*, 618–628, and in *Herbrand 1971*, 282–298.
1968   *Écrits logiques*, edited by Jean van Heijenoort (Paris: Presses Universitaires de France).
1971   *Logical writings*, English translation of *Herbrand 1968* by Warren Goldfarb (Cambridge, MA: Harvard University Press; Dordrecht: Reidel).

Herken, Rolf
 1988   (ed.) *The universal Turing machine: a half-century survey*
         (Oxford: Oxford University Press).

Heyting, Arend
 1930   Die formalen Regeln der intuitionistischen Logik, *Sitzungs-
         berichte der Preussischen Akademie der Wissenschaften, Phy-
         sikalisch-mathematische Klasse*, 42–56.
 1930a   Die formalen Regeln der intuitionistischen Mathematik, *ibid.*,
         57–71.
 1930b   Sur la logique intuitionniste, *Académie royale de Belgique, Bul-
         letins de la classe des sciences (5) 16*, 957–963.
 1931   Die intuitionistische Grundlegung der Mathematik, *Erkenntnis
         2*, 106–115; English translation by Erna Putnam and Gerald
         J. Massey in *Benacerraf and Putnam 1964*, 42–49.
 1934   *Mathematische Grundlagenforschung. Intuitionismus. Be-
         weistheorie*, Ergebnisse der Mathematik und ihrer Grenzge-
         biete 3 (Berlin: Springer).
See also Hahn et alii.

Hilbert, David
 1900   Mathematische Probleme. Vortrag, gehalten auf dem interna-
         tionalen Mathematiker-Kongreß zu Paris, 1900, *Nachrichten
         von der Königlichen Gesellschaft der Wissenschaften zu
         Göttingen*, 253–297; translated into English by Mary Winston
         Newson as *Hilbert 1902a*.
 1902a   Mathematical problems (English translation of *Hilbert 1900*),
         Bulletin of the American Mathematical Society 8, 437–479;
         reprinted in *Browder 1976*, 1–34.
 1926   Über das Unendliche, *Mathematische Annalen 95*, 161–190;
         English translation by Stefan Bauer-Mengelberg in *van Hei-
         jenoort 1967*, 367–392.
 1928   Die Grundlagen der Mathematik, *Abhandlungen aus dem
         mathematischen Seminar der Hamburgischen Universität 6*,
         65–85; English translation by Stefan Bauer-Mengelberg and
         Dagfinn Føllesdal in *van Heijenoort 1967*, 464–479.
 1931   Die Grundlegung der elementaren Zahlenlehre, *Mathematische
         Annalen 104*, 485–494; reprinted in part in *Hilbert 1935*, 192–
         195.
 1935   *Gesammelte Abhandlungen*, vol. 3 (Berlin: Springer).

Hilbert, David, and Wilhelm Ackermann
 1928   *Grundzüge der theoretischen Logik* (Berlin: Springer).

1938   Second, revised edition of *Hilbert and Ackermann 1928*; translated into English by Lewis M. Hammond, George G. Leckie and F. Steinhardt as *Hilbert and Ackerman 1950*.

1950   *Principles of mathematical logic*, English translation of *Hilbert and Ackermann 1938* (New York: Chelsea).

Hilbert, David, and Paul Bernays

1934   *Grundlagen der Mathematik*, vol. I (Berlin: Springer).

1939   *Grundlagen der Mathematik*, vol. II (Berlin: Springer).

1970   Second edition of *Hilbert and Bernays 1939*.

Hintikka, Jaakko

1965   Kant's "new method of thought" and his theory of mathematics, *Ajatus 27*, 37–47; reprinted in *Hintikka 1974*, 126–134.

1967   Kant on the mathematical method, *The monist 51*, 352–375; reprinted in *Hintikka 1974*, 160–183.

1974   *Knowledge and the known. Historical perspectives in epistemology* (Dordrecht: Reidel).

Hoensbroech, Franz G.

1931   Beziehungen zwischen Inhalt und Umfang von Begriffen, *Erkenntnis 2*, 291–300.

Hook, Sidney

1960   (ed.) *Dimensions of mind: a symposium* (New York: New York University Press).

Hopkins, Jasper, and Herbert Richardson

1974   (eds.) *Anselm of Canterbury*, volume one: *monologion, proslogion, debate with Gaunilo, and a meditation on human redemption* (Toronto and New York: Edwin Mellen Press).

Huntington, Edward V.

1932   A new set of independent postulates for the algebra of logic with special reference to Whitehead and Russell's *Principia mathematica*, *Proceedings of the National Academy of Sciences, U.S.A. 18*, 179–180.

1934   Independent postulates related to C. I. Lewis's theory of strict implication, *Mind (n.s.) 43*, 181–198.

Husserl, Edmund

1900   *Logische Untersuchungen*. Erster Teil: *Prolegomena zur reinen Logik* (Halle: Max Niemeyer); reprinted in *Husserl 1950–*, XVIII.

1901   *Logische Untersuchungen.* Zweiter Teil: *Untersuchungen zur Phänomenologie und Theorie der Erkenntnis* (Halle: Max Niemeyer); reprinted in *Husserl 1950–*, XIX.

1911   Philosophie als strenge Wissenschaft, *Logos 1*, 289–341; reprinted in *Husserl 1950–*, XXV; English translation by Quentin Lauer in *Husserl 1965*, 71–147.

1913   *Ideen zu einer reinen Phänomenologie und phänomenologischen Philosophie.* Erstes Buch: *Allgemeine Einführung in die reine Phänomenologie* (Halle: Max Niemeyer); reprinted in *Husserl 1950–*, III; translated into English by F. Kersten as *Husserl 1982*.

1913a   Second, revised edition of *Husserl 1900* and Investigations I–V of *Husserl 1901*; reprinted in *Husserl 1950–*, XIX, part 1; translated into English by J. N. Findlay in *Husserl 1970*.

1921   Second edition of Investigation VI of *Husserl 1901*; reprinted in *Husserl 1950–*, XIX, part 2; translated into English by J. N. Findlay in *Husserl 1970*.

1929   *Formale und transzendentale Logik. Versuch einer Kritik der logischen Vernunft* (Halle: Niemeyer); reprinted in *Husserl 1950–*, XVII; translated into English by Dorion Cairns as *Husserl 1969*.

1950–   *Husserliana. Edmund Husserl. Gesammelte Werke*, edited by the Husserl Archive (Leuven) on the basis of the *Nachlaß* (The Hague: Martinus Nijhoff; later Dordrecht: Kluwer).

1960   *Cartesian meditations*, English translation of *Kartesische Meditationen* (main text of *Husserl 1950–*, I) by Dorion Cairns (The Hague: Martinus Nijhoff).

1964   *The Paris lectures*, English translation of "Die Pariser Vorträge" (*Husserl 1950–*, I, 3–39) by Peter Koestenbaum (The Hague: Martinus Nijhoff).

1965   *Phenomenology and the crisis of philosophy*, English translation, with an introduction by Quentin Lauer, of *Husserl 1911* together with "Die Krisis des europäischen Menschentums und die Philosophie" (*Husserl 1950–*, VI, 314–348) (New York: Harper Torchbooks).

1968   *Logische Untersuchungen*, fifth edition of *Husserl 1900* and *1901*, unaltered from the second (*1913a*, *1921*) (Tübingen: Max Niemeyer).

1969   *Formal and transcendental logic*, English translation of *Husserl 1929* (The Hague: Martinus Nijhoff).

1970   *Logical investigations* (two volumes), English translation of *Husserl 1913a* and *1921* by J. N. Findlay (London: Routledge and Kegan Paul).

1970a   *The crisis of European sciences and transcendental phenom-
        enology. An introduction to phenomenological philosophy*,
        English translation of the main text and a selection of the ad-
        ditional texts in *Husserl 1950–*, VI, by David Carr (Evanston,
        Ill.: Northwestern University Press).
1982   *Ideas pertaining to a pure phenomenology and to a phenomeno-
        logical philosophy*, First book: *general introduction to a pure
        phenomenology*, English translation of *Husserl 1913* (The
        Hague: Martinus Nijhoff).

Jeans, James
1936   Man and the universe, in *Scientific progress* (Sir Halley Stewart
        lecture[s], 1935) (New York: MacMillan; London: Allen and
        Unwin), 11–38.

Jech, Thomas
1978   *Set theory* (New York: Academic Press).

Jensen, Ronald Björn
1972   The fine structure of the constructible hierarchy, *Annals of
        mathematical logic 4*, 229–308.

Jensen, Ronald Björn, and Robert M. Solovay
1970   Some applications of almost disjoint sets, in *Bar-Hillel 1970*,
        84–104.

Jones, James P.
1981   Classification of quantifier prefixes over diophantine equations,
        *Zeitschrift für mathematische Logik und Grundlagen der
        Mathematik 27*, 403–410.
1982   Universal diophantine equation, *The journal of symbolic logic
        47*, 549–571.

Kaczmarz, Stefan
1932   Axioms for arithmetic, *The journal of the London Mathemat-
        ical Society 7*, 179–182.

Kadets, Mikhail Iosifovich (Кадец, Михаил Иосифович), and Krzystof
Wozniakowski
1989   On series whose permutations have only two sums, *Bulletin of
        the Polish Academy of Sciences: mathematics 37*, 15–21.

Kadets, Vladimir Mikhailovich (Кадец, Владимир Михайлович)
   1986   A problem of S. Banach (problem 106 from the "Scottish book")
           (Russian), *Akademiia Nauk S.S.S.R. Funktsional'nyi analiz i
           ego prilozheniya 20 (4)*, 74–75; English translation in *Func-
           tional analysis and its applications 20* (1986), 317–319.

Kalmár, László
   1932   Ein Beitrag zum Entscheidungsproblem, *Acta litterarum ac
           scientiarum Regiae Universitatis Hungaricae Francisco-Jose-
           phinae, sectio scientiarum mathematicarum 5*, 222–236.
   1933   Über die Erfüllbarkeit derjenigen Zählausdrücke, welche in der
           Normalform zwei benachbarte Allzeichen enthalten, *Mathema-
           tische Annalen 108*, 466–484.
   1934   Über einen Löwenheimschen Satz, *Acta litterarum ac scien-
           tiarum Regiae Universitatis Hungaricae Francisco-Josephinae,
           sectio scientiarum mathematicarum 7*, 112–121.

Kaluza, Theodor Franz Eduard
   1921   Zum Unitätsproblem der Physik, *Sitzungsberichte der Preussi-
           schen Akademie der Wissenschaften*, 966–972.

Kant, Immanuel
   1781   *Critik der reinen Vernunft* (Riga: Hartknoch).
   1783   *Prolegomena zu einer jeden künftigen Metaphysik, die als Wis-
           senschaft wird auftreten können* (Riga: Hartknoch).
   1786   *Metaphysische Anfangsgründe der Naturwissenschaft* (Riga:
           Hartknoch).
   1787   *Kritik der reinen Vernunft*, second, revised edition of *Kant
           1781*.
   1902–   *Kants gesammelte Schriften*, edited by the Prussian Academy
           of Sciences, later the German Academy of Sciences (Berlin:
           Georg Reimer, later Walter de Gruyter).
   1929   *Critique of pure reason*, English translation of *Kant 1781* and
           1787 by Norman Kemp Smith (London: Macmillan).
   1933   Second impression, with corrections, of *Kant 1929*.

Katznelson, Yitzhak, and O. Carruth McGehee
   1974   Conditionally convergent series in R^∞, *The Michigan mathe-
           matical journal 21*, 97–106.

Kern, Iso
   1964   *Husserl und Kant. Eine Untersuchung über Husserls Verhält-
           nis zu Kant und zum Neukantianismus*, Phenomenologica 16
           (The Hague: Martinus Nijhoff).

Kino, Akiko
See Myhill, John, Akiko Kino and Richard E. Vesley.

Kleene, Stephen C.
1936   General recursive functions of natural numbers, *Mathematische Annalen 112*, 727–742; reprinted in *Davis 1965*, 236–252; for an erratum, a simplification and an addendum, see *Davis 1965*, 253.
1981   Origins of recursive function theory, *Annals of the history of computing 3*, 52–67; corrections, *Davis 1982*, footnotes 10 and 12.

Klein, Fritz
1931   Zur Theorie der abstrakten Verknüpfungen, *Mathematische Annalen 105*, 308–323.
1932   Über einen Zerlegungssatz in der Theorie der abstrakten Verknüpfungen, *ibid. 106*, 114–130.

Klibansky, Raymond
1968   (ed.) *Contemporary philosophy, a survey. I, Logic and foundations of mathematics* (Florence: La Nuova Italia Editrice).

Köhler, Eckehart
See Schimanovich et alii.

Kolmogorov, Andrei Nikolayevich (Kolmogoroff; Колмогоров, Андрей Николаевич)
1932   Zur Deutung der intuitionistischen Logik, *Mathematische Zeitschrift 35*, 58–65.

Kondô, Motokiti
1938   Sur les opérations analytiques dans la théorie des ensembles et quelques problèmes qui s'y rattachent, I, *Journal of the Faculty of Science, Hokkaido Imperial University. Series I. Mathematics 7*, 1–34.

Kornilov, P. A. (Корнилов, П. А.)
1988   On the set of sums of a conditionally convergent series of functions (Russian), *Matematicheskii sbornik 137*, 114-127; English translation in *Mathematics of the U.S.S.R. Sbornik 65* (1990), 119–131.

Kowalewski, Arnold
1924   (ed.) *Die philosophischen Hauptvorlesungen Immanuel Kants. Nach den neu aufgefunden Kollegheften des Grafen Heinrich zu Dohna-Wundlacken* (Munich: Rösl et Cie.); reprinted in 1965 (Hildesheim: Georg Olms).

Kreisel, Georg
1951   On the interpretation of non-finitist proofs—Part I, *The journal of symbolic logic 16*, 241–267.
1959c   Proof by transfinite induction and definition by transfinite induction in quantifier-free systems, *ibid. 24*, 322–323.
1962a   Foundations of intuitionistic logic, in *Nagel, Suppes and Tarski 1962*, 198–210.
1965   Mathematical logic, in *Saaty 1965*, vol. 3, 95–195.
1968   A survey of proof theory, *The journal of symbolic logic 33*, 321–388.
1971   Review of *Gentzen 1969*, *The journal of philosophy 68*, 238–265.
1987   Gödel's excursions into intuitionistic logic, in *Weingartner and Schmetterer 1987*, 65–186.

Kundt, Wolfgang
1956   Trägheitsbahnen in einem von Gödel angegebenen kosmologischen Modell, *Zeitschrift für Physik 145*, 611–620.

Lathrop, J. and R. Teglas
1978   Dynamics in the Gödel universe, *Il nuovo cimento 43, B*, 162–172.

Leibniz, Gottfried W.
1923–   *Sämtliche Schriften und Briefe*, edited by the Prussian Academy of Sciences, later the German Academy of Sciences (Darmstadt: O. Reichl).
1956   *Philosophical papers and letters*, translated and edited by Leroy E. Loemker (Chicago: University of Chicago Press).
1969   Second edition of *Leibniz 1956* (Dordrecht and Boston: Reidel).

Leivant, Daniel
1985   Syntactic translations and provably recursive functions, *The journal of symbolic logic 50*, 682–688.

Leonardi, Paolo, and Marco Santambrogio
199?   (eds.) *On Quine* (Cambridge: Cambridge University Press).

Lévy, Azriel
1970   Definability in axiomatic set theory II, in *Bar-Hillel 1970*, 129–145.

Lévy, Paul M.
1905   Sur les séries semi-convergentes, *Nouvelles annales de mathématiques (4) 5*, 506–511.

Lewis, Clarence I.
1932   Alternative systems of logic, *The monist 42*, 481–507.

Lewis, David K.
1970   Anselm and actuality, *Noûs 4*, 175–188; reprinted with postscripts in *Lewis 1983*, 10–25.
1983   *Philosophical papers*, vol. I (New York and Oxford: Oxford University Press).

Liang, Can-bin
See Chakrabarti, Sandip K., Robert Geroch and Can-bin Liang.

Lindenbaum, Adolf, and Alfred Tarski
1926   Communication sur les recherches de la théorie des ensembles, *Comptes rendus des séances de la Société des Sciences et des Lettres de Varsovie, Classe III, 19*, 299–330.

Lipschitz, Leonard
See Denef, Jan, and Leonard Lipschitz.

Łoš, Jerzy
1954   On the categoricity in power of elementary deductive systems and some related problems, *Colloquium mathematicum 3*, 58–62.

Lucas, J. R.
1961   Minds, machines and Gödel, *Philosophy 36*, 112–127.

Łukasiewicz, Jan, and Alfred Tarski
1930   Untersuchungen über den Aussagenkalkül, *Sprawozdania z posiedzeń Towarzystwa Naukowego Warszawskiego, wydział III, 23*, 30–50; English translation by Joseph H. Woodger in *Tarski 1956*, 38–59.

Malcolm, Norman
1960   Anselm's ontological arguments, *The philosophical review 69*, 41–62; reprinted in *Malcolm 1963*, 141–162.
1963   *Knowledge and certainty, essays and lectures* (Englewood Cliffs, N. J.: Prentice-Hall).

Martin-Löf, Per
    1975   An intuitionistic theory of types: predicative part, in *Rose and
            Shepherdson 1975*, 73–118.
    1984   *Intuitionistic type theory* (Naples: Bibliopolis).

Mathias, Adrian R. D., and Hartley Rogers
    1973   (eds.) *Cambridge summer school in mathematical logic*, Lec-
            ture notes in mathematics, vol. 337 (Berlin: Springer).

Matiyasevich, Yuri Vladimirovich (Matijacevič; Матиясевич, Юрий
Владимирович)
    1970   Enumerable sets are diophantine (Russian), *Doklady Akademii
            Nauk S.S.S.R. 191*, 279–282; English translation, with revi-
            sions, in *Soviet mathematics Doklady 11* (1970), 354–358.
    See also Davis, Martin, Yuri Matiyasevich and Julia Robinson.

Matiyasevich, Yuri, and Julia Robinson
    1975   Reductions of an arbitrary diophantine equation to one in 13
            unknowns, *Acta arithmetica 27*, 521–553.

Mauldin, R. Daniel
    1981   (ed.) *The Scottish book, mathematics from the Scottish Cafe*
            (Boston: Birkhauser).

McAloon, Kenneth
    1971   Consistency results about ordinal definability, *Annals of math-
            ematical logic 2*, 449–467.

McGehee, O. Carruth
    See Katznelson, Yitzhak, and O. Carruth McGehee.

McGuinness, Brian
    1987   (ed.) *Unified science: the Vienna Circle monograph series*
            (Dordrecht: D. Reidel).

McKinsey, John C. C., and Alfred Tarski
    1948   Some theorems about the sentential calculi of Lewis and Heyt-
            ing, *The journal of symbolic logic 13*, 1–15.

McTaggart, J. McTaggart Ellis
    1908   The unreality of time, *Mind* (n.s.) *17*, 457–474.

Menger, Karl
    1930a   Der Intuitionismus, *Blätter für deutsche Philosophie 4*, 311–
            325.

1981   Recollections of Kurt Gödel, private memoir, to appear in *Schimanovich et alii 199?*.

See also *Gödel 1933h*.

Mints, Gregory E. (Mints, Grigori E.; Минц, Григорий Е.)
1971   Quantifier-free and one-quantifier systems (Russian), *Zapiski nauchnyk seminarov Leningradskogo otdeleniya Matematischeskogo Instituta im. V. A. Steklova, Akademii nauk S.S.S.R. (Leningrad) 20*, 115–133; English translation in *Journal of soviet mathematics 1* (1973), 71–84.

Moore, Gregory H.
1982   *Zermelo's axiom of choice: its origins, development, and influence*, Studies in the history of mathematics and physical sciences, vol. 8 (New York: Springer).

Müller, Gert H., and Dana S. Scott
1978   (eds.) *Higher set theory. Proceedings, Oberwohlfach, Germany, April 13–23, 1977*, Lecture notes in mathematics, vol. 669 (Berlin: Springer).

Myhill, John, Akiko Kino and Richard E. Vesley
1970   (eds.) *Intuitionism and proof theory* (Amsterdam: North-Holland).

Nagel, Ernest, and James R. Newman
1958   *Gödel's proof* (New York: New York University Press).

Nagel, Ernest, Patrick Suppes and Alfred Tarski
1962   (eds.) *Logic, methodology, and philosophy of science. Proceedings of the 1960 International Congress* (Stanford, Calif.: Stanford University Press).

Neder, Ludwig
1931   Über den Aufbau der Arithmetik, *Jahresbericht der Deutschen Mathematiker-Vereinigung 40*, 22–37.

Newman, James R.
1956   (ed.) *The world of mathematics* (New York: Simon and Schuster).

See also Nagel, Ernest, and James R. Newman.

Notcutt, Bernard
   1934   A set of axioms for the theory of deduction, *Mind* (n.s.) *43*, 63–77.

O'Neill, Barrett
   1983   *Semi-Riemannian geometry, with applications to relativity* (New York: Academic Press).

Parry, William T.
   1933   Ein Axiomensystem für eine neue Art von Implikation (analytische Implikation), *Ergebnisse eines mathematischen Kolloquiums 4*, 5–6.

Parsons, Charles D.
   1970   On a number-theoretic choice schema and its relation to induction, in *Myhill, Kino and Vesley 1970*, 459–473.
   1972   On *n*-quantifier induction, *The journal of symbolic logic 37*, 466–482.
   199?   Quine and Gödel on analyticity, in *Leonardi and Santambrogio 199?*, to appear.

Pears, David F.
   1972   (ed.) *Bertrand Russell: a collection of critical essays* (Garden City, N.Y.: Anchor-Doubleday).

Penrose, Roger
   1989   *The emperor's new mind* (Oxford and New York: Oxford University Press).

Perzanowski, Jerzy
   1991   Ontological arguments II: Cartesian and Leibnizian, in *Burkhardt and Smith 1991*, 625–633.

Pheidas, Thanases
   1988   Hilbert's tenth problem for a class of rings of algebraic number fields, *Proceedings of the American Mathematical Society 104*, 611–620.
   1991   Hilbert's tenth problem for fields of rational functions over finite fields, *Inventiones mathematicae 103*, 1–8.

Plantinga, Alvin
   1965   (ed.) *The ontological argument, from St. Anselm to contemporary philosophers* (Garden City, N.Y.: Anchor-Doubleday).
   1974   *The nature of necessity* (Oxford: Clarendon Press).

Podolsky, Boris
    See Einstein, Albert, Boris Podolsky and Nathan Rosen.

Pohlers, Wolfram
    1989   *Proof theory: an introduction*, Lecture notes in mathematics, vol. 1407 (Berlin: Springer).
    See also Buchholz et alii.

Poincaré, Henri
    1902   *La science et l'hypothèse* (Paris: Bibliothèque de philosophie scientifique); translated into English by William John Greenstreet as *Poincaré 1905*.
    1905   *Science and hypothesis*, edited by Sir Joseph Larmor, English translation of *Poincaré 1902* (London and Newcastle-on-Tyne: Walter Scott).

Popper, Karl
    1976   *Unended quest, an intellectual autobiography* (La Salle, Ill.: Open Court).

Putnam, Hilary
    1960   Minds and machines, in *Hook 1960*, 148–179; reprinted in *Putnam 1975*, 362–385.
    1975   *Mind, language and reality: philosophical papers*, vol. 2 (Cambridge, U.K.: Cambridge University Press).
    See also Benacerraf, Paul, and Hilary Putnam.
    See also Davis, Martin, and Hilary Putnam.
    See also Davis, Martin, Hilary Putnam and Julia Robinson.

Quine, Willard van Orman
    1933   A theorem in the calculus of classes, *The journal of the London Mathematical Society 8*, 89–95.
    1936   Truth by convention, in *Philosophical essays for Alfred North Whitehead* (London and New York: Longmans, Green), 90–124; reprinted with corrections in *Benacerraf and Putnam 1964*, 322–345, and in *Quine 1966*, 70-99.
    1951   Two dogmas of empiricism, *The philosophical review 60*, 20–43; reprinted in *Quine 1953*, 20–46, and in *Benacerraf and Putnam 1964*, 346–365.
    1953   *From a logical point of view. 9 logico-philosophical essays* (Cambridge, Mass.: Harvard University Press).
    1960   Carnap and logical truth, *Synthese 12*, 350–374; also in *Schilpp 1963*, 385–406; reprinted in *Quine 1966*, 100–125.
    1960a   Posits and reality, in *Uyeda 1960*, 391–400; reprinted in *Quine 1966*, 233–241.

1966   *The ways of paradox and other essays* (New York: Random House).

1976   Second, enlarged edition of *Quine 1966* (Cambridge, Mass.: Harvard University Press).

Ramsey, Frank P.

1926   The foundations of mathematics, *Proceedings of the London Mathematical Society (2) 25*, 338–384; reprinted in *Ramsey 1931*, 1–61.

1931   *The foundations of mathematics and other logical essays*, edited by Richard B. Braithwaite (London: Kegan Paul).

1978   *Foundations: essays in philosophy, logic, mathematics and economics*, expanded and revised edition of *Ramsey 1931*, edited by D. H. Mellor (Atlantic Highlands, N. J.: Humanities Press; London: Routledge and Kegan Paul).

Rathjen, Michael

1991   Proof-theoretic analysis of KPM, *Archive for mathematical logic 30*, 377–403.

Reichenbach, Hans

1924   *Axiomatik der relativistischen Raum-Zeit-Lehre* (Braunschweig: Friedrich Vieweg); translated into English by Maria Reichenbach as *Reichenbach 1969*.

1928   *Philosophie der Raum-Zeit-Lehre* (Berlin: Walter de Gruyter); translated into English by Maria Reichenbach and John Freund as *Reichenbach 1957*.

1957   *The philosophy of space and time*, English translation of *Reichenbach 1928* (New York: Dover).

1969   *Axiomatization of the theory of relativity*, English translation of *Reichenbach 1924* (Berkeley: University of California Press).

Reidemeister, Kurt
See Hahn et alii.

Richardson, Herbert
See Hopkins, Jasper, and Herbert Richardson.

Ricketts, Thomas

1994   Carnap's principle of tolerance, empiricism, and conventionalism, in *Clark and Hale 1994*, in press.
See also Goldfarb, Warren, and Thomas Ricketts.

Robinson, Abraham
 1966   *Non-standard analysis* (Amsterdam: North-Holland).
 1974   Second edition of *Robinson 1966*.

Robinson, Julia
 1952   Existential definability in arithmetic, *Transactions of the American Mathematical Society 72*, 437–449.
 See also Davis, Martin, Yuri Matiyasevich and Julia Robinson.
 See also Davis, Martin, Hilary Putnam and Julia Robinson.
 See also Matiyasevich, Yuri, and Julia Robinson.

Rogers, Hartley
 See Mathias, Adrian R. D., and Hartley Rogers.

Rose, Harvey E.
 1984   *Subrecursion: functions and hierarchies* (Oxford: Clarendon Press).

Rose, Harvey E., and John C. Shepherdson
 1975   (eds.) *Logic colloquium '73* (Amsterdam: North-Holland).

Rosen, Nathan
 See Einstein, Albert, Boris Podolsky and Nathan Rosen.

Rosser, J. Barkley
 1936   Extensions of some theorems of Gödel and Church, *The journal of symbolic logic 1*, 87–91; reprinted in *Davis 1965*, 230–235.
 1937   Gödel theorems for non-constructive logics, *ibid. 2*, 129–137.

Rubin, Herman, and Jean E. Rubin
 1985   *Equivalents of the axiom of choice, II* (Amsterdam: North-Holland).

Russell, Bertrand
 1908   Mathematical logic as based on the theory of types, *American journal of mathematics 30*, 222–262; reprinted in *Russell 1956*, 59–102, and in *van Heijenoort 1967*, 150–182.
 1956   *Logic and knowledge: essays 1901–1950*, edited by Robert Charles Marsh (London: Allen and Unwin).
 See also Whitehead, Alfred North, and Bertrand Russell.

Saaty, Thomas L.
 1965   (ed.) *Lectures on modern mathematics* (New York: Wiley).

508 *References*

Santambrogio, Marco
See Leonardi, Paolo, and Marco Santambrogio.

Schilpp, Paul A.
1944   (ed.) *The philosophy of Bertrand Russell*, Library of living philosophers, vol. 5 (Evanston, Ill.: Northwestern University); third edition (New York: Tudor, 1951).
1949   (ed.) *Albert Einstein, philosopher-scientist*, Library of living philosophers, vol. 7 (Evanston, Ill.: Library of living philosophers); third edition (New York: Tudor, 1951).
1955   (ed.) *Albert Einstein als Philosoph und Naturforscher*, German translation by Hans Hartman (with additions) of *Schilpp 1949* (Stuttgart: Kohlhammer).
1963   (ed.) *The philosophy of Rudolf Carnap*, Library of living philosophers, vol. 11 (La Salle, Ill.: Open Court; London: Cambridge University Press).

Schimanovich, Werner, Bernd Buldt, Eckehart Köhler and Peter Weibel
199?   (eds.) *Wahrheit und Beweisbarkeit. Leben und Werk Kurt Gödels* (Vienna: Hölder-Pichler-Tempsky).

Schlick, Moritz
1918   *Allgemeine Erkenntnislehre* (Berlin: Springer); translated into English by Albert E. Blumberg as *Schlick 1985*.
1938   *Gesammelte Aufsätze 1926-1936* (Vienna: Gerold).
1985   *General theory of knowledge*, English translation of *Schlick 1918* (La Salle, Ill.: Open Court).

Schmetterer, Leopold
See Weingartner, Paul, and Leopold Schmetterer.

Scholz, Arnold
See Hahn et alii.

Scholz, Heinrich
See Hasse, Helmut, and Heinrich Scholz.

Schücking, Engelbert L.
See Heckmann, Otto, and Engelbert L. Schücking.

Schütte, Kurt
1977   *Proof theory* (Berlin: Springer).
See also Diller, Justus, and Kurt Schütte.

Schwichtenberg, Helmut
1977   Proof theory: some applications of cut-elimination, in *Barwise 1977*, 867–895.

Scott, Dana S.
1987   Gödel's ontological proof, in *Thomson 1987*, 257–258.
See also Müller, Gert H., and Dana S. Scott.

Sevast'yanov, Sergei V. (Севастьянов, Сергей В.)
See Grinberg, Victor S. and Sergei V. Sevast'yanov.

Shapiro, Harold N., and Alexandra Shlapentokh
1989   Diophantine relationships between algebraic number fields, *Communications on pure and applied mathematics 42*, 1113–1122.

Shapiro, Stewart
1985   (ed.) *Intensional mathematics* (Amsterdam: North-Holland).

Shepherdson, John C.
See Rose, Harvey E., and John C. Shepherdson.

Shlapentokh, Alexandra
1989   Extensions of Hilbert's tenth problem to some algebraic number fields, *Communications on pure and applied mathematics 42*, 939–962.
1992   Hilbert's tenth problem for rings of algebraic functions of characteristic 0, *Journal of number theory 40*, 218–236.
1992a   Hilbert's tenth problem for rings of algebraic functions in one variable over fields of constants of positive characteristic, *Transactions of the American Mathematical Society 333*, 275–298.
See also Shapiro, Harold N., and Alexandra Shlapentokh.

Sieg, Wilfried
1985   Fragments of arithmetic, *Annals of pure and applied logic 28*, 33–71.
1990   Relative consistency and accessible domains, *Synthese 84*, 259–297.
1991   Herbrand analyses, *Archive for mathematical logic 30*, 409–441.
See also Buchholz et alii.
See also Feferman, Solomon, and Wilfried Sieg.

Sierpiński, Wacław

1947   L'hypothèse généralisée du continu et l'axiome du choix, *Fundamenta mathematicae 34*, 1–5; reprinted in *Sierpiński 1976*, 484–488.

1976   *Oeuvres choisies*, Tome III: *Théorie des ensembles et ses applications. Travaux des années 1930–1966* (Warsaw: PWN).

Simpson, Stephen G.

1988   Partial realizations of Hilbert's program, *The journal of symbolic logic 53*, 349–363.

Skolem, Thoralf

1923a   Einige Bemerkungen zur axiomatischen Begründung der Mengenlehre, *Matematikerkongressen i Helsingfors den 4–7 Juli 1922, den femte skandinaviska matematikerkongressen, Redogörelse* (Helsinki: Akademiska Bokhandeln), 217–232; reprinted in *Skolem 1970*, 137–152; English translation by Stefan Bauer-Mengelberg in *van Heijenoort 1967*, 290–301.

1931   Über einige Satzfunktionen in der Arithmetik, *Skrifter utgitt av Det Norske Videnskaps-Akademi i Oslo, I. Matematisk-naturvidenskapelig klasse*, no. 7, 1–28; reprinted in *Skolem 1970*, 281–306.

1932   Über die symmetrisch allgemeinen Lösungen im identischen Kalkül, *Skrifter utgitt av Det Norske Videnskaps-Akademi i Oslo, I. Matematisk-naturvidenskapelig klasse*, no. 6, 1–32; also appeared in *Fundamenta mathematicae 18*, 61–76; reprinted in *Skolem 1970*, 307–336.

1933   Ein kombinatorischer Satz mit Anwendung auf ein logisches Entscheidungsproblem, *Fundamenta mathematicae 20*, 254–261; reprinted in *Skolem 1970*, 337–344.

1933a   Über die Unmöglichkeit einer vollständigen Charakterisierung der Zahlenreihe mittels eines endlichen Axiomensystems, *Norsk matematisk forenings skrifter, series 2*, no. 10, 73–82; reprinted in *Skolem 1970*, 345–354.

1934   Über die Nicht-charakterisierbarkeit der Zahlenreihe mittels endlich oder abzählbar unendlich vieler Aussagen mit ausschließlich Zahlenvariablen, *Fundamenta mathematicae 23*, 150–161; reprinted in *Skolem 1970*, 355–366.

1970   *Selected works in logic*, edited by Jens Erik Fenstad (Oslo: Universitetsforlaget).

Smith, Barry
See Burkhardt, Hans, and Barry Smith.

Sobel, Jordan Howard
  1987   Gödel's ontological proof, in *Thomson 1987*, 241–261.

Solovay, Robert M.
  1970   A model of set theory in which every set of reals is Lebesgue measurable, *Annals of mathematics (2) 92*, 1–56.
  See also Jensen, Ronald Björn, and Robert M. Solovay.

Spector, Clifford
  1962   Provably recursive functionals of analysis: a consistency proof of analysis by an extension of principles formulated in current intuitionistic mathematics, in *Dekker 1962*, 1–27.

Spiegelberg, Herbert
  1960   *The phenomenological movement: a historical introduction*, Phenomenologica 5, Volumes I and II (The Hague: Martinus Nijhoff).
  1965   Second edition of *Spiegelberg 1960*.

Staal, J. Frits
  See van Rootselaar, Bob, and J. Frits Staal.

Steinitz, Ernst
  1913   Bedingt konvergente Reihen und konvexe Systeme, *Journal für die reine und angewandte Mathematik 143*, 128–175.
  1914   Bedingt konvergente Reihen und konvexe Systeme (Fortsetzung), *ibid. 144*, 1–40.
  1916   Bedingt konvergente Reihen und konvexe Systeme (Schluß), *ibid. 146*, 1–52.

Stern, Jacques
  1982   (ed.) *Proceedings of the Herbrand symposium. Logic colloquium '81* (Amsterdam: North-Holland).

Sundholm, Göran
  1983   Constructions, proofs and the meaning of logical constants, *Journal of philosophical logic 12*, 151–172.

Suppes, Patrick
  See Nagel, Ernest, Patrick Suppes and Alfred Tarski.

Tait, William W.
  1967   Intensional interpretations of functionals of finite type. I, *The journal of symbolic logic 32*, 198–212.

1968   Constructive reasoning, in *van Rootselaar and Staal 1968*, 185–199.

1981   Finitism, *The journal of philosophy 78*, 524–546.

Takeuti, Gaisi

1975   *Proof theory* (Amsterdam: North-Holland).

1978   Gödel numbers of product spaces, in *Müller and Scott 1978*, 461–471.

1978a   *Two applications of logic to mathematics*, Publications of the Mathematical Society of Japan 13 (Tokyo: Iwanami Shoten; Princeton: Princeton University Press).

1987   Second edition of *Takeuti 1975*.

Tarski, Alfred

1933a   Pojecie prawdy w jezykach nauk dedukcyjnych (The concept of truth in the languages of deductive sciences), *Prace Towarzystwa Naukowego Warszawskiego, wydział III*, no. 34; translated into German by L. Blaustein as *Tarski 1935*; English translation by Joseph H. Woodger in *Tarski 1956*, 152–278.

1935   Der Wahrheitsbegriff in den formalisierten Sprachen, *Studia philosophica* (Lemberg) *1*, 261–405; German translation of *Tarski 1933a*.

1956   *Logic, semantics, metamathematics: papers from 1923 to 1938*, translated into English and edited by Joseph H. Woodger (Oxford: Clarendon Press).

See also Lindenbaum, Adolf, and Alfred Tarski.

See also Łukasiewicz, Jan, and Alfred Tarski.

See also McKinsey, John C. C., and Alfred Tarski.

See also Nagel, Ernest, Patrick Suppes and Alfred Tarski.

Teglas, R.

See Lathrop, J., and R. Teglas.

Thomson, Judith Jarvis

1987   (ed.) *On being and saying: essays for Richard Cartwright* (Cambridge, Mass.: MIT Press).

Troelstra, Anne S.

1973   (ed.) *Metamathematical investigation of intuitionistic arithmetic and analysis*, Lecture notes in mathematics, vol. 344 (Berlin: Springer).

Troelstra, Anne S., and Dirk van Dalen

1988   *Constructivism in mathematics*, volumes I and II (Amsterdam: North-Holland).

Troyanski, Stanimir (Троянски, Станимир)
 1967   Conditionally converging series and certain F-spaces (Russian),
         Teoriya funktsii funktsional'nyi analiz i ikh prilozheniya 5,
         102–107.

Uyeda, S.
 1960   (ed.) *Basis of the contemporary philosophy*, vol. 5 (Tokyo:
         Waseda University Press).

van Dalen, Dirk
 See Troelstra, Anne S., and Dirk van Dalen.

van Heijenoort, Jean
 1967   (ed.) *From Frege to Gödel: a source book in mathematical
         logic, 1879–1931* (Cambridge, Mass.: Harvard University
         Press).
 1982   L'oeuvre logique de Jacques Herbrand et son contexte his-
         torique, in *Stern 1982*, 57–85; English translation in *van Hei-
         jenoort 1985*, 99–121.
 1985   *Selected essays* (Naples: Bibliopolis).

van Rootselaar, Bob, and J. Frits Staal
 1968   (eds.) *Logic, methodology and philosophy of science III. Pro-
         ceedings of the Third International Congress for Logic, Method-
         ology and Philosophy of Science, Amsterdam 1967* (Amster-
         dam: North-Holland).

Vaught, Robert L.
 1954   Applications of the Löwenheim-Skolem-Tarski theorem to
         problems of completeness and decidability, *Koniklijke Neder-
         landse Akademie van Wetenschappen, Proceedings, Series A:
         Mathematical sciences 57*, 467–472; also *Indagationes mathe-
         maticae 16*, 467–472.

Vesley, Richard E.
 See Myhill, John, Akiko Kino and Richard E. Vesley.

von Juhos, Béla
 1930   *Das Problem der mathematischen Wahrscheinlichkeit* (Munich:
         Reinhardt).

von Neumann, John
 1927   Zur Hilbertschen Beweistheorie, *Mathematische Zeitschrift 26*,
         1–46; reprinted in *von Neumann 1961*, 256–300.

1929   Über eine Widerspruchfreiheitsfrage in der axiomatischen Men-
        genlehre, *Journal für die reine und angewandte Mathematik
        160*, 227–241; reprinted in *von Neumann 1961*, 494–508.

1931   Die formalistische Grundlegung der Mathematik, *Erkenntnis
        2*, 116–121; reprinted in *von Neumann 1961a*, 234–239; En-
        glish translation by Erna Putnam and Gerald J. Massey in
        Benacerraf and Putnam 1964, 50–54.

1961   *Collected works*, vol. I: *logic, theory of sets and quantum me-
        chanics*, edited by A. H. Taub (New York and Oxford: Perga-
        mon).

1961a   *Collected works*, vol. II: *operators, ergodic theory and almost
        periodic functions in a group*, edited by A. H. Taub (New York
        and Oxford: Pergammon).
See also Hahn et alii.

Vossenkuhl, Wilhelm
See Bell, David, and Wilhelm Vossenkuhl.

Wald, Abraham
1933   Vereinfachter Beweis des Steinitzschen Satzes über Vektorrei-
        hen im R^n, *Ergebnisse eines mathematischen Kolloquiums 5*,
        10–13.

1933a   Bedingt konvergente Reihen von Vektoren im R^ω, *ibid. 5*, 13–
        14.

1936   Über die Produktionsgleichungen der ökonomischen Wertlehre
        (II. Mitteilung), *ibid. 7*, 1–6.
See also *Gödel 1933h*.

Wald, Robert M.
1984   *General relativity* (Chicago: University of Chicago Press).

Wang, Hao
1974   *From mathematics to philosophy* (London: Routledge and
        Kegan Paul; New York: Humanities Press).

1978   Kurt Gödel's intellectual development, *The mathematical in-
        telligencer 1*, 182–184.

1981   Some facts about Kurt Gödel, *The journal of symbolic logic
        46*, 653–659.

1987   *Reflections on Kurt Gödel* (Cambridge, Mass.: MIT Press).

Weibel, Peter
See Schimanovich et alii.

Weingartner, Paul, and Leopold Schmetterer
1987   (eds.) *Gödel remembered. Salzburg 10–12 July 1983* (Naples:
        Bibliopolis).

Weinstein, Scott
  1983   The intended interpretation of intuitionistic logic, *Journal of philosophical logic 12*, 261–270.

Weizman, Karl Ludwig
  1915   *Lehr- und Übungsbuch der Gabelsbergerschen Stenographie*, twelfth edition (Vienna: Manzsche k.u.k. Hof-, Verlags- und Universitäts-Buchhandlung).

Weyl, Hermann
  1918   *Das Kontinuum. Kritische Untersuchungen über die Grundlagen der Analysis* (Leipzig: Veit); translated into English by Stephen Pollard and Thomas Bole as *Weyl 1987*.
  1918a   *Raum Zeit Materie: Vorlesungen über allgemeine Relativitätstheorie* (Berlin: Springer).
  1987   *The continuum. A critical examination of the foundation of analysis*, English translation of *Weyl 1918* (Kirksville, Mo.: Thomas Jefferson University Press).

Whitehead, Alfred North, and Bertrand Russell
  1910   *Principia mathematica*, vol. 1 (Cambridge, U. K.: Cambridge University Press).
  1912   *Principia mathematica*, vol. 2 (Cambridge, U. K.: Cambridge University Press).
  1913   *Principia mathematica*, vol. 3 (Cambridge, U. K.: Cambridge University Press).
  1925   Second edition of *Whitehead and Russell 1910*.

Wittgenstein, Ludwig
  1921   Logisch-philosophische Abhandlung, *Annalen der Naturphilosophie 14*, 185–262; English translation by C. K. Ogden in *Wittengenstein 1922*.
  1922   *Tractatus logico-philosophicus*, English translation of *Wittgenstein 1921*, with corrected German text (New York: Harcourt, Brace; London: Kegan Paul).

Woodin, W. Hugh
  See Dales, H. Garth, and W. Hugh Woodin.

Wozniakowski, Krzystof
  See Kadets, Mikhail Iosifovich, and Krzystof Wozniakowski.

Wright, James P.
  See Chandrasekhar, Subrahmanyan, and James P. Wright.

Zermelo, Ernst
  1908   Untersuchungen über die Grundlagen der Mengenlehre. I.,
          Mathematische Annalen 65, 261–281; English translation by
          Stefan Bauer-Mengelberg in *van Heijenoort 1967*, 199–215.
  1930   Über Grenzzahlen und Mengenbereiche: Neue Untersuchun-
          gen über die Grundlagen der Mengenlehre, *Fundamenta math-
          ematicae 16*, 29–47.

Zilsel, Edgar
  1976   *Die sozialen Ursprünge der neuzeitlichen Wissenschaft*, edited
          by Wolfgang Krohn (Frankfurt: Suhrkamp).
  199?   *The sociological roots of science*, edited by Robert S. Cohen
          (Dordrecht: Kluwer).

Addenda and corrigenda to Volumes I and II
of these *Collected Works*

Volume I – Further Corrigenda

The following errata in Volume I were noted after Volume II went to press. (In counting line numbers, titles in the middle of a page should be included.)

	Text as printed	*Correction*
25, 31	"objective"	"objective"
36, 7	zum 14 und 15	vom 14 auf 15
39, 15–16	American Mathematical Society	Mathematical Association of America
39, 34	May 4	May 9
41, 11	5 February - 4 March	February 20 - March 4
41, 30	1945 (ca.)	1951 [move entry to next page]
41, 41	May ... lectures	May 7 ... lecture
77, 6	(x_k)	(x_r)
80–81, 23	A	A_n
82–83, 19 and 23	[Misplaced brackets in displayed formulas]	Place left bracket at beginning of formula, right bracket after $(P_n)A_n$
83, 21	$P(A)$	$(P)A$
84, 29	Abschnitt 3	Abschnitt 2
85, 27	Section 3	Section 2
87, fn. 22	z	z_1
93, 2	&	
121, 6	expression	expressions
148, 9	$\overline{\text{Bew}}$	$\overline{Bew}$
186, fn. 55, 3	(in der	(in den
198, 19	645	654
227, 19	if undecidable	is undecidable
275, 1	743	748
298, fn. a, 2	Muzavitski	Muravitskii
352, 6	be substituting	by substituting
369, 4	equation	equations
386, 18	auf eine	auf einen
388, 12	solchen begriff	solchen Begriff
469, 24	Muzavitski	Muravitskii

In addition to the typographical errors listed above, the following more contentual errors should be noted:

(1) It now seems doubtful that Gödel's paper *1931* first appeared in January of that year, as stated in the Chronology (page 38): remarks by Gödel in a letter to Tarski of 20 January 1931 suggest that it more likely appeared in February or March, and a receipt in Gödel's *Nachlass* shows that 100 offprints of the paper were sent to him by the publisher on 25 March.

(2) The completeness proof described on page 57, lines 32 to 37, is not Henkin's original one, but rather the improvement of it due to Hasenjaeger.

(3) The remark (page 267, line 2) that Gödel's result in *1932* showed how many truth values are needed to model Heyting's intuitionistic propositional calculus is not quite correct: Gödel proved that there is no characteristic finite matrix for that calculus, but that it actually contains infinitely many inequivalent formulas in one variable was not demonstrated until 1949 (by Rieger; rediscovered, independently, by Nishimura in 1960).

(4) The commentary on pages 274–275 requires revision: Birkhoff showed only that every complemented modular lattice *of finite length* is isomorphic to the direct product of a finite Boolean algebra and a finite number of projective geometries (a restriction of which Gödel was clearly aware). The "important further question" was thus, in effect, whether the variety of complemented modular lattices is generated by its finite-dimensional members; and that question was later answered affirmatively by Christian Herrmann. (The editors are indebted to Alisdair Urquhart for this and the preceding two corrections.)

Volume II – Corrigenda

	Text as printed	*Correction*
xiii, 1	xi	xv
xiii, 18	*Gïdel 1947*	*Gödel 1947*
xv		The illustrations on pages 188 and 218 were inadvertently switched.
25, 6	*198?*	*1982*
98, fn. 23	⟦note omitted⟧	[23]See footnote 14.
114, 11	⟦last ten words are misaligned⟧	Align with start of line

120, 9	it so	it is so
172, 37	Woodin found	Woodin (*1991*) found
204–5, text and footnotes	[All commas, apostrophes and quotation marks appear as dots]	Replace dots by correct punctuation marks, as required in each instance.
205, fn. 11, 14	uncertainty principle:	uncertainty principle;
241, 29	the proofs	them
317, 35	130, 10	139, 10
317, 37	wie	*dasselbe wie*
318, 34	*tierbaren*	*tiven*
337, 25	Dodd, Anthony J.	Dodd, Antony J.
337, 26	198?	1982
337, 27	Dodd, Anthony J.	Dodd, Antony J.
353, 37	Dodd, Anthony J.	Dodd, Antony J.
389, 35	β_1	B_1
393–407		All prefatory page references (lower-case Roman numerals) should be increased by four.
395, 16	Dodd, Anthony J.	Dodd, Antony J.
back flap of dustjacket	Steven C. Kleene	Stephen C. Kleene

Addenda to References

The unpublished typescript cited in Volume I (394, 34–35) and listed in the index as *Solomon 1981* has since appeared in print. That reference should accordingly be changed to:

Solomon, Martin K.
1987 A connection between Blum speedable sets and Gödel's speed-up theorem, *Zeitschrift für mathematische Logik und Grundlagen der Mathematik 33*, 417–421.

Likewise, the following articles, cited in Volume II as "to appear", have since been published: *Foreman, Magidor and Shelah 198?*, *Foreman and Woodin 198?*, *Mitchell 198?*, and *Martin and Steel 198?*. Revised bibliographic data for them are:

Foreman, Matthew, Menachem Magidor and Saharon Shelah
1988   Martin's maximum, saturated ideals, and non-regular ultrafilters. Part I, *Annals of mathematics (2) 127*, 1–47.

Foreman, Matthew, and W. Hugh Woodin
1991   The generalized continuum hypothesis can fail everywhere, *Annals of mathematics (2) 133*, 1–35.

Martin, Donald A. and John R. Steel
1988   Projective determinacy, *Proceedings of the National Academy of Sciences, U.S.A. 85*, 6582–6586.
1989   A proof of projective determinacy, *Journal of the American Mathematical Society 2*, 71–125.

Mitchell, William J.
1984   The core model for sequences of measures. I, *Mathematical proceedings of the Cambridge Philosophical Society 95*, 229–260.

Citations to these papers in the text (24, 10; 25, fn. zz; 168, 33; and 171, 16) and in the index should be revised accordingly.

Added in press to Volume III

As the present volume was going to press, the editors learned that the unpublished paper of Petr Hájek cited on pages 398 and 402 is to be published in revised form as *Hajek 199?*.

Index